Sustainable Development Goals Series

The **Sustainable Development Goals Series** is Springer Nature's inaugural cross-imprint book series that addresses and supports the United Nations' seventeen Sustainable Development Goals. The series fosters comprehensive research focused on these global targets and endeavours to address some of society's greatest grand challenges. The SDGs are inherently multidisciplinary, and they bring people working across different fields together and working towards a common goal. In this spirit, the Sustainable Development Goals series is the first at Springer Nature to publish books under both the Springer and Palgrave Macmillan imprints, bringing the strengths of our imprints together.

The Sustainable Development Goals Series is organized into eighteen subseries: one subseries based around each of the seventeen respective Sustainable Development Goals, and an eighteenth subseries, "Connecting the Goals", which serves as a home for volumes addressing multiple goals or studying the SDGs as a whole. Each subseries is guided by an expert Subseries Advisor with years or decades of experience studying and addressing core components of their respective Goal.

The SDG Series has a remit as broad as the SDGs themselves, and contributions are welcome from scientists, academics, policymakers, and researchers working in fields related to any of the seventeen goals. If you are interested in contributing a monograph or curated volume to the series, please contact the Publishers: Zachary Romano [Springer; zachary.romano@springer.com] and Rachael Ballard [Palgrave Macmillan; rachael.ballard@palgrave.com].

J. Nicolás Urbina-Cardona
Carlos A. Navas
Alessandro Catenazzi
Carla Piantoni
Editors

Andean Herpetofauna

Explorations of Diversity, Ecology, and Conservation

Springer

Editors
J. Nicolás Urbina-Cardona
Pontificia Universidad Javeriana
Facultad de Estudios Ambientales y Rurales
Departamento de Ecología y Territorio
Carrera, Bogotá, Colombia

Carlos A. Navas
Department of Physiology
Biosciences Institute
University of São Paulo
São Paulo, SP, Brazil

Alessandro Catenazzi
Department of Biological Sciences
Florida International University
Miami, FL, USA

Carla Piantoni
School of Life Sciences
University of Hawai´i at Mānoa
Honolulu, HI, USA

ISSN 2523-3084 ISSN 2523-3092 (electronic)
Sustainable Development Goals Series
ISBN 978-3-032-00073-6 ISBN 978-3-032-00074-3 (eBook)
https://doi.org/10.1007/978-3-032-00074-3

This Springer imprint is published by the registered company Springer Nature Switzerland AG
The registered company address is: Gewerbestrasse 11, 6330 Cham, Switzerland

Contents

About the Editors

J. Nicolás Urbina-Cardona completed his undergraduate studies in Ecology at the Pontificia Universidad Javeriana in Colombia, followed by a Ph.D. in Biological Sciences at UNAM in Mexico. His research primarily focuses on the responses of amphibians and reptiles to anthropogenic landscape transformations, exploring their functional traits. During his time with a conservation NGO, he played a key role in shaping public policy on biodiversity conservation, including the implementation of biodiversity offsets and systematic conservation planning for watershed management. For the past 14 years, as an associate professor at the Department of Ecology and Territory within the Faculty of Environmental and Rural Studies at Pontificia Universidad Javeriana, he has contributed to both research and teaching in biodiversity conservation. Additionally, for a decade, he served as the regional chair for Colombia of the IUCN Amphibian Specialist Group, overseeing the extinction risk assessment of over 850 species and advocating for their adoption by the Colombian government. He has also been instrumental in establishing the National Program for Amphibian Conservation with the Ministry of Environment and contributed to the update of the Global Amphibian Conservation Action Plan. With 90 scientific articles, 30 book chapters, and 6 books published, he currently serves as an associate editor for three indexed scientific journals. Recently, he has expanded his interests to include scientific communication and outreach.

Carlos A. Navas is a biologist from Universidad de los Andes in Colombia, where he also obtained his master's degree. From the outset of his academic journey, he has focused on researching amphibian adaptation to high-altitude environments, a central theme in his Ph.D. thesis at the University of Connecticut. He later moved to Brazil, initially to take up a postdoctoral position at the University of São Paulo, before joining the Department of Physiology at the Biosciences Institute. In Brazil, he broadened his research interests to include other models and systems, such as arthropods and reptiles, developing interdisciplinary projects that converge behavior, physiology, and ecology. With nearly 200 scientific publications, including peer-reviewed articles, natural history notes, and book chapters, he has also advised or co-advised approximately 25 doctoral students, both nationally and internationally. He has held administrative roles at his institution, particularly in graduate studies, and has served on the editorial boards of several journals in herpetology, physiology, and general science. While maintaining his focus on Physiological Ecology, he is currently exploring other facets of science, including the challenges and benefits of interdisciplinary work and the communication of science to younger audiences.

Alessandro Catenazzi research focuses on the systematics and conservation of Neotropical amphibians and reptiles, as well as the ecological dimensions of biodiversity. His work primarily concentrates on the Andes and the Amazon, where he collaborates with colleagues in museums and other research institutions to describe new species and develop phylogenies of frogs from the biodiverse eastern slopes of the Andes. He is particularly interested in using phylogenies as a framework to compare physiological traits across species. His research also includes studying thermal preferences, heat tolerance, and the influence of temperature on physiological functions to better understand how climate change may impact these animals. A major aspect of his current work involves exploring the ecological effects of fungal diseases on amphibians, a group experiencing significant biodiversity losses globally. Having documented the collapse of a species-rich amphibian assemblage, he is focused on developing strategies to mitigate the impact of the fungal disease chytridiomycosis on surviving species.

Carla Piantoni holds a bachelor's degree in Biological Sciences from Universidad Nacional del Comahue in San Carlos de Bariloche, Argentina, and a Ph.D. in Sciences from the University of São Paulo in Brazil. Between her undergraduate and doctoral studies, she worked as a research assistant at the National Museum of Natural History at the Smithsonian Institution. Her herpetological research intersected with the study of marine sponges under the supervision of Dr. Klaus Rützler, Research Zoologist and Curator of Porifera. During this period, she also edited the Annual Reports for the Caribbean Coral Reef Ecosystem Program (CCRE) and co-authored several manuscripts on the ecology and systematics of marine sponges and other invertebrates. Currently, she balances her research with supervising graduate students and teaching first-year Biology at the School of Life Sciences, University of Hawai'i at Mānoa. Initially focused on life history traits and thermal ecology of lizards, her research has expanded to encompass a variety of experimental approaches aimed at understanding individual behavior. Her recent work centers on comparative and evolutionary ecophysiology, particularly in ectothermic animals, and the plasticity in growth and behavior related to thermal biology. Through her experimental methodologies, she seeks to understand how individual variability in behavior and physiological preferences, as well as interactions among individuals, influence the adjustment processes of lizards to new environments.

Prologue

Nora R. Ibargüengoytía

The Andes uplift occurred about 100 My ago and has significantly impacted the climate and environment, not only in the Andes but also the surroundings, extending to regions such as the Amazon basin and the Pacific Ocean coasts. This orographic process created new habitats, promoting isolation and diversification, resulting in a variety of environments with biodiversity hotspots and endemism. The most remarkable consequence was the formation of a dramatic barrier to atmospheric circulation, creating a rain shadow effect. This resulted in a distinct asymmetric topography and contrasting landscapes on either side of the Andes. These spatial heterogeneity and altitudinal gradients have given rise to some of the most unique and rich montane ecosystems and driven adaptive and plastic responses, shaping the evolutionary history and biodiversity patterns of both plant and animal species.

These pristine environments attracted early explorers such as the young Charles Darwin, who, at the age of 23, stayed in the Andes Mountains between 1832 and 1836, traveling through Chile and Argentina. At the beginning of the twentieth century, numerous investigators from Europe and North America followed in his footsteps to explore the Andean landscapes and discover various species of reptiles and amphibians. These explorers included notable naturalists such as Cei, Etheridge, Laurent, and Donoso-Barros, among others. They shared a deep curiosity, adventurous spirit, and extensive knowledge not only of taxonomy but also of geology and evolution. These pioneers laid the foundation of herpetological knowledge by describing and classifying the diverse array of species and writing the first books and articles in South America. Their invaluable legacy established the basis for current studies and sparked numerous questions about the physiology, ecology, behavior, and evolution of montane species living in unique conditions. As a result, many of these research lines began to be revealed and integrated at the end of the last century and continue to this day, pursued by several of the authors in this book.

The Editors have gathered second- and third-generation researchers from various universities across South America to share their most remarkable pioneering studies on the behavior, physiology, ecology, and evolution of reptiles and amphibians. These studies represent a complex puzzle, showcasing a variety of methods and approaches from different laboratories in Argentina, Australia, Brazil, Canada, Colombia, Chile, Ecuador, France, Mexico, Poland, Portugal, Spain, Sweden, Switzerland, and United States of America. They unveil the interplay among environmental constraints,

N. R. Ibargüengoytía (✉)
Instituto de Investigaciones en Biodiversidad y Medioambiente, Consejo Nacional de Investigaciones Científicas y Técnicas, Universidad Nacional del Comahue, San Carlos de Bariloche, Argentina

J. N. Urbina-Cardona et al. (eds.), *Andean Herpetofauna*, Sustainable Development Goals Series,
https://doi.org/10.1007/978-3-032-00074-3_1

opportunities, plasticity, and adaptive responses that allow amphibians and reptiles to thrive in some of the harshest conditions in the Andes complex.

Throughout this book, the authors present the results, conclusions, and open questions regarding the distributional patterns, adaptations, and resilience of species inhabiting the diverse landscapes along the nearly 8900 km stretch from Colombia to Patagonia. The herpetofauna of the cold, humid Páramo at high altitudes in the northern Andes, the arid and dry Puna in the central Andes, the subtropical humid Yungas on the eastern slopes of the Andes, and the coastal plains are studied. Additionally, this book includes research conducted in the southern Andes, where the Patagonian Steppes, characterized by their cold, dry climate and predominantly shrub and grassland vegetation, sharply transition into the humid temperate forests of *Cupressus* and *Nothofagus*, marked by a temperate climate with high precipitation. There is a common thread among the book chapters that emphasize the understanding of the intricate web of behavior, physiology, ecology, and evolution of species that exist in these regions. This approach aims to deepen our understanding of their distribution and life history patterns to infer about the possible responses to environmental changes. Finally, most chapters underscore the goal of preserving these unique ecosystems for future generations.

This book is organized into four parts, Part I, "Diversity and Distribution Patterns," delves into studies that analyze the functional and species diversity of amphibians and reptiles. Discusses the biogeographic patterns of endemism in one of the world's most diverse lizard families, the Family Liolaemidae, and explores the niche evolution of Andean lizards. In these studies, the authors analyzed how behavior and physiology have been shaped by geological, geomorphological, climatic, and anthropogenic disturbances to evolve in the actual biodiversity and ecosystems. Part II, "Evolutionary Histories and Life Strategies," investigates the evolutionary history of montane biota, the reproductive patterns of Patagonian anurans and invites the reader to know the development of a line of studies on reproduction, thermal biology, performance integrated to discover how lizards respond to chronic and acute disturbances. It also analyses the physiological responses within the context of Rapoport's rule as applied to Liolaemidae lizards integrating diet, metabolic rate, thermal biology, and body size. Part III, "Ecological and Physiological Adaptations," explores the resilience and adaptation of amphibians and reptiles to the extreme hot and cold temperatures found in the highlands of the Andes. Part IV, "Conservation Challenges and Emerging Threats," focuses on the impact of climate change and land use changes induced by human activities on amphibians and snakes in the Andes and how these modified environments can contribute to the rise of infectious diseases. They also highlight experiences in conservation practices.

This book adopts an integrative approach, addressing the Andean ecosystems that are facing significant challenges due to climate change, human activities such as urbanization and deforestation, natural drought, fires, and volcanic eruptions. These factors threaten the delicate balance that has been maintained for millennia. The diverse studies result in invaluable contributions to advance in the reconstruction of evolutionary paths leading to current biodiversity helping to predict the possible outcomes resulting from different anthropogenic and natural threats. By understanding these ecological and physiological strategies, we can better appreciate the importance of conserving these remarkable ecosystems in the face of ongoing environmental changes. This knowledge can also serve to inspire current and future researchers to develop effective conservation strategies.

The Andes: The Physical Setting

Juan C. Benavides and María Cristina Arenas-Bautista

1 Introduction

The Andes are the longest continental mountain range in the world, extending approximately 8900 km along the western portion of South America. The Andes is home to a unique biodiversity and to the most species rich biodiversity hotspot in the world due to a combination of its relatively recent geological age with a recent uplift in the last five million years, its location across over 75° of latitude from 10° north to 65° south with multiple macroclimates (Espinoza et al. 2020), the presence of biologically isolated areas connected and reconnected during the glacier and interglacial periods, unique geomorphological and climatic conditions on its valleys and mountains, and numerous microhabitats (Myers et al. 2000; Graham 2009; Mora et al. 2010). The Andes cordillera spans seven countries, from Argentina to Venezuela, passing through Chile, Bolivia, Peru, Ecuador, and Colombia with a variety of ecosystems and climates each with a unique biodiversity that is further enriched by the climatic and geological history that has disconnected and reconnected multiple populations (Hazzi et al. 2018; Esquerré et al. 2019). The major changes in species composition and geomorphology start around 500 m in elevation in the tropical and subtropical regions of the Andes (Hazzi et al. 2018; Morales-Rozo et al. 2024) (Fig. 1a).

Geological processes are the main factor behind the Andean biodiversity (see the geology and geomorphology section on this chapter). The configuration of the Andes as one of the planet's major tectonic formations is directly linked to the subduction of the Nazca oceanic plate beneath the South American continental plate (Armijo et al. 2015). This geological process has resulted in the formation of discontinuities in the geomorphology of the Andes, which have led to variations in their altitude and topography. The subduction of the Nazca plate is primarily observed beneath the central regions of Peru, Bolivia, and northern Chile, with a subsidence angle of 30° at its most oblique position between latitudes 15°S and 28°S (Montgomery et al. 2001; Armijo et al. 2015). This tectonic push has resulted in the formation of the Moho discontinuity, which is particularly deep in this region and represents one of the most distinctive features of Andean geology. The uplift of the mountain range because of this process has resulted in the formation of the Altiplano, a region characterized by volcanic and sedimentary rocks situated at high altitudes, predominantly in Bolivia and Peru (Graham 2009). This uplift has been accompanied by the depression of surrounding areas to the East, such as the Pastaza-Marañon and El Chaco and Pantanal regions, as part of the process of isostatic compensation, in which the Earth's crust adjusts to

J. C. Benavides (✉) · M. C. Arenas-Bautista
Pontificia Universidad Javeriana, Bogota, Colombia
e-mail: jubenavides@javeriana.edu.co

J. N. Urbina-Cardona et al. (eds.), *Andean Herpetofauna*, Sustainable Development Goals Series,
https://doi.org/10.1007/978-3-032-00074-3_2

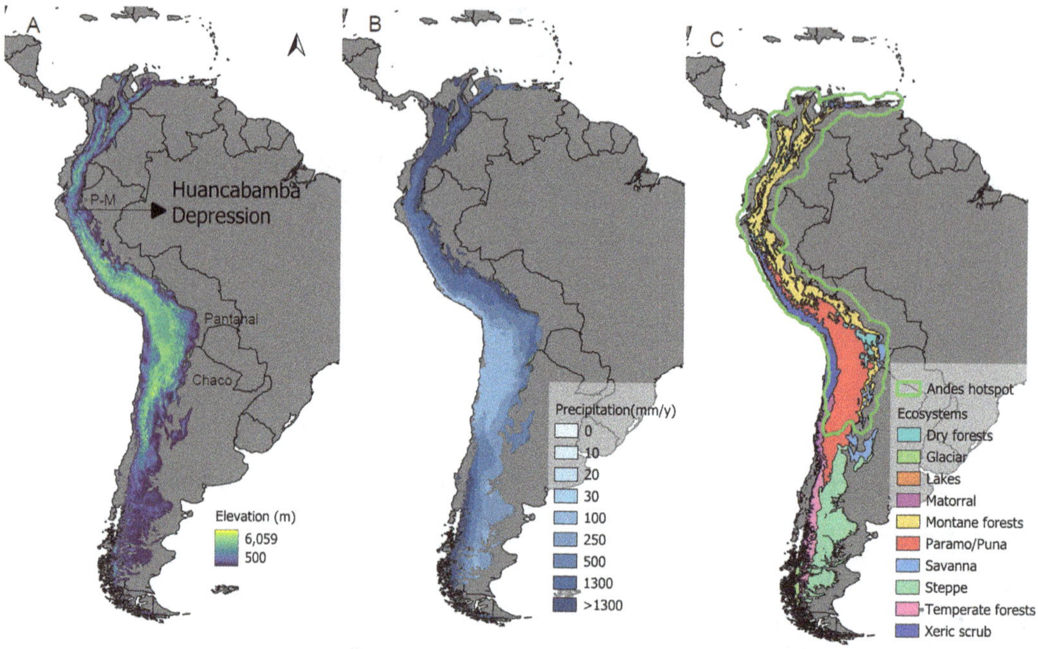

Fig. 1 (**a**): Elevation patterns across the Andes cordillera outline, displaying the Andes above 500 m a.s.l. (Danielson and Gesch 2011), (**b**): cumulated annual precipitation averages from 1970 to 2000 (Fick and Hijmans 2017), and (**c**): main ecosystems across the Andes (Yu et al. 2019), with the green outline showing the extent of the Andean hotspot (Myers et al. 2000). (P-M Pastaza Marañon foreland basin)

balance the material displaced by the mountain uplift. This geotectonic dynamic has not only shaped the physical landscape but also impacted the climatic patterns and ecosystems that characterize the Andes, thereby creating a unique climatic and biological diversity (Montgomery et al. 2001; Armijo et al. 2015) (Fig. 1a).

Overall, the biota of the Andes is divided into several main ecosystems, from the wet Patagonian forests, the dry areas in the central part, to the wet páramos in the north. The alpine areas of the Andes, areas without trees because of the elevation, are divided into three main groups: páramos in the north, covering the north of Perú, Ecuador, Colombia, and Venezuela; the Jalca, a drier region in northern Peru; and the Puna, which extends to the south to the Central part of the Chilean and Argentinean Andes (Josse et al. 2011; Pérez-Escobar et al. 2022) (Fig. 1b). The main ecosystems in the region respond to the elevational and climatic gradients with humid montane forests below the alpine regions covering mostly the eastern flank of the Andes from its northern limit to the south of Bolivia (Yu et al. 2019). Treeless areas in the south are part of the Austral Steppe, a unique biogeographic region transitional between the Andes mountains and the temperate low-lying areas (Roig-Juñent et al. 2018) (Fig. 1c).

The Andes have been inhabited by humans since pre-Columbian times, with a strong agriculture in the wet north and camelid and pastoral communities more common in the drier central and south portion (Guengerich and Berquist 2020; Duchicela et al. 2024). The oldest evidence of human habitation can be traced back to 14,000 ybp, but cultivation and expanded communities start to become more evident in the early Holocene, in the last 10,000 y (Rademaker et al. 2014). Modern, the last 200 y, changes in the land cover are associated with expansion of agriculture in the north, with nearly 60% of the area lost to agriculture at rates of change from natural areas to cultivated areas above 1.5% annually (Tovar et al. 2013; López et al. 2021). Thus, the Andes are particularly vulnerable to changes in land cover since those changes bring with them limitations on the species connectivity, climate, and hydrology (Clerici et al. 2019; Mosquera et al. 2022).

1.1 Geographic Characteristics

One of the most distinctive features of Andean geomorphology with regard to elevation gradients is the occurrence of diverse thermal floors, which undergo a transformation as altitude rises (Montgomery et al. 2001; Bobadilla et al. 2024). A thermal floor is defined as a geographical area where the climate and vegetation are primarily determined by altitude and the associated environmental conditions. As one ascends in the Andes, a transition in climatic conditions is observed, from warm and humid conditions in the lower zones to cold and dry conditions in the higher regions. In the northern Andes, particularly in areas situated near the equator, elevations are relatively low, and temperatures are higher (Masiokas et al. 2020) (Fig. 2b, c). The lowlands of countries such as Colombia and Ecuador are characterized by a tropical or subtropical climate. In this region, the lower thermal floors are typified by the warm and humid conditions of the tropical rain forests. In contrast, the southern Andes experience a significant increase in elevation, resulting in a notable shift toward a colder and drier climate (Arias et al. 2021) (Fig. 2d). In the highlands of Chile, such as the Atacama Desert, temperatures are low, and the environment is marked by extreme aridity. In these regions, the interplay between the topographical features and the main climate and wind patterns gives rise to an abrupt decline in precipitation, resulting in the formation of areas where temperatures are markedly extreme, and rainfall is virtually absent (Espinoza et al. 2020; Masiokas et al. 2020) (Fig. 2d, e).

The Andes also exert a significant influence on the climate patterns of neighboring regions. The mountain range functions as a natural barrier between the climatic and atmospheric flow patterns of the eastern and western regions of South America. On the eastern side, the Andes intercept the humid winds from the Atlantic Ocean, which particularly affect the northern Andes and regions such as the Chocó and Amazon. These winds carry considerable humidity (>70%), and when they meet the high peaks of the Andes, they are forced to rise. As the air rises, it cools and condenses, resulting in rainfall of over 5000 mm per year in the northern and central Andean regions. This leads to the formation of cloud cover at the canopy level, creating the montane forest and Páramo ecosystems, with the northern Andes being particularly affected by this process (Espinoza et al. 2020; Pérez-Escobar et al. 2022). In contrast, on the western side of the Andes, the blocking effect of the mountains becomes more pronounced in the central and southern Andean regions, where moist winds from the east are obstructed. This phenomenon provides an explanation for the existence of some of the world's driest deserts, such as the Atacama Desert in Chile, where precipitation is practically non-existent, with an annual rainfall of only approximately 15 mm (Bobadilla et al. 2024). The contrast in precipitation levels is also a contributing factor to the formation of expansive plateaus and arid valleys that are situated at elevated altitudes within the mountain range.

However, the geomorphology of the Andes not only exerts an influence on contemporary climatic patterns but has also left a discernible imprint on climate variations over time. These historical climatic shifts provide invaluable insights into the comprehension of contemporary climate change and its potential ramifications on landscapes and ecosystems. The Andes serve as a natural archive that documents these paleoclimatic changes over thousands of years, particularly during the Last Glacial Maximum (LGM), around 21,000 years ago. This period was distinguished by markedly reduced temperatures, with an estimated decline of 4 °C–6 °C in the high Andes (Vimeux 2009). The advent of colder conditions precipitated a descent of the snow line by as much as 1000 meters, thereby enabling glaciers to expand to altitudes that are no longer populated by such formations in the present day.

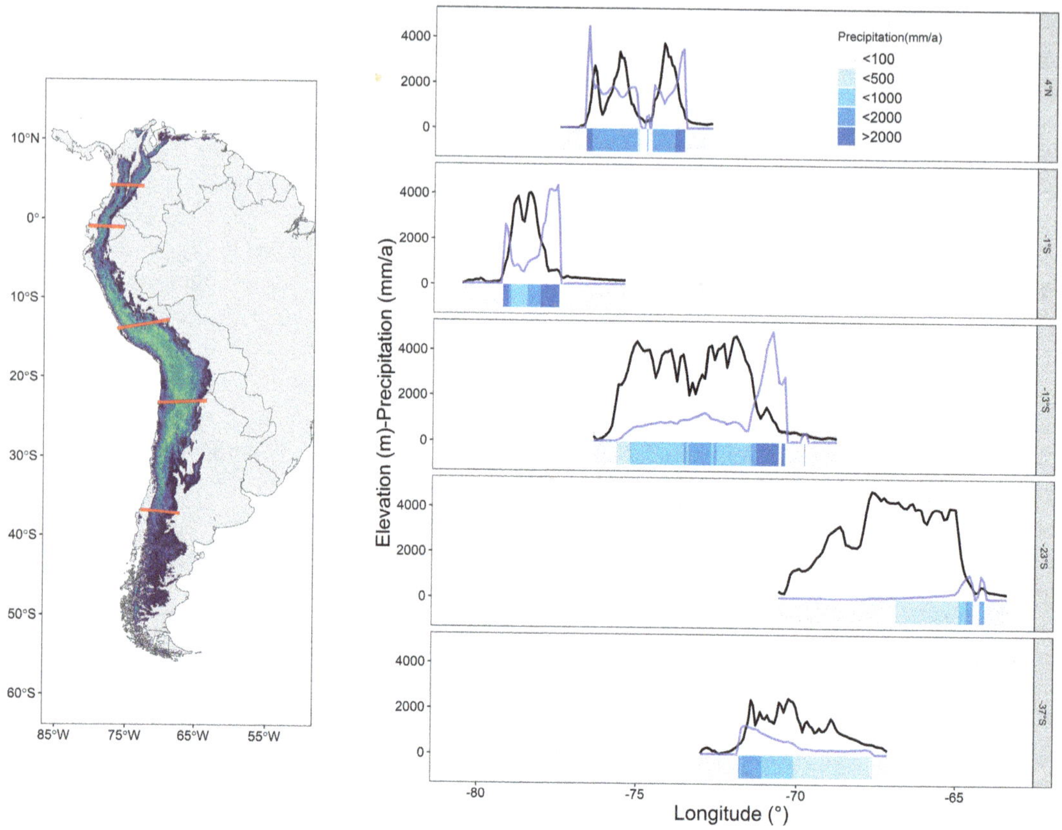

Fig. 2 (**a**): Digital elevation model of the Andes cordillera above 500 m in elevation (Danielson and Gesch 2011), with the red lines marking the location of five latitudinal transects. (**b–f**): Elevation (black lines) and cumulated annual precipitation averages from 1970 to 2000 (Fick and Hijmans 2017) (blue lines) cross section showing the changes in precipitation across the elevation and aspect gradients at five different latitudes from North to South

In addition to influencing the landscape, these glaciers affected the availability of water in river basins and left behind geological deposits that are now studied to reconstruct the environmental conditions of the time (Martini et al. 2017). The presence of moraines and lacustrine sediments in regions such as the Bolivian Altiplano and the Sierra Nevada de Santa Marta in Colombia offers compelling evidence that high-altitude ecosystems were predominantly characterized by cold and arid conditions. These climatic factors have been identified as exerting a considerable influence on the biodiversity observed in these environments (Graham 2009; Pérez-Escobar et al. 2022).

During the glacial periods, precipitation dynamics in the Andes also underwent significant changes, driven primarily by the interaction between elevation gradients and global climate system (Reato et al. 2021). As Hoffmann et al. (2003) have demonstrated, the northward shift of the Intertropical Convergence Zone (ITCZ) during colder periods resulted in a significant alteration of the distribution of moisture. This displacement resulted in a reduction of precipitation on the eastern slopes of the Andes, which in turn affected humid ecosystems such as Montane forests. Concurrently, the arid conditions that already prevailed on the western slopes intensified. These variations in moisture patterns, in conjunction with the influence of the trade winds, resulted in the establishment of a discernible moisture gradient that ultimately shaped the region's ecological conditions. Paleolimnological studies of high Andean lakes, such as Titicaca and Junín (Fiorella et al. 2015), corroborate this hypothesis, indicating reduced water levels and less optimal conditions for species reliant on humid environments.

A significant aspect of the Andean paleoclimate is the influence of elevation gradients as climatic refuges during glacial and interglacial cycles (Reato et al. 2021; Agustina et al. 2024). In response to global climate change, species adapted to warm and humid conditions migrated to lower elevations during interglacial periods, while they ascended to find suitable habitats during glacial periods. Lyu et al. (2024) emphasize that this altitudinal dynamic enabled species to persist over time, contributing substantially to the high biodiversity that characterizes the Andes today. For example, paramo ecosystems served as critical refuges for plants and animals during colder periods, while cloud forests played a similar role at intermediate altitudes. However, Landshuter et al. (2024) proposed that climatic variations also resulted in the fragmentation of ecosystems, thereby limiting connectivity between refuges and causing localized extinctions in some instances. This analysis of Andean paleoclimate serves two purposes: it helps to reconstruct past environmental conditions, and it provides a framework for understanding the interactions between climate, biodiversity, and altitudinal gradients.

Climate and Paleoclimate of the Andes

Holocene-scale variations indicate that the current climate is part of a cyclical process, characterized by periods of colder and warmer (Betancourt et al. 2000). Considering this variability framework, the climatic dynamics of the Andes are shaped by a complex interplay of global and local factors. One of the primary drivers of these changes is the El Niño-Southern Oscillation (ENSO) phenomenon, which operates on an interannual scale and causes significant fluctuations in temperature and precipitation. ENSO manifests in two principal phases: El Niño is distinguished by an anomalous warming of the eastern Pacific Ocean, which forces a reduction in precipitation in the tropical Andes. Conversely, La Niña is characterized by cooling in the same oceanic region, resulting in increased precipitation, particularly in areas such as Colombia. However, the impact of ENSO is not uniform throughout the Andes, as its effects are modulated by interactions with the Andean topography, resulting in different climatic responses in various regions (Ortega et al. 2019; Hagemans et al. 2022).

Currently, the climate of the Andes is a combination of the elevation effect with the main tropical and subtropical wind patterns. The northern part of the Andes from northern Peru to Venezuela are influenced by the latitudinal movement of the ITCZ that brings peaks of precipitation twice a year as the zone moves northward or southward, overall precipitation is relatively high with short dry periods. We have selected five transects that cross the Andes cordillera from west to east at five different latitudes, each of those latitudes has unique climatic conditions ranging from the wet tropical north to the dry areas in the center to the slightly wetter and colder regions in the south (Figs. 2 and 3). In the tropical region from 15°S to 8°N, the temperature is stable (Fig. 2a–c), and the presence of watershed-scale climatic variations is common with anomalous precipitation patterns that respond more to the turbulence processes of the mountain shadow than to the moisture transported by the trade winds (Fig. 3b, c) (Restrepo et al. 2009; Espinoza et al. 2020). Precipitation is relatively high given the prevalence of the ITCZ in the Northern Hemisphere, since there is evidence of a northward transport of heat from the Atlantic Ocean. The humid areas extend to the north, while to the south they reach only to 15°S latitude (Fig. 3c, d) (Grosjean et al. 2007).

The central part of the Andes is dominated by the westerlies influence on the subtropical pacific anticyclone that brings dry air from the cold pacific water in central and south Peru and northern Chile over the high elevations of the Altiplano

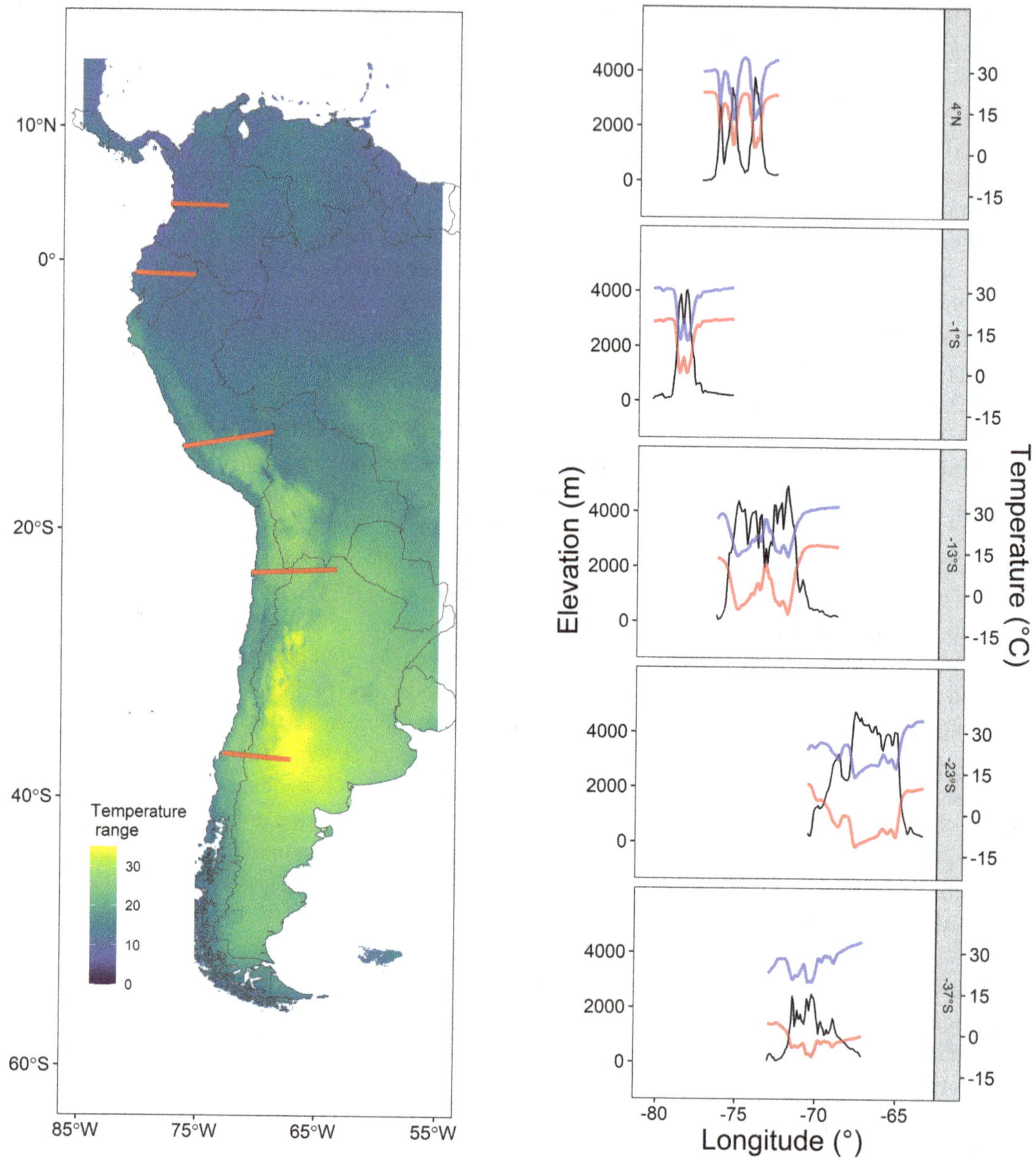

Fig. 3 (**a**) Temperature range: Difference in temperature between the lowest temperature of the coldest month and the highest temperature of the warmest month for the period 1970–2000 (Fick and Hijmans 2017). Right panel (**b–f**): minimal temperature of the coldest month (red) and maximum temperature of the warmest month (blue) (Fick and Hijmans 2017), and elevation (black) across five latitudinal transects (red lines in panel **a**) (Danielson and Gesch 2011)

(Figs. 2d and 3d). Moisture is brought to this region by the southern summer monsoon that is highly weakened by the elevation of the mountains and redirects the precipitation to the southern part of the Amazon and the Pantanal region (Fig. 1a) (Grosjean et al. 2007). More than 50%, and in some cases 70%, of the precipitation of the central Andes is provided by the summer monsoon concentrating the precipitation in the 1–2 months of activity. On the central part of Chile and the Andean portion of Argentina, the climate is dry dominated by a permanent presence of the westerlies with little influence from the summer monsoon creating one of the most arid places on earth (Grosjean et al. 2007) (Fig. 2e, f).

The weakening of the southern monsoon is also a consequence of the high elevation of the Altiplano region that prevents the humid winds to pass the mountain barrier and are diverted to the south, reducing even more the precipitation on this area and at the same time forcing the southern end of the ITCZ to the north on its eastern end and to the south on the western end over the Brazilian amazon (Bird et al. 2011). The divergent paths of the ITCZ in the northern part of the Andes are caused by the interaction of the topographic alignment of the Andes cordillera with the main direction of the trade winds by keeping the winds on an east-west trajectory during the northern summer and a more northeast-southwest trajectory during the southern summer. Bringing additional precipitation to Southern Colombia and Ecuador during the southern summer monsoon (Viveen et al. 2019; Jara et al. 2020). The southern portion of the Andes, in Patagonia, south from the 30°S latitude, lies south of the main path of the westerlies and the interaction of the westerlies with the Antarctic fronts brings moisture creating a humid belt in the Patagonia region. The humid belt is also known as the subtropical southern storm track and brings wet and dry precipitation to the Andes mountains. Precipitation is highly seasonal depending on the position of the westerlies with higher precipitation in southern Patagonia during the summer and higher precipitation in the north during the winter with a predominance of wet precipitation (Espinoza et al. 2020) (Fig. 2e, f).

Over the past three decades, climate variability has intensified these pre-existing climatic dynamics, exacerbating their impact on Andean ecosystems. Among the most obvious manifestations of this phenomenon are rising temperatures and shifts in precipitation patterns (Hagemans et al. 2022). The retreat of tropical glaciers, driven by the upward shift of the 0 °C isotherm, is perhaps the most visible indicator of global warming in the Andes. This retreat has drastically reduced water reservoirs at higher altitudes, directly affecting communities that rely on these glaciers for their water supply. At the same time, precipitation patterns have become more heterogeneous, with general decreases in many areas and local increases in regions such as the Western Cordillera of Colombia and southern Ecuador (Vuille et al. 2018). The combined effects of climate variability and human activities are increasing the pressure on Andean ecosystems. In response to diminishing resources and growing populations, traditional agricultural practices have been replaced by more intensive systems, leading to the expansion of farmland, deforestation, and the loss of native vegetation. These changes not only reduce the soil's capacity to retain water and nutrients but also contribute to erosion and landscape degradation. At medium and high altitudes, activities such as mining exacerbate these problems, creating additional pressure on natural resources and accelerating ecosystem degradation (Young 2009; Tovar et al. 2013; Vuille et al. 2018).

Similarly, altitudinal shifts of species, both flora and fauna, driven by rising temperatures are altering ecological dynamics and threatening to destabilize entire biological communities (Colwell et al. 2008; Huey et al. 2009), in particular those species unable to regulate their own temperatures (Navas et al. 2008). These altitudinal gradients are characterized by changes in temperature, atmospheric pressure, and humidity with increasing altitude, creating sharp ecological transitions. For every 1000 meters of altitude gain, temperatures drop by about 6.5 °C, creating a series of thermal ecological zones ranging from tropical rainforests in the lowlands to glaciers at the highest peaks. This thermal gradient not only influences the distribution of species but also determines their ability to adapt to changing environmental conditions (Graham et al. 2018; Pérez-Escobar et al. 2022).

Climatic patterns in mountain environments are complex and a relevant secondary pattern occurs at intermediate elevations, between 1800 and 3000 meters, where the combination of moderate temperatures and high humidity supports unique ecosystems such as montane forests and cloud forests. These habitats are particularly sensitive to changes in rainfall and temperature. At higher altitudes, in the paramo regions, extreme conditions limit biodiversity, favoring species adapted to cold temperatures, low atmospheric pressure, and freeze-thaw cycles. Meanwhile, latitudinal gradients add another layer of complexity, creating a mosaic of habitats that have driven the evolution and adaptation of countless species (Josse et al. 2011; Graham et al. 2018).

The interaction between latitudinal and altitudinal gradients is particularly evident at the transition from the tropical to the subtropical Andes. At tropical latitudes, most precipitation falls on the eastern slopes due to moisture from the Amazon basin, while the western slopes remain relatively dry. Further south, the influence of the Amazon decreases, and rainfall becomes more seasonal, particularly in the subtropical Andes, where rainfall is concentrated during the austral summer. In the Andes of Chile and Argentina, lowland areas are dominated by arid vegetation such as shrubs, while higher altitudes support temperate forests and eventually alpine zones. These variations are determined not only by altitude, but also by climatic factors such as moisture from the Pacific Ocean and the rain-shadow effect of the mountain range (Viale et al. 2019). The western slope at the northern part of the Andes shows a unique climate with one of the wettest places in the region due to the presence of equatorial westerlies associated to the tropical ocean countercurrent that brings precipitation from the west, making it a unique place across the Andes with extreme high precipitation and low seasonality (Fig. 2a) (Arias et al. 2021; Mejía et al. 2021).

Latitudinal and altitudinal gradients also play a critical role in the hydrological cycles of the Andes. In tropical regions, north of 8°S, high rainfall on the eastern slopes feeds numerous rivers that flow into the Amazon basin. In contrast, the western slopes rely on snow and glacier melt to maintain water flow in the river basins. Further south, precipitation patterns are shifting, and arid regions are heavily dependent on snow and ice for their water supply. Climate change is disrupting these patterns by reducing snow accumulation and accelerating glacier retreat, threatening the hydrological balance in many Andean regions. In subtropical areas, where water availability is already limited, these changes disproportionately affect communities that depend on water for agriculture and consumption (Young 2009; Viale et al. 2019).

Furthermore, latitudinal and altitudinal gradients in the Andes have a significant impact on the region's hydrological cycles, with implications for water availability and ecosystem health that extend beyond the immediate area. In the tropical regions of the Andes, the eastern slopes, which receive abundant precipitation due to moist winds from the Amazon, give rise to numerous rivers that flow eastward into the Amazon basin. These rivers are vital for maintaining the biodiversity of the Amazon and providing water for local communities. In contrast, the western slopes of the Andes are considerably drier, relying predominantly on the meltwater from glaciers and snowfields to sustain the flow of rivers toward the Pacific Ocean. As one progresses in a southward direction and moves further away from the equator, the climate becomes progressively more arid. Consequently, the hydrological balance of the region is increasingly dependent on the availability of snow and ice. The reliance on these high-altitude water sources is becoming increasingly vulnerable due to the ongoing effects of climate change, which are reducing snow accumulation and accelerating glacier retreat. This has significant implications for water security, particularly in regions where the availability of freshwater is already constrained, and the population is heavily reliant on these seasonal sources for drinking, agriculture, and hydropower (Espinoza et al. 2020).

Geology and Geomorphology

The Andes uplift is relatively recent, occurring approximately within a range of 40–120 Ma during the Late Cretaceous and Cenozoic periods, with variable formation times across different regions, as each area has experienced distinct formation processes and timelines (Graham 2009; Pérez-Escobar et al. 2022). This series of uplifts has induced significant landscape changes, such as the contraction and expansion of forests, increased hydraulic pressure, erosion, sedimentation, and fluvial dynamics in adjacent areas (Doan and Schargel 2003; Graham 2009).

The origin of the Andes is relatively recent with initial development in the late cretaceous nearly 120 Ma caused by the subduction of the Nazca plate below the South American Plate (Quiero et al. 2022). The thrust is the result of the collision between the south American continental plate with the Nazca oceanic plate along most if its length except for the southernmost limit where the Antarctic plate pushes the South American plate to the East (Fig. 1a). The thrust from the Nazca plate is not homogeneous and this asymmetry produces the most important discontinuity in the Cordillera. The main thrust of the Nazca plate has been observed under the central parts of Peru, Bolivia, and northern Chile with a subsidence angle of 30° on its most oblique position between 15° and 28°S latitude, occurring with the deepest occurrence of the Moho discontinuity in South America (Fig. 1a, b) (Schellart 2017). The result of this thrust is the uplifting of the Altiplano region, a large area of volcanic and sedimentary rocks, with the consequent depression of the Chaco and Pantanal region as part of the isostatic compensation process (Mora et al. 2010). A particular weak point in the subduction of the Nazca plate occurs at 5° S latitude originated by the rotation of the Carnegie Ridge that prevented the uplifting of the South American plate in that region and is known as the Huancabamba Depression (Fig. 1a) (Mitouard et al. 1990; Pérez-Escobar et al. 2022). The Huancabamba Depression marks the separation between the north and central Andes from a geological perspective but has large implications on the biogeographical patterns since it is an east-west depression that prevents the north-south migration of high elevation species (Fig. 1a) (Weigend 2002).

The Andes uplifting process has been the result of episodic thrust periods followed by relatively stable periods on the rate of convergence varying from 3 to 4 cm/year up to 15.25 cm/year (Mora et al. 2010). At the same time, the rates of convergence and thrust are not homogeneous across the latitudinal span of the western margin of the South American continent, with a diversity of mountain building process, including volcanic arcs, development of plutonic rocks, rock deformation by tectonic pressures, and marine and estuarine sediment uplifting (Frutos 1990; Quiero et al. 2022). The diversity of geological formations, rock types, and climates affects the geomorphological development, drainage patterns and edaphic conditions across the Andes, thus influencing the evolution of the different species (Graham et al. 2018). At the same time, diversity is affected by the geomorphological evolution of the topographic elevations where mountain building and uplifting isolates species and creates new opportunities as diverse climatic conditions develop as the mountains raise with cooler conditions but also particular climatic conditions in the mountain valleys (Rahbek et al. 2019). This long-term control of the habitats available interacts with more short-term climatic process related to the glacial and interglacial process mostly in the last 2 Ma that disconnects and reconnects mountain ranges as the glacial and ecosystems move downward or upward following the changing climatic conditions (Rahbek et al. 2019).

Although biogeography, topography, and geological and climatic evolution have shaped biological diversity in the Andes, the geomorphological has played a hidden role in shaping the current biodiversity patterns in the Andes (Mutke et al. 2014; Clark et al. 2016). Diversity patterns across the Andes are affected by landslides, drainage networks, stream transport, and geomorphological evolution, influencing habitat availability, nutritional constraints, limiting or increasing connectivity across habitats and regions among others (Doan and Schargel 2003; Kattan et al. 2004). Particularly, the role of landslides in offering a heterogeneity of habitats otherwise unavailable in more stable environments allows small populations of early successional species to be maintained in the landscapes (Restrepo et al. 2009). The history of the Andes has a large influence on the diversity and on the large proportions of endemic species, while not all the parts of the Andes were formed uniformly, and many small geological, climatic, or geomorphological changes created a multitude of vicariance and dispersal events (Hazzi et al. 2018).

Hydrology

In subtropical zones, where water availability is constrained by both the geography and the climate, the impacts of these changes are particularly pronounced. For example, in the central and southern Andes, where precipitation is already scarce, the retreat of glaciers and reduced snowmelt are intensifying the scarcity of water resources. These changes have a cascading effect on agricultural practices, as the lack of water threatens crop yields and the livelihood of farming communities. Furthermore, the rise in temperatures has resulted in alterations to precipitation patterns, frequently culminating in more extreme meteorological events, such as droughts followed by intense precipitation. These changes not only render water availability less predictable but also create conditions conducive to soil erosion and the depletion of groundwater reserves (Mettier et al. 2009; Espinoza et al. 2020).

Hydrology and Provision of Ecosystem Services

The hydrology of the Andes is characterized by a remarkable diversity, influenced by latitudinal and altitudinal gradients, seasonal climatic dynamics, and distinct geographical features. These factors create a complex mosaic of precipitation and water availability that varies dramatically from the tropical regions in the north to the temperate and arid zones further south (Young 2009; Viale et al. 2019).

In the northern Andes, from about 10°N to 5°S (Fig. 2a), precipitation is dominated by the interaction of the easterly trade winds and moisture transported from the Amazon, resulting in a hydrological asymmetry between the eastern and western slopes. Regions such as the Amazon Piedmont in Colombia and Ecuador experience intense rainfall for much of the year. These areas can receive more than 4000 mm of rain annually, with peaks between April and November, when the ITCZ is most active. This rainfall regime promotes the recharge of major rivers such as the Putumayo, Napo, and Caquetá, which are essential to the Amazon basin. On the other hand, coastal regions such as Tumaco in Colombia and Esmeraldas in Ecuador experience an orographic shadow effect that reduces rainfall. Although these areas are not as dry as the southern part of the continent, their annual totals range between 1500 and 2000 mm, favoring the presence of drier tropical forests (Espinoza et al. 2015). However, local phenomena such as the Chocó biogeographical region in western Colombia are notable exceptions, as this region, located near the northern limit of the Andes, combines high humidity with annual precipitation that can exceed 8000 mm, making it one of the rainy regions in the world (Espinoza et al. 2020).

The relationship between temperature and altitude also plays a crucial role in the hydrological processes of this Andean region. In the lowlands, where temperatures range from 20 °C to 30 °C, solar radiation increases evaporation and promotes the formation of frequent convective rains that feed tropical rainforest ecosystems and major river systems. In contrast, at altitudes above 3000 m, temperatures drop to 5 °C or lower, reducing evaporation by up to 50% compared to the lowlands and creating ideal conditions for water accumulation and retention. In regions such as the Páramos of Sumapaz and Chingaza in Colombia, these altitude features are key to hydrological regulation (Arias et al. 2021). During the driest months, between December and March, rainfall on the eastern slopes is reduced by up to 40% compared to the wetter months, but Andean ecosystems compensate for this reduction by gradually releasing water from glaciers and páramos.

As one moves southward, this hydrological dynamic evolves. In the central Andes, which extends roughly from 5°S to 27°S (Fig. 2—latitudinal transect −1°S, −13°S, and −23°S), rainfall is seasonal and influenced by the monsoon, especially between December and March. During these months, the region is affected by moist winds from the Atlantic, which concentrate rainfall on the eastern slopes of the mountain range (Junquas et al. 2018). Places such as Cusco in Peru and Lake Titicaca on the border between Peru and Bolivia receive between 600 and 1200 mm of annual rainfall (Fig. 2c), which is unevenly distributed throughout the year. This seasonal regime creates a wet-dry cycle in which montane forest ecosystems depend on water recharge during this season to maintain their biodiversity. In contrast, the western slopes of the central Andes are much drier due to the orographic barrier created by the mountains (Rabatel et al. 2013). In this region, water scarcity limits vegetation and fauna, favoring desert ecosystems (Figs. 1c and 2d). Temperature also plays a fundamental role in the hydrological dynamics of the central Andes. At altitudes above 3000 m, daily temperature variations are particularly extreme: temperatures can drop to −5 °C at night and rise to 15 °C during the day (Fig. 3c, d). This thermal amplitude limits evaporation, allowing the region's wetlands and lagoons to retain water even during the dry season (Arias et al. 2021).

Recent changes in climate and precipitation have a cascading effect on agricultural practices, as the lack of water threatens crop yields and the livelihood of farming communities. Furthermore, the rise in temperatures has resulted in alterations to precipitation patterns, frequently culminating in more extreme meteorological events, such as droughts followed by intense precipitation. These changes not only render water availability less predictable but also create conditions conducive to soil erosion and the depletion of groundwater reserves (Mettier et al. 2009; Espinoza et al. 2020).

In the southern Andes, which extend from about 27°S to 55°S (Fig. 2e, f), hydrological dynamics are influenced by westerly winds that transport moisture from the Pacific and cyclonic fronts. In northern Patagonia, the western slopes receive between 1500 and 3000 mm of annual precipitation, which favors the growth of temperate forests (Fig. 1c). As the Andes turn to the east, climatic conditions change significantly due to the orographic shadow effect. The eastern slopes receive less than 600 mm of total annual precipitation, creating a much drier environment. This contributes to the expansion of steppe ecosystems (Fig. 1c), as seen in the Argentine pampas and the Patagonian steppe, where vegetation is sparser and adapted to the lack of water (Mettier et al. 2009; Espinoza et al. 2020).

Additional gradients also significantly influence the ecosystem services provided by the Andes, from regulating the hydrological cycle to providing agricultural and medicinal resources (Fonte et al. 2012; Rolando et al. 2017). In the lowlands, tropical ecosystems support major crops such as cocoa and palm, while montane forests at mid-altitude play a crucial role in water capture and storage (Fonte et al. 2012). At higher altitudes, páramos act as natural sponges, regulating the flow of water into river basins and ensuring a steady supply during the dry season. However, climate-induced changes in gradients are undermining these services. For example, the loss of páramos due to agricultural expansion and changing rainfall patterns is reducing the ability of watersheds to store water, increasing the vulnerability of communities to droughts (Young 2009; Anderson et al. 2011).

The latitudinal and altitudinal gradients of the Andes also exert a profound influence on the ecosystem services provided by the region. These gradients contribute to the formation of a diverse array of habitats, each of which plays a specific role in maintaining ecological balance. In mid-altitude regions, montane forests play a pivotal role in regulating water flow (Lozano-Povis et al. 2021). These forests serve as a reservoir for precipitation, gradually releasing it into rivers and groundwater reserves. This consistent water release plays an essential role in maintaining river basins and ensuring a continuous water supply, particularly in regions experiencing seasonal variations in precipitation.

In the context of high-altitude alpine ecosystems, they play a vital role as natural water reservoirs, a function that is becoming increasingly important in the face of climate change and human-driven pressures (Barnes and Pelletier 2006; Young 2009). These ecosystems absorb moisture from the atmosphere, storing it in their unique soils and vegetation. During the dry season, this stored water is gradually released, ensuring a steady supply of water to the surrounding river systems. This hydrological service is crucial for both local communities and biodiversity, especially in regions that depend on consistent water flow for agricultural activities and ecosystem stability. However, the combined effects of climate change, such as rising temperatures, altered precipitation patterns, and glacier retreat, and human activities, including deforestation, agricultural expansion, and urbanization, are significantly threatening the sustainability of these ecosystems (Lovejoy and Nobre 2018; Sierra et al. 2022). The degradation of Páramo, Puna, and Jalca ecosystems reduces their ability to retain water, leading to more pronounced water scarcity and increasing vulnerability to droughts, particularly in areas already struggling with limited water resources (Vento et al. 2023). Moreover, the transformation of these alpine ecosystems has broader ecological implications, as these ecosystems are home to numerous endemic species that rely on their specific conditions for survival, further underlining the critical importance of preserving these areas (Sklenář et al. 2021).

Concluding Remarks

The transformation of both hydrological and ecological systems in the Andes due to climate change represents a formidable challenge for the region. The unpredictability of water availability, once a relatively stable resource, is now a growing concern for both agricultural production and the overall health of ecosystems. The consequences of these shifts are particularly alarming for vulnerable communities, where the destabilization of water sources threatens livelihoods, local economies, and social cohesion. As agricultural practices become less reliable and water scarcity intensifies, these communities are increasingly faced with the urgent task of adapting to the new realities of a changing (Lozano-Povis et al. 2021; Marchant Santiago et al. 2021). This may involve rethinking traditional farming practices, adopting more sustainable water management strategies, and enhancing conservation efforts to protect crucial ecosystems, such as the paramos.

This ongoing crisis underscores the need for a comprehensive understanding of the interconnectedness between climate patterns, water cycles, and human development in the Andes. The challenges faced by this region extend beyond the domains of hydrology and ecosystem services; they also have profound implications for agriculture, biodiversity conservation, and social well-being. The ability to address these interrelated concerns, through a holistic approach that integrates scientific knowledge, policy action, and community involvement, will determine the future stability and sustainability of one of the world's most ecologically diverse and culturally rich mountain ranges. The decisions made in the coming years will shape the region's resilience to the inevitable challenges brought by climate change.

References

Agustina R, Martínez OA, Cottescu A, Serrat D (2024) Glacial and periglacial geomorphology of the eastern flank of the north Patagonian Andes: quaternary morphoclimatic evolution. J S Am Earth Sci 149: 105211

Anderson EP, Marengo J, Villalba R, Halloy S, Young B, Cordero D, Gast F, Jaimes E, Ruiz D, Herzog SK (2011) Consequences of climate change for ecosystems and ecosystem services in the tropical Andes. In: Climate change and biodiversity in the tropical Andes, vol 1. IAI, pp 1–18

Arias PA, Garreaud R, Poveda G, Espinoza JC, Molina-Carpio J, Masiokas M, Viale M, Scaff L, Van Oevelen PJ (2021) Hydroclimate of the Andes part II: hydroclimate variability and sub-continental patterns. Front Earth Sci 8:505467

Armijo R, Lacassin R, Coudurier-Curveur A, Carrizo D (2015) Coupled tectonic evolution of Andean orogeny and global climate. Earth Sci Rev 143:1–35

Barnes JB, Pelletier JD (2006) Latitudinal variation of denudation in the evolution of the Bolivian Andes. Am J Sci 306:1–31

Betancourt J, Latorre C, Rech J, Quade J, Rylander K (2000) A 22,000-year record of monsoonal precipitation from northern Chile's Atacama Desert. Science 289:1542–1546

Bird BW, Abbott MB, Vuille M, Rodbell DT, Stansell ND, Rosenmeier MF (2011) A 2,300-year-long annually resolved record of the South American summer monsoon from the Peruvian Andes. Proc Natl Acad Sci 108:8583–8588

Bobadilla A, Stehr A, Toro N (2024) Evaluation of climate-change impacts on the temporal and spatial behaviour of drought in South-Central Chile. Hydrol Sci J 69:165–184

Clark KE, West AJ, Hilton RG, Asner GP, Quesada CA, Silman MR, Saatchi SS, Farfan-Rios W, Martin RE, Horwath AB (2016) Storm-triggered landslides in the Peruvian Andes and implications for topography, carbon cycles, and biodiversity. Earth Surf Dyn 4:47–70

Clerici N, Cote-Navarro F, Escobedo FJ, Rubiano K, Villegas JC (2019) Spatio-temporal and cumulative effects of land use-land cover and climate change on two ecosystem services in the Colombian Andes. Sci Total Environ 685:1181–1192

Colwell RK, Brehm G, Cardelús CL, Gilman AC, Longino JT (2008) Global warming, elevational range shifts, and lowland biotic attrition in the wet tropics. Science 322:258–261

Danielson JJ, Gesch DB (2011) Global multi-resolution terrain elevation data 2010 (GMTED2010). US Geological Survey

Doan TM, Schargel WE (2003) Bridging the gap in Proctoporus distribution: a new species (Squamata: Gymnophthalmidae) from the Andes of Venezuela. Herpetologica 59:68–75

Duchicela SA, Llambí LD, Bonnesoeur V, Román-Dañobeytia F (2024) Pastoralism in the high tropical Andes: a review of the effect of grazing intensity on plant diversity and ecosystem services. Appl Veg Sci 27:e12791

Espinoza JC, Chavez S, Ronchail J, Junquas C, Takahashi K, Lavado W (2015) Rainfall hotspots over the southern tropical Andes: spatial distribution, rainfall intensity, and relations with large-scale atmospheric circulation. Water Resour Res 51:3459–3475

Espinoza JC, Garreaud R, Poveda G, Arias PA, Molina-Carpio J, Masiokas M, Viale M, Scaff L (2020) Hydroclimate of the Andes part I: main climatic features. Front Earth Sci 8:64

Esquerré D, Brennan IG, Catullo RA, Torres-Pérez F, Keogh JS (2019) How mountains shape biodiversity: the role of the Andes in biogeography, diversification, and reproductive biology in South America's most species-rich lizard radiation (Squamata: Liolaemidae). Evolution 73:214–230

Fick SE, Hijmans RJ (2017) WorldClim 2: new 1-km spatial resolution climate surfaces for global land areas. Int J Climatol 37:4302–4315

Fiorella RP, Poulsen CJ, Pillco Zola RS, Barnes JB, Tabor CR, Ehlers TA (2015) Spatiotemporal variability of modern precipitation δ18O in the Central Andes and implications for paleoclimate and paleoaltimetry estimates. J Geophys Res Atmos 120:4630–4656

Fonte SJ, Vanek SJ, Oyarzun P, Parsa S, Quintero DC, Rao IM, Lavelle P (2012) Pathways to agroecological intensification of soil fertility management by smallholder farmers in the Andean highlands. Adv Agron 116:125–184

Frutos J (1990) The Andes Cordillera: a synthesis of the geologic evolution. In: Stratabound ore deposits in the Andes. Springer, pp 3–35

Graham A (2009) The Andes: a geological overview from a biological perspective. Ann Mo Bot Gard 96:371–385

Graham CH, Parra M, Mora A, Higuera C (2018) The interplay between geological history and ecology in mountains. In: Mountains, climate and biodiversity. John Wiley & Sons, pp 231–243

Grosjean M, Santoro CM, Thompson LG, Núñez L, Standen VG (2007) Mid-Holocene climate and culture change in the South Central Andes. In: Climate change and cultural dynamics. Elsevier, pp 51–115

Guengerich A, Berquist S (2020) Earthen terrace technologies and environmental adaptation in the montane forests of pre-Columbian northeastern Peru. J Field Archaeol 45:153–169

Hagemans K, Urrego D, Gosling W, Rodbell D, Wagner-Cremer F, Donders T (2022) Intensification of ENSO frequency drives forest disturbance in the Andes during the Holocene. Quat Sci Rev 294:107762

Hazzi NA, Moreno JS, Ortiz-Movliav C, Palacio RD (2018) Biogeographic regions and events of isolation and diversification of the endemic biota of the tropical Andes. Proc Natl Acad Sci 115:7985–7990

Hoffmann G, Ramirez E, Taupin J-D, Francou B, Ribstein P, Delmas R, Dürr H, Gallaire R, Simões J, Schotterer U (2003) Coherent isotope history of Andean ice cores over the last century. Geophys Res Lett 30. https://doi.org/10.1029/2002GL014870

Huey RB, Deutsch CA, Tewksbury JJ, Vitt LJ, Hertz PE, Álvarez Pérez HJ, Garland T Jr (2009) Why tropical forest lizards are vulnerable to climate warming. Proc R Soc B Biol Sci 276:1939–1948

Jara IA, Maldonado A, de Porras ME (2020) Late Holocene dynamics of the south American summer monsoon: new insights from the Andes of northern Chile (21 S). Quat Sci Rev 246:106533

Josse C, Cuesta F, Navarro G, Barrena V, Becerra MT, Cabrera E, Chacón-Moreno E, Ferreira W, Peralvo M, Saito J (2011) Physical geography and ecosystems in the tropical Andes. In: Climate change and biodiversity in the tropical. MacArthur Foundation, IAI, SCOPE, pp 152–169

Junquas C, Takahashi K, Condom T, Espinoza J-C, Chávez S, Sicart J-E, Lebel T (2018) Understanding the influence of orography on the precipitation diurnal cycle and the associated atmospheric processes in the central Andes. Clim Dyn 50:3995–4017

Kattan GH, Franco P, Rojas V, Morales G (2004) Biological diversification in a complex region: a spatial analysis of faunistic diversity and biogeography of the Andes of Colombia. J Biogeogr 31:1829–1839

Landshuter N, Aemisegger F, Mölg T (2024) Stable water isotope signals and their relation to stratiform and convective precipitation in the tropical Andes. J Geophys Res Atmos 129:e2023JD040630

López S, López-Sandoval MF, Jung J-K (2021) New insights on land use, land cover, and climate change in human–environment dynamics of the equatorial Andes. Ann Am Assoc Geogr 111:1110–1136

Lovejoy TE, Nobre C (2018) Amazon tipping point. American Association for the Advancement of Science, p eaat2340

Lozano-Povis A, Alvarez-Montalván CE, Moggiano N (2021) Climate change in the Andes and its impact on agriculture: a systematic review. Sci Agropecu 12:101

Lyu Z, Vuille M, Goosse H, Orrison R, Novello VF, Cruz FW, Stríkis NM, Cauhy J (2024) South American monsoon intensification during the last millennium driven by joint Pacific and Atlantic forcing. Sci Adv 10:eado9543

Marchant Santiago C, Rodriguez Diaz P, Morales-Salinas-L, Paz Betancourt L, Ortega Fernandez L (2021) Practices and strategies for adaptation to climate variability in family farming. An analysis of cases of rural communities in the Andes mountains of Colombia and Chile. Agriculture 11:1096

Martini MA, Kaplan MR, Strelin JA, Astini RA, Schaefer JM, Caffee MW, Schwartz R (2017) Late Pleistocene glacial fluctuations in Cordillera oriental, subtropical Andes. Quat Sci Rev 171:245–259

Masiokas MH, Luckman BH, Vuille M, Villalba R, Poveda G (2020) Observed and predicted changes in climate in the Andes Cordillera. Front Environ Sci 8: 591996

Mejía JF, Yepes J, Henao JJ, Poveda G, Zuluaga MD, Raymond DJ, Fuchs-Stone Ž (2021) Towards a mechanistic understanding of precipitation over the far eastern tropical Pacific and western Colombia, one of the rainiest spots on earth. J Geophys Res Atmos 126: e2020JD033415

Mettier R, Schlunegger F, Schneider H, Rieke-Zapp D, Schwab M (2009) Relationships between landscape morphology, climate and surface erosion in northern Peru at 5 S latitude. Int J Earth Sci 98:2009–2022

Mitouard P, Kissel C, Laj C (1990) Post-Oligocene rotations in southern Ecuador and northern Peru and the formation of the Huancabamba deflection in the Andean Cordillera. Earth Planet Sci Lett 98:329–339

Montgomery DR, Balco G, Willett SD (2001) Climate, tectonics, and the morphology of the Andes. Geology 29:579–582

Mora A, Baby P, Roddaz M, Parra M, Brusset S, Hermoza W, Espurt N (2010) Tectonic history of the Andes and sub-Andean zones: implications for the development of the Amazon drainage basin. In: Amazonia, landscape and species evolution: a look into the past. Wiley-Blackwell, pp 38–60

Morales-Rozo C, Acuña-Rodríguez IS, Benavides JC, Rincón-Parra VJ, Diazgranados M (2024) Prioritization of important plant areas for conservation of frailejones (Espeletiinae, Asteraceae) in the Northern Andes. Biodivers Conserv 33:1–26

Mosquera GM, Marin F, Stern M, Bonnesoeur V, Ochoa-Tocachi BF, Roman-Danobeytia F, Crespo P (2022) Progress in understanding the hydrology of high-elevation Andean grasslands under changing land use. Sci Total Environ 804:150112

Mutke J, Jacobs R, Meyers K, Henning T, Weigend M (2014) Diversity patterns of selected Andean plant groups correspond to topography and habitat dynamics, not orogeny. Front Genet 5:351

Myers N, Mittermeier RA, Mittermeier CG, Da Fonseca GA, Kent J (2000) Biodiversity hotspots for conservation priorities. Nature 403:853–858

Navas CA, Gomes FR, Carvalho JE (2008) Thermal relationships and exercise physiology in anuran amphibians: integration and evolutionary implications. Comp Biochem Physiol A Mol Integr Physiol 151: 344–362

Ortega C, Vargas G, Rojas M, Rutllant JA, Muñoz P, Lange CB, Pantoja S, Dezileau L, Ortlieb L (2019) Extreme ENSO-driven torrential rainfalls at the southern edge of the Atacama Desert during the Late Holocene and their projection into the 21th century. Glob Planet Chang 175:226–237

Pérez-Escobar OA, Zizka A, Bermúdez MA, Meseguer AS, Condamine FL, Hoorn C, Hooghiemstra H, Pu Y, Bogarín D, Boschman LM (2022) The Andes through time: evolution and distribution of Andean floras. Trends Plant Sci 27:364–378

Quiero F, Tassara A, Iaffaldano G, Rabbia O (2022) Growth of Neogene Andes linked to changes in plate convergence using high-resolution kinematic models. Nat Commun 13:1339

Rabatel A, Francou B, Soruco A, Gomez J, Cáceres B, Ceballos JL, Basantes R, Vuille M, Sicart J-E, Huggel C (2013) Current state of glaciers in the tropical Andes: a multi-century perspective on glacier evolution and climate change. Cryosphere 7:81–102

Rademaker K, Hodgins G, Moore K, Zarrillo S, Miller C, Bromley GR, Leach P, Reid DA, Álvarez WY, Sandweiss DH (2014) Paleoindian settlement of the high-altitude Peruvian Andes. Science 346:466–469

Rahbek C, Borregaard MK, Antonelli A, Colwell RK, Holt BG, Nogues-Bravo D, Rasmussen CM, Richardson K, Rosing MT, Whittaker RJ (2019) Building mountain biodiversity: geological and evolutionary processes. Science 365:1114–1119

Reato A, Carol ES, Cottescu A, Martínez OA (2021) Hydrological significance of rock glaciers and other periglacial landforms as sustenance of wet meadows in the Patagonian Andes. J S Am Earth Sci 111:103471

Restrepo C, Walker LR, Shiels AB, Bussmann R, Claessens L, Fisch S, Lozano P, Negi G, Paolini L, Poveda G (2009) Landsliding and its multiscale influence on mountainscapes. Bioscience 59:685–698

Roig-Juñent SA, Griotti M, Cecilia Domínguez M, Agrain FA, Campos-Soldini P, Carrara R, Cheli G, Fernández-Campón F, Flores GE, Katinas L (2018) The Patagonian Steppe biogeographic province: Andean region or South American transition zone? Zool Scr 47:623–629

Rolando JL, Turin C, Ramírez DA, Mares V, Monerris J, Quiroz R (2017) Key ecosystem services and ecological intensification of agriculture in the tropical high-Andean Puna as affected by land-use and climate changes. Agric Ecosyst Environ 236:221–233

Schellart W (2017) Andean mountain building and magmatic arc migration driven by subduction-induced whole mantle flow. Nat Commun 8:2010

Sierra JP, Junquas C, Espinoza JC, Segura H, Condom T, Andrade M, Molina-Carpio J, Ticona L, Mardoñez V, Blacutt L (2022) Deforestation impacts on Amazon-Andes hydroclimatic connectivity. Clim Dyn 58:1–28

Sklenář P, Romoleroux K, Muriel P, Jaramillo R, Bernardi A, Diazgranados M, Moret P (2021) Distribution changes in Páramo plants from the equatorial high Andes in response to increasing temperature and humidity variation since 1880. Alp Bot 131:201–212

Tovar C, Seijmonsbergen AC, Duivenvoorden JF (2013) Monitoring land use and land cover change in mountain regions: an example in the Jalca grasslands of the Peruvian Andes. Landsc Urban Plan 112:40–49

Vento B, Cuesta F, Cavieres L (2023) Carbon sequestration and climate change in crops, natural vegetation, and wetland dynamics in the high Andes. Front Plant Sci 14:1338577

Viale M, Bianchi E, Cara L, Ruiz LE, Villalba R, Pitte P, Masiokas M, Rivera J, Zalazar L (2019) Contrasting climates at both sides of the Andes in Argentina and Chile. Front Environ Sci 7:69

Vimeux F (2009) Similarities and discrepancies between andean ice cores over the last deglaciation: climate implications. In: Past climate variability in South America and surrounding regions: from the last glacial maximum to the holocene. Springer Science & Business Media, pp 239–255

Viveen W, Zevallos-Valdivia L, Sanjurjo-Sanchez J (2019) The influence of centennial-scale variations in the South American summer monsoon and base-level fall on Holocene fluvial systems in the Peruvian Andes. Glob Planet Chang 176:1–22

Vuille M, Carey M, Huggel C, Buytaert W, Rabatel A, Jacobsen D, Soruco A, Villacis M, Yarleque C, Timm OE (2018) Rapid decline of snow and ice in the tropical Andes–Impacts, uncertainties and challenges ahead. Earth Sci Rev 176:195–213

Weigend M (2002) Observations on the biogeography of the Amotape-Huancabamba zone in northern Peru. Bot Rev 68:38–54

Young KR (2009) Andean land use and biodiversity: humanized landscapes in a time of change. Ann Mo Bot Gard 96:492–507

Yu D, Liu Y, Shi P, Wu J (2019) Projecting impacts of climate change on global terrestrial ecoregions. Ecol Indic 103:114–123

Part I

Diversity and Distribution Patterns

Beyond Species Richness: A Behavioural Perspective of the Anuran Diversity in the Andes of Colombia

Fernando Vargas-Salinas, María Paula Toro-Gómez, Natalia Vergara-Herrera, Elkin A. Tenorio, and Luis Alberto Rueda-Solano

1 Introduction

The exploration of the diversity of anurans in Colombia has focused on taxonomy and related research areas (Urbina-Cardona et al. 2023). This bias can be attributed to several factors. First, taxonomy serves as the fundamental basis for organism classification and is indispensable for biological research, facilitating the understanding of species distribution patterns, ecological relationships, and evolutionary processes. Moreover, taxonomy serves as a vital tool for communication among scientists, policymakers, and conservationists (Kaiser et al. 2013; Thomson et al. 2018). Second, the primary research focus of researchers associated with the Institute of Natural Sciences at the National University of Colombia (ICN) emphasized taxonomy and systematics of anurans; these studies have been fundamental in documenting anuran species richness in various areas and ecosystems in Colombia, and in identifying knowledge gaps. Nevertheless, taxonomy is not the only approach through which biological diversity should be studied; among other approaches is the study of behaviour (Sutherland 1998; Devictor et al. 2010).

From the early stages of Colombian herpetology, there was a recognized need to delve deeper into behavioral studies (Dunn 1944a). The study of animal behaviour offers a unique window into appreciating the diversity and beauty of the natural world. Besides, behavioural studies are crucial for understanding the complexity of ecosystems, the interactions among organisms, and the dynamics of populations (Sutherland 1996; Mittelbach and McGill 2019); they provide critical insights into species ecology and evolution that, in turn, help to predict how animals respond to environmental changes and how we can better manage natural resources (Gosling and Sutherland 2000; Candolin and Wong 2012; Weis and Sol 2016).

Anurans are an important component of vertebrate diversity, with countless behavioural traits that make them unique and fascinating. They

Supplementary Information: The online version contains supplementary material available at https://doi.org/10.1007/978-3-032-00074-3_3.

F. Vargas-Salinas (✉)
Evolución, Ecología y Conservación EECO, Programa de Biología, Facultad de Ciencias Básicas y Tecnologías, Universidad del Quindío, Armenia, Colombia
e-mail: fvargas@uniquindio.edu.co

M. P. Toro-Gómez
Evolución, Ecología y Conservación (EECO), Programa de Biología, Facultad de Ciencias Básicas y Tecnologías, Universidad del Quindío, Armenia, Colombia

N. Vergara-Herrera · L. A. Rueda-Solano
Fundación Atelopus, and Grupo de Investigación en Biodiversidad y Ecología Aplicada (GIBEA), Programa de Biología, Facultad de Ciencias Básicas, Universidad del Magdalena, Santa Marta, Colombia

E. A. Tenorio
Graduate Center, City University of New York & American Museum of Natural History, American Museum of Natural History, New York, NY, USA

J. N. Urbina-Cardona et al. (eds.), *Andean Herpetofauna*, Sustainable Development Goals Series,
https://doi.org/10.1007/978-3-032-00074-3_3

exhibit complex life cycles, a wide range of communication systems and reproductive modes, and are an integral part of nutrient flow in ecosystems as predators and prey in terrestrial and aquatic habitats (Wells 2007; Valencia-Aguilar et al. 2013). They are abundant, with small body size, and their phylogenetic relationships are well known (Pyron and Wiens 2011; Portik et al. 2023), facilitating their use in experiments and modern comparative analyses for understanding the origin and maintenance of biological diversity (see Gomez-Mestre et al. 2012; Pyron and Burbrink 2013; Delia et al. 2017a, 2020; Tonini et al. 2020; Carvajal-Castro et al. 2021).

In this chapter, we present an overview of the main studies on animal behaviour of anurans belonging to the Hylidae, Dendrobatoidea (Dendrobatidae + Aromobatidae) and Centrolenidae in the Andes of Colombia. We focus on these families because they are the most frequently used groups in behavioural studies in the country. We also include behavioural studies conducted with species of the genus *Atelopus* in the Sierra Nevada of Santa Marta, a mountain range located at the Caribbean region of Colombia, because their ecology is likely similar to the ecology of Andean species, whose populations have declined dramatically (La Marca et al. 2005; Lötters et al. 2023). We do not intend to disregard the contributions from studies of other anuran families in Colombia (e.g. Craugastoridae and Strabomantidae: Vargas-Salinas et al. 1999; Rojas-Rivera et al. 2011; Escobar-Vargas et al. 2016; Rios-Soto and Ospina 2018; Cárdenas-Ortega 2022; Cárdenas-Ortega and Herrera-Lopera 2022; Londoño-Quiceno and Gutiérrez-Cárdenas 2023; Pisso-Florez et al. 2023a). However, most of these publications are not the result of studies based on a continuous monitoring of individuals or with an experimental design focused on testing a specific hypothesis. Finally, we present a preliminary analysis of the geographic component of some anuran traits (e.g. body size, reproductive mode) in the Andes of Colombia. This analysis can generate insights into the ecological and evolutionary causes of the behavioural diversity of anurans, helping explain the exceptional biodiversity of frogs and toads in Colombia.

2 Behavioural Studies with Anurans in the Andes of Colombia

The northern Andes, recognized as one of the most diverse regions on the planet, extend from the Amotape-Huancabamba depression in Peru and Ecuador to the Mérida Mountain range of Venezuela (Graham 2009; Smith et al. 2014). In its northern range, located in Colombia, the Andes hold the highest diversity of fauna (e.g. birds: Kattan et al. 2006: mammals: Solari et al. 2013; fish: Maldonado-Ocampo et al. 2005; amphibians: Bernal and Lynch 2008; Armesto and Señaris 2017). The topographic complexity in the Andes of Colombia and their subregions (inter-Andean valleys) creates a very high variety of climatic conditions in relation to elevation and latitude. This climate variability is reflected in a spatial replacement of species and ecosystem types ranging from humid and dry forests in the lowlands to elfin forest and paramo on the top of mountains. One can thus envision the Andes of Colombia as a vast natural laboratory in which there is still a lot of biological diversity to be discovered.

Next, we include studies conducted with specific anuran clades. While we used a 900 m a.s.l. elevation criterion (sensu Lynch 1999) as a reference to illustrate general trends and gaps in Colombian anuran behavioural studies, we acknowledge the limitations of such arbitrary elevation limits, as highlighted by Herrera-Lopera et al. (2023) in their critique of altitudinal cutoffs used for differentiating Andean species from lowlands species. We summarize studies on animal behaviour conducted in the Colombian Andes with species from Hylidae, Dendrobatoidea (Aromobatidae + Dendrobatidae) and Centrolenidae by following a chronological order of research performed over the last five decades. Notably, species belonging to these clades exhibit striking contrasts in key aspects of their natural history and ecology. For instance, species of Dendrobatoidea are primarily diurnal, terrestrial, and many lineages display vibrant coloration as conspicuous warning signals to potential predators (Rojas 2016; Rojas and Vargas-Salinas 2024). In contrast, species of Centrolenidae are nocturnal, arboreal, and rely predominantly on camouflage as their primary defence mechanism (Kubicki 2007; Barnett et al. 2020). Therefore, the behavioural patterns found in one of these clades expand and complement our knowledge of the behavioural ecology in anurans obtained from studies conducted on the other clades.

2.1 Hylidae (Tree Frogs)

Hylidae, or tree frogs, is a clade composed of 760 species distributed worldwide (Frost 2025). Tree frogs exhibit nocturnal habits, along with high interspecific variability in body size and behaviours related to thermoregulation, antipredator strategies, communication, breeding and parental care (Wells 2007; Taboada et al. 2017; Warkentin et al. 2019; Lee et al. 2021). They also possess streamlined bodies and adhesive structures on the terminal part of their phalanges, allowing them to climb many types of substrates, and are a prominent component of amphibian assemblages, especially in the lowlands (Duellman 2001, 2005). Three genera of Hylidae are particularly diverse in the Andes, extending into the Amazon region: *Boana* with 101 spp., *Hyloscirtus* with 44 spp., and *Dendropsophus* with 106 spp. (Frost 2025). Like most studies with anurans in the northern Andes, mainly in Colombia, tree frog research focused on describing new species and resolving inter- and intra-generic taxonomy (e.g. Faivovich et al. 2005, 2010; Orrico et al. 2021).

A tree frog species distributed in the Colombian Andes whose behaviour has been studied is *Dendropsophus molitor* (Fig. 1). It can be found in humid forests, sub-paramos and paramos, and has been the subject of multiple taxonomic, ecological, eco-physiological and behavioural studies (Navas 1996a, b; Lüddecke 1997; Amézquita and Lüddecke 1999; Lüddecke 2001; Gutierrez and Lüddecke 2002; Lüddecke and Sínchez 2002; Higuera-Rojas and Carvajal-Cogoyo 2021; Cortés-Suárez 2021; Saboyá-Acosta and Urbina-Cardona 2023; Arias-Cárdenas et al. 2024). This species is particularly abundant in the Bogotá savanna, where body coloration patterns, mating, egg-laying behaviour and tadpole and adult body sizes have been documented (Dunn 1944b; Hunter and Valdivieso 1962). Moreover, aspects of its reproductive biology have been studied extensively (Lüddecke 1997; Amézquita and Lüddecke 1999; Lüddecke 2001; Gutierrez and Lüddecke 2002; Lüddecke and Sínchez 2002). At present, most of our knowledge comes from studies led by Dr. Horst Lüddecke, who dedicated over 13 years of systematic observations on *D. molitor* in the Chingaza paramo. Lüddecke laid the groundwork for studies of behavioural ecology with Colombian anurans using both correlative and experimental approaches, through his studies on *D. molitor*, and the dendrobatoids *Rheobates palmatus* and *Hyloxalus subpunctatus* (see section below), but also by the mentoring of students who eventually became leading researchers of behavioural studies with anurans in and out of Colombia (Molina 2023).

Most females of *D. molitor* in the Chingaza paramo spawn once per year or over longer intervals. The interval between female breeding events is unrelated to body size or body condition but is associated with environmental factors, such as temperature, which influence females' physiological recovery rate (Lüddecke 1997). This study highlights the importance of understanding reproductive patterns at both the population and individual levels. Regarding mating patterns, a 4-year monitoring period revealed a significant disparity in sex ratio, which was variable but always male biased (Lüddecke 2001). When male availability is high, female choice is more important than male-male competition in determining mating pattern. Larger males do not exhibit a clear advantage in mating success than smaller males (Lüddecke 2001), as is common in other frogs (Wells 2007). Under controlled laboratory experiments, larger female laid larger egg clutches than smaller females, and females

Fig. 1 Images showing basic characteristics of the reproductive ecology of the Colombian Andean endemic tree frog *Dendropsophus molitor*. (**a**) calling male, (**b**) amplectant pair and some black eggs in floating clutches, (**c**) free swimming larva (note the differential coloration between body and the posterior part of the tail), (**d**) metamorphic individual. (Pictures by Fernando Vargas-Salinas (**a**, **c**, **d**) and Giovanny Díaz (**b**))

amplexed by proportionally larger males had a higher percentage of hatching success (Gutiérrez and Lüddecke 2002). In other words, assortative mating by body size improves fertilization rates in *D. molitor*, possibly because of an optimal cloaca-cloaca positioning during spawning. Overall, results of these studies challenged established notions about the importance of male size in female mate selection and serves as a relevant example of mating mechanisms in Andean frog populations. Moreover, Gutiérrez and Lüddecke (2002)'s study was one of the first to include quantitative analyses of advertisement calls emitted by males to test specific behavioural hypotheses. Their results suggest that female mate choice in *D. molitor* is partially based on call features related to male body size.

Published behavioural studies on tree frogs in the Andes of Colombia other than *D. molitor* are few. Among them, the tree frogs *Hyloscirtus antioquia* and *H. larinopygion* are two closely related species that breed alongside streams, for which distress call and defensive body posture have been described (Duarte-Marín et al. 2019; Tovar et al. 2022). Also, features of their advertisement calls were quantified by Rivera-Correa et al. (2017), who attributed the limited call differentiation to their allopatric distribution. The tadpoles of *H. antioquia* and *H. larinopygion* have been described (Coloma et al. 2012; Rivera-Correa and Faivovich 2021), but their reproductive behaviours and basic aspects such as type of amplexus (Carvajal-Castro et al. 2020), oviposition site, and postures are not known. In fact, these aspects are unknown for all *Hyloscirtus* (Rivera-Correa M. Comm. Pers.). Other tree frogs in the Andes of Colombia for which some natural history traits are known include *Dendropsophus norandinus* (Gutiérrez-Cárdenas et al. 2013) and *D. columbianus* (Gómez-Hoyos et al. 2012; Bedoya et al. 2014; Díaz-Ayala et al. 2014; Méndez-Narvaez et al.

2014; Zornosa-Torres and Hoyos 2016). *Dendropsophus* reproduces in ponds in open and forested areas where calling males frequently congregate, but details of communication and breeding behaviour in species and populations in the Andes of Colombia are unknown.

2.2 Dendrobatoidea (Poison Frogs)

The neotropical super family Dendrobatoidea (Aromobatidae + Dendrobatidae), hereafter dendrobatoids, is composed of around 353 species (Frost 2025). Most dendrobatoids are diurnal, with males calling from territories, and females ovipositing small clutches (1–40 eggs) in humid terrestrial microhabitats in the leaf litter or on rocky substrates (Pröhl 2005; Lötters et al. 2007; Carvajal-Castro et al. 2021). After some days, the eggs hatch and, depending on the species, the tadpoles are transported on the back of the mother, father, or both parents, to slow moving water creeks, temporal pools on the ground or phytotelmata (Weygoldt 1987; Summers and Tumulty 2014). In some species (e.g. *Oophaga pumilio*), mothers feed tadpoles with unfertilized trophic eggs and possibly, with nutrients in their skin (Weygoldt 1980; Stynoski et al. 2024). At least 100 species of dendrobatoids have evolved aposematism as an antipredator strategy (Carvajal-Castro et al. 2021). Such aposematic species exhibit conspicuous coloration that warn potential predators about their skin toxins, which they synthetize or sequester from dietary items (Santos et al. 2003; Saporito et al. 2012; Waters et al. 2024). Because of the toxicity of these skin secretions, dendrobatoids are commonly referred as poison frogs. In fact, some dendrobatoids have some of the strongest toxins found in nature (Myers et al. 1978; Gonzalez and Carazzone 2023). These specialized behaviours in reproduction, parental care and strategies to avoid attack by predators, together with their beautiful and conspicuous coloration, make dendrobatoids highly interesting from an academic perspective (Summers and Tumulty 2014; Rojas 2016; Rojas et al. 2023; Vargas-Salinas and Rojas 2024; Rojas and Vargas-Salinas 2024). Their bright coloration, however, have made some dendrobatoids susceptible to illegal pet trafficking (Rueda-Almonacid et al. 2004; Kahn et al. 2016; Betancourth-Cundar et al. 2020).

In Colombia, where at least 17 species of Aromobatidae and 96 species of Dendrobatidae live (Frost 2025), most behavioural studies have been conducted with species distributed below 900 m a.s.l. (e.g. De Luna et al. 2010; Vargas-Salinas and Amézquita 2013a; Vargas-Salinas and Amézquita 2013b; Gil-Guevara and Amézquita 2020; González-Santoro et al. 2021, 2023; Vergara-Herrera et al. 2024), following trends seen in other countries (e.g. Summers 1999; Saporito et al. 2010; Brown et al. 2010; Schulte 2014; Rojas et al. 2020; Fouilloux et al. 2024). Generally, what is known about the ecology and behaviour of dendrobatoids in the Colombian Andes is mainly limited to descriptions of habitat use, diet, predation and advertisement call features, in some cases published as part of new species descriptions (e.g. Bernal et al. 2007; Amézquita et al. 2013a; Meza-Joya et al. 2015; Marin et al. 2018; Ramos et al. 2018; Chaves-Portilla et al. 2021) or natural history observations (e.g. Valderrama-Vernaza et al. 2009; Gómez-Hoyos et al. 2014; Duarte-Marín et al. 2020; Bobadilla-Molina et al. 2021; Pisso-Florez et al. 2023b). Studies testing behavioural hypotheses in Colombian dendrobatoids occurring >900 m a.s.l. are limited to a few of species (e.g. *Rheobates palmatus*, *Andinobates bombetes*; Fig. 2).

2.2.1 Behavioural Studies of Colombian Andean Dendrobatoids

Beyond early work in the mid-twentieth century (e.g. Dunn 1944a; Stebbins and Hendrickson 1959), studies quantifying dendrobatoid behaviour in the Colombian Andes are rare, with those on *Rheobates palmatus* and *Hyloxalus subpunctatus* being notable exceptions. *Rheobates palmatus* is a cryptically coloured species distributed from 250 to 2520 m a.s.l., and the most detailed studies about its behaviour have been conducted on highland populations (Lüddecke 1976, 1993, 1999, 2003; Bernal et al. 2005). These studies cover aspects of habitat use,

Fig. 2 Photographs of dendrobatoids that have been most studied regarding diverse aspects of their ecology and behaviour in the Andes of Colombia. (**a**) *Hyloxalus subpunctatus*, (**b**) *Andinobates bombetes*, (**c**) *Rheobates palmatus*, and (**d**) *Leucostethus brachistriatus*. (Pictures by Mauricio Rivera-Correa (**a**), Cristian González-Acosta (**b**), Khristian Venegas-Valencia (**c**), and Fernando Vargas-Salinas (**d**))

territorial behaviour, cognitive abilities of individuals to navigate space, communication, breeding behaviour, and parental care. Further ecological and behavioural aspects have been complemented with natural history records about parental care, diet, and predation, especially in populations below 900 m a.s.l. (Astwood-Romero et al. 2016; Marín-Martínez and Rojas-Morales 2016; Zuluaga-Isaza et al. 2020; Rivera-Pérez et al. 2020; Bobadilla-Molina et al. 2021).

Hyloxalus subpunctatus is another cryptic species endemic to the Andes of Colombia with interesting behavioural aspects related to acoustic interactions between territorial males, female choice and parental care (Fandiño et al. 1997; Lüddecke et al. 1999, 2000; Lüddecke 2002). For instance, males transport up to 21 tadpoles on their backs simultaneously and deposit them in small temporary water bodies formed on the ground after rains. When the carrying male introduces his body in water, some tadpoles detach themselves, while other tadpoles remain attached until the male visits another puddle (Stebbins and Hendrickson 1959; Fandiño et al. 1997). This behaviour opens interesting questions regarding tadpole puddle selection and how this decision is influenced by inter-puddle differences in relative location and quality.

From a physiological perspective, *H. subpunctatus* is particularly interesting because of their activity at low temperatures. Despite experiencing high levels of thermal fluctuation from 6 to 25 °C, males occurring above 3000 m a.s.l. call from dawn to dusk throughout the year (Navas 1996a, b; Navas and Bevier 2001). Including environmental temperature in studies of frog calling behaviour is necessary because they are ectothermic organisms and hence, temperature greatly influences their metabolism and functioning of muscles associated to air pumping for call production (Gerhardt and Huber 2002; Wells 2007). Temporal call features, but not spectral, are temperature-dependent in *H. subpunctatus*, as expected in most anurans (Navas and Bervier 2001; Wells 2007). Regarding acoustic signalling, *R. palmatus* also has been studied; Bernal et al. (2005) explores patterns and causes, including drift, of the intraspecific geographic variation in the advertisement call among seven populations. These studies are important because understanding factors promoting variability in call features, and behavioural traits in general, within a population or between populations of a given species, allows us to explore the earliest stages of evolutionary divergence, which in some cases generate speciation (Rundle and Boughman 2010; Wilkins et al. 2013).

2.2.2 Dendrobatoids Calling in Noisy Environments

Recent studies on acoustic communication with dendrobatoids in the Andes of Colombia have focused on documenting intraspecific geographic variation in features of the advertisement call and its relationship with habitat use. Abiotic noise can limit the effectiveness of acoustic communication among individuals by masking their signals, particularly those transmitted at lower frequencies (Brumm and Slabbekoorn 2005; Slabbekoorn and Ripmeester 2008). This is because abiotic noise tends to be more pronounced at lower frequencies compared to higher frequencies (Brumm and Slabbekoorn 2005). Consequently, it is expected that individuals alter their behaviours and features of their calls in a way that increase the signal-to-noise ratio (Fig. 3).

According to the acoustic adaptation hypothesis (Morton 1975; Ey and Fischer 2009), the emission of higher-frequency signals by species engaging in vocalizations and breeding in noisy environments, such as those near streams, has often been suggested or demonstrated as an evolutionary adaptation in anuran communication (Feng et al. 2006; Arch et al. 2008; Boeckle et al. 2009). This adaptive response, however, cannot be generalized to all anuran species calling alongside streams (Vargas-Salinas and Amézquita 2014; Carvajal-Castro and Vargas-Salinas 2016; Vargas-Salinas et al. 2024; Escalona et al. 2025). Among anurans, dendrobatoids offer a good opportunity for testing ecological and evolutionary changes in calling behaviour of individuals exposed to abiotic noise. This is because dendrobatoids call from territories scattered throughout their habitat, and they do not form dense choruses as do many species of Bufonidae, Hylidae and Ranidae. In anurans that gather to reproduce around lentic water bodies and form dense chorus, the calling behaviour of males may be primarily shaped by intrasexual competition and noise generated by nearby calling males, rather than by the background abiotic noise (Gerhardt and Huber 2002; Vargas-Salinas et al. 2014a). In contrast, in dendrobatoids, male mating success is largely influenced by the qualities of their calls and the amount of time and energy they invest in calling (Pröhl and Hödl 1999; Souza et al. 2021). The Colombian Andean endemic *Andinobates bombetes* is one such species (Fig. 2b).

The rubí poison frog *A. bombetes* is a species in which males establish territories in the leaflitter and call away from streams, both in the interior and on the edge of humid montane forest (Fig. 4). However, males in a tropical dry forest establish territories and call from leaf litter and rocky crevices alongside streams (Vargas-Salinas et al. 2014b). After statistically controlling for the potential effects of temperature and body size of males, these authors found that males in populations distributed away from streams call at lower frequencies than those in populations distributed alongside streams (Fig. 4). Similar results have been found with the lowland Colombian dendrobatoids *Oophaga histrionica* and *O. lehmanni* (Vargas-Salinas and Amézquita 2013b). It would be very interesting to use playbacks to perform signal recognition tests in *A. bombetes* and other dendrobatoids (Gerhard 1998; Amézquita et al. 2011). If individuals from non-stream populations approach and respond to a speaker broadcasting a stream-population male's call, it would indicate recognition of the call as a conspecific signal. On the contrary, a weak or absent response would suggest lack of recognition of the call as a conspecific signal (Gerhardt 1998; Gannon et al. 2022), suggesting some level of reproductive isolation as a possible indirect effect of calling divergence

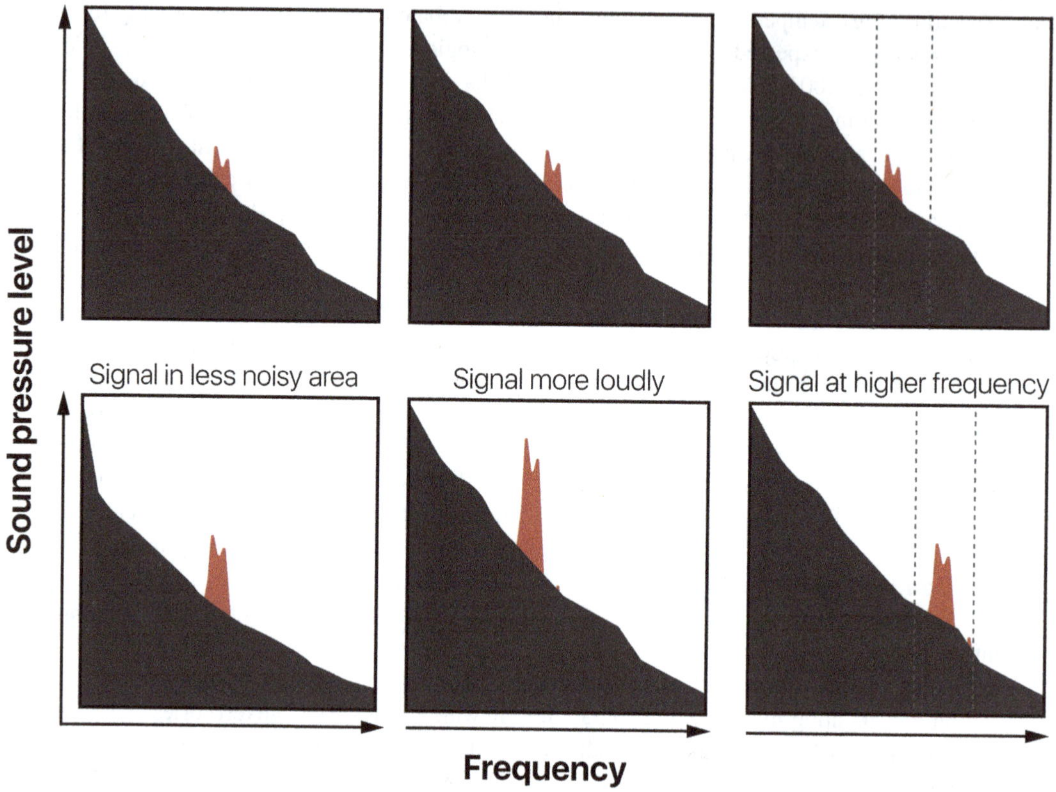

Fig. 3 Study scheme of ecological and evolutionary effects of abiotic noise in the emission of auditory signals in animals. The diagram illustrates the power spectra of a signal (depicted in red) and the background acoustic environment (depicted in grey) alongside streams or highways. In the top row, the scenarios show how the acoustic signal is masked by background abiotic noise, rendering the signal almost undetectable (i.e. low signal-to-noise ratio). In contrast, the bottom row depicts three scenarios where signallers could respond to mitigate the masking effects of noise. Left column: escaping the noise by signalling in a less noisy area. Middle column: signalling with greater intensity. Right column: signalling at a higher frequency. In each case, these adjustments would increase the signal-to-noise ratio. Figure adapted from Gomes et al. (2021) and Francis et al. (2023). Another behavioural alternative to communicate in noisy habitats include emitting highly repetitive calls, long pulsed calls or altering calling emission rate (Dubois and Martens 1984; Vélez et al. 2013), and even, performing visual displays (Hödl and Amézquita 2001)

promoted by noise on streams. To our knowledge, no studies have tested this potential role of stream noise as a factor promoting speciation in frogs.

Similar to abiotic noise alongside streams, anthropogenic noise tends to be more pronounced at lower frequencies than at higher frequencies (Brumm and Slabbekoorn 2005). Consequently, numerous authors have proposed that calling at higher frequencies represents an adaptation that mitigates acoustic masking for individuals communicating in urban habitats or near roadways (Slabbekoorn and Peet 2003; Halfwerk and Slabbekoorn 2009; Pohl et al. 2012; Römer 2013). Males of *A. bombetes* from a population exposed to anthropogenic noise from a highway for over 40 years alter their calling rate according to local variations in the level of anthropogenic noise (Vargas-Salinas and Amézquita 2013a) (Fig. 4). They also vocalize at higher frequencies than individuals from populations in habitats without abiotic noise, whether anthropogenic or natural (Vargas-Salinas et al. 2014b). Such shift in call frequency does not necessarily imply an

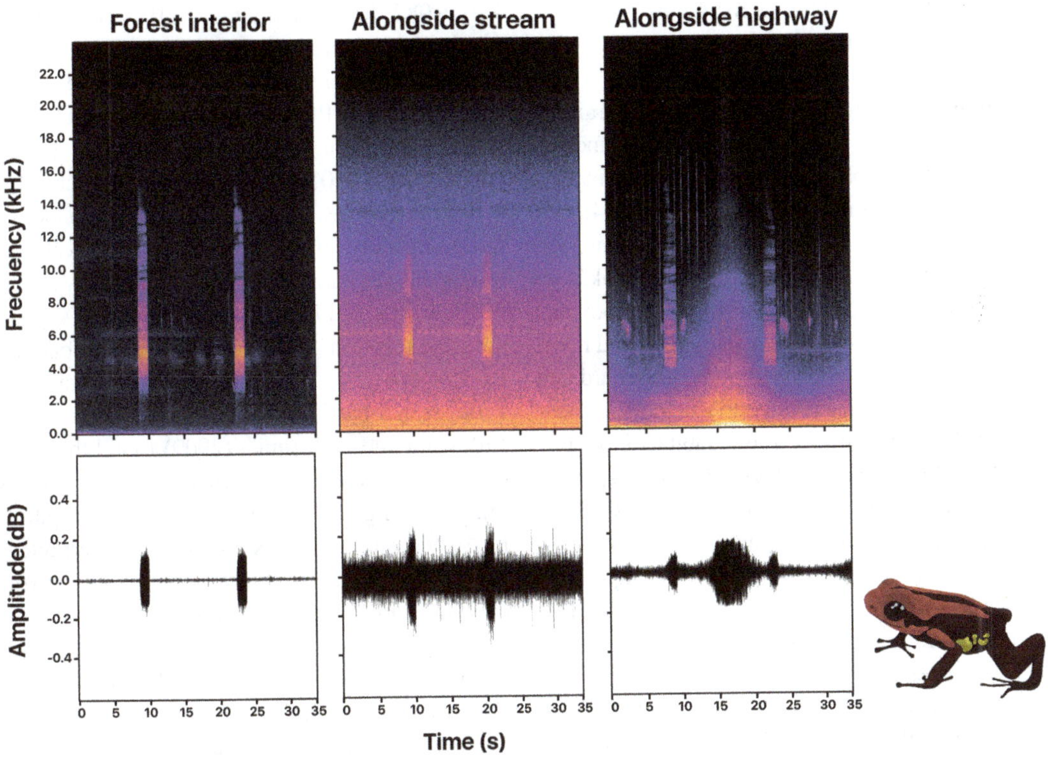

Fig. 4 Spectrograms (top row) and oscillograms (bottom row) of the advertisement call of the rubí poison frog *Andinobates bombetes* across three levels of background noise. Left column: two calls of a male in a habitat with low levels of abiotic noise (e.g. places in a forest away from streams and highways). Middle column: two calls emitted by a male signalling alongside a stream. Right column: scenario in which the male is alternating the emission of their calls with variation in the level of anthropogenic noise produced by trucks on a highway. Note that the call frequency is higher in the frog signalling in habitats with higher level of abiotic noise. Panel showing trends documented by Vargas-Salinas and Amézquita (2013a) and Vargas-Salinas et al. (2014b). (Drawing of the frog by María Paula Toro-Gómez)

evolutionary change, as it can be a reversible behavioural change (Warren et al. 2006; Cunnington and Fahrig 2010) or may result from an indirect effect of temporarily calling at higher amplitude (Nemeth and Brumm 2009, 2010; Nemeth et al. 2013). Subsequent studies using playback experiments with anthropogenic noise (Jiménez-Vargas and Vargas-Salinas 2021) and comparing call features among 12 populations of *A. bombetes* (González-Acosta et al. 2024) provide additional support for the adaptive change in call frequency in relation to anthropogenic noise, but more evidence is necessary to reach robust conclusions.

The Andes is the region of Colombia with the highest density of human population, and with the highest development rate of cities and highways. Since almost all dendrobatoids are diurnal, they call when the level of anthropogenic noise tends to be highest. Studies about the effects of anthropogenic noise with anurans in the Colombian Andes are limited to those with *A. bombetes*. It is important to perform more studies to determine how much anthropogenic noise can alter habitat use by anurans, affect population size, or promote evolutionary changes. Some models considering studies performed with species in several regions of the world (e.g. Fahrig and Rytwinski 2009; Rytwinski and Fahrig 2013) offer baseline data for establishing some predictions, but the high diversity of natural history traits in anurans hinders generalizations.

2.2.3 Selection and Drift: Calls in Dendrobatoids as a Multicomponent Signal

Communication signals consist of several traits or components, even within a single sensory modality, which can evolve in a semi-independent way to the selective forces (Grether et al. 2004; Ord and Martins 2006). In the case of acoustic signals, there are spectral components such as peak, low and high frequency, and temporal components such as call duration, call emission rate, and number and duration of notes and pulses (Gerhard and Huber 2002). Studies about acoustic communication of dendrobatoids in noisy habitats have primarily addressed potential adaptations in spectral components of male advertisement calls. However, noise in streams also may promote the evolution of calls with specific temporal components (Dubois and Martens 1984; Vélez et al. 2013). Evidence in dendrobatoids and other anuran clades suggests that the presence of constant abiotic noise near streams and waterfalls may have influenced the evolution of spectral call components without greatly affecting its temporal components (Vargas-Salinas and Amézquita 2013b; Goutte et al. 2016; Vargas-Salinas et al. 2024). For instance, in the species *Oophaga histrionica* and *O. lehmanni*, both with populations in noisy (because streams) and less-noisy environments, the temporal components of the call reflect taxonomic affinity, as expected, but the spectral components of the call reflect similarities in the acoustic environment (Vargas-Salinas and Amézquita 2013b). Clearly, more studies are needed to determine the influence of abiotic noise on the decoupling of spectral and temporal components of anuran vocalizations before generalizations can be made. Such studies should consider physiological/morphological constraints on the evolution of specific acoustic traits; for example, the production of long pulsed calls is energetically expensive, and the peak frequency of calls is related in a negative way to body size in most species (Wells 2001; Bevier 2016; Tonini et al. 2020).

Biologists might incorrectly interpret phenotypic traits as the result of selection, especially those traits associated with a function (Gould and Vrba 1982; Fox and Westneat 2010). Since communication signals mediate biotic interactions, which can determine the fitness of individuals, it is assumed that the evolution of communication signals has been under strong selective forces (Enquist et al. 2010; Bradbury and Vehrencamp 2011). However, stochasticity may also determine diversity patterns at different levels of biological organization (Kimura 1968; Vellend 2010; Losos 2017), and communication signals are no exception. In this context, dendrobatoids have been a good study model. Using a phylogenetic signal-based approach with 21 species, Erdtmann and Amézquita (2009) found that several advertisement call components exhibited a relatively high phylogenetic signal, indicating that stochastic processes might have determined, at least in part, the structure of the advertisement calls in those anurans (for a similar tendency in tree frogs genus *Hyloscirtus* and glass frogs, see Vargas-Salinas et al. 2024). Most dendrobatoids included in this study are from the lowlands (i.e. <900 m a.s.l.), but one can expect similar trends with dendrobatoids distributed at higher elevations. In agreement with this hypothesis, a positive relationship between the geographical distance and the level of call differentiation between populations has been recorded in some Andean dendrobatoids (*R. palmatus*: Bernal et al. 2005; *A. bombetes*: González-Acosta et al. 2024).

2.2.4 What Role Does the Receiver Perspective Play?

To understand the evolutionary ecology of communication in anurans, it is fundamental to study the receiver's perspective (Narins et al. 2006; Bradbury and Vehrencamp 2011). The mechanistic aspect of the auditory system in anurans has been studied for decades (Ehret et al. 1990; Fay and Popper 1999; Smotherman and Narins 2004; Narins et al. 2024), and the ecological and evolutionary approach of the auditory perception in anurans is an active research area (Feng et al. 2006; Lee et al. 2021; Narins et al. 2024). As mentioned previously, dendrobatoids are good study models because males are territorial and increase their rate and intensity of acoustic signals, or even engage in physical attacks, against an intruding male (Pröhl 1997, 2005).

Since these responses are conspicuous and can be easily quantified, dendrobatoids have served to understand which components of the calls stimulate the sensory system of receptors and elicit behavioural responses (Narins et al. 2003; Mayer et al. 2014; Ringler et al. 2017; Protti-Sánchez et al. 2023). Studies using this experimental approach with dendrobatoids in Colombia have been performed with lowland species (e.g. Amézquita et al. 2011; Betancourth-Cundar et al. 2016, 2024; Gil-Guevara and Amézquita 2020; González-Santoro et al. 2023). Nevertheless, they offer a framework and valuable insights into the dynamics of communication that could be extrapolated to Andean species.

2.2.5 Not Just Acoustic Signalling: Multimodal Communication in Dendrobatoids

Multimodal communication occurs when signals emitted by a sender simultaneously stimulate distinct sensory systems and responses in the receiver (Rowe 1999; Partan and Marler 2005). In the Colombian Andes, no studies have focused directly on testing multimodal communication in dendrobatids. However, many colourful dendrobatoids (e.g. genus *Andinobates*) are considered aposematic (Brown et al. 2011; Kahn et al. 2016; Gonzalez and Carazzone 2023). Aposematism is an antipredator strategy that implies multimodal communication given that it includes the use of acoustic or visual warning signalling by the prey with a chemical signalling that confers an unpleasant experience to the predator (Poulton 1890; Cott 1940; Ruxton et al. 2018; Rojas et al. 2018). The few studies that have been performed about aposematism in dendrobatoids of the Andes of Colombia, and therefore in a multisensory communication system, have addressed the origin and maintenance of variation in warning visual signals (Casas-Cardona et al. 2018; Márquez et al. 2020).

Research conducted with the polytypic dendrobatid *A. bombetes* aimed to determine whether different colour morphs are equally effective as visual predator deterrents. Most known populations of this species have individuals displaying red dorsolateral stripes on a dark brown background (Fig. 2b). However, in some populations, individuals display yellow dorsolateral stripes (Vargas-Salinas et al. 2020). Using clay models that imitate frogs, researchers tested whether visual predators (e.g. birds) accustomed to one warning signal (e.g. red colour) can generalize their learned avoidance when exposed to a new, unknown warning signal (e.g. yellow colour) (Casas-Cardona et al. 2018). These authors found that both red and yellow colours effectively serve as warning signals. Thus, there is little support for the hypothesis that divergent local predation pressures from primarily visual animals are driving the red/yellow polytypism in *A. bombetes*. Similar results have been obtained in lowland dendrobatoids displaying warning signals (e.g. species of the genera *Oophaga*, *Dendrobates* and *Ranitomeya*; Chouteau and Angers 2011; Amézquita et al. 2013b; Hegna et al. 2013; Rojas et al. 2014; Stuckert et al. 2014). Further studies with *A. bombetes* could include genetic analyses to estimate patterns of gene flow between populations (Brusa et al. 2013; Ebersbach et al. 2019; Márquez et al. 2020) or the role of imprinting in mate selection (Ewer 1956; Ten Cate et al. 2006; Yang et al. 2016, 2019), shedding light on the origin of the polytypism in this species.

Márquez et al. (2020) studied five aposematic dendrobatoids of the *Phyllobates* genus, some of them reaching what traditionally has been considered an Andean distribution (e.g. *P. bicolor* at >900 m a.s.l.). Those authors found that the colour patterns among *Phyllobates* populations in Colombia exhibit a leapfrog distribution, in which populations phenotypically similar (e.g. solid-yellow: *P. terribilis*, *P. bicolor*) are separated by populations with different aposematic patterns (e.g. striped: *P. aurotaenia*) (see Supplementary Material 3.1). This leapfrog distribution pattern represents recurring occurrences of similar phenotypes in different locations and offers a valuable opportunity to understand what processes could promote the origin and colour pattern variability among species. There is gene flow between lowlands populations of *P. aurotaenia* and *P. terribilis* (<500 m a.s.l.) and one population of *P. bicolor* located at ±600 m a.s.l., but not with a population of *P. bicolor* at higher elevation (1500 m a.s.l.).

According to Márquez et al. (2020), it is possible that mid-elevation habitats exert distinct selective pressures compared to lowland habitats, potentially favouring solid-yellow patterns over striped ones in mid-elevation habitats.

Multimodal communication has also been studied with dendrobatoids in contexts of territorial defence (Narins et al. 2003; Crothers et al. 2010; Crothers and Cummings 2015; Sonnleitner et al. 2020), mate choice (Dreher and Pröhl 2014), and parental care (Schulte et al. 2011). No similar studies exist for the Colombian Andes, but there are some notable studies with lowland species. For instance, in *Allobates femoralis*, a territorial non-toxic dendrobatid in the Amazon, the probability of attack by a focal male was strongly linked to body movement and body size of a clay frog model, with body coloration having a weak correlation with the likelihood of attack. These findings emphasize the significance of body movement as a nonredundant component shaping male aggression within the context of multimodal signalling. Moreover, this result challenges prior assumptions about the significance of body coloration in conspecific recognition in dendrobatoids (Reynolds and Fitzpatrick 2007; Maan and Cummings 2008; Richards-Zawacki et al. 2012; Siddiqi et al. 2004). In the polytypic species *Oophaga histrionica*, and *O. lehmanni*, mtDNA and playback experiments with acoustic and visual signals indicate that hybridization may contribute to create polytipism (Medina et al. 2013). A similar conclusion was found with another polytypic congeneric species (*O. sylvatica*) based on mitochondrial and nuclear genes obtained from 13 populations in Ecuador (Roland et al. 2017).

There are no studies about multisensory communication in the context of parental care behaviour with dendrobatoids in Colombia. However, the presence of dark-coloured tadpoles on the dorsum of parents might compromise the efficacy of their warning coloration as a defence strategy against visually oriented predators (Toro-Gómez et al. 2022). These authors conducted an experiment employing chicks as a predator that has been trained to recognize and avoid specific warning signals on printed frog models with tadpoles and without tadpoles on the back. Overall, trained chicks do not attack frog models randomly; frog models featuring warning coloration and without tadpoles on the back experienced a reduced risk of attack by the chicks compared to similar models with tadpoles. This result suggests that at short distances from predators, aposematic frog could maintain the warning function of conspicuous coloration if it was situated on parts of the parent's body not covered by the tadpoles during transportation. Toro-Gómez et al. (2022) also suggest that at large distances from predators, black tadpoles on the back of parents could hinder the detection of the parental frog because they could break the body contour of the parent in a natural complex environment (Cuthill et al. 2005; Stevens and Cuthill 2006; Honma et al. 2015). More studies are necessary to corroborate these findings and hypotheses, but in dendrobatoids, conspicuous coloration occurs toward the head and other parts of the body outside the dorsum. In line with this hypothesis, species climbing trees to deposit their tadpoles in arboreal phytotelmata, which implies high detectability by predators, are those with conspicuous warning coloration and a reduction in the number of tadpoles simultaneously transported on the back (Carvajal-Castro et al. 2021; Ringler et al. 2023).

2.2.6 Other Behavioural Lines of Research with Dendrobatoids in the Colombian Andes

In addition to the ones mentioned above, there are many interesting behavioural studies that could be conducted with Andean dendrobatoids in Colombia; see, for example, Vargas-Salinas and Rojas (2024) and Rojas and Vargas-Salinas (2024) for an overview based on manuscripts submitted to a recent special issue covering evolutionary ecology of poison frogs. Below, we mention three developing lines of research: the use of pheromones during courtship, homing, and laterality.

The use of chemical signals and cues in context of predator-prey interaction, mating behaviour and parental care has been hypothesized or demonstrated for several anuran lineages (Vences et al. 2007; Starnberger et al. 2013; Brunetti et al. 2014; Schulte et al. 2021,

2023a, b; Zheng et al. 2024) including dendrobatoids (Schulte et al. 2011, 2023b; Schulze et al. 2024). Nonetheless, the realm of sexual chemical communication in anurans has rarely been tested (Starnberger et al. 2013; Brunetti et al. 2014). One of the few exceptions is a study performed with two dendrobatoids, one of them the endemic *Leucostethus brachistriatus* from the Colombian Andes (Abondano-Almeida et al. 2024). A transcriptomic sequencing approach found that swollen fingers in the hands of males have a larger number of transcripts (60 transcripts) encoding sodefrin precursor-like factors (SPFs), a well-known pheromone system in other amphibians, than non-swollen fingers.

Homing is the ability by an animal to return to a specific place (e.g. territory) after it has been displaced (Papi 1992). Homing has been studied in several lowland dendrobatoids: *Allobates femoralis*, *A. talamancae*, *Ameerega trivittata*, *Dendrobates auratus*, *Oophaga pumilio* and *O. histrionica* (McVey et al. 1981; Nowakowski et al. 2013; Pašukonis et al. 2013, 2018, 2019; Pichler et al. 2017; Nothacker et al. 2018; Gómez-Consuegra and Amézquita 2024), but only in one dendrobatid of the Colombian Andes: *Andinobates bombetes* (Arcila-Pérez et al. 2020). This study stands out because it includes many females who, like the males, exhibited homing. Most homing studies in anurans have been conducted with males, or the authors have not differentiated the sex of translocated individuals. Since a controversial question in evolutionary biology is whether selective pressures have led to sex differences in cognitive processes related to spatial memory (Olmstead and Kuhlmeier 2015), the said sex bias limits our understanding of processes shaping the evolution of homing in anurans. Studying homing in a greater number of Colombian dendrobatoids is necessary because they exhibit diverse life history characteristics that could influence this behaviour. For instance, how do homing ability and motivation differ among dendrobatoids based on body size, antipredator strategies (aposematism, crypsis) and microhabitat of tadpole development (small ponds at ground vs arboreal phytotelmata)?

Dendrobatoids can also serve as excellent models for studying laterality, which is the tendency of animals to use structures on one side of the body more frequently or efficiently than the other side (Bianki 1988). In fact, two of the three studies on laterality with neotropical anurans include dendrobatoids (Anderson et al. 2021; Rojas-Montoya and Vargas-Salinas 2023; Keuroghlian-Eaton et al. 2024). In their study with Colombian anurans, Rojas-Montoya and Vargas-Salinas (2023) conducted experiments covering the head and eyes of individuals with a damp cloth and subsequently recorded which limb they used to remove it. In a population of *A. bombetes* (Fig. 2b), most of individuals (87%) used their right front hand, while in *L. brachistriatus* (Fig. 2d), individuals were mostly left-handed (83%). Given that the right hemisphere of the brain of vertebrates is associated to the regulation of quick responses to stimuli such as attacks by competitors or predators (left-side body dominance), and the left hemisphere is associated with more analytical and planned task behaviours (right-side body dominance) (Robins 2006; Csermely and Regolin 2013), it is intriguing that these two dendrobatoids exhibit opposite patterns of laterality. Could differences in predation risk between aposematic and non-aposematic species have driven opposite trends in lateralization across these groups? (Rojas-Montoya and Vargas-Salinas 2023).

2.2.7 Conclusions of Behavioural Studies with Dendrobatoids in the Colombian Andes

The early behavioural studies were conducted with cryptic species that do not exhibit the conspicuous and striking coloration that has driven much research on dendrobatoids (Fandiño et al. 1997; Vargas-Salinas and Rojas 2024; Rojas and Vargas-Salinas 2024). Including less conspicuously coloured species of dendrobatoids in research is important for several reasons. First, it allows for comparative studies within the same family, which has shed light on the influence of aposematism on species diversification rates and the evolution of complex parental care behaviours (Santos et al. 2014; Carvajal-Castro et al. 2021;

Medina et al. 2024). Second, in well-studied aposematic species, new and previously unknown behaviours for dendrobatoids continue to be discovered (Stynoski et al. 2024; Dugas et al. 2024a, b), and the same is true for non-aposematic species (e.g. Fouilloux et al. 2024; Abondano-Almeida et al. 2024). Third, although aposematic species are charismatic and can serve as flagship species for conservation (Amézquita et al. 2016), less showy species also contribute to the biological diversity of this clade and, therefore, need to be valued in conservation initiatives (Buitrago-Marulanda et al. 2025). In general, behavioural research on dendrobatoids has multiplied in recent decades, but clearly, many Colombian Andean species have been overlooked.

2.3 Centrolenidae (Glass Frogs)

The family Centrolenidae, commonly referred to as glass frogs due to their distinctive translucent dorsal coloration and the transparent nature of their belly, comprises over 167 arboreal and nocturnal species (Frost 2025). Glass frogs reproduce near streams, where males emit calls to attract females. Egg clutches are deposited on rocks, leaves, and overhanging stems above the water's surface. Upon hatching, the embryos drop into the water and complete their development as gastromyzophorous tadpoles (Kubicki 2007). In some glass frogs, parental care is undertaken by males, while in others, females assume such role (Delia et al. 2017a, b).

In Colombia, there are at least 81 species of glass frogs. Most of what we know about the behaviour of glass frogs that live in the Andes of Colombia comes from records mentioned in the descriptions of new species and from natural history observations made in populations of Colombia (Lynch et al. 1983; Guayara-Barragan et al. 2010; Escobar-Lasso and Rojas-Morales 2012; Hutter et al. 2013; Guevara-Molina and Vargas-Salinas 2014; Arcila-Pérez et al. 2017; Rios-Soto et al. 2017a, b; Prado-Sañudo et al. 2018; Villegas-Mora et al. 2020) and other countries (e.g. Rojas-Runjaic and Cabello 2011; Delia et al. 2017a, b). Next, we present an overview of the topics that have been most studied with Andean glass frogs distributed in Colombia: acoustic signalling, mating behaviour, parental care and, recently, homing (Fig. 5).

2.3.1 Evolution of Acoustic Signal Structure

Extensive efforts have been made to elucidate the phylogenetic relationships among glass frogs (Guayasamin et al. 2009; M. Rada, unpublished data) and to document their advertisement calls (Duarte-Marín et al. 2022; Rodríguez-Correa et al. 2024). Consequently, glass frogs, much like their dendrobatid counterparts, have emerged as valuable models for performing macroecological studies investigating evolutionary patterns of acoustic communication signals.

The evolution of advertisement calls in glass frogs appears to have been shaped by physical characteristics of the habitat and energetically constraints rather than by competition and predation pressures. Using comparative analyses based on 72 species of glass frogs, Escalona et al. (2019) found that peak frequency of the advertisement call was inversely related with body size, as expected, whereas temporal components of the call were associated with calling site and paternal care. These authors hypothesized that (1) the emission of pulsed calls can be easier and energetically less demanding if emitted from the upper surface of the leaf, while emitting tonal calls can be less demanding for males positioned on the underside of leaves, and (2) males in species with paternal care (another energetically expensive behaviour) tend to produce tonal calls, while males in species with maternal care tend to produce pulsed calls. A posterior study found that the structure of the advertisement calls in glass frogs exhibits a convergent pattern, as expected under a habitat filtering effect, possibly because of habitat similarities among species (Mendoza-Henao et al. 2023). The phylogenetic signal of spectral components of advertisement calls of 94 species of Centrolenidae indicate that, contrary to what was found in some dendrobatoids (see above), abiotic noise in streams does not seem to promote the evolution of high-frequency calls in Centrolenidae (Vargas-

Fig. 5 Photographs of some of the glass frogs in the Andes of Colombia for which there is more detailed information about behaviour. (**a**) *Centrolene savagei*, (**b**) *Espadarana prosoblepon*, (**c**) *Hyalinobatrachium tatayoi* and (**d**) *Nymphargus grandisonae*. (Pictures by Fernando Vargas-Salinas (**a**, **c**) and Cristian González-Acosta (**b**, **d**))

Salinas et al. 2024). Apparently, the small body size (average SVL = 23.8 mm; range: 18.4–75.2 mm; Duarte-Marin et al. 2022) and the concomitant high call frequency of most glass frogs make their acoustic signals less susceptible to masking by abiotic noise.

2.3.2 Breeding Behaviour and Sexual Selection

Sexual selection theory predicts that males compete to gain access to females, and females select males based on their monopolized resources, phenotypic traits or displays (Andersson 1994; but see Schlupp 2021 for many examples of competition among females and choosy males). Studies of reproductive behaviour in glass frogs from the Colombian Andes (e.g. Vargas-Salinas et al. 2017; Ospina et al. 2017; Basto-Riascos et al. 2017; Rojas-Montoya et al. 2024b) and other countries in Central (e.g. Jacobson 1985; Mangold et al. 2015) and South America (e.g. Noronha and Rodriguez 2018; Salgado and Guayasamin 2018) report a non-random mating pattern.

The mating success of males in glass frogs has been related to their mate search effort (i.e. chorus tenure). This mating pattern is in accordance with sexual selection acting upon males, promoting the capacity to remain reproductively active during longer periods of time in the breeding season (Andersson and Iwasa 1996). Males that spend more nights calling would have a greater mating success simply because they have a greater probability of being found by a female when she approaches the stream to reproduce. The height of the calling site also may be positively associated male mating success in some species (Greer and Wells 1980; Wells and Schwartz 1982; Muñoz and Halfwerk 2022). Regarding body size, it does not seem to play an important role in the mating success of males in glass frogs (but see Ospina et al. 2017) as it does in species from other anuran families (Sullivan et al. 1995; Wells 2007).

In glass frogs, little research has focused on female choice, but the elaborate courtship behaviour described in some Colombian Andean species (Vargas-Salinas et al. 2014c; Rios-Soto et al. 2017b; Rojas-Montoya et al. 2024b) suggests female choice can be important. Moreover, calling is energetically costly in anurans (Pough et al. 1992; Wells 2001; Bevier 2016) and implies risks of predation and parasitism (Gerhardt and Huber 2002; Bernal et al. 2006). Therefore, when females mate with males that call during numerous nights (see examples above), they could be mating with males of high genetic quality, good physical condition, or with good foraging capacity and fitness recovery throughout the breeding season (Bateson 1983; Rosenthal 2017). This hypothesis agrees with a process of passive female choice (Parker 1982; Parker and Waite 1997). Among Colombian glass frogs, *Centrolene savagei* (Fig. 6) is the only species where male body size correlates with mating success. However, while male body size relates mating in *C. savagei*, it does not affect parental care (egg brooding) performance (Ospina et al. 2017). In contrast, studies of lowland Amazonian (*Hyalinobatrachium cappellei*) and Central American/Ecuadorian Pacific coast (*H. valerioi*) glass frogs have revealed that females play a significant role in non-random mating patterns. In these species, females prefer to mate with males already caring for clutches laid by other females than with single males (Valencia-Aguilar et al. 2020; Garrido-Priego et al. 2024). Whether female mate choice results from social copying decisions by females or own decisions based on call features emitted by males and related to their parental performance deserves further studies (Garrido-Priego et al. 2024).

2.3.3 Parental Care: Who Cares and How Much?

Our understanding of the evolution of parental care behaviour and its adaptive advantages in glass frogs has significantly increased over the past 15 years. Traditionally, it was believed that parental care in glass frogs was solely carried out by males, which can extend until the embryos hatch (Vockenhuber et al. 2009; Valencia-Aguilar et al. 2012; Vargas-Salinas et al. 2014c). The discovery of maternal care in an endemic glass frog to the Sierra Nevada de Santa Marta (SNSM) in the north of Colombia was considered novel when first reported (Cisneros-Heredia and McDiarmid 2007; Bravo-Valencia and Delia 2016). However, Delia et al. (2017a) demonstrated that the traditional concept of exclusive paternal care in Centrolenidae is incorrect. In many species, females remain with the egg masses after oviposition, covering them with their body; since females only attend their offspring for a few hours, this behaviour has not been properly recorded. Subsequently, Delia et al. (2020) confirmed that the structure of the egg masses is associated with the sex that cares for the offspring. Jelly-rich clutches, which are less susceptible to desiccation, tend to be cared by females (but see Goyes Vallejos et al. 2024), while extended male care is associated with simple clutches (i.e. with reduction in egg-jelly), which are more susceptible to desiccation. Other factors such as oviposition site characteristics are also related to desiccation risk of clutches and hence play a role in the evolution of female or male paternal care in glass frogs (Delia et al. 2020).

Macroecological studies of parental care in glass frogs (Delia et al. 2017a, 2020; Valencia-Aguilar et al. 2024) include species that are present in the Colombian Andes (e.g. *N. grandisonae*, *C.* aff. *savagei*, *E prosoblepon*; Fig. 5) or other mountain systems in the country (e.g. *I. tayrona*), but studies about parental care in glass frogs performed in Colombia are few. Males of *Hyalinobatrachium aureoguttatum*, a lowland species, stayed near eggs, touching or sitting on the clutches throughout the day, and during most of embryo development (Valencia-Aguilar et al. 2012). These authors also highlighted that such parental behaviour included hydration, cleaning, and defence of the embryos against predators. In *Centrolene savagei*, there is behavioural plasticity in egg attendance between day and night, and throughout embryo development (Ospina et al. 2020). Males of this Andean glass frog attend egg clutches more at night than at day, and the time invested in egg brooding reduces as embryos

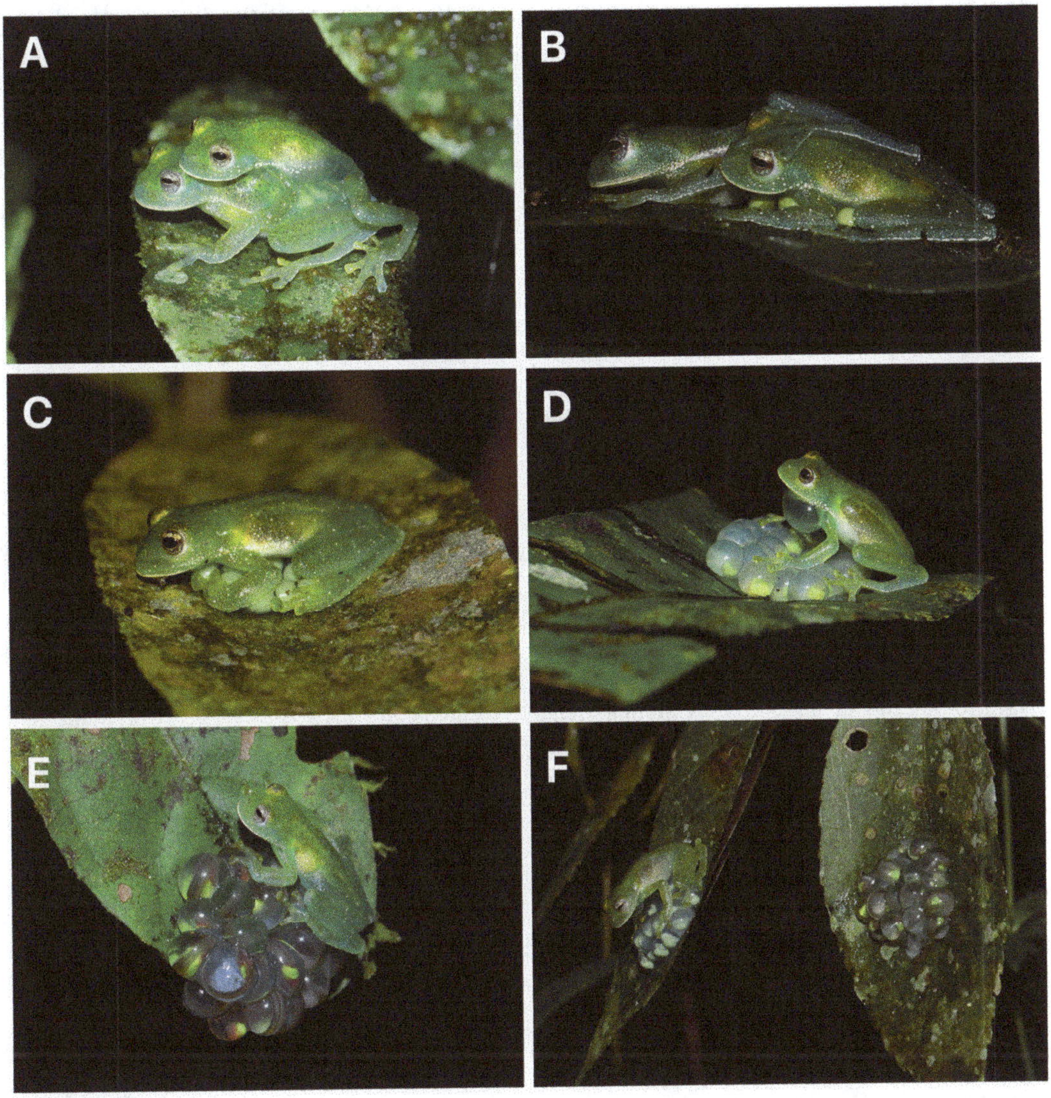

Fig. 6 Sequence of images showing breeding in the glass frog *Centrolene savagei*. (**a**) Axillary amplexus. (**b**) Coordinated movements between male and female after oviposition; with these movements, the male begins to cover the egg clutch with its body. (**c**) While the female moves away from the oviposition site towards adjacent vegetation, the male permanently covers the eggs with its body. (**d**) After a few days and when the embryos have progressed in their development, the male begins to call again. (**e**) Egg attendance by the male can extend until the embryos in the clutch have developed sufficiently to hatch from the egg and fall in the water where they complete their development until metamorphosis. (**f**) If the male attracts a new female while he is attending an egg clutch, he will start caring the new clutch more frequently than the previous one

develop and males begin to call again (Fig. 6). Behavioural plasticity has also been reported in *H. fleischmanni*, a Central American lowland species in which males increase or decrease brooding of clutches according to precipitation levels and the concomitant probability of offspring mortality by desiccation (Delia et al. 2013). The behavioural plasticity in parental care that has been recorded both in *C. savagei* and *H. fleischmanni* may be widespread among glass frogs, but more studies are necessary to test patterns and evolutionary implications. The Andes of Colombia and its high diversity of glass frogs provide many opportunities to perform such studies.

2.3.4 Homing: Where Is Home and Why Return?

As mentioned previously, dendrobatoids have been extensively studied regarding patterns and mechanisms associated with homing. However, research in other neotropical anurans than dendrobatoids will help elucidate the extent and patterns of this behaviour. Recently, montane Colombian glass frogs have served as valuable models for testing hypotheses about the relationship between homing, parental care and chorus membership.

Centrolene savagei is a glass frog in which parental care is performed by males (Vargas-Salinas et al. 2007, 2014c). Embryo-brooding males translocated by researchers to places away from clutches exhibit a higher probability of homing than translocated non-brooding males (Navarro-Salcedo et al. 2021). A similar trend was recorded in males of *H. valeriori,* a glass frog species in Central America (Garrido-Prieto et al. 2025), but not found for females in *I. tayrona*, a glass frog in which extended parental care is performed by females (Navarro-Salcedo et al. 2022). The reasons for these differences in homing motivation are unknown. On the other hand, Gómez-Murcia et al. (2024) investigated homing in *E. prosoblepon* (Fig. 5b), whose males gather in choruses located at specific sites along streams (hotspots sensu Angeli et al. 2015) and display remarkable calling site fidelity (Jacobson 1985; Robertson et al. 2008). After been translocated to another chorus, most males of this species prefer to return to their original chorus. Such homing response is intriguing, because habitat attributes do not seem to differ between chorus sites and independently of the presence of choruses (Rodríguez-Correa et al. unpublished data). Therefore, physical attributes in the habitat does not seem to influence homing motivation in *E. prosoblepon*. Instead, it is possible that males prefer to return to a chorus where they are familiar with the location of refuge sites or territorial limits of conspecific males with which they have had previous agonistic interactions (Rios-Soto et al. 2017b; Tumulty and Bee 2021). Acoustic playback experiments provide a robust methodology to examine if homing males of *E. prosoblepon* experience a reduction in fitness costs when interacting with unfamiliar males (dear enemy effect; Tumulty 2022).

Homing motivation and ability in dendrobatoids can depend on previous experience, visual and acoustic cues, or true navigation (Pašukonis et al. 2022). How similar mechanisms could apply to glass frogs requires designing hypothesis-driven experiments. Differences in the reproductive mode among dendrobatoids can influence the evolution of different homing abilities (Pašukonis et al. 2019, 2022), but since glass frogs do not exhibit such high diversity of reproductive modes, they offer a relatively simplified study system of homing with fewer confounding variables. Although traditionally overlooked in studies of homing, glass frogs offer many opportunities to develop this line of research in the Andes of Colombia and other regions or countries.

2.3.5 Conclusions of Behavioural Studies with Glass Frogs in the Colombian Andes

Glass frogs are an excellent study model system in anuran behavioural ecology. However, there are still many gaps that need to be addressed. For instance, it is unknown whether the intraspecific geographical variation in acoustic (and even visual) signals that has been frequently reported in other anuran clades (e.g. dendrobatoids) occurs in glass frogs (Wilczynski and Ryan 1999). A preliminary comparison of advertisement call between ten populations of the glass frog *H. tatayoi* distributed in Panama, Colombia, Venezuela, and Ecuador suggests little intraspecific geographic variation in advertisement call and courtship call features (Rojas-Montoya et al. 2024a; see also results by Mendoza-Henao et al. 2020; Mendoza-Henao et al. 2023). A similar tendency might be happening with *E. prosoblepon* (Rodríguez-Correa et al. 2024). Regarding variation in behavioural aspects other than acoustic signalling, some of the few available studies are those with species having populations in the Colombian Andes:

H. fleischmanni complex (Delia et al. 2014), *H. tatayoi* (Rojas-Montoya et al. 2024b) and *E. prosoblepon* (Basto-Riascos et al. 2017; Goyes Vallejos et al. 2021). Clearly, more studies about intraspecific geographic variation in glass frogs are necessary.

At least 35 species of glass frogs are endemic to Colombia, some of them with unique behaviours, hence providing opportunities to test behavioural hypotheses that cannot be tested with other species. Some examples of unique behaviours are *Ikakogi tayrona* and *I. ispacue*, the only known glass frogs with prolonged maternal care (Bravo-Valencia et al. 2016; Rada et al. 2019), and *Centrolene geckoideum*, endemic to the Colombian and Ecuadorian Andes, which has an extreme morphology compared to other glass frogs (Lynch et al. 1983; Rueda-Almonacid 1994; Grant et al. 1998). It is expected that the number of studies covering behavioural aspects in glass frogs will increase and diversify in the coming years. Hopefully, a percentage of these studies will be conducted with Colombian Andean species, which may be particularly susceptible to high rates of deforestation and disturbance of natural habitats, as well as large-scale processes such as climate change.

3 Beyond the Andes: Behavioural Studies with Anurans in Another Mountain Ranges of Colombia

The Sierra Nevada de Santa Marta (SNSM), located in northern Colombia and isolated from the Andes, is considered as the world's highest coastal mountain range, reaching elevations of over 5700 m a.s.l. (Rangel-Ch and Garzón 1995). A wide range of ecosystems, from tropical rainforests to paramos, occur within a relatively short distance, and host numerous endemic plant and animal species (Wurdack 1976; Lynch et al. 1997; Carbono and Lozano-Contreras 1997; Sánchez-Pacheco et al. 2018; Arroyo et al. 2022; Botero-Delgadillo et al. 2024). The harlequin frogs of the genus *Atelopus* (Anura: Bufonidae) are prominent among the anuran lineages that live in the SNSM (Ruiz-Carranza et al. 1994); their populations survive in apparent abundance despite their Andean counterparts suffering population declines or possible extinctions (La Marca et al. 2005; Lötters 2007; Lötters et al. 2023).

The Santa Marta harlequin toad (*Atelopus laetissimus*) has attracted behavioural research. In this species, the reproductive effort of males (measured as loss of body weight) varies from 25 to 30% depending on climatic conditions, which, in turn, influences its reproductive phenology (Rocha-Usuga et al. 2017). In dry years, females opt for quick spawning in streams, resulting in lower reproductive effort for males; conversely, in rainy years, streams increase their flow and optimal spawning sites are harder to find for females, prolonging amplexus duration and considerably increasing the reproductive effort of males. *Atelopus laetissimus* also has been a study model for understanding factors behind the evolution of prolonged amplexus (several months duration) in the genus *Atelopus*, a distinctive behavioural feature of these toads (Dole and Durant 1974; Lötters 1996). Male-male competition and an unequal operational sex ratio in *A. laetissimus* populations may have promoted mate-guarding behaviour through prolonged amplexus (Rueda-Solano et al. 2022). These authors found evidence supporting such hypothesis through in situ behavioural experiments, which included comparisons of sexual dimorphism, test of functionality of nuptial embraces and the development of forelimbs in amplexed and non-amplexed males. This study is the only one offering experimental evidence about the evolution of prolonged amplexus in anurans and the basis for further studies looking for testing a possible relationship between sexual selection, mate-guarding behaviour, and two divergent morphological tendencies between "Andean" and "lowland" *Atelopus* species (Rueda-Solano et al. in prep).

Other behavioural studies with *Atelopus* in the SNSM covered nocturnal activity of adults (Rueda-Solano et al. 2016; Rueda-Solano and Warkentin 2016) and the preference for specific microhabitats by tadpoles in streams (Pérez-

González et al. 2020; Rueda-Solano et al. 2015). Therefore, *Atelopus* species in the SNSM offer a window into the ecology of their Andean congeners, whose populations are decimated and for whom hypothesis-testing studies are not feasible.

4 A Macroecological Exploration of Behavioural Patterns in "Andean Anurans"

As described before, the landscape of Colombia is composed of lowlands interrupted by three main mountain ranges (Andean region) and other mountains isolated systems, covering a broad elevational range (sea level to >5000 m a.s.l.; Fig. 7a) which translates into a broad range of temperature (Fig. 7b) and precipitation regimes (Fig. 7c). The combination of these environmental variables influences the spatial patterns of species richness and behavioural and ecological traits (although recognizing the importance of many other factors). For example, the western slope of the cordillera Occidental, the northern portion of the Magdalena valley and the eastern slope of the cordillera Oriental present the highest levels of species richness (Fig. 7d). These regions share a common influence from the highly humid Amazon and the Chocó biogeographic region.

Although species richness decreases with elevation on the Andes region of Colombia (Fig. 8a), following common macroecological trends in other taxa (McCain and Grytnes 2010; Peters et al. 2016), precipitation exerts a stronger influence on species richness than elevation (Fig. 8b). The well-documented dependence of species richness to moisture in anurans (Buckley and Jetz 2007), apparently covaries with other geographic patterns of behavioural ecology that have been less explored (but see Jiménez-Vargas et al. 2021, and Acevedo et al. 2022). For instance, sexual size dimorphism does not vary with elevation (Fig. 8c) but seem to increase with precipitation (Fig. 8d), despite a negative relationship between precipitation and body size (measured as maximum body size per species) (Fig. 8f). This pattern holds only when the entire elevational gradient is considered (see grey dots in bivariate plots; Fig. 8), meaning that assemblages in lowland rainforest drive these patterns.

The diversity of reproductive modes of anurans in the Andes (i.e. >900 m a.s.l.) showed a stronger trend with precipitation (Fig. 8h) than with elevation (Fig. 8g). Although this trend may be influenced by the number of species across regions, our metric implementing a diversity index should partially control for this effect (see details in the legend of Fig. 7). This means that areas with higher precipitation provide more ecological opportunities (macro- and microenvironments), which anurans are taking advantage of. For instance, high precipitation levels promote wet environments facilitating the establishment of frogs with direct development (e.g. genus *Strabomantis*), and the presence of ponds and streams for species with reproductive modes involving free swimming tadpoles (Wells 2007; Vargas-Salinas et al. 2019). Moreover, regions with high precipitation level are those in which forests have a high complexity in the vertical stratification, with dense understories, high level of epiphytes, high canopies and a high diversity of plans that create a diversity of substrates available for diverging reproductive modes. This may be evidenced in Fig. 9, where toward lowlands regions, mainly involving the Pacific coast and the northern portion of the Magdalena Valley (which has an influence from the Chocó biogeographic region), harbour a high proportion of diverging reproductive modes, from species laying eggs and tadpoles in ponds or streams to species with direct development.

Although it is not possible to assess the directional causality between species richness and size dimorphism, body size and diversity of reproductive modes, we can formulate some hypotheses. It is possible that diverging pre-existent reproductive modes facilitate the accumulation of more species through a mechanism of complementary, or alternatively, where occurring species using different spatial recourses reduce competition (Jops and O'Dwyer 2023). Further studies covering the historical context of biogeographic colonization and the evolution of morphological (e.g. body size and sexual dimorphism) and behavioural traits (e.g. reproductive modes) are

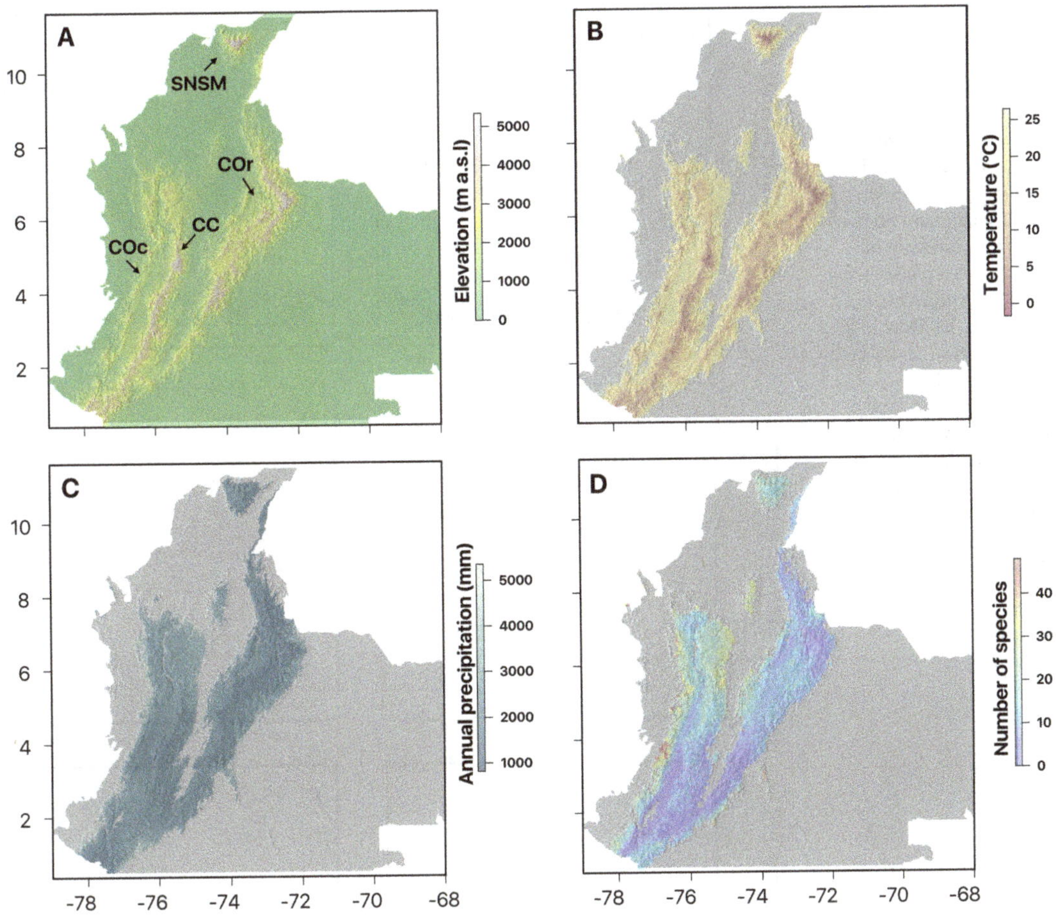

Fig. 7 Geographic patterns of anuran species richness in the Andean region of Colombia (COc cordillera Occidental, CC cordillera Central, COr cordillera Oriental) associated with topography and precipitation (note than other mountain ranges in Colombia are shown, e.g. SNSM: the Sierra Nevada of Santa Marta). (**a**) Topographic map derivates from the SRTM digital elevation model V4.1 (http://www2.jpl.nasa.gov/srtm/). (**b**) Temperature gradients as characterized by the Bio1 bioclime variable from WorldClim (http://www.worldclim.org). (**c**) Precipitation patterns as characterized by the Bio12 bioclime variable from WorldClim. (**d**) Spatial variation of the number of species across the Colombian mountain ranges. Species richness was estimated by overlapping species range polygons of 653 species that are present in at least one pixel of the study area, defined as any area above 900 m elevation in Colombia. Species range maps were obtained from the IUCN (version 2022). (Maps are displayed using a WGS84 coordinates system)

necessary to interpret the spatial divergence of traits in anurans of the Colombian Andes.

5 General Conclusions and Perspectives

Relatively few studies on animal behaviour have been conducted with anurans in the Andes of Colombia. This is concerning given that Colombia is also one of the countries with more threatened species of anurans (Stuart et al. 2008; Luedtke et al. 2023), and behavioural studies can inform the conservation and management of populations (Sutherland 1996; Gosling and Sutherland 2000). Moreover, animal behaviour, with its diversity and complexity, has the power to captivate and enchant people (Castillo-Velandia et al. 2019). Sharing aspects of Colombian Andean frog behaviour with diverse

Fig. 8 Bivariate relationships between elevation and precipitation level in the Andes of Colombia with species richness (**a**, **b**), size dimorphism: female body size minus male body size (**c**, **d**), maximum body size (**e**, **f**) and diversity of reproductive modes sensu Jiménez-Vargas et al. (2021) (**g**, **h**). Dots in grey correspond to pixels below 900 m a.s.l. Lines correspond to lowess functions. Analysis performed in R software using the *terra* and *sf* packages. Each dot in plots corresponds to a cell from the species richness map

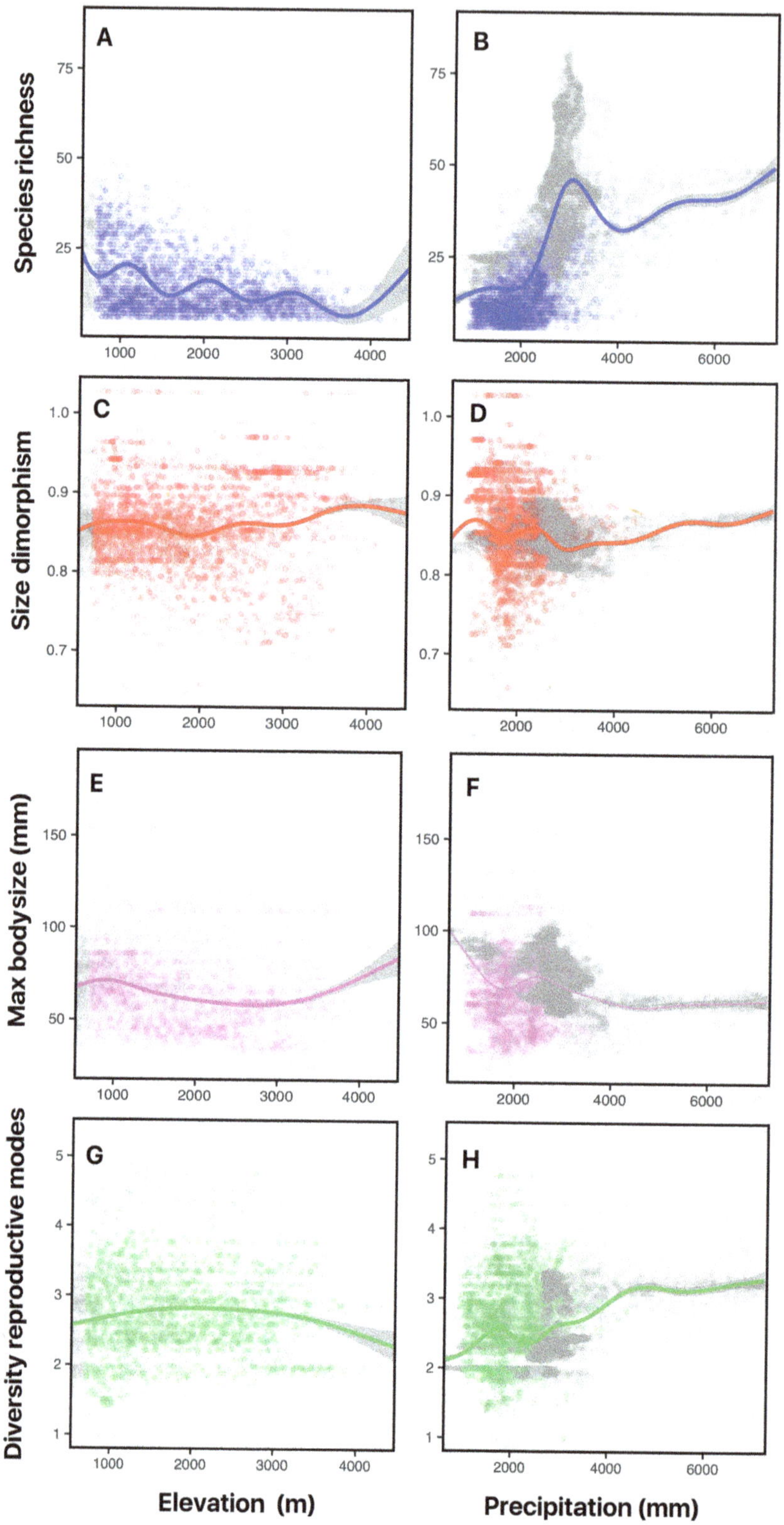

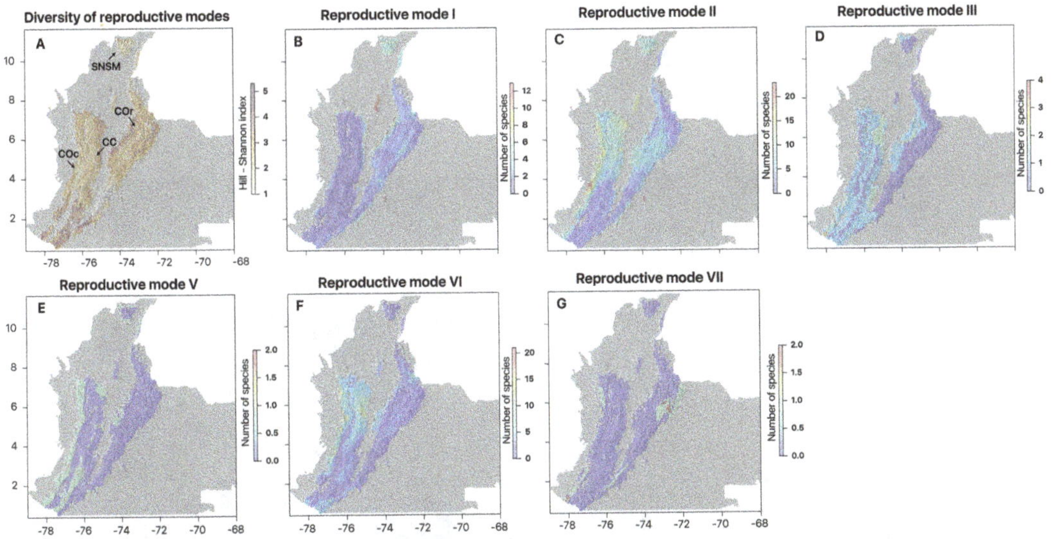

Fig. 9 Spatial distribution of reproductive modes of anurans in the Andes of Colombia (>900 m a.s.l.). Note that other mountain ranges in Colombia are shown. The first panel (**a**) represents the diversity of reproductive modes of the 653 species present in the species richness map (see Fig. 7d). To estimate the diversity of reproductive modes, we adapted the Shannon-Weaver index transformed to true diversity (sensu Jost 2006), where instead of species, we used the reproductive modes present in a community, and instead of abundance, we used the number of species having each reproductive mode. For reproductive modes, we used the seven broad categories proposed by Jiménez-Vargas et al. (2021) which are represented in panels (**b**–**g**), corresponding to: Mode I—eggs and tadpoles in ponds or streams; Mode II—eggs out of water (terrestrial, arboreal or foam nest) with tadpoles in ponds, streams or small pools in ground cavities; Mode III—eggs out of water with tadpoles in phytotelma; Mode V—eggs out of water that hatch into non-aquatic tadpoles or froglets by direct development; Mode VI—eggs carried by the adult, with tadpoles in ponds or streams; Mode VII—eggs carried by the adult hatch into froglets by direct development. The mode IV—eggs and tadpoles in phytotelma—is not present in a species included in this analysis. For details of map elaboration, see the legend of Fig. 7. SNSM: Sierra Nevada of Santa Marta, COc: cordillera Occidental, CC: cordillera Central, COr: cordillera Oriental

audiences, including children (Fig. 10), has consistently yielded positive and motivating experiences, demonstrating the value of studying animal behaviour to understand and conserve the biodiversity that surrounds us.

While most behavioural studies with Colombian Andean anurans focus on acoustic communication and breeding behaviour in a handful of species (e.g. *D. molitor*, *R. palmatus*, *A. bombetes*, *C. savagei*), many more questions remain unanswered with these and other species. Below, we highlight notable gaps and promising lines of research with anurans in the Andes of Colombia. First, most studies have been conducted with postmetamorphic individuals, especially adults. A quick systematic review of the literature on embryos and tadpoles in Colombia (Fig. 11) indicates that most studies have been conducted with lowland species, and there is a preponderance of morphological descriptions, either for identification during inventories (Lynch 2006; Lynch and Suárez-Mayorga 2011), for taxonomy and systematics (Anganoy-Criollo 2013; Ospina-Sarria et al. 2011; Rueda-Solano et al. 2015; Rada et al. 2019; Duarte-Marín et al. 2020; Pérez-González et al. 2020; Rivera-Correa and Faivovich 2021; Montilla et al. 2023; dos Santos Diaz and Anganoy-Criollo 2024), or to document ontogenetic changes (e.g. Vargas-Salinas and Gutiérrez-Cárdenas 2005; Rodríguez-Suárez and Corredor-Londoño 2012). Other studies have documented the effects of contaminants on the development and survival of embryos and tadpoles (Bernal et al. 2009; Triana Velásquez et al. 2013; Riaño et al. 2020; Henao et al. 2021; López-Florez et al.

Fig. 10 Charismatic species, such as dendrobatids with striking colours and elaborate parental care behaviours, offer wonderful opportunities to change misconceptions (e.g. ugly and nasty animals) about frogs among people. The dendrobatid *Andinobates supatae* (upper drawing) is a great example of how frogs can become a symbol of conservation initiatives (Amézquita et al. 2016; Meneses-Pelayo 2024). Misconceptions can be changed from early ages in children through speeches and playful activities, as shown in the picture. Drawing of the poison frog by María José Tovar-Gil. (Photo: María Paula Toro-Gómez)

2023), and to a much lesser extent, physiology (Méndez-Narváez et al. 2015), community ecology (Camacho-Rozo and Urbina-Cardona al 2021), social behaviour (Gutiérrez-Cárdenas 2006a, b) and aspects of natural history (e.g. Guayara-Barragan et al. 2010; Villegas-Mora et al. 2020).

Second, an important component of the anuran fauna in the Andes of Colombia are the species of the family Strabomantidae, which are characterized by having a reproductive mode that includes oviposition outside of water and direct development of the embryo; that is, upon hatching from the egg, the individual is fully developed, a miniature version of adult individuals (Bernal and Lynch 2008; Saboyá-Acosta and Urbina-Cardona 2023). Publications of Colombian Andean strabomantids cover aspects of diet (Arroyo et al. 2008; García et al. 2012, 2015; Moreno-Barbosa and Hoyos-Hoyos 2014), parental care and reproduction (Vargas-Salinas and Castro-Herrera 1999; Rojas-Rivera et al. 2011; Granados-Pérez and Ramirez-Pinilla 2020; Falcón-Espitia et al. 2023), evolution of body size (Acevedo et al. 2022) and habitat use and vocal characteristics aiming to contribute to understanding species coexistence (Rincón-Franco and Castro-Herrera 1998; García et al. 2007; Cortés et al. 2008; Méndez-Narváez and Bolívar-García 2016; Arriaga-Jaramillo et al. 2023; Cuellar-Valencia et al. 2023). These studies are very few if we consider the high taxonomic diversity of this group from the Colombian Andes. Hence, the prospects for behavioural studies with these frogs are immense and much remains to be done.

Finally, in this chapter, emphasis is placed on the contribution of amphibians in the Andes of Colombia, from a behavioural perspective to the biological diversity in the country. However, the Andes Mountain range also contribute to the biological diversity of Colombia through its role as a geographical barrier to species dispersal or as a determining factor in the asynchrony of spatial climatic regimes that could promote divergence among populations (Quintero et al. 2014;

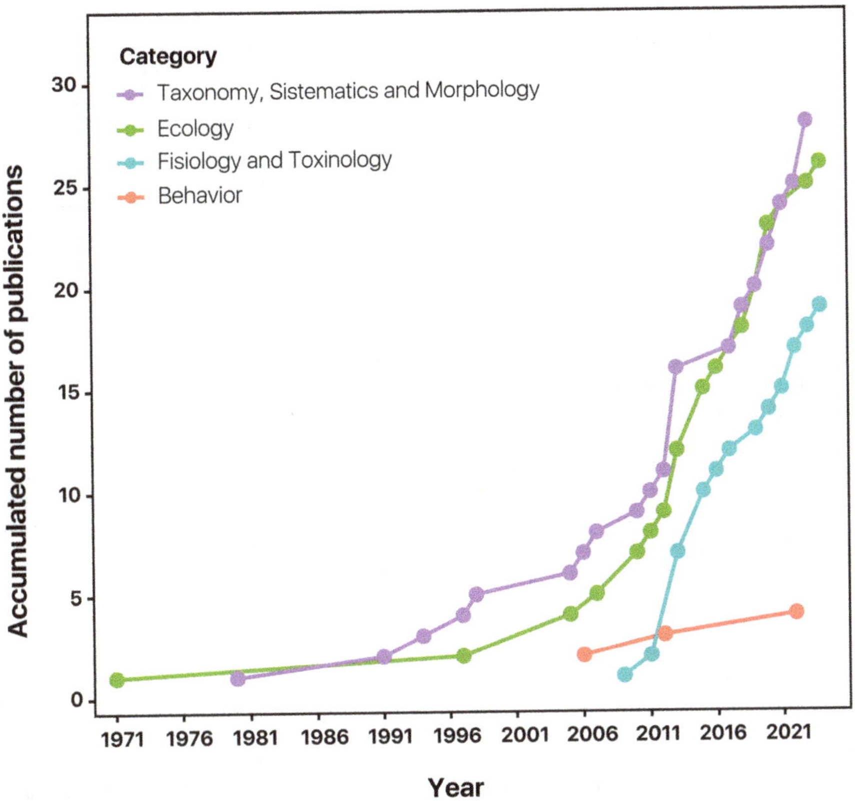

Fig. 11 Summary of results from a quick systematic review of studies on embryos and tadpoles in Colombia. Seventy-seven publications were identified, covering aspects related to taxonomy (including systematics and morphology), ecology, physiology and toxicology and behaviour. The systematic review covered publications made between 1971 and August 2024 and was conducted using five search terms: [Tadpoles], [Embryos], [Anurans], [Larvae], and [Colombia], in Google Scholar, PubMed, and Elsevier

Guarnizo et al. 2022). We hope, with this chapter, to encourage studies that contribute to understand the origin, maintenance and conservation of the behavioural diversity of anurans in Colombia.

Acknowledgements We thank the editors for the invitation to write this chapter. We express our gratitude to everyone who contributed to the completion of this chapter through interchange of literature and sharing photographs and drawings (credits in each figure legend). The University of Quindío, the University of Magdalena, Colombia, New York University, and the American Museum of Natural History, USA, supported logistically our contributions. Commentaries by A. Catenazzi greatly improved previous versions of this manuscript.

Competing Interests The authors have no conflicts of interest to declare that are relevant to the content of this chapter.

References

Abondano D, Twomey E, Vargas-Salinas F, Meyer C, Schulte LM (2024) Sexy fingers: pheromones in the glands of male dendrobatid frogs. Mol Ecol 33:e17476

Acevedo AA, Palma RE, Olalla-Tárraga MA (2022) Ecological and evolutionary trends of body size in *Pristimantis* frogs, the world's most diverse vertebrate genus. Sci Rep 12:18106

Amézquita A, Lüddecke H (1999) Correlates of intrapopulational variation in size at metamorphosis of the high-Andean frog *Hyla labialis*. Herpetologica 55(3):295–303

Amézquita A, Flechas SV, Lima AP et al (2011) Acoustic interference and recognition space within a complex assemblage of dendrobatid frogs. PNAS 108(41): 17058–17063

Amézquita A, Márquez R, Medina R et al (2013a) A new species of Andean poison frog, *Andinobates* (Anura: Dendrobatidae), from the northwestern Andes of Colombia. Zootaxa 3620:163–178

Amézquita A, Castro L, Arias M, González M, Esquivel C (2013b) Field but not lab paradigms support generalisation by predators of aposematic polymorphic prey: the *Oophaga histrionica* complex. Evol Ecol 27(4):769–782

Amézquita A, Kahn TR, Kraus K et al (2016) Amphibian conservation in the tropical Andes and Amazonian basin. In: Kahn T, La Marca E, Lötters S, Brown JL, Twomey E, Amézquita A (eds) Aposematic poison frogs (Dendrobatidae) of the Andean countries: Bolivia, Colombia, Ecuador, Perú and Venezuela. Conservation International, Arlington, pp 21–65

Amézquita A, Vargas-Salinas F, Ramos I, Palacios-Rodréguez P, Salazar EN, Quiroz M, Bolévar W, Galindo-Uribe DM, Mazariegos-H LA (2024) Molecular phylogenetics uncovers two new species in the genus *Phyllobates* (Anura, Dendrobatidae): the terrible frog gets two new sisters. Zookeys 1212:217-240

Anderson HM, Fisher DN, McEwen BL et al (2021) Episodic correlations in behavioural lateralization differ between a poison frog and its mimic. Anim Behav 174:207–215

Andersson M (1994) Sexual selection. Princeton University Press, Princeton

Andersson M, Iwasa Y (1996) Sexual selection. Trends Ecol Evol 11(2):53–58

Anganoy-Criollo MA (2013) Tadpoles of the High-Andean *Hyloxalus subpunctatus* (Anura: Dendrobatidae) with description of larval variation and species distinction by larval morphology. Pap Avulsos Zool 53:221–224

Angeli NF, DiRenzo GV, Cunha A, Lips KR (2015) Effects of density on spatial aggregation and habitat associations of the glass frog (Centrolene) *Espadarana prosoblepon*. J Herpetol 49(3):388–394

Arch VS, Grafe TU, Narins PM (2008) Ultrasonic signalling by a Bornean frog. Biol Lett 4(1):19–22

Arcila-Pérez LF, Rios-Soto JA, Montilla SO et al (2017) Vocalization and natural history in populations of a glassfrog assigned to *Nymphargus griffithsi* in the Central Andes of Colombia. Herpetol Rev 48:275–280

Arcila-Pérez LF, Atehortua-Vallejo MA, Vargas-Salinas F (2020) Homing in the rubí poison frog *Andinobates bombetes* (Dendrobatidae). Copeia 108(4):948–956

Arias-Cárdenas A, Barrientos LS, Pardo-Diaz C et al (2024) Taxonomic inflation and a reconsideration of speciation in the Andes: the case of the high-elevation tree frog *Dendropsophus molitor* (Anura: Hylidae). Zool J Linnean Soc 200(3):763–775

Armesto LO, Señaris JC (2017) Anuros del norte de los Andes: patrones de riqueza de especies y estado de conservación. Pap Avulsos Zool 57:491–526

Arriaga-Jaramillo FG, Cuellar-Valencia OM, García-Gómez I et al (2023) Acoustic segregation of five sympatric and syntopic species of genus *Pristimantis* (Anura: Strabomantidae) from Western Colombia. Stud Neotropical Fauna Environ 58:275–282

Arroyo SB, Serrano-Cardozo VH, Ramírez-Pinilla MP (2008) Diet, microhabitat and time of activity in a *Pristimantis* (Anura, Strabomantidae) assemblage. Phyllomedusa 7:109–119

Arroyo S, Targino M, Rueda-Solano LA, Daza JM, Grant T (2022) A new genus of terraranas (Anura: Brachycephaloidea) from northern South America, with a systematic review of *Tachiramantis*. Syst Biodivers 20(1):1–25

Astwood-Romero JA, Álvarez-Perdomo N, Parra-Torres MF et al (2016) Contenidos estomacales de especies de anuros en reservas naturales del municipio de Villavicencio, Meta, Colombia. Caldasia 38(1): 165–181

Barnett JB, Michalis C, Anderson HM et al (2020) Imperfect transparency and camouflage in glass frogs. PNAS 117(23):12885–12890

Bravo Valencia L, Delia J (2016) Maternal care in a glassfrog: care function and commitment to offspring in *Ikakogi tayrona*. Behav Ecol Sociobiol, 70, 41-48.

Basto-Riascos MC, López-Caro J, Vargas-Salinas F (2017) Reproductive ecology of the glass frog *Espadarana prosoblepon* (Anura: Centrolenidae) in an urban forest of the Central Andes of Colombia. J Nat Hist 51(41–42):2535–2550

Bateson PG (1983) Mate choice. Cambridge University Press, Cambridge

Bedoya SC, Mantilla-Castaño JC, Pareja-Márquez IM (2014) Necrophiliac and interspecific amplexus in *Dendropsophus columbianus* (Anura: Hylidae) in the Central Cordillera of Colombia. Herpetol Notes 7(1): 16

Bernal MH, Lynch JD (2008) Review and analysis of altitudinal distribution of the Andean anurans in Colombia. Zootaxa 1826(1):1–25

Bernal XE, Guarnizo C, Lüddecke H (2005) Geographic variation in advertisement call and genetic structure of *Colostethus palmatus* (Anura, Dendrobatidae) from the Colombian Andes. Herpetologica 61(4):395–408

Bernal XE, Rand AS, Ryan MJ (2006) Acoustic preferences and localization performance of blood-sucking flies (*Corethrella coquillett*) to túngara frog calls. Behav Ecol 17(5):709–715

Bernal MH, Luna-Mora VF, Gallego O, Quevedo A (2007) A new species of poison frog (Amphibia: Dendrobatidae) from the Andean mountains of Tolima, Colombia. Zootaxa 1638:59–68

Bernal MH, Solomon K, Carrasquilla G (2009) Toxicity of formulated glyphosate (Glyphos) and Cosmo-Flux to larval and juvenile Colombian frogs 2. Field and laboratory microcosm acute toxicity. J Toxicol Environ Health Part A 72(15–16):966–973

Betancourth-Cundar M, Lima AP, Hödl W, Amézquita A (2016) Decoupled evolution between senders and receivers in the Neotropical *Allobates femoralis* frog complex. PLoS One 11(6):e0155929

Betancourth-Cundar M, Palacios-Rodríguez P, Mejía-Vargas D, Paz A, Amézquita A (2020) Genetic differentiation and overexploitation history of the critically endangered Lehmann's poison frog: *Oophaga lehmanni*. Conserv Genet 21:453–465

Betancourth-Cundar M, Amézquita A, Cadena CD (2024) Novel and classical methods similarly describe variation in territory size among males in Neotropical poison frogs with contrasting reproductive and behavioral strategies. Evol Ecol 38(5):799–823

Bevier CR (2016) Physiological and biochemical correlates of calling behavior in anurans with different calling strategies. In: Amphibian and reptile adaptations to the environment. CRC Press, Boca Raton, pp 63–80

Bianki VLV (1988) The right and left hemispheres of the animal brain: cerebral lateralization of function. Gordon and Breach, New York

Bobadilla-Molina JS, Acosta-Ortiz JM, Montoya-Cruz A (2021) Observaciones sobre el cuidado parental de *Rheobates palmatus* (Anura: Aromobatidae) en Villavicencio Colombia. Rev Latinoam Herpet 4(1): 178–180

Boeckle M, Preininger D, Hödl W (2009) Communication in Noisy environments I: acoustic signals of *Staurois latopalmatus* Boulenger 1887. Herpetologica 65(2): 154–165

Botero-Delgadillo E, Meza-Angulo E, Bayly NJ (2024) Abundance models of endemic birds of the Sierra Nevada de Santa Marta, northern South America, suggest small population sizes and dependence on montane elevations. Ornithol Appl 126(2):duae002

Bradbury JW, Vehrencamp SL (2011) Principles of animal communication. Sinauer Associates, Sunderland

Bravo-Valencia L, Delia J (2016) Maternal care in a glassfrog: care function and commitment to offspring in *Ikakogi tayrona*. Behav Ecol Sociobiol 70:41–48

Brown JL, Morales V, Summers K (2010) A key ecological trait drove the evolution of biparental care and monogamy in an amphibian. Am Nat 175(4):436–446

Brown JL, Twomey E, Amezquita A et al (2011) A taxonomic revision of the Neotropical poison frog genus *Ranitomeya* (Amphibia: Dendrobatidae). Zootaxa 3083(1):1–120-121–120

Brumm H, Slabbekoorn H (2005) Acoustic communication in noise. Adv Stud Behav 35:151–209

Brunetti AE, Taboada CA, Faivovich J (2014) The reproductive biology of *Hypsiboas punctatus* (Anura: Hylidae): male territoriality and the possible role of different signals during female choice. Salamandra 50(4):215–224

Brusa O, Bellati A, Meuche I et al (2013) Divergent evolution in the polymorphic granular poison-dart frog, *Oophaga granulifera*: genetics, coloration, advertisement calls and morphology. J Biogeogr 40(2): 394–408

Buckley LB, Jetz W (2007) Environmental and historical constraints on global patterns of amphibian richness. Proc R Soc B Biol Sci 274:1167–1173

Buitrago-Marulanda BB, Vargas-Salinas F, Martínez V, Buitrago-Torres DL (2025) What is essential is invisible to the eye: a conservation approach for poison frogs (Dendrobatidae) based on evolutionary distinctiveness. Biodivers. Conserv 4(6): 965-984.

Camacho-Rozo CP, Urbina-Cardona N (2021) Tadpoles inhabiting natural and anthropogenic temporary water bodies: which are the environmental factors that affect the diversity of the assemblages? Front Environ Sci 9: 667448

Candolin U, Wong BB (2012) Behavioural responses to a changing world: mechanisms and consequences. Oxford University Press, Oxford

Carbono E, Lozano-Contreras G (1997) Endemisms and other singularities; Sierra Nevada of Santa Marta, Colombia. Possible origin causes and necessity to conserve them. Revista Acad Colomb Ci Exact 21(81): 409–419

Cárdenas-Ortega MS (2022) *Pristimantis achatinus* predation. Herpetol Rev 53(1):107

Cárdenas-Ortega MS, Herrera-Lopera JM (2022) *Pristimantis thectopternus* (Northern cordilleras robber frog) predation. Herpetol Rev 53:298–299

Carvajal-Castro JD, Vargas-Salinas F (2016) Stream noise, habitat filtering, and the phenotypic and phylogenetic structure of Neotropical anuran assemblages. Evol Ecol 30(3):451–469

Carvajal-Castro JD, López-Aguirre Y, Ospina-L AM, Santos JC, Rojas B, Vargas-Salinas F (2020) Much more than a clasp evolutionary patterns of amplexus diversity in anurans. Biol J Linn Soc 20:1–12

Carvajal-Castro JD, Vargas-Salinas F, Casas-Cardona S, Rojas B, Santos JC (2021) Aposematism facilitates the diversification of parental care strategies in poison frogs. Sci Rep 11:19047

Casas-Cardona S, Márquez R, Vargas-Salinas F (2018) Different colour morphs of the poison frog *Andinobates bombetes* (Dendrobatidae) are similarly effective visual predator deterrents. Ethology 124(4): 245–255

Castillo-Velandia DR, Dueñas Valderrama F, Muñoz Avila JA et al (2019) Educación ambiental comunitaria. In: Vargas-Salinas F, Muñoz-Avila J, Morales ME (eds) Biología de anfibios y reptiles en el Bosque seco tropical del norte de Colombia. Editorial UPTC, Tunja, pp 414–449

Chaves-Portilla GA, Salazar EN, Gil-Acero J, Dorado-Correa A et al (2021) A new species of Andean golden poison frog (Andinobates, Dendrobatidae) from the Eastern Andes of Colombia. Zootaxa 5047(5):531–546

Chouteau M, Angers B (2011) The role of predators in maintaining the geographic organization of aposematic signals. Am Nat 178(6):810–817

Cisneros-Heredia DF, Mcdiarmid RW (2007) Revision of the characters of Centrolenidae (Amphibia: Anura: Athesphatanura), with comments on its taxonomy and the description of new taxa of glassfrogs. Zootaxa 1572(1):1–82

Coloma LA, Carvajal-Endara S, Duenas JF et al (2012) Molecular phylogenetics of stream treefrogs of the *Hyloscirtus larinopygion* group (Anura: Hylidae), and description of two new species from Ecuador. Zootaxa 3364(1):1–78

Cortés AM, Ramírez-Pinilla MP, Suárez HA, Tovar E (2008) Edge effects on richness, abundance and diversity of frogs in Andean cloud forest fragments. S Am J Herpetol 3(3):213–222

Cortés-Suárez JE (2021) Microhabitat use of the Savannah Treefrog *Dendropsophus luddeckei* (Anura: Hylidae) in a pasture agro-ecosystem in Villa de Leyva, Colombia. Herpetol Notes 14:1193–1198

Cott HB (1940) Adaptive coloration in animals. Methuen & Co. London.

Crothers LR, Cummings ME (2015) A multifunctional warning signal behaves as an agonistic status signal in a poison frog. Behav Ecol 26(2):560–568

Crothers L, Gering E, Cummings M (2010) Aposematic signal variation predicts male-male interactions in a polymorphic poison frog. Evolution 65(2):599–605

Csermely D, Regolin L (2013) Behavioral lateralization in vertebrates. Springer, Berlin

Cuellar-Valencia OM, Bolaños-Molina BE, Bolívar-García W (2023) Exploring the acoustic diversity of Andean frogs in the Farallones de Cali National Natural Park, southwestern Colombia. Stud Neotropical Fauna Environ 59:1–19

Cunnington GM, Fahrig L (2010) Plasticity in the vocalizations of anurans in response to traffic noise. Acta Oecol 36(5):463–470

Cuthill IC, Stevens M, Sheppard J, Maddocks T, Párraga CA, Troscianko TS (2005) Disruptive coloration and background pattern matching. Nature 434(7029):72–74

De Luna AG, Hödl W, Amézquita A (2010) Colour, size and movement as visual subcomponents in multimodal communication by the frog *Allobates femoralis*. Anim Behav 79(3):739–745

Delia JR, Ramírez-Bautista A, Summers K (2013) Parents adjust care in response to weather conditions and egg dehydration in a Neotropical glassfrog. Behav Ecol Sociobiol 67:557–569

Delia JR, Ramírez-Bautista A, Summers K (2014) Glassfrog embryos hatch early after parental desertion. Proc R Soc B Biol Sci 281(1785):20133237

Delia J, Bravo-Valencia L, Warkentin K (2017a) Patterns of parental care in Neotropical glassfrogs: fieldwork alters hypotheses of sex-role evolution. J Evol Biol 30(5):898–914

Delia J, Bravo-Valencia L, McDiarmid RW (2017b) Notes on paternal behavior in *Hyalinobatrachium* glassfrogs (Anura: Centrolenidae). Phyllomedusa 16(1):101–107

Delia J, Bravo-Valencia L, Warkentin KM (2020) The evolution of extended parental care in glassfrogs: do egg-clutch phenotypes mediate coevolution between the sexes? Ecol Monogr 90(3):e01411

Devictor V, Mouillot D, Meynard C et al (2010) Spatial mismatch and congruence between taxonomic, phylogenetic and functional diversity: the need for integrative conservation strategies in a changing world. Ecol Lett 13(8):1030–1040

Díaz-Ayala RF, Hurtado LF, Sierra-Giraldo JA (2014) First record of Anura (*Dendropsophus columbianus* Anura: Hylidae) as floral visitors of Araceae (*Zantedeschia aethiopica* (L.) Spreng). Aroideana 37E(1):4–10

Dole JW, Durant P (1974) Movements and seasonal activity of *Atelopus oxyrhynchus* (Anura: Atelopodidae) in a Venezuelan cloud forest. Copeia 1974:230–235

dos Santos Dias PH, Anganoy-Criollo M (2024) Harlequin frog tadpoles comparative buccopharyngeal morphology in the gastromyzophorous tadpoles of the genus *Atelopus* (Amphibia, Anura, Bufonidae), with discussion on the phylogenetic and evolutionary implication of characters. Sci Nat 111(1):3

Dreher CE, Pröhl H (2014) Multiple sexual signals: calls over colors for mate attraction in an aposematic, color-diverse poison frog. Front Ecol Evol 2:22

Duarte-Marín S, López-Molina KJ, Arcila-Pérez LF (2019) Defensive behavior in *Hyloscirtus larinopygion* Duellman, 1973 (Anura: Hylidae). Herpetol Notes 12:509–511

Duarte-Marín S, González-Acosta C, Santos Dias PH, Arias-Álvarez GA, Vargas-Salinas F (2020) Advertisement call, tadpole morphology, and other natural history aspects of the threatened poison frog *Andinobates daleswansoni* (Dendrobatidae). J Nat Hist 54(45–46):3005–3030

Duarte-Marín S, Rada M, Rivera-Correa M et al (2022) Tic, Tii and Trii calls: advertisement call descriptions for eight glass frogs from Colombia and analysis of the structure of auditory signals in Centrolenidae. Bioacoustics 32(2):143–180

Dubois A, Martens J (1984) A case of possible vocal convergence between frogs and a bird in Himalayan torrents. J Ornithol 125(4):455–463

Duellman WE (2001) Hylid Frogs of middle America, 2nd edn. Society for the Study of Amphibians and Reptiles, Contributions to Herpetology, No. 18

Duellman WE (2005) Cusco Amazónico, The lives of amphibians and reptiles in an Amazonian rainforest. Comstock Publishing Associates, Cornell University, New York

Dugas MB, Brooks OL, Saporito RA et al (2024a) Adult poison frogs can capture and consume aquatic tadpoles. Evol Ecol 38:69–75

Dugas MB, Stevens TC, Cosman ME et al (2024b) Proximate determinants of family size in an obligate egg-feeding frog. Evol Ecol 38:91–102

Dunn ER (1944a) Los géneros de anfibios y reptiles de Colombia. Caldasia 10(2):497–529

Dunn ER (1944b) Herpetology of the Bogotá area. Rev Acad Colomb Cienc Exactas, Físicas y Nat 6(21):68–81

Ebersbach J, Posso-Terranova A, Bogdanowicz S et al (2019) Complex patterns of differentiation and gene flow underly the divergence of aposematic phenotypes in *Oophaga* poison frogs. Mol Ecol 29(11):1944–1956

Ehret G, Tautz J, Schmitz B, Narins P (1990) Hearing through the lungs: lung-eardrum transmission of sound in the frog *Eleutherodactylus coqui*. Naturwissenschaften 77(4):192–194

Enquist M, Hurd PL, Ghirlanda S (2010) Signaling. In: Fox CW, Wesneat DF (eds) Evolutionary behavioral ecology. Oxford University Press Inc, Oxford, pp 266–284

Erdtmann L, Amézquita A (2009) Differential evolution of advertisement call traits in Dart-Poison frogs (Anura: Dendrobatidae). Ethology 115(9):801–811

Escalona Sulbaran MD, Ivo Simões P, Gonzalez-Voyer A, Castroviejo-Fisher S (2019) Neotropical frogs and mating songs: the evolution of advertisement calls in glassfrogs. J Evol Biol 32(2):163–176

Escalona M, Simões PI, Gonzalez-Voyer et al (2025) Allometric constraint predominates over the acoustic adaptation hypothesis in a radiation of Neotropical treefrogs. Integr. Zool, 20, 763–773.

Escobar-Lasso S, Rojas-Morales JA (2012) Antipredatory behaviors of the Colombian endemic glassfrog *Centrolene savagei* (Anura: Centrolenidae). Bol Cient Centro Mus Mus Hist Nat 16:226–232

Escobar-Vargas S, Marín-Martínez M, Zuluaga-Isaza JC (2016) First report of flesh fly eggs (Diptera: Sarcophagidae) in *Pristimantis achatinus* (Anura: Craugastoridae) from Colombia. Bol Cient Centro Mus Mus Hist Nat 20(2):231–234

Ewer R (1956) Imprinting in animal behaviour. Nature 177(4501):227–228

Ey E, Fischer J (2009) The "acoustic adaptation hypothesis": a review of the evidence from birds, anurans and mammals. Bioacoustics 19(1–2):21–48

Fahrig L, Rytwinski T (2009) Effects of roads on animal abundance: an empirical review and synthesis. Ecol Soc 14(1):1–20

Faivovich J, Haddad CF, Garcia PC et al (2005) Systematic review of the frog family Hylidae, with special reference to Hylinae: phylogenetic analysis and taxonomic revision. Bull Am Mus Nat Hist 294:1–240

Faivovich J, Haddad CF, Baêta D et al (2010) The phylogenetic relationships of the charismatic poster frogs, Phyllomedusinae (Anura, Hylidae). Cladistics 26(3): 227–261

Falcón-Espitia N, Ríos-Orjuela JC, Martínez-Botero MC et al (2023) Notes on reproductive behavior and vocalizations of *Pristimantis taeniatus* (Anura: Strabomantidae). Phyllomedusa 22(1):57–61

Fandiño MC, Lüddecke H, Amézquita A (1997) Vocalisation and larval transportation of male *Colostethus subpunctatus* (Anura: Dendrobatidae). Amphibia-Reptilia 18(1):39–48

Fay RR, Popper A (1999) Hearing in fishes and amphibians: an introduction. In: Fay RR, Popper AN (eds) Comparative hearing: fish and amphibians, vol 11. Springer Science & Business Media, New York, pp 1–15

Feng AS, Narins PM, Xu C-H et al (2006) Ultrasonic communication in frogs. Nature 440(7082):333–336

Fouilloux CA, Goyes Vallejos J, Tumulty JP (2024) Home is where the high-quality resources are: nursery characteristics and territory distribution suggest reproductive resource defense in golden rocket frogs. Evol Ecol 38:223–240

Fox CW, Westneat DF (2010) Adaptation. In: Fox CW, Westneat DF (eds) Evolutionary behavioral ecology. Oxford University Press, New York, pp 16–31

Francis CD, Phillips JN, Barber JR (2023) Background acoustics in terrestrial ecology. Annu Rev Ecol Evol Syst 54:351–373

Frost DR (2025) Amphibian Species of the World: an online reference. Version 5.4. http://research.amnh.org/vz/herpetology/amphibia/

Gannon WL, Dunlop R, Hawkins A, Thomas JA (2022) Collecting, documenting, and archiving bioacoustical data and metadata. In: Erbe C, Thomas JA (eds) Exploring animal behavior through sound, vol 1. Springer, Cham, pp 87–110

García RJC, Cárdenas-H H, Castro-Herrera F (2007) Relación entre la diversidad de anuros y los estados sucesionales de un bosque muy húmedo montano bajo del Valle del Cauca, suroccidente colombiano. Caldasia 29:363–374

García JC, Lucas-Velásquez L, Cárdenas Henao H, Posso Gómez CE (2012) Ecología alimentaria de la rana de lluvia endémica *Pristimantis jubatus* (Craugastoridae) en el Parque Nacional Natural Munchique, Colombia. Acta Biol Colomb 17(2):411–420

García RJC, Posso-Gómez CE, Cardenas-Henao H (2015) Diet of direct-developing frogs (Anura: Craugastoridae: *Pristimantis*) from the Andes of western Colombia. Acta Biol Colomb 20(1):79–87

Garrido-Priego M, Tószeghi M, Angiolani-Larrea FN, Valencia-Aguilar A et al (2024) Clutch attendance and call parameters are linked to mating sucess in a glassfrog with paternal care. Behav Ecol 35(6): arae078

Garrido-Priego M, Angiolani-Larrea FN, Valencia-Aguilar A et al (2025) Previous mating success motivates homing of male glassfrogs Hyalinobatrachium valerioi. Behav Ecol Sociobiol 79, 75

Gerhardt HC (1998) Acoustic signals of animals: recording, field measurements, analysis and description. In: Hopp SL, Owren MJ, Evans CS (eds) Animal acoustic communication. Springer, Berlin, Heidelberg, pp 1–25

Gerhardt HC, Huber F (2002) Acoustic communication in insects and anurans: common problems and diverse solutions. The University of Chicago Press, Chicago

Gil-Guevara O, Amézquita A (2020) Adjusted phonotactic reactions to sound amplitude and pulse number mediate territoriality in the harlequin poison frog. Behav Process 181:104249

Gomes DG, Francis CD, Barber JR (2021) Using the past to understand the present: coping with natural and anthropogenic noise. Bioscience 71(3):223–234

Gómez-Consuegra J, Amézquita A (2024) Ecological and behavioral correlates of homing performance in the egg-feeding poison frog *Oophaga histrionica*. Evol Ecol 38:241–253

Gómez-Hoyos DA, Marín-Gómez OH, Vanegas Guerrero J (2012) Unusual amplexus in *Dendropsophus columbianus* (Anura: Hylidae). Herpetol Notes 5: 497–498

Gómez-Hoyos DA, López-García MM, Soto-Garzón CA et al (2014) Geographic variation in the diet of the Cauca Poison Frog *Andinobates bombetes* (Anura: Dendrobatidae) in the Andes of Colombia. Herpetol Notes 7:559–564

Gomez-Mestre I, Pyron RA, Wiens JJ (2012) Phylogenetic analyses reveal unexpected patterns in the evolution of reproductive modes in frogs. Evolution 66(12): 3687–3700

Gómez-Murcia DA, Bedoya-Ospina MM, Arcila-Pérez LF, Vargas-Salinas F (2024) Nothing like home: most males of *Espadarana prosoblepon* (Anura Centrolenidae) exhibit homing to calling site despite the availability of alternative suitable sites for calling and mating. Ethol Ecol Evol 36(1):70–85

Gonzalez M, Carazzone C (2023) Eco-metabolomics applied to the chemical ecology of poison frogs (Dendrobatoidea). J Chem Ecol 49(9–10):570–598

González-Acosta C, Amézquita A, Vargas-Salinas F (2024) Variation in the advertisement call of the poison frog *Andinobates bombetes* relates to geographic distance between populations, but not to divergence in warning coloration. Stud Neotropical Fauna Environ 59(3):686–698

González-Santoro M, Hernández-Restrepo J, Palacios-Rodríguez P (2021) Aggressive behaviour, courtship and mating call description of the neotropical poison frog *Phyllobates aurotaenia* (Anura: Dendrobatidae). Herpetol Notes 14:1145–1149

González-Santoro M, Palacios-Rodríguez P, Gonzalez M et al (2023) Decoupled auditory perception from acoustic signal divergence hinders species recognition in territorial poison frogs. Behav Ecol Sociobiol 77(1), 8:1–11

Gosling LM, Sutherland WJ (2000) Behaviour and conservation. Cambridge University Press, Cambridge

Gould SJ, Vrba ES (1982) Exaptation—a missing term in the science of form. Paleobiology 8(1):4–15

Goutte S, Dubois A, Howard SD et al (2016) Environmental constraints and call evolution in torrent-dwelling frogs. Evolution 70(4):811–826

Goyes Vallejos J, Siles JS, Calero V, Rodriguez N, Machado G (2024) Not enough time: short-term female presence after oviposition does not improve egg survival in the emerald glass frog. Anim Behav 213, 161–171

Goyes Vallejos J, Gomez J, Hernández-Figueroa AD, Vera R, Green DM (2021) Fertilization success suggests random pairing in frogs with regard to body size. Behav Ecol Sociobiol, 75(10), 140

Graham A (2009) The Andes: a geological overview from a biological perspective. Ann Mo Bot Gard 96(3): 371–385

Granados-Pérez Y, Ramírez-Pinilla MP (2020) Reproductive phenology of three species of *Pristimantis* in an Andean cloud forest. Rev Acad Colomb Cienc Exactas Fis Nat 44(173):1083–1098

Grant T, Bolivar-G W, Castro F (1998) The advertisement call of *Centrolene geckoideum*. J Herpetol 32(3): 452–455

Greer BJ, Wells KD (1980) Territorial and reproductive behavior of the tropical American frog *Centrolenella fleischmanni*. Herpetologica 36:318–326

Grether GF, Kolluru GR, Nersissian K (2004) Individual colour patches as multicomponent signals. Biol Rev 79(3):583–610

Guarnizo C, Montoya P, Quintero I, Cadena CD (2022) Population divergence associated with spatial asynchrony in precipitation in Neotropical frogs. J Biogeogr 49:2169–2180

Guayara-Barragan M, Bernal M, Hernández-Cuadrado E (2010) *Centrolene prosoblepon* (glass frog). Fungal Infect Herpetol Rev 41:61–62

Guayasamin JM, Castroviejo-Fisher S, Trueb L et al (2009) Phylogenetic systematics of glassfrogs (Amphibia: Centrolenidae) and their sister taxon *Allophryne ruthveni*. Zootaxa 2100:1–100

Guevara-Molina SC, Vargas-Salinas F (2014) *Nymphargus grandisonae* (Anura: Centrolenidae). Reprod Behav Herpetol Bull 128:29–30

Gutiérrez G, Lüddecke H (2002) Mating pattern and hatching success in a population of the Andean frog *Hyla labialis*. Amphibia-Reptilia 23(3):281–292

Gutiérrez-Cárdenas PDA (2006a) Embryonic olfactory learning in larvae of *Smilisca phaeota* (Cope, 1862) (Anura: Hylidae). Actual Biol 28(84):59–65

Gutiérrez-Cárdenas PDA (2006b) *Smilisca phaeota* (Cope, 1862) (Anura: Hylidae) tadpoles do not discriminate between siblings and non-siblings. Actual Biol 28:59–65

Gutiérrez-Cárdenas PA, Rivera-Correa M (2013) *Dendropsophus norandinus* (north Andean treefrog). Diet Herpetol Rev 44:120–121

Halfwerk W, Slabbekoorn H (2009) A behavioural mechanism explaining noise-dependent frequency use in urban birdsong. Anim Behav 78(6):1301–1307

Hegna RH, Saporito RA, Donnelly MA (2013) Not all colors are equal: predation and color polytypism in the aposematic poison frog *Oophaga pumilio*. Evol Ecol 27:831–845

Henao LM, Mendez JJ, Bernal MH (2021) UVB radiation enhances the toxic effects of three organophosphorus insecticides on tadpoles from tropical anurans. Hydrobiologia 849(1):141–153

Herrera-Lopera JM, Castaño VA, Cultid-Medina CA (2023) What are the Andean Colombian anurans? Empirical regionalization proposals vs. observed patterns of compositional dissimilarity. PeerJ 11: e15217

Higuera-Rojas DF, Carvajal-Cogollo JE (2021) Diet of *Dendropsophus molitor* (Anura: Hylidae) in a High-Andean agricultural ecosystem, Colombia. Univ Sci 26(1):119–137

Hödl W, Amézquita A (2001) Visual signaling in anuran amphibians. In: Ryan MJ (ed) Anuran communication.

Smithsonian Institution Press, Washington, pp 121–141

Honma A, Mappes J, Valkonen JK (2015) Warning coloration can be disruptive: aposematic marginal wing patterning in the wood tiger moth. Ecol Evol 5(21): 4863–4874

Hunter AS, Valdivieso B (1962) La reproducción de la rana *Hyla labialis*. Caldasia 8:573–583

Hutter CR, Escobar-Lasso S, Rojas-Morales JA, Gutiérrez-Cárdenas PDA, Imba H, Guayasamin JM (2013) The territoriality, vocalizations and aggressive interactions of the red-spotted glassfrog, *Nymphargus grandisonae*, Cochran and Goin, 1970 (Anura: Centrolenidae). J Nat Hist 47(47–48):3011–3032

Jacobson SK (1985) Reproductive behavior and male mating success in two species of glass frogs (Centrolenidae). Herpetologica 41:396–404

Jiménez-Vargas GM, Vargas-Salinas F (2021) Does anthropogenic noise promotes advertisement call adjustments in the rubí poison frog *Andinobates bombetes*? Behaviour 158(7):565–583

Jiménez-Vargas GM, Atehortua-Vallejo MA, Arcila-Pérez LF, Carvajal-Castro JD, Vargas-Salinas F (2021) Does abiotic noise promote segregation of functional diversity in Neotropical anuran assemblages? Biol J Linn Soc 132(4):847–860

Jops K, O'Dwyer JP (2023) Life history complementarity and the maintenance of biodiversity. Nature 618:986–991

Jost L (2006) Entropy and diversity. Oikos 113:363–375

Kahn TR, La Marca E, Lötters S et al (2016) Aposematic poison frogs (Dendrobatidae) of the Andean countries: Bolivia, Colombia, Ecuador, Perú and Venezuela. Conservation International

Kaiser H, Crother BI, Kelly CM et al (2013) Best practices: in the 21st century, taxonomic decisions in herpetology are acceptable only when supported by a body of evidence and published via peer-review. Herpetol Rev 44(1):8–23

Kattan GH, Franco P, Saavedra-Rodríguez CA et al (2006) Spatial components of bird diversity in the Andes of Colombia: implications for designing a regional reserve system. Conserv Biol 20(4):1203–1211

Keuroghlian-Eaton S, dos Santos SP, Marrie Pinheiro Müller M, Cavalheri D, Bolochio B, Santana D (2024) Escape behavior lateralization of pointed-belly frog (*Leptodactylus podicipinus*) (Anura, Leptodactylidae) in the southern Pantanal. Biotropica 56:e13326

Kimura M (1968) Evolutionary rate at the molecular level. Nature 217(5129):624–626

Kubicki B (2007) Glass frogs, Costa Rica. Editorial INBio, Costa Rica

La Marca E, Lips KR, Lötters S, Puschendorf R, Ibáñez DR et al (2005) Catastrophic population declines and extinctions in Neotropical harlequin frogs (Bufonidae: *Atelopus*). Biotropica 37(2):190–201

Lee N, Christensen-Dalsgaard J, White LA et al (2021) Lung mediated auditory contrast enhancement improves the signal-to-noise ratio for communication in frogs. Curr Biol 31(7):1488–1498.e1484

Londoño-Quiceno C, Gutiérrez-Cárdenas PDA (2023) Advertisement call of *Pristimantis taeniatus* (Boulenger, 1912) (Anura: Strabomantidae) from a population in the Magdalena River valley of Colombia. Zootaxa 5319(3):443–450

López-Flórez C, Ruíz MAO, Gómez-Ramírez E (2023) Effect of sublethal concentrations of glyphosate-based herbicides (Roundup Active®) on skin of the tropical frog (*Dendropsophus molitor*). Environ Sci Pollut Res 30(50):109618–109626

Losos J (2017) Improbable destinies: how predictable is evolution? Penguin UK

Lötters S (1996) The Neotropical toad genus *Atelopus*, checklist, biology and distribution. M Vences & Glaw Verlags GbR, Köln, pp 1–143

Lötters S (2007) The fate of the harlequin toads–help through a synchronous multi-disciplinary approach and the IUCN Amphibian Conservation Action Plan. Zoosyst Evol 83(S1):69–73

Lötters S, Jungfer K, Henkel F, Schmidt W (2007) Poison frogs-biology, species & captive husbandry. Chiamara, Frankfurt am Main

Lötters S, Plewnia A, Catenazzi A et al (2023) Ongoing harlequin toad declines suggest the amphibian extinction crisis is still an emergency. Commun Earth Environ 4(1):412

Lüddecke H (1976) Einige ergebnisse aus feldbeobachtungen and *Phyllobates palmatus* (Amphibia: Ranidae) in Kolumbien. Mitt Inst Colombo-Alemán Invest Cient 8:157–163

Lüddecke H (1993) Gruppenhaltung des Raketenfrosches *Colostethus palmatus* (Dendrobatidae). Elaphe 1:14–16

Lüddecke H (1997) Field reproductive potential of tropical high mountain *Hyla labialis* females: direct and indirect evidence from mark-recapture data. Amphibia-Reptilia 18(4):357–368

Luddecke H (1999) Behavioral aspects of the reproductive biology of the Andean frog *Colostethus palmatus* (Amphibia: Dendrobatidae). Rev Acad Colomb Cienc Exactas Fis Nat 23:S303–S303

Lüddecke H (2001) Variation in mating pattern in a population of the Andean frog *Hyla labialis*. Amphibia-Reptilia 22(2):199–207

Lüddecke H (2002) Male and female responses to call playback in the Andean frog *Colostethus subpunctatus*. Amphibia-Reptilia 23(2):141–150

Lüddecke H (2003) Space use, cave choice, and spatial learning in the dendrobatid frog *Colostethus palmatus*. Amphibia-Reptilia 24(1):37–46

Lüddecke H, Sínchez OR (2002) Are tropical highland frog calls cold–adapted? The case of the Andean frog *Hyla labialis*. Biotropica 34(2):281–288

Lüddecke H, Amézquita A, Bernal X, Guzmán F, Agudelo L, Miaud C, Guyétant R (1999) Comparison of environmental conditions and their effects on call characteristics of five frog species in a Colombian highland community. Curr Stud Herpetol:285–294

Luedtke JA, Chanson J, Neam K et al (2023) Ongoing declines for the world's amphibians in the face of emerging threats. Nature 622(7982):308–314

Lüddecke H, Amézquita A, Bernal X, Guzmán F (2000). Partitioning of vocal activity in a Neotropical highland-frog community. Stud Neotrop Fauna Environ, 35(3), 185-194.

Lynch JD (1999) Ranas pequeñas, la geometría de evolución, y la especiación en los Andes colombianos. Rev Acad Colomb Cienc Exactas Fis Nat 23(86): 143–160

Lynch JD (2006) The tadpoles of frogs and toads found in the lowlands of Northern Colombia. Rev Acad Colomb Cienc Exact Fis Nat 30(116):443–457

Lynch JD, Suárez Mayorga AM (2011) Clave ilustrada de los renacuajos en las tierras bajas al oriente de los Andes, con énfasis en Hylidae. Caldasia 33(1): 235–270

Lynch JD, Ruiz PM, Rueda JV (1983) Notes on the distribution and reproductive biology of *Centrolene geckoideum* Jimenez de la Espada in Colombia and Ecuador (Amphibia: Centrolenidae). Stud Neotropical Fauna Environ 18(4):239–243

Lynch JD, Ruiz-Carranza PM, Ardila-Robayo MC (1997) Biogeographic patterns of Colombian frogs and toads (Patrones biogeográficos de las ranas y los sapos de Colombia). Rev Acad Colomb Cienc Exactas Fis Nat 21(80):237–248

Maan ME, Cummings ME (2008) Female preferences for aposematic signal components in a polymorphic poison frog. Evolution 62(9):2334–2345

Maldonado-Ocampo JA, Ortega-Lara A, Usma J et al (2005) Peces de los Andes de Colombia. Instituto de Investigación de Recursos Biológicos Alexander von Humboldt, Bogotá, DC, Colombia

Mangold A, Trenkwalder K, Ringler M, Hödl W, Ringler E (2015) Low reproductive skew despite high male-biased operational sex ratio in a glass frog with paternal care. BMC Evol Biol 15(1):1–13

Marin CM, Molina-Zuluaga C, Restrepo A, Cano E, Daza JM (2018) A new species of Leucostethus (Anura: Dendrobatidae) from the eastern versant of the Central Cordillera of Colombia and the phylogenetic status of *Colostethus fraterdanieli*. Zootaxa 4461(3):359–380

Marín-Martínez M, Rojas-Morales JA (2016) Predation by a mygalomorphic spider (Araneae: Theraphosidae) on a stream-dwelling frog, *Rheobates palmatus* (Anura: Aromobatidae). Rept Amphib 23(3):175–177

Márquez R, Linderoth TP, Mejía-Vargas D, Nielsen R, Amézquita A, Kronforst MR (2020) Divergence, gene flow, and the origin of leapfrog geographic distributions: the history of colour pattern variation in *Phyllobates* poison-dart frogs. Mol Ecol 29(19): 3702–3719

Mayer M, Schulte LM, Twomey E, Lötters S (2014) Do male poison frogs respond to modified calls of a Müllerian mimic? Anim Behav 89:45–51

McCain CM, Grytnes JA (2010) Elevational gradients in species richness. In: Encyclopedia of Life Sciences (ELS). Wiley, Chichester

McVey ME, Zahary RG, Perry D, MacDougal J (1981) Territoriality and homing behavior in the poison dart frog (*Dendrobates pumilio*). Copeia 1:1–8

Medina I, Wang IJ, Salazar C, Amézquita A (2013) Hybridization promotes color polymorphism in the aposematic harlequin poison frog, *Oophaga histrionica*. Ecol Evol 3(13):4388–4400

Medina I, Dong C, Marquez R, Perez DM, Wang IJ, Stuart-Fox D (2024) Anti-predator defences are linked with high levels of genetic differentiation in frogs. Proc R Soc B 291(2015):20232292

Méndez-Narváez J, Bolívar-G W (2016) Complementary ecological approaches to understand decreases in frog diversity in altered Andean ecosystems. South Am J Herpetol 11(1):1–11

Méndez-Narváez J, Ospina-Sarria J, Bolívar-G W (2014) Diet and trophic ecology of *Leptodactylus fragilis* (Leptodactylidae) and *Dendropsophus columbianus* (Anura: Hylidae) in a disturbed area in southwestern Colombia. Herpetol Notes 7:299–305

Méndez-Narváez J, Flechas S, Amézquita A (2015) Foam nests provide context-dependent thermal insulation to embryos of three leptodactylid frogs. Physiol Biochem Zool 88(3):246–253

Mendoza-Henao AM, Arias E, Townsend JH, Parra-Olea G (2020) Phylogeny-based species delimitation and integrative taxonomic revision of the *Hyalinobatrachium fleischmanni* species complex, with resurrection of *H. viridissimum* (Taylor, 1942). Syst Biodivers 18(5):464–484

Mendoza-Henao AM, Zamudio KR, Guayasamin JM, Escalona M, Parra-Olea G (2023) Environment rather than character displacement explains call evolution in glassfrogs. Evolution 77(2):355–369

Meneses-Pelayo E (2024) Una guía para los anfibios y reptiles de Santander. Bucaramanga, Santander, Colombia. Available online at http://www.santanderherps.com

Meza-Joya FL, Ramos-Pallares E, Hernández-Jaimes C (2015) Use of an agroecosystem by the threatened dart poison frog *Andinobates virolinensis* (Dendrobatidae). Herpetol Rev 46(2):171–176

Mittelbach GG, McGill BJ (2019) Community ecology. Oxford University Press, Oxford.

Molina J (2023) La generación generadora de nuevas generaciones: Los maestros extranjeros que formaron a los actuales y futuros profesores del DCB. Universidad de los Andes. Revista Hipótesis, No 25

Montilla SO, Arcila-Pérez LF, Toro-Gómez MP, Vargas-Salinas F, Rada M (2023) A multidisciplinary approach reveals a new species of glassfrog from Colombia (Anura: Centrolenidae: *Nymphargus*). Zootaxa 5271(1):1–48

Moreno-Barbosa SE, Hoyos-Hoyos J (2014) Ontogeny of the diet in anurans (Amphibia) collected at La Vieja

river basin in the departmento of Quindío (Colombia). Caldasia 36(2):365–372
Morton ES (1975) Ecological sources of selection on avian sounds. Am Nat 109(965):17–34
Muñoz MI, Halfwerk W (2022) Amplification of frog calls by reflective leaf substrates: implications for terrestrial and arboreal species. Bioacoustics 31(4):490–503
Myers CW, Daly JW, Malkin B (1978) A dangerously toxic new frog (*Phyllobates*) used by Emberá Indians of western Colombia, with discussion of blowgun fabrication and dart poisoning. Bull Am Mus Nat Hist 161(2):309–365
Narins PM, Hödl W, Grabul DS (2003) Bimodal signal requisite for agonistic behavior in a dart-poison frog, *Epipedobates femoralis*. PNAS 100(2):577–580
Narins P, Feng AS, Fay R (2006) Hearing and sound communication in amphibians. In: Handbook of auditory research. Springer, New York, pp 1–12
Narins P, Gerhardt H, Christensen-Dalsgaard J (2024) A nasty, brutish, and short history of amphibian bioacoustics. In: Ketten CA, Coffin A, Fay R, Popper A (eds) A history of discoveries on hearing. Springer, Cham, pp 75–112
Navarro-Salcedo P, Duarte-Marín S, Rada M, Vargas-Salinas F (2021) Parental status is related to homing motivation in males of the glassfrog *Centrolene savagei*. Ethol Ecol Evol 33(5):528–542
Navarro-Salcedo P, Arcila-Pérez L, Pérez-González J, Rueda-Solano L, Rada M, Vargas-Salinas F (2022) Sex difference in homing: males but not females return home despite offspring mortality in *Ikakogi tayrona*, a glassfrog with prolonged maternal care. Behav Ecol Sociobiol 76(1):18
Navas CA (1996a) Implications of microhabitat selection and patterns of activity on the thermal ecology of high elevation neotropical anurans. Oecologia 108:617–626
Navas CA (1996b) Thermal dependency of field locomotor and vocal performance of high-elevation anurans in the tropical Andes. J Herpetol 30(4):478–487
Navas CA, Bevier CR (2001) Thermal dependency of calling performance in the eurythermic frog *Colostethus subpunctatus*. Herpetologica 57(4): 384–395
Nemeth E, Brumm H (2009) Blackbirds sing higher-pitched songs in cities: adaptation to habitat acoustics or side-effect of urbanization? Anim Behav 78(3): 637–641
Nemeth E, Brumm H (2010) Birds and anthropogenic noise: are urban songs adaptive? Am Nat 176(4): 465–475
Nemeth E, Pieretti N, Zollinger SA et al (2013) Bird song and anthropogenic noise: vocal constraints may explain why birds sing higher-frequency songs in cities. Proc R Soc B 280(1754):20122798
Noronha JC, Rodrigues DJ (2018) Reproductive behaviour of the glass frog *Hyalinobatrachium cappellei* (Anura: Centrolenidae) in the Southern Amazon. J Nat Hist 52(3–4):207–224
Nothacker JA, Neu CP, Mayer M, Wagner N, Lötters S (2018) Homing behavior in the Neotropical poison frog *Ameerega trivittata*. Salamandra 54(1):29–34
Nowakowski AJ, Otero Jiménez B, Allen M, Diaz-Escobar M, Donnelly MA (2013) Landscape resistance to movement of the poison frog, *Oophaga pumilio*, in the lowlands of northeastern Costa Rica. Anim Conserv 16(2):188–197
Olmstead MC, Kuhlmeier VA (2015) Comparative cognition. In: Olmstead MC, Kuhlmeier VA (eds) The Routledge international handbook of comparative psychology, Cambridge University Press, United Kingdom, pp 84–95
Ord TJ, Martins EP (2006) Tracing the origins of signal diversity in anole lizards: phylogenetic approaches to inferring the evolution of complex behaviour. Anim Behav 71(6):1411–1429
Orrico VG, Grant T, Faivovich J et al (2021) The phylogeny of Dendropsophini (Anura: Hylidae: Hylinae). Cladistics 37(1):73–105
Ospina-L AM, Rios-Soto JA, Vargas-Salinas F (2017) Size, endurance, or parental care performance? Male–male competition, female choice, and non-random mating pattern in the Glassfrog *Centrolene savagei*. Copeia 105(3):575–583
Ospina-L AM, Navarro-Salcedo P, Rios-Soto JA, Duarte-Marín S, Vargas-Salinas F (2020) Temporal patterns, benefits, and defensive behaviors associated with male parental care in the glassfrog *Centrolene savagei*. Ethol Ecol Evol 32(2):162–174
Ospina-Sarria JJ, Bolívar-G W, Mendez-Narvaez J, Burbano-Yandi C (2011) The tadpole of *Nymphargus grandisonae* (Anura, Centrolenidae) from Valle del Cauca, Colombia. South Am J Herpetol 6(2):79–86
Papi F (1992) Animal homing. Chapman & Hall, London
Parker GA (1982) Why are there so many tiny sperm? Sperm competition and the maintenance of two sexes. J Theor Biol 96(2):281–294
Parker P, Waite T (1997) Mating systems, effective population size, and conservation of natural populations. In: Behavioral approaches to conservation in the wild. Cambridge University Press, Cambridge, pp 243–261
Partan SR, Marler P (2005) Issues in the classification of multimodal communication signals. Am Nat 166(2): 231–245
Pašukonis A, Ringler M, Brandl HB, Mangione R, Ringler E, Hödl W (2013) The homing frog: high homing performance in a territorial dendrobatid frog *Allobates femoralis* (Dendrobatidae). Ethology 119(9): 762–768
Pašukonis A, Loretto M-C, Hödl W (2018) Map-like navigation from distances exceeding routine movements in the three-striped poison frog (*Ameerega trivittata*). J Exp Biol 221(2):jeb169714
Pašukonis A, Loretto M-C, Rojas B (2019) How far do tadpoles travel in the rainforest? Parent-assisted dispersal in poison frogs. Evol Ecol 33(4):613–623
Pašukonis A, Serrano-Rojas SJ, Fischer M-T et al (2022) Contrasting parental roles shape sex differences in

poison frog space use but not navigational performance. elife 11:e80483

Pérez-González JL, Rada M, Vargas-Salinas F, Rueda-Solano LA (2020) The tadpoles of two *Atelopus* species (Anura: Bufonidae) from the Sierra Nevada de Santa Marta, Colombia, with notes on their ecology and comments on the morphology of *Atelopus* larvae. South Am J Herpetol 2020(15):47–62

Pichler C, Weinlein S, Kopeinig L, Pašukonis A (2017) Homing performance in a territorial dendrobatid frog, *Allobates talamancae*. Salamandra 53(2):309–313

Pisso-Florez GA, Liévano-Bonilla AF, Mazepa G, Mulder KP, Ramírez-Chaves HE (2023b) Distribution update, advertisement call, and phylogenetic position of the Malvasá Rocket Frog, *Hyloxalus pinguis* (Anura: Dendrobatidae). South Am J Herpetol 28(1):78–88

Pisso-Florez GA, Liévano-Bonilla AF, Mendoza-Henao AM, Duarte-Marín S (2023a) The distress call and range extension of the Hermosas Robber Frog, *Pristimantis racemus* (Anura: Strabomantidae). Revista Latinoamericana de Herpetología 6(1):23–31

Pohl NU, Leadbeater E, Slabbekoorn H, Klump GM, Langemann U (2012) Great tits in urban noise benefit from high frequencies in song detection and discrimination. Anim Behav 83(3):711–721

Portik DM, Streicher JW, Wiens JJ (2023) Frog phylogeny: a time-calibrated, species-level tree based on hundreds of loci and 5,242 species. Mol Phylogenet Evol 188:107907

Pough F, Magnusson W, Ryan M, Wells K, Taigen T (1992) Behavioral energetics. In: Feder ME, Burggren WW (eds) Environmental physiology of the amphibians, University of Chicago Press, Chicago, pp 395–436

Poulton EB (1890) Colours of animals. D. Appleton and Company, New York

Prado-Sañudo ML, Giraldo A, Garcia WB (2018) Parental attendance and site selection of *Centrolene savagei* (Ruiz and Lynch 1991) in the western and central Andes of Colombia. Herpetol Notes 11:293–299

Pröhl H (1997) Territorial behaviour of the strawberry poison-dart frog, *Dendrobates pumilio*. Amphibia-Reptilia 18(4):437–442

Pröhl H (2005) Territorial behavior in Dendrobatid frogs. J Herpetol 39(3):354–365

Pröhl H, Hödl W (1999) Parental investment, potential reproductive rates, and mating system in the strawberry dart-poison frog, *Dendrobates pumilio*. Behav Ecol Sociobiol 46:215–220

Protti-Sánchez F, García-Rodríguez A, Barrantes G, Sandoval L (2023) Does variation in call rate affect the response of territorial males in the strawberry poison frog (*Oophaga pumilio*)? Ichthyol Herpetol 111(2):248–253

Pyron RA, Burbrink FT (2013) Phylogenetic estimates of speciation and extinction rates for testing ecological and evolutionary hypotheses. Trends Ecol Evol 28(12): 729–736

Pyron RA, Wiens JJ (2011) A large-scale phylogeny of Amphibia including over 2800 species, and a revised classification of extant frogs, salamanders, and caecilians. Mol Phylogenet Evol 61(2):543–583

Quintero I, González-Caro S, Zalamea P-C, Cadena CD (2014) Asynchrony of seasons: genetic differentiation associated with geographic variation in climatic seasonality and reproductive phenology. Am Nat 184: 352–363

Rada M, Dias PHDS, Pérez-Gonzalez JL, Anganoy-Criollo M, Rueda-Solano LA, Pinto-E MA, Quintero LM, Vargas-Salinas F, Grant T (2019) The poverty of adult morphology: bioacoustics, genetics, and internal tadpole morphology reveal a new species of glassfrog (Anura: Centrolenidae: Ikakogi) from the Sierra Nevada de Santa Marta, Colombia. PLoS One 14(5): e0215349

Ramos E, Meza-Joya FL, Hernandez-Jaimes C (2018) Distribution and conservation status of *Andinobates virolinensis* (Dendrobatidae), a threatened Andean poison frog endemic to Colombia. Herpetol Conserv Biol 13(1):58–69

Rangel-Ch O, Garzón A (1995) Sierra Nevada de Santa Marta (Colombia). Colombia diversidad biótica, Instituto de Ciencias Naturales, Bogotá, DC, pp 155–170

Reynolds RG, Fitzpatrick BM (2007) Assortative mating in poison-dart frogs based on an ecologically important trait. Evolution 61(9):2253–2259

Riaño C, Ortiz-Ruiz M, Pinto-Sánchez NR, Gómez-Ramírez E (2020) Effect of glyphosate (Roundup Active®) on liver of tadpoles of the Colombian endemic frog *Dendropsophus molitor* (Amphibia: Anura). Chemosphere 250:126287

Richards-Zawacki CL, Wang IJ, Summers K (2012) Mate choice and the genetic basis for colour variation in a polymorphic dart frog: inferences from a wild pedigree. Mol Ecol 21(15):3879–3892

Rincón-Franco F, Castro-Herrera F (1998) Aspectos ecológicos de una comunidad de *Eleutherodactylus* (Anura: Leptodactylidae) en un bosque de niebla del Occidente de Colombia. Caldasia:193–202

Ringler M, Szipl G, Hödl W, Khil L, Kofler B, Lonauer M, Provin C, Ringler E (2017) Acoustic ranging in poison frogs—it is not about signal amplitude alone. Behav Ecol Sociobiol 71:1–17

Ringler E, Rojas B, Stynoski JL, Schulte LM (2023) What amphibians can teach us about the evolution of parental care. Annu Rev Ecol Evol Syst 54:43–62

Rios-Soto JA, Ospina-L AM (2018) The advertisement call of *Pristimantis boulengeri* (Lynch, 1981) from a population in the Central Andes of Colombia (Anura: Craugastoridae). Herpetol Notes 11:719–723

Rios-Soto JA, Ospina-L AM, Vargas-Salinas F (2017a) The advertisement call and notes on the reproductive ecology of the glassfrog *"Centrolene" quindianum* (Anura: Centrolenidae). South Am J Herpetol 12(2): 117–127

Rios-Soto JA, Ospina-L AM, Basto-Riascos MC, López-Caro J, Vargas-Salinas F (2017b) Description of the distress call in *Espadarana prosoblepon* and the post-amplexus vocal display in *Centrolene savagei* (Anura: Centrolenidae). Herpetol Notes 10:27–29

Rivera Correa M, Vargas Salinas F, Grant T (2017) Statistical differences and biological implications: a comparative analysis of the advertisement calls of two Andean stream treefrogs (Hylidae: *Hyloscirtus*) and the evolution of acoustic characters. Salamandra 53(2):237–244

Rivera-Correa M, Faivovich J (2021) The tadpole of the stream treefrog *Hyloscirtus antioquia* (Anura: Hylidae: Cophomantini). Stud Neotropical Fauna Environ 56(3):203–209

Rivera-Pérez JM, Orrego-Meza JG, Escobar-Lasso S, Zuluaga-Isaza JC, Londoño-Quiceno C, Caicedo-Martínez LS (2020) Unexpected case of cancrivory in a lotic tropical ecosystem: first record of predation of *Strengeriana maniformis* by the frog *Rheobates palmatus*. Food Webs 25:e00163

Robertson JM, Lips KR, Heist EJ (2008) Fine scale gene flow and individual movements among subpopulations of *Centrolene prosoblepon* (Anura: Centrolenidae). Rev Biol Trop 56(1):13–26

Robins LJ (2006) Lateralized visual processing in anurans: new vistas through ancient eyes. In: Malashichev YB, Deckel AW (eds) Behavioral and morphological asymmetries in vertebrates. Landes Bioscience, Georgetown, pp 86–106

Rocha-Usuga AA, Vargas-Salinas F, Rueda-Solano LA (2017) Not every drought is bad: quantifying reproductive effort in the Harlequin Frog *Atelopus laetissimus* (Anura: Bufonidae). J Nat Hist 51(31–32):1913–1928

Rodríguez-Correa V, González-Alzate S, Rada M, Vargas-Salinas F (2024) The advertisement call of *Espadarana prosoblepon* (Anura: Centrolenidae) from a population in the Central Andes of Colombia. Phyllomedusa 23(2):165–178

Rodríguez-Suárez GE, Corredor-Londoño G (2012) Manejo y reproducción ex situ de la rana venenosa del Cauca *Ranitomeya bombetes* en el Zoológico de Cali, Colombia. Rev Biodivers Neotrop 2:113–125

Rojas B (2016) Behavioural, ecological, and evolutionary aspects of diversity in frog colour patterns. Biol Rev Camb Philos Soc 92(2):1059–1080

Rojas B, Vargas-Salinas F (2024) Advances in the study of poison frog evolutionary ecology II: decoding the hidden messages in coloration and unique behaviours. Evol Ecol 38(5):551–570

Rojas B, Devillechabrolle J, Endler JA (2014) Paradox lost: variable colour-pattern geometry is associated with differences in movement in aposematic frogs. Biol Lett 10(6):20140193

Rojas B, Burdfield-Steel E, De Pasqual C et al (2018) Multimodal aposematic signals and their emerging role in mate attraction. Front Ecol Evol 6:93

Rojas D, Lima AP, Momigliano P et al (2020) The evolution of polymorphism in the warning coloration of the Amazonian poison frog *Adelphobates galactonotus*. Heredity 124(3):439–456

Rojas B, Lawrence JP, Márquez R (2023) Amphibian coloration: proximate mechanisms, function, and evolution. In: Moreno-Rueda G, Comas M (eds) Evolutionary ecology of amphibians, CRC Press, Boca Raton, pp 219–258

Rojas-Montoya M, Vargas-Salinas F (2023) Right or left-handed: are locomotion type, body size, and microhabitat related to forelimb laterality in anurans? Evol Ecol 37(6):961–980

Rojas-Montoya M, López-Aguirre Y, González-Acosta C, Vargas-Salinas F (2024a) Repertorio de señales acústicas en la rana de cristal *Hyalinobatrachium tatayoi* (Anura: Centrolenidae). Rev Latinoam Herpet 7(1):760–782

Rojas-Montoya M, López-Aguirre Y, González-Acosta C, Vargas-Salinas F (2024b) Comportamiento reproductivo de la rana de cristal *Hyalinobatrachium tatayoi* (Anura: Centrolenidae) en una población de los andes colombianos. Rev Latinoam Herpet 7(3):95–118

Rojas-Rivera A, Cortés-Bedoya S, Gutiérrez-Cárdenas P, Castellanos J (2011) *Pristimantis achatinus* (Cachabi robber frog). Parental care and clutch size. Herpetol Rev 42:588–589

Rojas-Runjaic FJ, Cabello P (2011) *Centrolene daidaleum* (Ruiz-Carranza & Lynch, 1991) (Anura, Centrolenidae): a glassfrog with primitive and derived combat behavior. Zootaxa 2833(1):60–64

Roland AB, Santos JC, Carriker BC et al (2017) Radiation of the polymorphic Little Devil poison frog (*Oophaga sylvatica*) in Ecuador. Ecol Evol 7(22):9750–9762

Römer H (2013) Masking by noise in acoustic insects: problems and solutions. In: Brumm H (ed) Animal communication and noise. Animal signals and communication 2. Springer, Berlin, Heidelberg, pp 33–63

Rosenthal GG (2017) Mate choice: the evolution of sexual decision making from microbes to humans. Princeton University Press, Princeton

Rowe C (1999) Receiver psychology and the evolution of multicomponent signals. Anim Behav 58(5):921–931

Rueda-Almonacid J (1994) Estudio anatómico y relaciones sistemáticas de *Centrolene geckoideum* (Salientia: Anura: Centrolenidae). Trianea 5:133–187

Rueda-Almonacid JV, Lynch JD, Amézquita A (2004) Libro rojo de los anfibios de Colombia. Conservación Internacional Colombia

Rueda-Solano LA, Warkentin KM (2016) Foraging behavior with possible use of substrate-borne vibrational cues for prey localization in *Atelopus laetissimus* (Ruiz-Carranza, Ardila-Robayo, and Hernández-Camacho, 1994). Herpetol Notes 9:191–195

Rueda-Solano LA, Vargas-Salinas F, Rivera-Correa M (2015) The highland tadpole of the harlequin frog *Atelopus carrikeri* (Anura: Bufonidae) with an analysis of its microhabitat preference. Salamandra 51(1): 25–32

Rueda-Solano LA, Flechas SV, Galvis-Aparicio M et al (2016) Epidemiological surveillance and amphibian assemblage status at the Estación Experimental de San Lorenzo, Sierra Nevada de Santa Marta, Colombia. Amphib Reptil Conserv 10(1):7–19. (e114)

Rueda-Solano LA, Vargas-Salinas F, Pérez-González JL, Sánchez-Ferreira A, Ramírez-Guerra A, Navas CA, Crawford AJ (2022) Mate-guarding behaviour in anurans: intrasexual selection and the evolution of prolonged amplexus in the harlequin toad *Atelopus laetissimus*. Anim Behav 185:127–142

Ruíz-Carranza PM, Ardila-Robayo MC, Hernández-Camacho JA (1994) Tres nuevas especies de *Atelopus* (Amphibia: Bufonidae) de la Sierra Nevada de Santa Marta. Rev Acad Colomb Cienc Exactas Fis Nat 19: 153–163

Rundle DH, Boughman JW (2010) Behavioral ecology and speciation. In: Fox CW, Westneat DF (eds) Evolutionary behavioural ecology. Oxford University Press, New York, pp 471–487

Ruxton G, Allen WL, Sherrat TN, Speed MP (2018) Avoiding attack: the evolutionary ecology of crypsis, warning signals and mimicry, 2nd edn. Oxford University Press, Oxford

Rytwinski T, Fahrig L (2013) Why are some animal populations unaffected or positively affected by roads? Oecologia 173:1143–1156

Saboyá-Acosta LP, Urbina-Cardona JN (2023) Current state of knowledge of Páramo Amphibians in Colombia: Spatio temporal trends and information gaps to be strengthened for effective conservation. Trop Conserv Sci 16:1–23

Salgado AL, Guayasamin JM (2018) Parental care and reproductive behavior of the minute dappled glassfrog (Centrolenidae: *Centrolene peristictum*). South Am J Herpetol 13(3):211–219

Sánchez-Pacheco SJ, Torres-Carvajal O, Aguirre-Peñafiel V et al (2018) Phylogeny of *Riama* (Squamata: Gymnophthalmidae), impact of phenotypic evidence on molecular datasets, and the origin of the Sierra Nevada de Santa Marta endemic fauna. Cladistics 34(3):260–291

Santos JC, Coloma LA, Cannatella DC (2003) Multiple, recurring origins of aposematism and diet specialization in poison frogs. PNAS 100(22):12792–12797

Santos JC, Baquero M, Barrio-Amoros C, Coloma LA, Erdtmann LK, Lima AP, Cannatella DC (2014) Aposematism increases acoustic diversification and speciation in poison frogs. Proc R Soc Biol Sci 281(1796):20141761

Saporito RA, Donnelly MA, Madden AA, Garraffo HM, Spande TF (2010) Sex-related differences in alkaloid chemical defenses of the dendrobatid frog *Oophaga pumilio* from Cayo Nancy, Bocas del Toro, Panama. J Nat Prod 73(3):317–321

Saporito RA, Donnelly MA, Spande TF, Garraffo HM (2012) A review of chemical ecology in poison frogs. Chemoecology 22:159–168

Schlupp I (2021) Male choice, female competition, and female ornaments in sexual selection. Oxford University Press, Oxford UK

Schulte LM (2014) Feeding or avoiding? Facultative egg feeding in a Peruvian poison frog (*Ranitomeya variabilis*). Ethol Ecol Evol 26(1):58–68

Schulte LM, Yeager J, Schulte R et al (2011) The smell of success: choice of larval rearing sites by means of chemical cues in a Peruvian poison frog. Anim Behav 81(6):1147–1154

Schulte LM, Martel A, Cruz-Elizalde R, Ramírez-Bautista A, Bossuyt F (2021) Love bites: male frogs (*Plectrohyla*, Hylidae) use teeth scratching to deliver sodefrin precursor-like factors to females during amplexus. Front Zool 18(1):1–14

Schulte LM, Abondano Almeida D, Stadler S, Köhler J (2023a) Pheromones for amphibian conservation–considering new approaches in conservation breeding programmes. Anim Conserv 23:603–605

Schulte LM, Lipkowski K, Abondano Almeida D (2023b) Chemical communication and semiochemical recognition in frogs—from eggs to adults. In: Schaal B, Rekow D, Keller M (eds) Chemical signals in vertebrates, vol 15. Springer Nature, Switzerland, pp 75–111

Schulze E, Lipkowski K, Abondano Almeida D, Schulte LM (2024) Adult poison dart frogs avoid potential heterospecific competitors using their sense of smell. Ethology 130:e13486

Siddiqi A, Cronin TW, Loew ER, Vorobyev M, Summers K (2004) Interspecific and intraspecific views of color signals in the strawberry poison frog *Dendrobates pumilio*. J Exp Biol 207:2471–2485

Slabbekoorn H, Peet M (2003) Birds sing at a higher pitch in urban noise. Nature 424(6946):267–267

Slabbekoorn H, Ripmeester EA (2008) Birdsong and anthropogenic noise: implications and applications for conservation. Mol Ecol 17(1):72–83

Smith BT, McCormack JE, Cuervo AM et al (2014) The drivers of tropical speciation. Nature 515(7527): 406–409

Smotherman M, Narins P (2004) Evolution of the amphibian ear. In: Manley GA, Fay RR, Popper AN (eds) Evolution of the vertebrate auditory system. Springer handbook of auditory research 22. Springer, New York, pp 164–199

Solari S, Muñoz-Saba Y, Rodríguez-Mahecha JV, Defler TR, Ramírez-Chaves HE, Trujillo F (2013) Riqueza, endemismo y conservación de los mamíferos de Colombia. Mastozool Neotrop 20(2):301–365

Sonnleitner R, Ringler M, Loretto MC, Ringler E (2020) Experience shapes accuracy in territorial decision-making in a poison frog. Biol Lett 16(5):20200094

Souza JR, Kaefer IL, Lima AP (2021) Calling activity determines male mating success in a territorial frog with parental care. Ethology 127(4):359–365

Starnberger I, Poth D, Peram PS et al (2013) Take time to smell the frogs: vocal sac glands of reed frogs (Anura: Hyperoliidae) contain species-specific chemical cocktails. Biol J Linn Soc 110(4):828–838

Stebbins RC, Hendrickson JR (1959) Field studies of amphibians in Colombia, South America. University of California Press

Stevens M, Cuthill IC (2006) Disruptive coloration, crypsis and edge detection in early visual processing. Proc Biol Sci 273(1598):2141–2147

Stuart SN, Hoffmann M, Chanson J, Cox N, Berridge R, Ramani P et al (2008) Threatened amphibians of the world. Lynx Edicions, Spain

Stuckert AM, Venegas PJ, Summers K (2014) Experimental evidence for predator learning and Müllerian mimicry in Peruvian poison frogs (*Ranitomeya*, Dendrobatidae). Evol Ecol 28:413–426

Stynoski JL, Jones I-M, Brown J, Granados-Martínez S (2024) Mother strawberry poison frogs might supplement nutritive eggs with secretory provisioning. Evol Ecol 38:77–90

Sullivan BK, Ryan MJ, Verrell PA (1995) Female choice and mating system structure. Amphib Biol 2:469–517

Summers K (1999) The effects of cannibalism on Amazonian poison frog egg and tadpole deposition and survivorship in Heliconia axil pools. Oecologia 119:557–564

Summers K, Tumulty J (2014) Parental care, sexual selection, and mating systems in neotropical poison frogs. In: Macedo RH, Machado G (eds) Sexual selection: perspectives and models from the Neotropics, AcademicPress, San Diego, pp 289–320

Sutherland WJ (1996) From individual behaviour to population ecology. Oxford University Press, Oxford, USA

Sutherland WJ (1998) The importance of behavioural studies in conservation biology. Anim Behav 56(4): 801–809

Taboada C, Brunetti AE, Pedron FN et al (2017) Naturally occurring fluorescence in frogs. Proc Natl Acad Sci 114(14):3672–3677

Ten Cate C, Verzijden MN, Etman E (2006) Sexual imprinting can induce sexual preferences for exaggerated parental traits. Current Biology, 16(11), 1128–1132.

Thomson SA, Pyle RL, Ahyong ST et al (2018) Taxonomy based on science is necessary for global conservation. PLoS Biol 16(3):e2005075

Tonini JFR, Provete DB, Maciel NM et al (2020) Allometric escape from acoustic constraints is rare for frog calls. Ecol Evol 10(8):3686–3695

Toro-Gómez MP, Carvajal-Castro JD, Casas-Cardona S, Vargas-Salinas F (2022) Experimental evidence in a poison frog model suggests that tadpole transport on the dorsum may affects warning signal effectiveness in poison frogs. Evol Ecol 37(2):267–289

Tovar B, Pineda G, Clavijo-Garzón S, Vargas D, Medina R (2022) Defensive behavior of the Antioquia Chocolate frog (*Hyloscirtus antioquia*). Rev Latinoam Herpet 5(3):46–48

Triana Velásquez TM, Rojas CM, Bernal Bautista MH (2013) Lethal and sublethal effects of glyphosate (Roundup® Active) to embryos of Colombian anurans. Acta Biol Col 18(2):271

Tumulty JP (2022) Dear enemy effect. In: Encyclopedia of animal cognition and behavior. Springer, Switzerland, pp 1937–1940

Tumulty JP, Bee MA (2021) Ecological and social drivers of neighbor recognition and the dear enemy effect in a poison frog. Behav Ecol 32(1):138–150

Urbina-Cardona N, Saboyá-Acosta LP, Camacho-Rozo CP et al (2023) Producción científica sobre la herpetología en Colombia: perspectivas desde los temas de investigación hacia la conservación biológica. Caldasia 45(1):1–20

Valderrama-Vernaza M, Ramírez-Pinilla MP, Serrano-Cardozo VH (2009) Diet of the Andean Frog *Ranitomeya virolinensis* (Athesphatanura: Dendrobatidae). J Herpetol 43(1):114–123

Valencia-Aguilar A, Castro-Herrera F, Ramírez-Pinilla MP (2012) Microhabitats for oviposition and male clutch attendance in *Hyalinobatrachium aureoguttatum* (Anura: Centrolenidae). Copeia 4: 722–731

Valencia-Aguilar A, Cortés-Gómez AM, Ruiz-Agudelo CA (2013) Ecosystem services provided by amphibians and reptiles in Neotropical ecosystems. Int J Biodivers Sci Ecosyst Serv Manag 9(3):257–272

Valencia-Aguilar A, Zamudio KR, Haddad CF, Bogdanowicz SM, Prado CP (2020) Show me you care: female mate choice based on egg attendance rather than male or territorial traits. Behav Ecol 31(4): 1054–1064

Vargas-Salinas F, Amézquita A (2013a) Traffic noise correlates with calling time but not spatial distribution in the threatened poison frog *Andinobates bombetes*. Behaviour 150:569–584

Vargas-Salinas F, Amézquita A (2013b) Stream noise, hybridization, and uncoupled evolution of call traits in two lineages of poison frogs: *Oophaga histrionica* and *Oophaga lehmanni*. PLoS One 8(10):e77545

Vargas-Salinas F, Amézquita A (2014) Abiotic noise, call frequency and stream-breeding anuran assemblages. Evol Ecol 28:341–359

Vargas-Salinas F, Castro-Herrera F (1999) Cuidado parental en anuros del género *Eleutherodactylus* (Amphibia: Leptodactylidae) presentes en Colombia. Rev Acad Colomb Cienc Exactas Fis Nat 23:407–410

Vargas-Salinas F, Gutiérrez-Cárdenas PDA (2005) Cambios morfológicos y mortalidad en embriones y renacuajos de *Agalychnis spurrelli* Boulenger (Anura: Hylidae). Actual Biol 27(83):189–202

Vargas-Salinas F, Rojas B (2024) Developments in the study of poison frog evolutionary ecology I: social interactions, life history and habitat use across space and ontogeny. Evol Ecol 38(1):1–22

Vargas-Salinas F, López-Aranda F, Muñoz-G N (2007) Savage Glassfrog *Cochranella savagei* (reproduction). Herpetol Rev 38(4):436–437

Vargas-Salinas F, Cunnington GM, Amézquita A, Fahrig L (2014a) Does traffic noise alter calling time in frogs and toads? A case study of anurans in Eastern Ontario, Canada. Urban Ecosyst 17(4):945–953

Vargas-Salinas F, Dorado-Correa A, Amézquita A (2014b) Microclimate and stream noise predict geographic divergence in the auditory signal of a threatened poison frog. Biotropica 46(6):748–755

Vargas-Salinas F, Quintero-Ángel A, Osorio-Domínguez D, Rojas-Morales J, Escobar-Lasso S, Gutiérrez-Cárdenas PD, Rivera-Correa M, Amézquita A (2014c) Breeding and parental behavior in the glassfrog *Centrolene savagei*. J Nat Hist 48(27–28): 1689–1705

Vargas-Salinas F, Ospina-L A, Rios-Soto J, Rivera-Correa M (2017) *Centrolene savagei*. Catalog Am Amphib Rept 912:1–12

Vargas-Salinas F, Muñoz-Ávila JA, Morales Puentes ME (eds) (2019) Biología de los anfibios y reptiles en el bosque seco tropical del norte de Colombia. Editorial UPTC, Tunja

Vargas-Salinas F, Atehortua-Vallejo MA, Arcila-Pérez LF, Jiménez-Vargas GM, González-Acosta C, Casas-Cardona S, Grajales-Echeverry A (2020) *Andinobates bombetes*. Catalog Am Amphib Rept 926:1–16

Vargas-Salinas F, Londoño-Guarnizo CA, Duarte-Marin S, Torres-Suárez OL, Amézquita A (2024) Uncoupled evolutionary patterns in spectral and temporal components of acoustic signals in anurans associated to streams. Ethology 130(11):e13510

Valencia-Aguilar A, Ringler E, Lüpold, S, Guayasamin JM, Prado CP (2024). Evolutionary trade-offs between testes size and parenting in Neotropical glassfrogs. Proc. R. Soc. B Biol. Sci. 291(2016), 20240054

Vélez A, Schwartz JJ, Bee MA (2013) Anuran Acoustic Signal Perception in Noisy Environments. In: Brumm H (eds) Animal Communication and Noise vol 2. Springer, Berlin, Heidelberg, pp 133 -185.

Vellend M (2010) Conceptual synthesis in community ecology. Q Rev Biol 85(2):183–206

Vences M, Wahl-Boos G, Hoegg S (2007) Molecular systematics of mantelline frogs from Madagascar and the evolution of their femoral glands. Biol J Linn Soc 92(3):529–539

Vergara-Herrera N, Cocroft R, Rueda-Solano LA (2024) Eating to the beat of the drum: vibrational parameters of toe tapping behavior in *Dendrobates truncatus* (Anura: Dendrobatidae). Evol Ecol 38:1–17

Villegas-Mora D, Muñoz-Acevedo S, Guevara-Molina EC, Vargas-Salinas F (2020) Phenology and mortality of embryos in a Colombian population of *Nymphargus grandisonae* (Anura: Centrolenidae). Phyllomedusa 19(2):225–241

Vockenhuber EA, Hödl W, Amézquita A (2009) Glassy fathers do matter: egg attendance enhances embryonic survivorship in the glass frog *Hyalinobatrachium valerioi*. J Herpetol 43(2):340–344

Warkentin KM, Jung J, Rueda Solano LA, McDaniel JG (2019) Ontogeny of escape-hatching decisions: vibrational cue use changes as predicted from costs of sampling and false alarms. Behav Ecol Sociobiol 73: 1–14

Warren PS, Katti M, Ermann M, Brazel A (2006) Urban bioacoustics: it's not just noise. Anim Behav 71(3): 491–502

Waters KR, Dugas MB, Grant T, Saporito RA (2024) The ability to sequester the alkaloid epibatidine is widespread among dendrobatid poison frogs. Evol Ecol 38:1–15

Weis JS, Sol D (2016) Biological invasions and animal behaviour. Cambridge University Press, Cambridge

Wells KD (2001) The energetics of calling in frogs. Anuran communication, pp 45–60

Wells KD (2007) The ecology and behavior of amphibians. The University of Chicago Press, Chicago

Wells KD, Schwartz JJ (1982) The effect of vegetation on the propagation of calls in the neotropical frog *Centrolenella fleischmanni*. Herpetologica 38:449–455

Weygoldt P (1980) Complex brood care and reproductive behavior in captive poison frogs, *Dendrobates pumilio* O. Schmidt Behav Ecol Sociobiol 7:329–332

Weygoldt P (1987) Evolution of parental care in dart poison frogs (Amphibia: Anura: Dendrobatidae). J Zool Syst Evol Res 25(1):51–67

Wilczynski W, Ryan MJ (1999) Geographic variation in animal communication systems. In: Foster SA, Endler JA (eds) Geographic variation in behavior. Oxford University Press, New York, pp 234–261

Wilkins MR, Seddon N, Safran JR (2013) Evolutionary divergence in acoustic signals: causes and consequences. Trends Ecol Evol 28(3):156–166

Wurdack J (1976) Endemic Melastomataceae of the Sierra Nevada de Santa Marta, Colombia. Brittonia 28:138–143

Yang Y, Richards-Zawacki CL, Devar A, Dugas MB (2016) Poison frog color morphs express assortative mate preferences in allopatry but not sympatry. Evolution 70(12):2778–2788

Yang Y, Servedio MR, Richards-Zawacki CL (2019) Imprinting sets the stage for speciation. Nature 574(7776):99–102

Zheng P, Gong Y, Wang B et al (2024) Love hug functional validation of nuptial pad-secreted pheromone in anurans. Animals 14(11):1550

Zornosa-Torres C, Hoyos JM (2016) Relationship between body size and sexual maturity in *Dendropsophus columbianus* (Anura, Hylidae) and description of their sexual apparatus. Herpetol Rev 47:19–23

Zuluaga-Isaza JC, Escobar-Lasso S, Llano-Arias CA, Londoño C (2020) Predation on Colombian endemic frog *Rheobates palmatus* (Werner, 1899) (Anura: Aromobatidae) by a whip-spider *Heterophrynus* sp. (Amblypygi: Phrynidae). Herpetol Notes 13:231–233

Functional Diversity of Amphibians and Reptiles in Tropical Dry Forest Remnants and Anthropogenic Cover Types in the Lower Bogotá River Sub-Basin (Department of Cundinamarca, Colombia)

Felipe Veloza Romero and J. Nicolás Urbina-Cardona

1 Introduction

The anthropogenic transformation of tropical ecosystems is one of the main drivers of biodiversity loss worldwide (Foley et al. 2005; Gonthier et al. 2014). Currently, fragmented landscapes contain forest remnants embedded within anthropogenic matrices used for agriculture, livestock, mining, and urbanization (Fahrig 2003; Portillo-Quintero and Sánchez-Azofeifa 2010; Rivas and Navarro-Cerrillo 2024). These land cover changes degrade habitat quality for native species, reduce ecosystem service provision, and disrupt ecological processes at multiple scales (Ewers and Didham 2006; Fischer and Lindenmayer 2007).

Landscape transformation disproportionately affects species in tropical forests, which, unlike temperate ecosystems, have not historically been exposed to frequent and intense disturbances (Pickett et al. 2011). Consequently, species in temperate environments have evolved greater resilience to habitat changes, demonstrating higher reproductive adaptability and tolerance to disturbance regimes with extreme temperature and precipitation conditions (*sensu* extinction filter hypothesis; Balmford 1996). In contrast, tropical landscapes with a history of significant anthropogenic transformation have already undergone an ecological filtering process, resulting in the loss of species most vulnerable to edge effects and habitat fragmentation (Betts et al. 2019). Those landscape processes reshape the structure and composition of assemblages and alter resource availability in the remaining natural habitats (Hobbs et al. 2009).

Both natural and anthropogenic disturbances influence the dynamics of tropical forests, directly impacting ectothermic taxa such as amphibians and reptiles (Gardner et al. 2007; Tan et al. 2023). These disturbances affect behavior, habitat use, and foraging areas, often leading to local declines of habitat specialists and a subsequent simplification of assemblages (Thompson et al. 2016; Marroquín-Páramo et al. 2021). In the long term, this sequential species loss disrupts biotic interactions and may lead to extinction debts, where ecological consequences manifest over extended timescales (Figueiredo et al. 2019). Habitat loss has serious consequences for herpetofauna, particularly in tropical regions,

F. V. Romero (✉)
Pontificia Universidad Javeriana, Facultad de Estudios Ambientales y Rurales, Maestría en Conservación y Uso de Biodiversidad, Bogotá, Colombia
e-mail: fveloza@javeriana.edu.co

J. N. Urbina-Cardona
Pontificia Universidad Javeriana, Facultad de Estudios Ambientales y Rurales, Departamento de Ecología y Territorio. Carrera, Bogotá, Colombia

J. N. Urbina-Cardona et al. (eds.), *Andean Herpetofauna*, Sustainable Development Goals Series,
https://doi.org/10.1007/978-3-032-00074-3_4

where the effects are more pronounced (Bohm et al. 2013; Luedtke et al. 2023). Annual deforestation rates continue to increase due to agricultural and livestock expansion, intensifying ecosystem change (Hansen et al. 2013; Laurance et al. 2014). Furthermore, amphibians remain underrepresented in protected area networks in the tropical Andes (Nori et al. 2015), in one of the regions with the highest levels of amphibian endemism (Tobar-Suarez et al. 2022), further exacerbating conservation challenges.

Tropical dry forests (TDFs) have experienced the highest levels of anthropogenic habitat loss, with only 25% of their original global cover remaining (Rivas and Navarro-Cerrillo 2024). In Colombia, the rate of deforestation has reached 174,000 hectares per year as of 2019 (IDEAM 2018). Deforestation of Colombia's TDFs began well before the 1500s, prior to the arrival of the Spanish conquest (Etter et al. 2008), and today, only 8% of its original extent remains (García et al. 2014). Continued primary forest loss and urban expansion in Colombia are expected to homogenize species assemblages by favoring generalist species over habitat specialists (Solar et al. 2015; Echeverría-Londoño et al. 2016). This drastic reduction threatens amphibian and reptile populations that persist in these fragmented landscapes of the lowland and dry areas of the Northern Andes (Urbina-Cardona et al. 2014; Carvajal-Cogollo et al. 2019).

Given the spatial configuration of remaining TDF patches and the limited extent of TDF it is crucial to evaluate how different types of anthropogenic land use affect species assemblages (Pita et al. 2007; Kupfer et al. 2006; Didham et al. 2012). Previous research suggests that the size, structure, management, and heterogeneity of the surrounding matrix exert greater pressure on biodiversity than the spatial arrangement of native forest patches (Prugh et al. 2009; Phillips et al. 2018). Thus, matrix management plays a pivotal role in biodiversity conservation, with the potential to either mitigate or exacerbate the effects of habitat loss and fragmentation (Vandermeer and Perfecto 2007; Franklin and Lindenmayer 2009; Arroyo-Rodríguez et al. 2020). This process is closely tied to edge effects and mechanisms such as the "spillover effect" (Ries et al. 2004; Granda-Rodriguez et al. 2025) which explains the increased species richness in fragmented landscapes, where generalist and invasive exotic species often colonize new forest edge environments, exploiting the altered conditions. According to Driscoll et al. (2013), anthropogenic matrices influence species persistence in fragmented landscapes through three key factors:

1. *Mobility and dispersal ability:* Species movement is influenced by the structural similarity between anthropogenic covers and native forest patches, where greater similarity facilitates dispersal (López-Bedoya et al. 2022).
2. *Resource availability:* The matrix can provide resources for some species while negatively impacting forest specialists by altering resource distribution (Ewers and Didham 2006).
3. *Abiotic conditions:* Changes in microclimate and edge effects, such as increased light exposure and airflow, elevate local extinction risks for groups like amphibians and reptiles (Lehtinen et al. 2003; Schneider-Maunoury et al. 2016; Pfeifer et al. 2017). These conditions also facilitate the expansion of generalist and invasive species adapted to new microclimatic conditions (Watling et al. 2011; Galindo-Uribe et al. 2022).

Functional traits are key to understanding how species respond to these environmental changes (Vetter et al. 2011; Ernst et al. 2012; Ribeiro et al. 2017a, b). Physiological, morphological, behavioral, and life history traits influence species' adaptation to environmental shifts (Díaz and Cabido 2001; Violle et al. 2007). Amphibians and reptiles with specialized traits, such as thick, tuberculated skin in amphibians and keeled, imbricate scales in squamates, exhibit greater resistance to desiccation (Hernández-Ordóñez et al. 2019; Rincón-Aranguri et al. 2019, 2023). Conversely, amphibian species with limited dispersal ability and low biomass are particularly vulnerable to elevated temperatures and increased solar radiation (Pfeifer et al. 2017; Nowakowski et al. 2018).

Another important trait influencing species adaptability and survival in anthropogenic landscapes is reproductive timing and sexual maturation (Allen et al. 2017). Species with long reproductive cycles and late sexual maturation are more likely to be negatively affected and extirpated from transformed habitats. In contrast, highly fertile species with accelerated sexual maturity often thrive in modified environments, frequently becoming dominant (Van Wilgen and Richardson 2012). This suggests that within an assemblage, some species are inherently more sensitive to habitat changes (Stuart et al. 2004; Bohm et al. 2013). Species with specific physiological and ecological requirements, low dispersal capacity, and small home ranges face higher extinction risks in altered landscapes (Huey 1982).

One of the most measured functional traits in amphibians and reptiles is body size (Pfeifer et al. 2017), which influences desiccation tolerance in amphibians (Olalla and Rodríguez 2007) and thermoregulatory requirements in reptiles (Nowakowski et al. 2018). In this sense, small-bodied amphibians tend to inhabit native forests, while larger ones can adapt to anthropogenic matrices (Pfeifer et al. 2017; Galindo-Uribe et al. 2022). The opposite is true for reptiles, especially snakes, where larger body sizes require greater exposure to direct sunlight or warm surfaces for proper thermoregulation (Kjoss and Litvaitis 2001). This behavior increases their vulnerability to hunting by local communities at forest edges (Pfeifer et al. 2017) and the risk of road mortality near remaining native forests, ultimately affecting the functional diversity of assemblages (Rincón-Aranguri et al. 2023).

When considering ecosystem transformation as an environmental filter, only species with traits that provide greater tolerance and/or plasticity will persist in anthropogenic habitats, while others will face local extinction (Aronson et al. 2016; Vanessa et al. 2025). This process leads to functional homogenization, where the remaining species share traits that enhance plasticity and tolerance, ultimately disrupting ecosystem functioning (Clavel et al. 2011). Additionally, the loss of low-abundance species, which possess unique functional traits, further contributes to functional homogenization, particularly in ecosystems undergoing ecological succession and habitat degradation (Zabala-Forero et al. 2025). Recent studies, such as Álvarez-Grzybowska et al. (2020), have demonstrated that dry seasons in TDF act as environmental filters for amphibians, favoring species with similar functional traits, whereas in more water-abundant ecosystems, natural and anthropogenic land cover types support greater functional diversity following a limiting similarity mechanism (Zabala-Forero and Urbina-Cardona 2021; Vanessa et al. 2025).

For the TDF of the Caribbean of Colombia, Carvajal-Cogollo and Urbina-Cardona (2015) reported that the proportion of reptile species belonging to each functional group differed between pasture and forest habitats, with no significant variation found between the forest edge and interior. Four of the seven ecological groups inhabited both pasture and forest interior, while all ecological groups were represented in the forest edge. The functional group, which consists of small, diurnal, oviparous species, accounted for 50% of species in the forest interior, suggesting that functional roles may differ between habitat types and provide insights into how ecological functions could be disrupted by habitat fragmentation, loss, and degradation. Recently, Granda-Rodriguez et al. (2025) observed that taxonomic and functional diversity of amphibians varies along a habitat gradient from pasture to forest edge and interior in remnant TDF. In the interior of the forest, specialized species predominate, as evidenced by high functional evenness and a greater number of functional groups, each with low richness. The most resilient functional group (with terrestrial amphibian species with a body temperature greater than 26 °C, body size over 27.3 mm, and weight greater than 2.2 g) showed the highest abundance in pasture, forest edge, and forest interior. Evaluating species' morphological traits and life history is, therefore, essential to understanding how ecosystem transformation affects functional diversity and its implications for ecosystem processes (Luck et al. 2010; Tilman et al. 2014).

By measuring functional traits, it is possible to assess functional diversity, which serves as a proxy for how species within an assemblage exploit resources in an ecosystem (Díaz and Cabido 2001; Violle et al. 2007; Salgado-Negret and Paz 2015). Research has shown that the functional diversity of amphibians and reptiles increases in areas with higher levels of conservation, albeit with different species composition and functional groups (Ernst et al. 2006; Rosas-Espinoza et al. 2024). Functional richness of anurans tends to be higher in well-conserved areas compared to pastures or croplands (Ribeiro et al. 2017a) and increases near water bodies (Ribeiro et al. 2017b). Conversely, functional divergence and equity of amphibians and reptiles are often higher in highly transformed areas (Ribeiro et al. 2017a; Ramírez-Arce et al. 2025), while functional evenness is lower in lizard assemblages (Brasileiro et al. 2023). In contrast, the interior of TDF shows lower functional evenness and higher dominance of amphibian species (Granda-Rodriguez et al. 2025).

Amphibians remain largely understudied in transformed tropical landscapes, with even fewer studies focusing on reptiles (Gardner, Barlow, and Peres 2007). As global research on habitat transformation, land-cover change, and its effects on herpetofauna continues to grow (López-Bedoya et al. 2022; Tan et al. 2023), these studies provide crucial insights for developing conservation strategies to mitigate the decline of these taxa (Filippi and Luiselli 2006; Morse et al. 2009; Reading et al. 2010; Alford 2011). Consequently, the development of effective management and conservation plans for threatened species, as well as the implementation of landscape planning tools, must account for both the quality of the surrounding matrix and the remaining native forest fragments (Urbina-Cardona et al. 2024). Furthermore, evaluating species' responses to habitat gradients through their functional traits is essential for understanding the impacts of habitat transformation on biodiversity (Driscoll et al. 2013).

In this context, this study aims to compare the functional diversity of amphibians and reptiles between natural and anthropogenic land covers in a TDF ecosystem. Specifically, we aim to (1) analyze the structure of amphibian and reptile assemblages in different land cover types of the Lower Bogotá River sub-basin; (2) classify amphibians and reptiles into functional groups based on morphological and natural history (fiixed) traits, assessing their distribution across natural and anthropogenic land cover; and (3) determine the effect of land cover type on amphibian and reptile functional diversity indices.

2 Materials and Methods

2.1 Study Area

The Bogotá River basin, located in the Department of Cundinamarca, Colombia, is classified as a second-order watercourse with a total surface area of 589,143 hectares (CAR 2005). The Lower Bogotá River sub-basin accounts for 9.2% of the total basin area and includes the municipalities of Tocaima, Agua de Dios, Girardot, and Ricaurte. The terrain is predominantly flat, with temperatures ranging between 24 and 30 °C, where annual variation does not exceed 3 °C. The relative humidity averages 81%, and the average annual precipitation ranges 1200 mm and 1300 mm. The region has a bimodal rainfall regime, with rainy periods occurring from April to June and September to October, and two dry periods from January to March and July to August. The upper part of the sub-basin has a warm semi-arid climate, while the Girardot area experiences a warm arid climate (CAR 2005). According to Holdridge (1982)'s life zone classification system, this region corresponds to the tropical dry forest (TDF) life zone.

The area is undergoing rapid population growth, leading to a continuous expansion of residential zones, primarily for vacation and recreational purposes. Additionally, land use has shifted toward productive activities, with livestock farming and extensive cultivation of

Table 1 Characteristics of the 14 sites nested in 6 locations in the Lower Bogotá River sub-basin in the department of Cundinamarca (Colombia)

Location	L1		L2				L3		L4		L5		L6	
Site	V1	V2	V3	V4	V5	V6	V7	V8	V9	V10	V11	V12	V13	V14
Municipality	Tocaima	Tocaima	Tocaima	Tocaima	Tocaima	Tocaima	Tocaima	Tocaima	Girardot	Girardot	Girardot	Girardot	Ricaurte	Ricaurte
Altitude (m a.s.l.)	351	351	324	325	356	340	303	300	296	293	281	284	283	290
Coordinates (WGS)	N 04° 26′5.9″ W 74° 38′31″	N 04° 26′49″ W 74° 38′ 35.5″	N 04° 25′3.7″ W 74° 40′ 58.3″	N 04° 24′ 55.7″ W 74°41′ 9.5″	N 04° 25′ 32.8″ W 74°41′ 2″	N 04° 25′ 26.7″ W 74°41′ 9.6″	N 04° 23′27.6″ W 74° 43′54.8″	N 04° 23′ 23.9″ W 74°43′ 46″	N 04° 20′ 11.7″ W 74°47′ 22.4″	N 04° 20′ 16.3″ W 74°47′ 32.8″	N 04° 18′ 17.8″ W 74°46′ 32.6″	N 04° 18′ 10.5″ W 74°46′ 37.1″	N 04° 18′ 05.1″ W 74°46′ 48.5″	N 04° 18′18″ W 74° 46′ 46.8″
Distance to the urban center (Km)	2.11	0.74	5.11	5.46	4.61	4.92	11.22	11.3	0.7	0.7	0	0	0	0
Type of body water	rb	rb	rb	rb	q	q	q	q	q	q	rb	rb	rb	q

Two categories: stream (q) or Bogotá river (rb). The locations are indicated as L1,..., L6

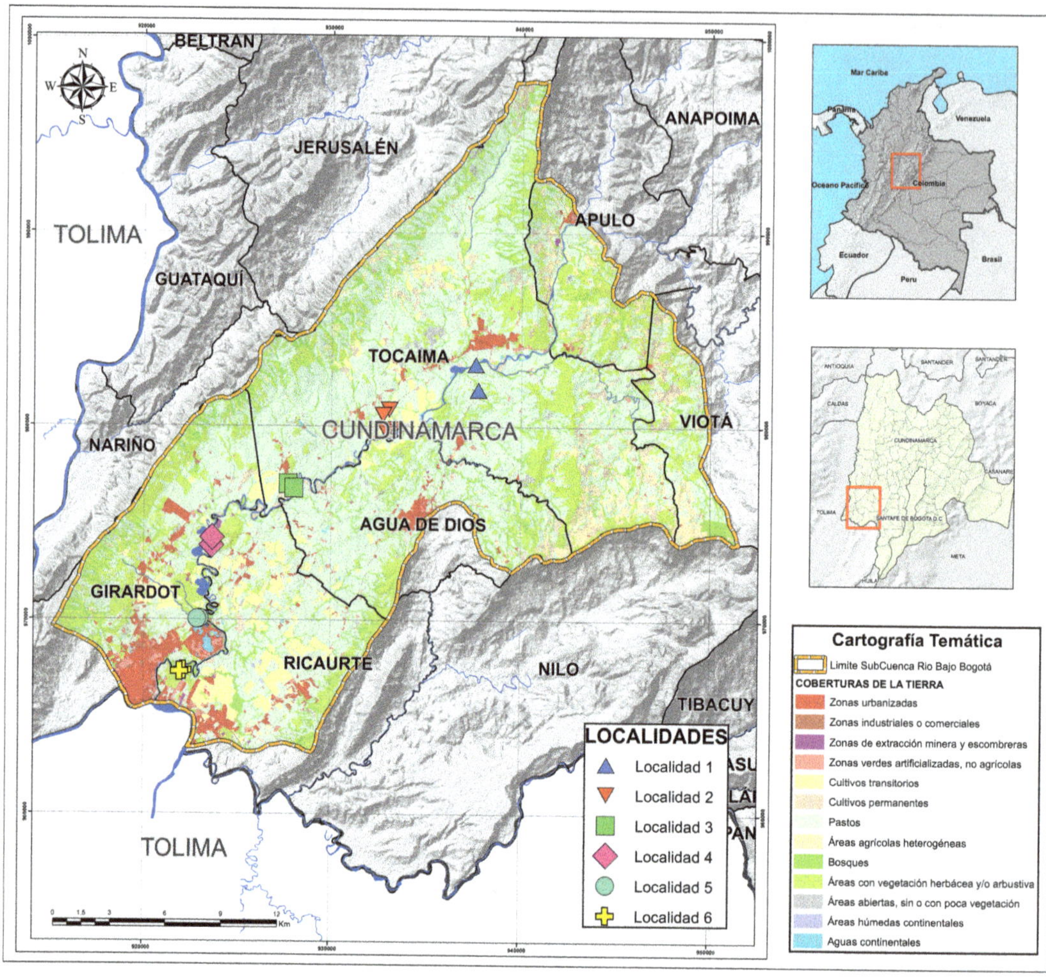

Fig. 1 Study area in the Lower Bogotá River sub-basin in the department of Cundinamarca (Colombia)

banana, rice, and kalica palm (also known as Huano) being the main economic activities in the region (CAR 2005) (Table 1).

2.2 Sampling Design

Using 2019 satellite imagery (Google Earth), and subsequent field verification, 14 sites were selected across six locations within the Lower Bogotá River sub-basin (Fig. 1 and Table 1). Each location presented three land cover types and was separated by a minimum linear distance of 1 km to avoid spatial pseudoreplication in herpetofauna studies (Mendenhall et al. 2014).

The natural land covers selected for herpetofauna samplings were as follows:

Riparian forest (Br): Patches of TDF composed of native tree vegetation located along the banks of the Bogotá River or adjacent streams. The dominant tree species included saman (*Pithecellobium saman*), diomate (*Astronium graveolens*), ceiba (*Ceiba pentandra*), olive (*Capparis odoratissima*), and cumulá (*Aspidosperma polyneuron*), all of which exhibited greater height and diameters at breast height than trees in secondary forest patches. No evidence of selective logging or cattle intrusion was observed in these areas;

however, there was visible accumulation of garbage and rubble.

Secondary forest (Bs): Patches of TDF characterized by a closed canopy with an average height of 5 m. The vegetation consists of tree and shrub species, most of which have diameters at breast height of less than 10 cm, along with the presence of broadleaf and creeping herbs. This area is undergoing secondary succession following the clearing of primary forest species, subsequent land use for anthropogenic activities, and eventual abandonment. The predominant species in this cover is forage acacia (*Leucaena leucocephala*), with additional individuals of saman (*Pithecellobium saman*), ceiba (*Ceiba pentandra*), diomate (*Astronium graveolens*), trupillo (*Prosopis juliflora*), and dividivi (*Caesalpinia coriaria*). Evidence of cattle intrusion was observed in these patches.

The anthropogenic land covers selected for herpetofauna samplings were as follows:

Rice field (AR): Extensive rice cultivation on flood-prone lands with constant agrochemical application. This land cover was found for sites 4, 5, and 6 (Table 1). During sampling, different crop phases, including post-harvest stages, were represented, where the soil remained uncovered.

Pasture (P): Extensive pastures with a permanent cattle presence. This land cover was found for sites 1, 2, 3, and 8 (Table 1).

Banana cultivation (Pl): Extensive banana plantations with evidence of burning and agrochemicals application, present in site 7 (Table 1).

Kalica palm (K) *Sabal mauritiiformis*, also known as Huano: Extensive kalica palm plantations with selective logging management, controlled burns, and permanent agrochemical use, present in sites 11 and 12 (Table 1).

Infrastructure (I): Abandoned and in-use constructions within sites 9, 10, 13, and 14 (Table 1).

In each land cover type, sampling was conducted at a minimum distance of 100 meters from any other cover to minimize potential edge effects (Schneider-Maunoury et al. 2016). The visual encounter survey (VES) technique (Crump and Scott 1994) was employed by a team of three researchers. Each survey covered a 50-meter transect, with a two-meter width on each side, and lasted 30 minutes per land cover type per field trip. All individuals were recorded from ground level, where plant and rocky material were carefully removed to locate specimens, up to 2 m high in the vegetation (Urbina-Cardona et al. 2015). Fieldwork was conducted from March to November 2019.

During each field trip, transects were surveyed at three different times (morning, afternoon, and night) between 5:00 a.m. and 11:30 p.m. In total, each land cover type per location was sampled four times: twice during the dry season and twice during the wet season. The total sampling effort was 1080 hours over 168 transects. Six functional traits (weight, snout-vent length, head width, tail length, crypticity, and habitat type) were recorded for each captured individual. In addition, four natural history (fixed) traits (habit, diet, reproductive strategy, and defensive mechanisms (venom or poison production)) were obtained from scientific literature; these traits influence the reproductive performance (fitness) of species and determine their responses to environmental changes (Tables 2, 3, and 4).

2.3 Data Analysis

2.3.1 Assemblage Structure

The species abundance matrix per transect was square root transformed, and the Bray-Curtis similarity index was calculated for 168 transects.

2.3.2 Functional Grouping

To study functional diversity, species can be classified into functional groups based on the values of their functional traits (Tilman et al. 1997; Blaum et al. 2011). Species within the same functional group are assumed to exhibit similar resource acquisition strategies and fulfill comparable ecological roles within their habitat (Gitay

Table 2 Functional traits measured and analyzed for amphibians and reptiles in the Lower Bogotá River sub-basin, in the department of Cundinamarca (Colombia)

Trait type	Complex	Traits	Attribute	Acquisition	Environmental response
Morphometric	Body	Weight	Continuous	Direct measurement	Relation to predation or protection against predation
		Snout-vent length	Continuous	Direct measurement	Relation to predation or protection against predation
		Head width	Continuous	Direct measurement	Relation to predation or protection against predation
		Tail length	Continuous	Direct measurement	Relation to predation or protection against predation
		Crypticity	Categorical (cryptic or not cryptic)	Direct measurement	Relation to predation strategies or protection against predation
Behavioral	Activity period	Habit	Categorical (crepuscular, diurnal, or nocturnal)	Literature	Flow of matter and energy over time
	Habitat use	Preferred habitat	Categorical (arboreal, aquatic, terrestrial, or fossorial)	Direct observation and literature	Flow of matter and energy in space. Relation to predation or protection against predation
Life history	Feeding	Diet	Categorical (great variability)	Literature	Flow of matter and energy, nutrient cycling
	Reproduction	Reproductive strategy	Categorical (viviparous, ovoviviparous, or oviparous)	Literature	Flow of matter and energy
	Defense mechanisms	Poison (anurans) or Venom (snakes)	Categorical (poisonous/ venomous or non-poisonous/ non-venomous)	Literature	Flow of matter and energy. Energy balance and nutrient cycling. Relationship with predation or protection against predation

et al. 1996). The impact of functional groups on ecosystem processes is considered more significant than that of species richness (Scherer-Lorenzen 2009; Srivastava and Vellend 2005; Balvanera et al. 2006). Furthermore, functional groups are useful for describing and elucidating assembly rules, such as environmental filtering or limiting similarity (Brown et al. 2000; Bellwood et al. 2002; Hoeinghaus et al. 2007).

Amphibian and reptile species were classified into functional groups based on the average values of their functional traits (Tables 3 and 4, respectively) using the k-means clustering method. This method partitions data into k groups by minimizing the sum of squared distances between data points and their assigned cluster centers, creating a cascade from the smallest to the largest number of groups (Ouellette et al. 2019). To determine the optimal number of functional groups, the Calinski-Harabasz criterion (Caliński and Harabasz 1974) was applied, and the corresponding dendrogram was generated. This analysis was conducted in R (v3.5.3; R Core Team 2014) using the vegan package (Oksanen et al. 2020).

2.3.3 Functional Diversity

For each of the 168 transects, five functional diversity indices of amphibians and reptiles were calculated using the FDiversity program (Casanoves et al. 2011):

Table 3 Average functional traits measured in amphibians from the Lower Bogotá River sub-basin

Species	Avg weight	Avg LRC	Avg AC	Crypticity	Habit	Habitat	Diet	Repro. Strat	Poison
Boana platanera	0.84	3.96	2.78	Yes	Nocturnal	Arboreal	Insectivorous	Oviparous	No
Dendrobates truncatus	0.79	3.26	2.20	Yes	Diurnal	Terrestrial	Insectivorous	Oviparous	Yes
Elachistocleis pearsei	2.40	3.83	2.77	Yes	Nocturnal	Terrestrial	Insectivorous	Oviparous	No
Engystomops pustulosus	0.49	3.50	2.33	Yes	Nocturnal	Terrestrial	Insectivorous	Oviparous	No
Leptodactylus colombiensis	0.47	3.73	2.44	Yes	Nocturnal	Terrestrial	Insectivorous	Oviparous	No
Leptodactylus fragilis	0.78	3.44	2.10	Yes	Nocturnal	Terrestrial	Insectivorous	Oviparous	No
Leptodactylus fuscus	1.85	3.84	2.51	Yes	Nocturnal	Terrestrial	Insectivorous	Oviparous	No
Rhaebo glaberrimus	0.10	1.79	1.10	Yes	Nocturnal	Terrestrial	Insectivorous	Oviparous	No
Rhinella humboldti	1.78	3.70	2.45	Yes	Nocturnal	Terrestrial	Insectivorous	Oviparous	No
Rhinella horribilis	2.89	4.00	2.83	Yes	Nocturnal	Terrestrial	Omnivorous	Oviparous	No
Trachycephalus typhonius	1.36	3.96	2.74	Yes	Nocturnal	Arboreal	Insectivorous	Oviparous	No

Functional richness (FRic) refers to the amount of functional space occupied by species within a community, regardless of their abundance (Villéger et al. 2008). Low functional richness indicates that some available resources remain unused, potentially leading to a decline in ecosystem productivity (Mason et al. 2005).

Functional divergence (FDiv) assesses the functional similarity among dominant species within a community (Villéger et al. 2008). High divergence reflects strong niche differentiation among dominant species, minimizing competition and maximizing resource utilization (Mason et al. 2005; Villéger et al. 2008).

Functional dispersion (FDis) evaluates the distribution of species within the functional trait space and quantifies how spread out species are based on their functional traits. Higher values indicate that species are utilizing distinct functional traits and occupying diverse ecological niches, which can enhance ecosystem stability and resilience by reducing functional overlap and increasing resource partitioning. FDis provides insights into the variability of ecological roles within a community and reflects the degree of niche differentiation (Laliberté and Legendre 2010).

RAOQ, on the other hand, balances functional richness and functional divergence by integrating both species presence and pairwise functional trait differences between species (Botta-Dukát 2005; Mason and de Bello 2013). Meanwhile, FAD 2 evaluates the degree of difference in functional traits among species, considering both the number of functional traits present and their relative distribution within the community (Ricotta 2005).

2.3.4 Statistical Analysis

The effect of land cover type on each functional diversity index was determined using PERMANOVA (Anderson et al. 2008). The design considered land cover type as a fixed factor with seven levels: riparian forest (Br), secondary forest (Bs), pasture (P), rice field (AR), kalica palm cultivation (K), banana cultivation (Pl), and infrastructure (I). Location was included as a

Table 4 Average functional traits measured in reptiles from the Lower Bogotá River sub-basin

Species	Avg weight	Avg LRC	Avg AC	Avg LC	Crypticity	Habit	Habitat	Diet	Repro. Str	Venom
Anolis auratus	1.14	147.90	17.50	181.47	Yes	Diurnal	Arboreal	Insectivorous	Oviparous	No
Basiliscus basiliscus	243.12	280.30	53.58	323.02	Yes	Diurnal	Arboreal	Insectivorous	Oviparous	No
Boa constrictor	12,000	930	150.00	900	Yes	Nocturnal	Arboreal	Carnivorous	Viviparous	No
Caiman crocodilus	1139.95	756.92	164.92	557.21	Yes	Nocturnal	Aquatic	Carnivorous	Oviparous	No
Cnemidophorus lemniscatus	1.34	42.27	8.71	78.73	No	Diurnal	Terrestrial	Omnivorous	Oviparous	No
Crocodylus acutus	150,000	160	600	1200	Yes	Nocturnal	Aquatic	Carnivorous	Oviparous	No
Crotalus durissus	700	850	55	100	Yes	Diurnal	Terrestrial	Carnivorous	Viviparous	Yes
Drymarchon melanurus	1800	1200	70	500	Yes	Diurnal	Terrestrial	Carnivorous	Oviparous	No
Gonatodes albogularis	0.75	46.43	6.01	38.03	Yes	Diurnal	Arboreal	Insectivorous	Oviparous	No
Hemidactylus brookii	0.10	26.63	4.54	17.46	Yes	Nocturnal	Arboreal	Insectivorous	Oviparous	No
Hemidactylus frenatus	2.79	44.72	7.48	25.98	Yes	Nocturnal	Arboreal	Insectivorous	Oviparous	No
Holcosus festivus	9.81	185.47	43.43	164.88	Yes	Diurnal	Terrestrial	Insectivorous	Oviparous	No
Iguana iguana	422.21	272.49	74.08	627.60	Yes	Diurnal	Arboreal	Herbivorous	Oviparous	No
Leptodeira annulata	7.79	287.12	10.76	196.37	Yes	Diurnal	Terrestrial	Carnivorous	Oviparous	No
Mabouya mabouya	1.00	70	5	60	Yes	Diurnal	Terrestrial	Insectivorous	Viviparous	No
Podocnemis lewyana	5636.66	76.96	82.58	120.98	Yes	Diurnal	Aquatic	Herbivore	Oviparous	No
Spilotes pullatus	2000	1300	60	500	No	Diurnal	Arboreal	Carnivore	Oviparous	No
Thecadactylus rapicauda	7.25	112.07	15.01	69.57	Yes	Nocturnal	Arboreal	Insectivore	Oviparous	No

covariate with six levels. The analysis was performed using a sequential partial sum of squares (Type I SS) and 9999 permutations of the residuals under a reduced model. Significant factors and their interactions were analyzed using a *post hoc* pairwise comparison, where the t-statistics were calculated with 9999 permutations. This routine was executed using PRIMER v7 & PERMANOVA+ add-on (Clarke et al. 2014).

3 Results

3.1 Assemblage Structure

A total of 29 herpetofauna species were recorded, including 11 amphibian species and 18 reptile species. The highest individual abundance was observed in secondary forest cover (Bs), with 185 individuals, out of a total of 486 individuals recorded (Table 5). For amphibians, significant differences were found between land cover types in assemblage structure (Bray-Curtis similarity; Pseudo-F = 1.71; p(perm) = 0.034). For reptiles, significant differences in assemblage structure (Bray-Curtis similarity) were observed between locations (Pseudo-F = 3.69; p(perm) = 0.003) and land cover types (Pseudo-F = 2.307; p (perm) = 0.0002).

3.2 Functional Grouping

The functional grouping of amphibians (Fig. 2) identified six species clusters (Calinski-Harabasz Criterion: 2 groups = 5.64; 3 groups = 7.2; 4 groups = 14.48; 5 groups = 24.44; 6 groups = 48.39), resulting in the following functional groups: (I) *Rhinella horribilis*; (II) *Engystomops pustulosus*, *Rhinella humboldti*, *Leptodactylus colombiensis*, *Leptodactylus fuscus*, and *Leptodactylus fragilis*; (III) *Elachistocleis pearsei;* (IV) *Rhaebo glaberrimus*; (V) *Boana platanera* and *Trachycephalus typhonius*; and (VI) *Dendrobates truncatus* (Table 6).

The NocTerInsOv group included five species, representing nearly half of the total amphibian richness. Four species belonged to the family Leptodactylidae, while the remaining species, *Rhinella humboldti*, belonged to the family Bufonidae. Across all land cover types, the NocTerInsOv group showed the highest individual abundance, representing 100% of individuals in banana cultivation (Pl) and exceeding 80% abundance in all other land covers, except in infrastructure (I), where its percentage was 64.3%.

Regarding species richness across land cover types, riparian forest (Br) was 60% dominated by NocTerInsOv species, with the NoctTerOmnOv and ArbInsOv groups each representing 20%. In secondary forest (Bs), the NocTerInsOv group dominated with 57.1%, while NoctTerOmnOv, NocTerInsOvOval, and DiurTox groups were evenly represented (14% each). Notably, banana cultivation (Pl) exclusively contained species from the NocTerInsOv group. No single vegetation cover contained all functional groups, with Bs hosting the highest number of functional groups (four), while Pl had the lowest, with only one group (Fig. 4a, b).

For amphibians, three functional groups were recorded in riparian forest, the same number observed in anthropogenic covers. Pastures and infrastructure also contained three groups, whereas kalica palm cultivation had two, and banana cultivation had only one. Secondary forest cover (Bs) contained the highest functional group diversity (four groups), lacking only NocTerInsOvPeq and ArbInsOv. NocTerInsOvPeq was exclusive to kalica palm cultivation, while ArbInsOv was composed of the only arboreal amphibians, *Boana platanera* (present in riparian forest and rice fields, inhabiting high trunks and open areas) and *Trachycephalus typhonius* (found only in infrastructure and capable of living in highly transformed habitats). The NocTerInsOvOval and DiurTox groups were exclusive to natural covers and were absent from transformed areas. These groups were monospecific, with NocTerInsOvOval containing *Elachistocleis pearsei* (a species of intermediate biomass but

Table 5 Registered species and their abundances by coverage type. Riparian forest (Br), secondary forest (Bs), pasture (P), rice field (Ar), kalica palm cultivation (K), banana cultivation (Pl), and infrastructure (I)

Class	Order	Family	Species	Coverages							Abundance	Previous reports in the region
				Br	Bs	P	Ar	K	Pl	I	total	
Amphibia	Anura	Bufonidae	*Rhaebo glaberrimus*	–	–	–	–	1	–	–	1	https://www.gbif.org/species/2423156
			Rhinella horribilis	3	2	1	4	–	–	2	12	https://www.batrachia.com/orden-anura/bufonidae-88-spp/rhinella-horribilis/
			Rhinella humboldti	8	7	7	4	–	2	–	28	https://www.batrachia.com/orden-anura/bufonidae-88-spp/rhinella-humboldti/
		Dendrobatidae	*Dendrobates truncatus*	–	1	–	–	–	–	–	1	https://www.batrachia.com/orden-anura/dendrobatidae-102-spp/dendrobates-truncatus/
		Hylidae	*Boana platanera*	3	–	4	–	–	–	–	7	https://www.batrachia.com/orden-anura/hylidae-154-spp/boana-platanera/
			Trachycephalus typhonius	–	–	–	–	–	–	3	3	https://www.batrachia.com/orden-anura/hylidae-154-spp/trachycephalus-typhonius/
		Leptodactylidae	*Engystomops pustulosus*	–	–	–	3	–	–	–	3	https://www.batrachia.com/orden-anura/leptodactylidae-40-spp/engystomops-pustulosus/
			Leptodactylus colombiensis	–	2	–	–	–	–	–	2	https://www.batrachia.com/orden-anura/leptodactylidae-40-spp/leptodactylus-colombiensis/
			Leptodactylus fragilis	1	8	1	8	1	–	2	21	https://www.batrachia.com/orden-anura/leptodactylidae-40-spp/leptodactylus-fragilis/
			Leptodactylus fuscus	28	38	16	25	4	–	7	118	https://www.batrachia.com/orden-anura/leptodactylidae-40-spp/leptodactylus-fuscus/
		Microhylidae	*Elachistocleis pearsei*	–	1	–	–	–	–	–	1	https://www.batrachia.com/orden-anura/microhylidae-21-spp/elachistocleis-pearsi/
Reptilia	Crocodilia	Alligatoridae	*Caiman crocodilus*	20	1	–	–	–	–	–	21	https://www.gbif.org/species/8953936
		Crocodylidae	*Crocodylus acutus*	1	–	–	–	–	–	–	1	https://www.gbif.org/species/2441321
	Testudines	Podocnemididae	*Podocnemis lewyana*	5	–	–	–	–	–	–	5	https://www.gbif.org/species/2442789
	Squamata	Boidae	*Boa constrictor*	1	–	–	–	–	–	–	1	https://www.gbif.org/species/2464899
		Colubridae	*Drymarchon melanurus*	–	1	–	–	–	–	–	1	https://www.gbif.org/species/2452321
			Spilotes pullatus	–	1	–	–	–	–	–	1	https://www.gbif.org/species/9195230
		Corytophanidae	*Basiliscus basiliscus*	18	–	–	–	–	–	–	18	https://www.gbif.org/species/5221813

		Dactyloidae	*Anolis auratus*	2	2	–	1	2	–	–	7	https://www.gbif.org/species/9129157
		Dipsadidae	*Leptodeira annulata*	1	–	2	1	1	–	–	5	https://www.gbif.org/species/2454954
		Gekkonidae	*Hemidactylus brookii*	–	1	–	1	–	–	5	7	https://www.gbif.org/species/5960063
		Gekkonidae	*Hemidactylus frenatus*	–	–	–	1	–	–	1	2	https://www.gbif.org/species/9537238
		Iguanidae	*Iguana iguana*	6	5	1	–	–	–	1	13	https://www.gbif.org/species/2459658
		Phyllodactylidae	*Thecadactylus rapicauda*	1	3	7	–	1	–	–	12	https://www.gbif.org/species/2445829
		Scindidae	*Mabuya mabouya*	–	1	–	–	–	–	–	1	https://www.gbif.org/species/5960489
		Sphaerodactylidae	*Gonatodes albogularis*	24	95	6	4	9	4	6	148	https://www.gbif.org/species/2447609
		Teiidae	*Cnemidophorus lemniscatus*	11	10	4	–	4	3	–	32	https://www.gbif.org/species/5227564
			Holcosus festivus	3	6	3	–	–	–	1	13	https://www.gbif.org/species/7836710
		Viperidae	*Crotalus durissus*	–	–	–	1	–	–	–	1	https://www.gbif.org/species/2444354
Richness				**17**	**18**	**11**	**11**	**8**	**3**	**9**	**29**	
Abundance				**136**	**185**	**52**	**53**	**23**	**9**	**28**	**486**	

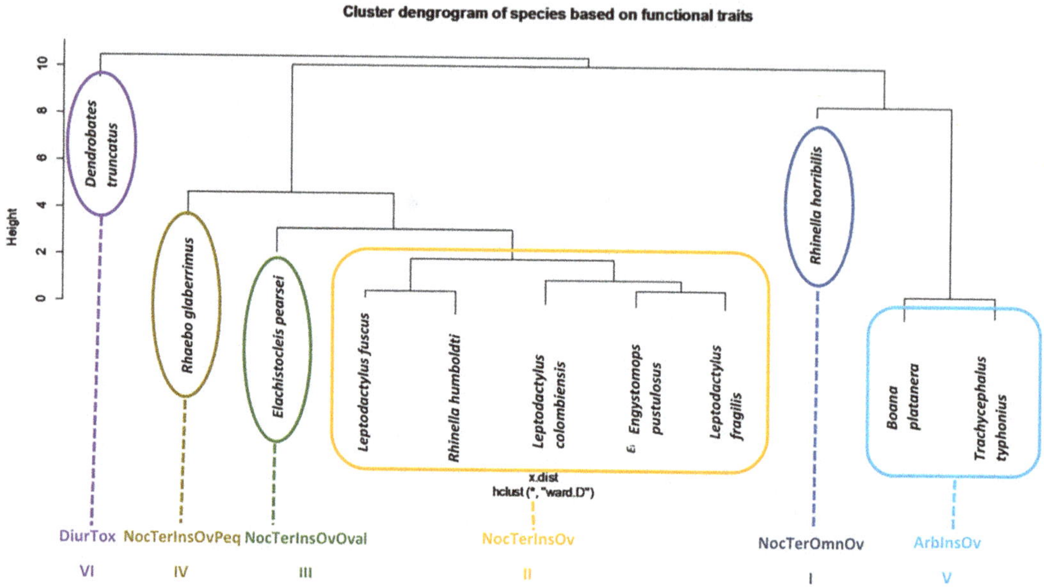

Fig. 2 Functional grouping of amphibian species based on ten functional traits. Each functional group is represented by different colored ovals grouping the species. Below each oval is the name of each functional group according to the acronyms in Table 6

Table 6 Composition of amphibian functional groups and their functional trait values

Functional group Name	Species that comprise it	Characteristic functional traits (S = species richness; n = total abundance)
NoctTerOmnOv	*Rhinella horribilis*	Nocturnal, terrestrial individuals, with large biomass, omnivorous diet and oviparous $S = 1$ $n = 12$
NocTerInsOv	*Engystomops pustulosus*, *Rhinella humboldti*, *Leptodactylus colombiensis*, *Leptodactylus fuscus*, and *Leptodactylus fragilis*	Nocturnal, terrestrial individuals, insectivorous diet, oviparous, with the highest biomass $S = 5$ $n = 172$
NocTerInsOvOval	*Elachistocleis pearsei*	Nocturnal, terrestrial individuals, insectivorous diet, oviparous, with intermediate biomass, higher surface/volume ratio (oval body) $S = 1$ $n = 1$
NocTerInsOvPeq	*Rhaebo glaberrimus*	Nocturnal, terrestrial individuals, insectivorous diet, oviparous, with the smallest biomass $S = 1$ $n = 1$
ArbInsOv	*Boana platanera* and *Trachycephalus typhonius*	Arboreal individuals, insectivorous diet and oviparous $S = 2$ $n = 10$
DiurTox	*Dendrobates truncatus*	Diurnal and poisonous individuals $S = 1$ $n = 1$

with a high surface/volume ratio due to its oval body), and DiurTox comprising *Dendrobates truncatus* (the only diurnal amphibian recorded and the only poisonous species).

The classification of reptiles (Fig. 3) identified six functional groups (Calinski-Harabasz Criterion: 2 groups = 4.36; 3 groups = 4.92; 4 groups = 5.12; 5 groups = 5.2; 6 groups = 5.66), resulting in the following functional grouping: i) *Basiliscus basiliscus*, *Anolis auratus* and *Gonatodes albogularis*; ii) *Boa constrictor*, *Caiman crocodilus* and *Crocodylus acutus*; iii)

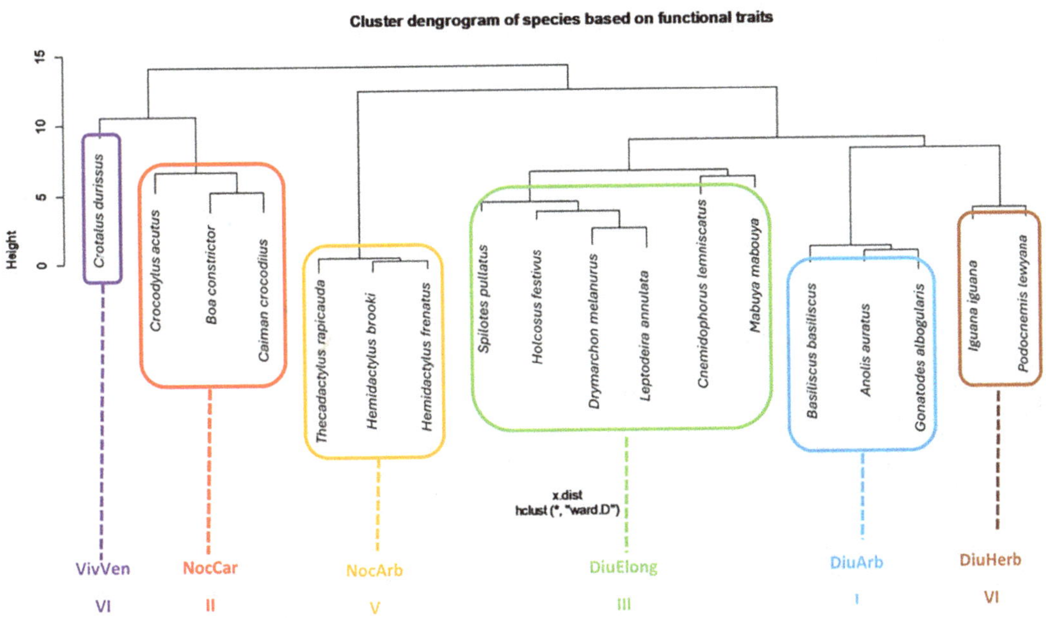

Fig. 3 Functional grouping of reptile species based on ten functional traits. Each functional group is represented by different colored ovals grouping the species. The name of each functional group, using the acronyms from Table 7, is shown below each oval

Table 7 Composition of reptile functional groups and their functional trait values

Functional group name	Species that comprise it	Characteristic functional traits (S = species richness; n = total abundance)
DiuArb	*Basiliscus basiliscus*, *Anolis auratus*, and *Gonatodes albogularis*	Diurnal and arboreal individuals $S = 3$ $n = 173$
NocCar	*Boa constrictor*, *Caiman crocodilus acutus*, and *Crocodylus acutus*	Individuals with the highest biomass, nocturnal habit, and carnivorous diet $S = 3$ $n = 23$
DiuElong	*Cnemidophorus lemniscatus*, *Drymarchon melanurus*, *Holcosus festivus*, *Spilotes pullatus*, *Leptodeira annulata*, and *Mabuya mabouya*	Diurnal individuals with the lowest surface/volume ratio (elongated body) S = 6 n = 53
VivVen	*Crotalus durissus*	Individuals with viviparous reproduction type and venomous $S = 1$ $n = 1$
NocArb	*Thecadactylus rapicauda*, *Hemidactylus brookii*, and *Hemidactylus frenatus*	Nocturnal and arboreal individuals $S = 3$ $n = 21$
DiuHerb	*Iguana iguana* and *Podocnemis lewyana*	Diurnal individuals with a herbivorous diet $S = 2$ $n = 18$

Cnemidophorus lemniscatus, *Drymarchon melanurus*, *Holcosus festivus*, *Spilotes pullatus*, *Leptodeira annulata*, and *Mabuya mabouya*; iv) *Crotalus durissus*; v) *Thecadactylus rapicauda*, *Hemidactylus brookii*, and *Hemidactylus frenatus*; and vi) *Iguana iguana* and *Podocnemis lewyana* (Table 7).

Across all land cover types, the DiuArb group had the highest abundance. Bs was dominated by this group (77%), while Br contained all groups except VivVen, and was also dominated by DiuArb. In P, DiuElong represented 39%, followed by NocArb (30%) and DiuArb (26%). Regarding species richness, Bs was dominated by

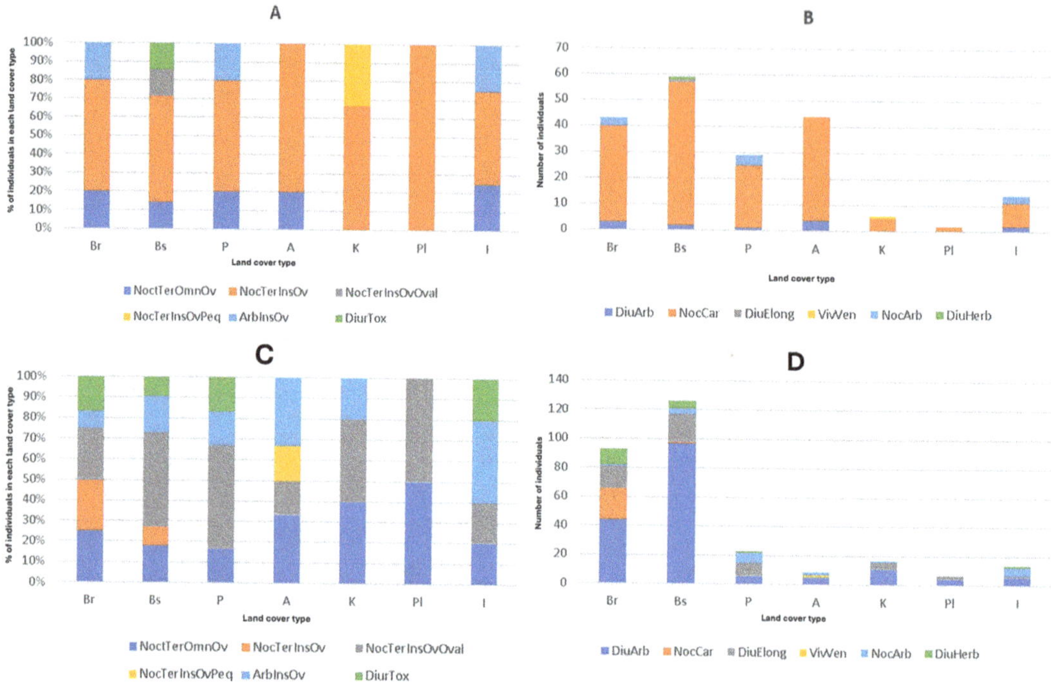

Fig. 4 Functional groups of anurans and reptiles in the land covers sampled in the Lower Bogotá River sub-basin in the department of Cundinamarca (Colombia). (**a**) Richness of anuran functional groups by land cover type; (**b**) Number of individuals per anuran functional group by land cover type; (**c**) Richness of reptiles functional groups by land cover type; (**d**) Number of individuals per reptiles functional group by land cover type. Br: Riparian forest, Bs: Secondary forest, P: Pasture, A: Rice field, K: Kalica palm cultivation, Pl: Banana cultivation, I: Infrastructure

DiuElong (46%), while in Br, the DiuArb, NocCar, and DiuElong groups each comprised 25% of recorded species. Pl showed equal representation (50%) of DiuArb and DiuElong (Fig. 4c, d).

For reptiles, riparian forest (Br) and secondary forest (Bs) contained five of the six functional groups, though in different proportions. Riparian forest exhibited greater homogeneity, with three groups comprising 75% of recorded species, whereas secondary forest was dominated by DiuElong. Both land covers lacked the VivVen group, represented only by *Crotalus durissus*, which was found exclusively in rice fields.

The NocCar and DiuHerb groups were restricted to natural habitats. NocCar (*Boa constrictor*, *Caiman crocodilus fuscus*, and *Crocodylus acutus*) included the largest-bodied species in the study area, all nocturnal carnivores. DiuHerb group, consisting of *Iguana iguana* and *Podocnemis lewyana*, was the only herbivorous reptile group recorded, with 90% of individuals found in natural habitats and only two individuals of *I. iguana* in transformed habitats (A and I covers).

3.3 Functional Diversity Indices

For amphibians, significant differences were found between land cover types in FDis (Pseudo-F = 2.91; p(perm) = 0.04) and RAOQ (Pseudo-F = 3.04; p(perm) = 0.04). For RAOQ, significant differences were found between rice fields and secondary forest (t = 3.37; p-perm = 0.001) and between rice fields and riparian forest (t = 2.21; p-perm = 0.03). Similarly, for FDis, significant differences were found

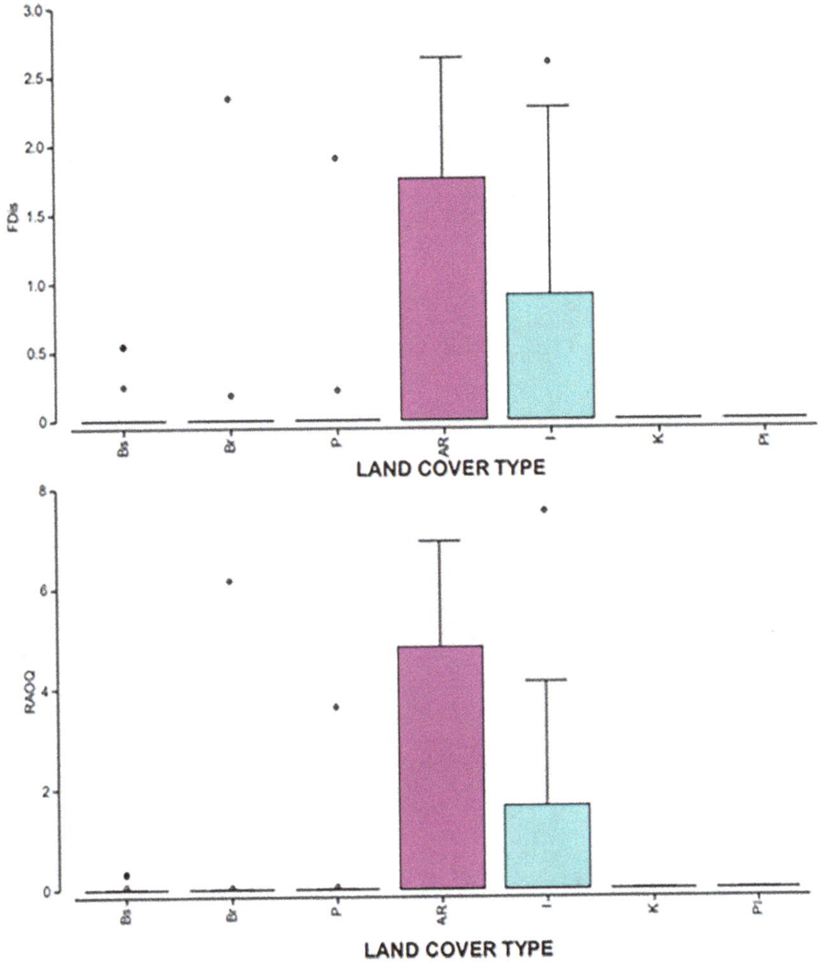

Fig. 5 Box plots of the functional diversity indices calculated for amphibians for the land cover types evaluated in the Lower Bogotá river sub-basin. FDis: functional dispersion, RaoQ: quadratic entropy based on multiple traits, Bs: secondary forest, Br: riparian forest, P: pastures, AR: rice field, I: infrastructure, K: kalica palm cultivation, and Pl: banana cultivation

between rice fields and secondary forest (t = 3.26; p-perm = 0.002) and between rice fields and riparian forest (t = 2.32; p-perm = 0.02) (Fig. 5). For reptiles, none of the functional diversity indices showed significant differences between locations or land cover types (Table 8).

4 Discussion

This study demonstrated that amphibian and reptile assemblages in the Lower Bogotá River sub-basin are predominantly composed of species that are highly tolerant to disturbances, with broad distributions and extensive geographical ranges across the tropical dry forest (TDF) of Colombia (Urbina-Cardona et al. 2014). The region is characterized by the absence of highly conserved TDF areas, with ecosystems instead being dominated by remnants in secondary successional stages within transformed landscapes (Etter and van Wyngaarden 2000). These findings align with studies on anthropogenic land covers in the Andes, which indicate that habitat structure

Table 8 Result of the PERMANOVA evaluating differences between the functional diversity indices of amphibians and reptiles across vegetation covers (fixed factor) and locations (covariable) in the Lower Bogotá River sub-basin, in the department of Cundinamarca (Colombia).

									Estimates of components of variation	
		Source	df	SS	MS	Pseudo-F	P(perm)	Unique perms	Estimate	Sq. root
Anurans	Assemblage structure	LO	5	4960.2	4960.2	2.0589	0.0911	9946	28.033	5.2946
		CO	6	24,728	4121.4	1.7107	**0.0345**	9908	152.04	12.33
		Res	79	2.00E+05	2409.2				2409.2	49.084
		Total	90	2.30E+05						
	Fric	LO	5	1.14E−02	1.14E−02	27.658	0.1615	9972	0.0001216	0.011027
		CO	6	−3.05E−02	−5.09E−03	Insufficient data			−0.00049193	−0.022179
		Res	79	3.37E−02	0.00041052	Insufficient data			0.00041052	0.020261
		Total	89	0.014477						
	Fdis	LO	5	6.00E−04	6.00E−04	0.0016496	0.9671	9830	−0.0040326	−0.063503
		CO	6	6.35E+00	1.06E+00	2.9125	**0.0483**	9945	0.062178	0.24935
		Res	79	2.98E+01	0.36353				0.36353	0.60294
		Total	89	36.163						
	Fdiv	LO	5	8.77E−05	8.77E−05	2.7705	0.4075	9964	6.23E−07	0.00078933
		CO	6	−2.16E−03	−3.60E−04	Insufficient data			−3.50E−05	−0.0059191
		Res	79	2.60E−03	3.17E−05	Insufficient data			3.17E−05	0.0056277
		Total	89	0.00052428						
	FAD2	LO	5	3.59E+05	3.59E+05	1.2285	0.275	9879	742.13	27.242
		CO	6	3.10E+06	5.16E+05	1.7658	0.0974	9910	20,017	141.48
		Res	79	2.40E+07	2.92E+05				2.92E+05	540.61
		Total	89	2.74E+07						
	RAO Q	LO	5	3.16E−02	3.16E−02	0.013251	0.9115	9821	−0.026154	−0.16172
		CO	6	4.36E+01	7.26E+00	3.0453	**0.0485**	9939	0.43633	0.66055
		Res	79	1.96E+02	2.3855				2.3855	1.5445
		Total	89	239.23						
Reptiles	Assemblage structure	LO	5	12,128	12,128	3.6992	**0.0033**	9942	78.316	8.8496
		CO	6	45,399	7566.5	2.3078	**0.0002**	9860	318.97	17.86
		Res	101	3.44E+05	3278.6				3278.6	57.259
		Total	112	4.02E+05						
	Fric	LO	5	3.70E−01	3.70E−01	Insufficient data			0.0032957	0.057409
		CO	6	1.33E+00	2.22E−01	Insufficient data			0.016674	0.12913
		Res	101	−2.02E−01	−1.93E−03				−0.0019266	−0.043893
		Total	112	1.50E+00						

	FDis	LO	5	0.040626	0.040626	0.044516	0.8321	9830	−0.0077168	−0.087845
		CO	6	5.1246	0.8541	0.93588	0.4525	9935	−0.0043532	−0.065979
		Res	101	95.825	0.91262				0.91262	0.95531
		Total	112	100.99						
	FDiv	LO	5	0.00066988	0.00066988	Insufficient data			5.94E−06	0.0024369
		CO	6	0.00061118	0.00010186	Insufficient data			7.67E−06	0.0027688
		Res	101	−0.0001255	−1.20E−06				−1.20E−06	−0.0010933
		Total	112	0.0011556						
	FAD2	LO	5	7.09E+08	7.09E+08	0.37515	0.731	9934	−1.04E+07	−3232.4
		CO	6	4.79E+09	7.99E+08	0.42259	0.5453	9862	−8.12E+07	−9009
		Res	101	1.98E+11	1.89E+09				1.89E+09	43,469
		Total	112	2.04E+11						
	RAO Q	LO	5	0.0038861	0.0038861	0.00046242	0.9814	9845	−0.074336	−0.27265
		CO	6	41.062	6.8437	0.81436	0.5055	9929	−0.11605	−0.34067
		Res	101	882.4	8.4038				8.4038	2.8989
		Total	112	923.47						

LO = Effects of the location, CO = Effects of the vegetation cover, Res = Residuals and management play crucial roles in shaping amphibian diversity (Roach et al. 2020; Galindo-Uribe et al. 2022). Similarly, Zabala-Forero and Urbina-Cardona (2021) highlighted that the composition and functional groups of amphibian assemblages are influenced by the synergistic effects of vegetation cover type and climatic season. Additionally, Granda-Rodriguez et al. (2025) also suggests that, for TDF, elevation plays an important role in shaping the functional diversity of anurans.

The impact of land use on amphibian and reptile assemblages is further supported by recent reviews, which suggest that while pastures negatively affect the abundance and richness of these species, mixed forest plantations can aid herpetofaunal conservation in tropical regions, provided they maintain a diverse shrub layer with native species (López-Bedoya et al. 2022). Specifically, in the TDF of the Colombian Caribbean, palm plantations have been shown to provide suitable habitat for some amphibian species by maintaining permanent water bodies and preserving canopy structure, thereby mitigating edge effects on the remaining native forest patches (Silvera-Chimá et al. 2024). In a similar vein, edge effects significantly influence amphibian diversity in TDF fragments, with higher species abundance and richness found at the forest edges and adjacent pastures, while the forest interior showed lower evenness and higher dominance. These findings underscore the critical importance of conserving the forest interior and managing edge environments to maintain functional diversity and enhance the resilience of TDF ecosystems (Granda-Rodriguez et al. 2025).

The mechanisms that determine the structure of species assemblages tolerant to anthropogenic disturbance are closely linked to the organisms' natural history and are primarily influenced by small-scale factors, such as microhabitat conditions (Carvajal-Cogollo et al. 2019). Blanco-Torres (2009) found that certain anthropogenic activities alter the landscape, creating a mosaic of humid environments that facilitate the establishment of anuran species while reducing habitat competition in forest fragments, as observed in rice field habitats. This is further supported by Pianka (1966), who emphasized that the structural complexity of vegetation is essential for taxonomic diversity, with species richness being higher in heterogeneous habitats. In line with these findings, Granda-Rodriguez et al. (2025) showed that edge effects in TDF influence anuran taxonomic diversity through changes in leaf litter and understory density, while functional diversity increases with altitude in forest fragments, highlighting the role of habitat structure and elevation in shaping biodiversity. Additionally, species with specific functional traits may be better adapted to tolerate the conditions generated by disturbances in human-modified ecosystems, whereas others, whose functional traits limit their adaptability (e.g., size, diet, and habit), may struggle to persist in these disturbed environments (Berriozabal et al. 2017).

4.1 Functional Traits of Amphibians and Reptiles

Body size has been proposed as one of the traits most strongly associated with extinction risk in amphibians and reptiles (Purvis et al. 2005; Böhm et al. 2016; Meiri 2008; Vargas-Santamaría et al. 2011; Tingley et al. 2013). In this study, large-bodied species, such as *Crocodylus acutus*, *Boa constrictor*, *Drymarchon melanurus*, *Spilotes pullatus*, *Podocnemis lewyana*, *Iguana iguana*, and even *Caiman crocodilus*, were found to be restricted to natural vegetation covers, particularly riparian forests. In anthropogenic habitats, their abundances were lower compared to those recorded in natural areas. Pfeifer et al. (2017) highlight that larger-bodied reptile species require greater exposure to direct sunlight or warm surfaces to maintain proper thermoregulation and this behavior increases their vulnerability to hunting by local communities at forest edges and heightens the risk of road mortality near remaining native forests, ultimately impacting the functional diversity of assemblages (Rincón-Aranguri et al. 2023).

A similar pattern was observed in smaller species, which are generally more sensitive to habitat transformation and loss due to their limited mobility. Almost all reptile species experienced a drastic decline in abundance in anthropogenic habitats, except for the invasive species *Hemidactylus brookii* and *H. frenatus*, as well as generalist species with high disturbance tolerance, such as *Leptodeira annulata*, *Anolis auratus*, and *Holcosus festivus*. A recent review identifies several key functional traits that contribute to the success of the invasive species (such as *H. brookii* and *H. frenatus)* in colonizing new environments. Fast life-history traits, such as high reproductive output and early sexual maturity, are essential characteristics of successful invaders; these traits enable rapid population growth, facilitating establishment even in disturbed habitats (Allen et al. 2017). For instance, *H. frenatus* has the ability to store sperm for extended periods, increasing its reproductive efficiency and enabling it to quickly populate newly invaded areas (Weterings and Vetter 2018). Its capacity to reproduce year-round under tropical conditions further enhances its invasive potential. Additionally, these species are highly adaptable, thriving in both urban and agricultural land covers, as well as in natural habitats, and exhibit high thermal and foraging efficiency, which allows them to exploit available food sources more effectively than native species (Rödder et al. 2008; Lapwong et al. 2021). *H. frenatus* has demonstrated its ability to outcompete other geckos through aggressive behavior and efficient foraging strategies, including high consumption rates and a tendency to actively pursue prey, rather than passively waiting for it to approach (Hoskin 2011). These traits, combined with their capacity for rapid range expansion via human-mediated transport, make them formidable invaders with significant implications for local gecko species, particularly those from the Phyllodactylidae and Sphaerodactylidae families.

Regarding reproductive strategies, no significant pattern was observed in the richness and abundance of functional groups between natural and anthropogenic habitats for oviparous reptile species. Oviparous species are generally more vulnerable, as they are directly affected by the loss of potential nesting sites and notable microclimatic changes, which can disrupt temperature regulation during incubation (Doherty et al. 2020). In this study, oviparous species were widely distributed, while viviparous species were primarily restricted to natural vegetation habitats. Viviparity allows gravid females to regulate embryonic temperatures more effectively than egg-laying species (Andrews and Mathies 2000). However, viviparous species tend to reproduce more slowly than oviparous species (Böhm et al. 2016), making them particularly vulnerable due to their specific habitat requirements (Richardson 2023). In their study, Böhm et al. (2016) found that viviparous species were more likely to be at risk of extinction, suggesting that their limited reproductive capacity and habitat specificity may contribute to their susceptibility in anthropogenic landscapes.

4.2 Functional Groups and Functional Diversity Indices

In contrast to the findings of Flynn et al. (2009), which observed a greater number of functional groups in natural areas compared to anthropogenic ones, the present study reveals a different pattern. The composition of amphibian functional groups was highly similar across riparian and secondary native forests, as well as pastures and infrastructure. This suggests that historical deforestation in the region has not only homogenized species assemblages but also led to functional homogenization. Such pattern is further supported by the similar values of functional diversity indices (e.g., FRic, FDiv, FAD2) between natural and anthropogenic covers mainly for both reptiles. Functional divergence, defined as the distance between the abundances of the most representative species and the center of the functional space (Mason et al. 2005), was assessed using the FDiv and RAOQ indices. Notably, the RAOQ index revealed significant differences among land cover types for amphibians, indicating a reduction in niche differentiation. This reduction could lead to a decline in ecosystem functionality due to the inefficient utilization of resources (Mason et al.

2005). Such a phenomenon may be attributed to the loss of species exclusive to natural covers and the increase of highly tolerant species, such as *Leptodactylus fuscus*, which exhibit lower niche differentiation. This pattern became evident when comparing natural forests with pastures, where the RAOQ index showed statistically significant differentiation. This suggests a potential decline in ecosystem function, likely resulting from reduced niche differentiation and the dominance of tolerant species with low specialization.

For amphibians, significant differences were observed only in RAOQ and FDis between natural forests (riparian and secondary) and rice fields. This indicates that the average distance of each species from the functional centroid of the assemblage (Mason et al. 2005; Córdova-Tapia and Zambrano 2015) directly responds to environmental stress and ecosystem functionality loss. Such patterns suggest that rice fields act as an environmental filter, selecting for generalist species that can tolerate the altered conditions, while excluding more specialized species. Similar studies have highlighted that rice fields, though often considered temporary wetlands, tend to favor species with generalist traits, which can thrive under disturbance but may lack the ability to support specialized, sensitive species in the long term (Ribeiro et al. 2019). In fact, rice fields have been shown to support species with a higher functional divergence, potentially limiting the overall functional richness of the community (Attademo et al. 2018). Furthermore, the conversion of natural habitats into agricultural systems, like rice fields, results in a reduction in species richness and functional complexity, potentially undermining ecosystem processes that depend on functional diversity. Rice fields are not homogeneous, but their dynamic hydrological regimes still offer certain microhabitats that amphibians utilize, though these habitats may not support the same diversity of species found in pristine wetlands (Moreira and Maltchik 2014). While some amphibians, especially those with shorter hydroperiod requirements, can adapt to rice fields, others, particularly those associated with more stable environments, face increased risk of local extinction due to the modified habitat structure and water management practices (Ribeiro et al. 2017a). Additionally, the loss of functional groups, particularly those requiring specific microhabitats or longer hydroperiods, further compounds the threat posed by habitat transformation (Moreira and Maltchik 2014).

Despite the loss of some species in transformed areas, others persist, such as *Leptodactylus fuscus* and *Rhinella horribilis* among amphibians, and *Gonatodes albogularis*, *Cnemidophorus lemniscatus*, *Hemidactylus brookii*, and *H. frenatus* among reptiles. Clearings in degraded TDF, whether caused by natural processes or anthropogenic activities, facilitate the movement of species from surrounding human-modified landscapes into the forest (Carvajal-Cogollo and Urbina-Cardona 2015; Silvera-Chimá et al. 2024; Granda-Rodriguez et al. 2025). This pattern was evident in the present study, where *Hemidactylus brookii* was recorded, albeit in low abundance, even within secondary forest. Martínez-Abraín and Jiménez (2016) found similar results, suggesting that these species may assume the functional roles of species lost due to disturbance. Invasive species can even replace native species in these roles. However, some functional groups were represented by only one species (e.g., *Rhinella horribilis*, *Elachistocleis pearsei*, *Rhaebo glaberrimus*, *Dendrobates truncatus*, and *Crotalus durissus*), indicating that these species are particularly sensitive to disturbances, potentially putting their ecological functions at risk (Martín-López et al. 2007).

The classification of *Rhinella horribilis* as a monospecific group does not indicate high vulnerability, as the species is highly tolerant to disturbances. Instead, it suggests that other native species sharing its functional group have been excluded due to a limiting similarity mechanism, hindering their persistence. Granda-Rodríguez et al. (2025) show that limiting similarity mechanisms explain the persistence of species like *R. horribilis*, which dominate forest edges in TDF. In the Colombian Caribbean, *R. horribilis* shares its functional group with *Leptodactylus savagei* and *L. insularum*, both of which thrive in disturbed environments.

However, in the TDF of the Bogotá River region, *R. horribilis* does not share a functional group with any other species, suggesting it occupies a unique ecological niche previously filled by other species in more stable environments. This pattern of functional exclusion aligns with the findings of Granda-Rodríguez et al. (2025), who suggest that edge effects in TDF favor generalist species like *R. horribilis*, which can fill vacant ecological roles. The absence of other species in the same functional group in the Bogotá River's TDF indicates that *R. horribilis* may have replaced other native amphibians, with implications for ecosystem processes, such as nutrient cycling and predator-prey dynamics, potentially affecting ecosystem functioning and resilience. Nevertheless, as noted by Petchey and Gaston (2006) and Berriozabal et al. (2017), these results may also be influenced by methodological decisions, such as the selection of cut-off values for functional grouping in dendrogram analyses. Therefore, interpretations should be made with caution and complemented in future studies with physiological and acoustic functional traits.

Consistent with studies by Vargas and Flores (2006) and Berriozabal et al. (2017), certain species were found exclusively in transformed areas, including *Hemidactylus brookii* and *Hemidactylus frenatus*, both of which are invasive species. In Colombia, *H. brookii* is projected to expand its geographical range by 43.4% under climate change scenarios by 2050 (Urbina-Cardona and Castro 2011). Despite their resilience to landscape transformation, the fact that these species belong to a distinct functional group suggests that they fulfill unique ecosystem roles within transformed habitats (Martínez and Jiménez 2016). Additionally, the atypical values in some functional traits of certain invasive species tend to increase the functional volume of the assemblage they inhabit, thereby elevating the value of some functional diversity indices (Galindo-Uribe et al. 2022). This could indicate that they either replace the ecological functions of native species that are absent or, alternatively, that their invasive status has led to the displacement of native species historically occupying those functional roles.

Both *H. brookii* and *H. frenatus* share a functional group with *Thecadactylus rapicauda*, making it the only potential competitor capable of regulating these invaders—or, conversely, a species at risk of exclusion by them. This phenomenon of species substitution has significant implications for ecosystems and functional diversity, as it reflects shifts in resource distribution based on species traits and abundances. Such shifts, in turn, influence functional evenness (FEve), which measures the balance of resource utilization within an ecosystem. High FEve values indicate optimal resource use, while low values suggest inefficiencies in resource exploitation (Mouchet et al. 2010).

Functional richness (FRic) and FAD2 indices did not vary significantly between land cover types, suggesting that species continue to use available resources efficiently, even in the most disturbed areas (Berriozabal et al. 2017). This finding aligns with Díaz-García et al. (2017) and Vanessa et al. (2025), who reported no significant differences in FRic, functional evenness (FEve), and functional divergence (FDiv) across transformation gradients for amphibians. Such results suggest a high level of functional redundancy across land cover types, implying that while habitat transformation and loss may lead to species declines, the overall functional roles of both amphibians and reptiles remain largely unaffected. Functional richness, which represents the amount of functional space occupied by amphibian assemblages across the evaluated land cover types (Mason et al. 2005), showed no significant differences between land cover types. This suggests that, despite habitat alterations and disturbances, species in both natural and anthropogenic cover types can occupy the available functional space to a similar extent.

4.3 Implications for Conservation

This study recorded two reptile species of high conservation priority: *Podocnemis lewyana* (a critically endangered endemic species according to the IUCN) and *Mabuya mabouya* (a critically endangered species requiring updated records, geographical distribution assessments, and periodic population studies by the IUCN; Hedges and Powell 2016). Additionally, the study documented the presence of invasive species, including *Hemidactylus brookii* and *H. frenatus*. A high abundance of generalist species was observed, reflecting their adaptation to historical disturbances in the region. This adaptation has led to a lack of significant differences in the species' functional diversity indices between natural and anthropogenic cover types, despite the high beta diversity, which is reflected in the change in the structure of the amphibian and reptile assemblages. This pattern is further reinforced by the extensive human intervention in the study area, where remaining patches of TDF are heavily influenced by anthropogenic activities. Practices such as logging, burning, and livestock intrusion create openings that facilitate the entry of highly resilient species, while simultaneously displacing forest-specialist species (López-Camacho and Rojas 2019). These findings suggest that extinction filters (Balmford 1996; Betts et al. 2019) may have already removed specialist species from the assemblage, resulting in biotic homogenization and an influx of generalist and invasive species through the "spillover effect" associated with edge dynamics (Ries et al. 2004).

Our results underscore the importance of incorporating multiple biodiversity dimensions—taxonomic, functional, and phylogenetic—into future studies to achieve a more comprehensive understanding of habitat transformation and its impact on amphibians and reptiles (Hernández-Ordóñez et al. 2019). While no significant differences were detected in some of the functional diversity indices, this should not be interpreted as a limitation of the study but rather as a reflection of the historical biodiversity decline in the region's herpetofauna. Future research should integrate ecophysiological functional traits, and, in the case of anurans, bioacoustic analyses enhance the representation of functional diversity.

Conservation and restoration efforts are essential to restoring ecosystem functionality and the ecological roles of species, while also improving connectivity between the remaining patches of TDF within increasingly expansive anthropogenic landscapes (Kanowski et al. 2006). The presence of species exclusive to anthropogenic cover types may also provide insights into the need for structural enrichment in these areas to increase habitat availability, resource diversity, and ecological opportunities for both generalist and specialist species. Furthermore, habitat management actions for amphibians and reptiles should be integrated into ecological restoration efforts, ensuring that restoration is not only approached from a botanical perspective (Urbina-Cardona et al. 2015). It is also crucial to monitor herpetofaunal functional diversity following restoration actions to detect changes in the assemblages that go beyond species richness, providing a more comprehensive understanding of ecosystem recovery (Díaz-García et al. 2017).

5 Conclusions

Human intervention, land-use changes, and, in general, ecosystem transformation have significant consequences for herpetofaunal assemblages, affecting both community structure and functional diversity. This study assessed the structure of amphibian and reptile assemblages across natural and anthropogenic cover types in the Lower Bogotá River sub-basin, Cundinamarca. For reptiles, functional diversity did not vary significantly among the seven land cover types evaluated, suggesting a high degree of homogenization and functional redundancy between natural and anthropogenic habitats. In contrast, amphibian assemblages exhibited significant differences between rice cultivation areas and native forests (riparian and secondary), with reductions in functional dispersion (FDis) and functional divergence (Rao Q) indices. This indicates that amphibians respond more directly to environmental stress, experiencing stronger environmental filtering and reduced niche differentiation, which may ultimately diminish ecosystem functionality due to inefficient resource utilization (Mason et al. 2005).

Species identity and trait composition revealed differentiation between land cover types through functional grouping. This classification was based primarily on fixed traits, such as reproductive modes, habitat use, and diet, while plastic traits, including weight, snout-vent length, and head width, contributed to the differentiation of certain groups, mainly due to biomass differences rather than inherent ecological roles. A total of six functional groups were identified for both amphibians and reptiles, with the highest species richness and abundance observed in natural covers. In contrast, anthropogenic covers exhibited both species loss and the disappearance of entire functional groups. This study represents the first assessment of changes in the functional and taxonomic diversity of herpetofaunal assemblages across natural and anthropogenic land cover types in the Bogotá River Basin and one of the few in the TDF of the Neotropics. This region, undergoing rapid population growth and urban expansion, remains understudied in terms of biodiversity. Integrating these findings into land-use and management plans and formulating conservation strategies that protect species, their habitats, and the ecological processes they support is critical.

Conflicts of Interest The authors have no conflicts of interest to declare relevant to this chapter's content.

References

Alford RA (2011) Bleak future for amphibians. Nature 480:461–462

Allen WL, Street SE, Capellini I (2017) Fast life history traits promote invasion success in amphibians and reptiles. Ecol Lett 20(2):222–230

Álvarez-Grzybowska E, Urbina-Cardona N, Córdova-Tapia F, García A (2020) Amphibian communities in two contrasting ecosystems: functional diversity and environmental filters. Biodivers Conserv 29:2457–2485

Andrews RM, Mathies T (2000) Natural history of reptilian development: constraints on the evolution of viviparity. Bioscience 50(3):227–238

Anderson MJ, Gorley RN, Clarke KR (2008) PERMANOVA+ for PRIMER: Guide to Software and Statistical Methods. Plymouth, UK: PRIMER-E Ltd

Aronson MF, Nilon CH, Lepczyk CA, Parker TS, Warren PS, Cilliers SS, Zipperer W (2016) Hierarchical filters determine community assembly of urban species pools. Ecology, 97(11):2952-2963

Arroyo-Rodríguez V, Fahrig L, Tabarelli M, Watling JI, Tischendorf L, Benchimol M et al (2020) Designing optimal human-modified landscapes for forest biodiversity conservation. Ecol Lett 23(9):1404–1420

Attademo AM, Lorenzón RE, Peltzer P, Lajmanovich RC (2018) Diversity of anurans in rice fields under organic and conventional management in Santa Fe province, Argentina. Herpetol Rev 49(4):632–635

Balmford A (1996) Extinction filters and current resilience: the significance of past selection pressures for conservation biology. Trends Ecol Evol 11(5): 193–196

Balvanera P, Pfisterer A, Buchmann N, He J, Nakashizuka T, Raffaelli D, Schmid B (2006) Quantifying the evidence for biodiversity effects on ecosystem functioning and services. Ecol Lett 9(10): 1146–1156. https://doi.org/10.1111/j.1461-0248.2006.00963.x

Bellwood DR, Wainwright PC, Fulton CJ, Hoey A (2002) Assembly rules and functional groups at global biogeographical scales. Funct Ecol 16:557–562. https://doi.org/10.1046/j.1365-2435.2002.00655.x

Berriozabal-Islas C, Badillo-Saldaña LM, Ramírez-Bautista R, Moreno CE (2017) Trop Conserv Sci 10: 1–11. https://doi.org/10.1177/1940082917704972

Betts MG, Wolf C, Pfeifer M, Banks-Leite C, Arroyo-Rodríguez V, Ribeiro DB, Ewers RM (2019) Extinction filters mediate the global effects of habitat fragmentation on animals. Science, 366(6470):1236–1239

Blanco-Torres A (2009) Distribution of microhabitats and trophic resources among species of Bufonidae and Leiuperidae (Amphibia: Anura) in areas with tropical dry forest of the Caribbean region-Colombia. Master's Thesis in Biology. Faculty of Sciences, National University of Colombia

Blaum N, Mosner E, Schwager M, Jeltsch F (2011) How functional is functional? Ecological groupings in terrestrial animal ecology: Towards an animal functional type approach. Biodivers Conserv 20:2333–2345

Böhm M, Collen B, Baillie J, Bowles P, Chanson J, Cox N, Hammerson G, Hoffmann M, Livingstone S, Ram M, Rhodin A, Stuart S et al (2013) The conservation status of the world's reptiles. Biol Conserv 157: 372–385

Böhm M, Williams R, Bramhall HR, McMillan KM, Davidson AD, Garcia A et al (2016) Correlates of extinction risk in squamate reptiles: the relative importance of biology, geography, threat and range size. Glob Ecol Biogeogr 25:391–405

Botta-Dukát Z (2005) Rao's quadratic entropy as a measure of functional diversity based on multiple traits. J Veg Sci 16(5):533–540

Brasileiro AC, Benício RA, Gonçalves-Sousa JG, Ávila RW (2023) Influence of vegetation regeneration and agricultural land use on lizard composition, taxonomic and functional diversity between different vegetation types in Caatinga domain, Brazil. Austral Ecol 48(7): 1274–1291

Brown JH, Fox BJ, Kelt DA (2000) Assembly rules: desert rodent communities are structured at scales from local to continental. Am Nat 156:314–321

Caliński T, Harabasz J (1974) A dendrite method for cluster analysis. Commun Stat 3(1):1–27

CAR (2005) Planning and management of the Bogotá River watershed. In: Preparation of the diagnosis, prospective and formulation of the Bogotá River Watershed, p 104

Carvajal-Cogollo JE, Urbina-Cardona N (2015) Ecological grouping and edge effects in tropical dry forest: reptile-microenvironment relationships. Biodivers Conserv 24:1109–1130

Carvajal-Cogollo JE, Castaño-Mora OV, Cárdenas-Arévalo G, Urbina-Cardona JN (2007) Reptiles of areas associated with wetlands of the plain of the department of Córdoba, Colombia. Caldasia 29(2): 427–438

Carvajal-Cogollo JE, Bernal-González V, Paternina-Hernandez A, Muñóz-Ávila JA, Urbina-Cardona JN, Vargas-Salinas F (2019) Habitat use and assembly rules: patterns and mechanisms. Chap VII. In: Vargas-Salinas F, Muñoz-Avila JA, Morales-Puentes ME, (Coord.) (eds) Biology of amphibians and reptiles in the tropical dry forest of northern Colombia. Editorial UPTC, Tunja, pp 297–337. ISBN 978-958-660-341-6

Casanoves F, Pla L, Di Rienzo JA, Díaz S (2011) FDiversity: a software package for the integrated analysis of functional diversity. Methods Ecol Evol 2(3): 233–237

Clarke KR, Gorley RN, Somerfield PJ, Warwick RM (2014) Change in marine communities: an approach to statistical and interpretation, 3rd edn. Primer-E, Ltd., Plymouth, UK

Clavel J, Julliard R, Devictor V (2011) Worldwide decline of specialist species: toward a global functional homogenization? Front Ecol Environ 9:222–228

Córdova-Tapia F, Zambrano L (2015) Functional diversity in community ecology. Ecosyst J 24(3):78–87. https://doi.org/10.7818/ECOS.2015.24-3.10

Crump ML, Scott NY (1994) Visual encounter surveys. In: Heyer W, Donnelley MA, McDiarmid RA, Hayek LC, Foster MC (eds) Measuring and monitoring biological diversity: standard methods for Amphibians. Smithsonian Institution, USA

Díaz S, Cabido M (2001) Vive la différence: plant functional diversity matters to ecosystem processes. Trends Ecol Evol 16(11):646–655

Díaz-García JM, Pineda E, López-Barrera F, Moreno CE (2017) Amphibian species and functional diversity as indicators of restoration success in tropical montane forest. Biodivers Conserv 26:2569–2589

Didham R, Kapos V, Ewers R (2012) Rethinking the conceptual foundations of habitat fragmentation research. Oikos 121:161–170. https://doi.org/10.1111/j.1600-0706.2011.20273.x

Doherty T, Balouch S, Bell K, Burns T, Feldman A, Fist C, Garvey T, Jessop T, Mei S, Driscoll D (2020) Reptile responses to anthropogenic habitat modification: a global meta-analysis. Glob Ecol Biogeogr 29: 1265–1279. https://doi.org/10.1111/geb.13091

Driscoll DA, Banks SC, Barton PS, Lindenmayer DB, Smith AL (2013) Conceptual domain of the matrix in fragmented landscapes. Trends Ecol Evol 28(10): 605–613

Durán E, Balvanera P, Lott E, Segura G, Pérez-Jiménez LA, Islas A, Franco M (2002) Structure, composition and dynamics of vegetation. In: Noguera FAJH, Vega R, García-Aldrete AN, Quesada MA (eds) Natural history of Chamela. Biology Institute. National Autonomous University of Mexico, Mexico, D. F., Mexico, pp 443–472

Echeverria-Londono S, Newbold T, Hudson L, Contu S, Hill SLL, Lysenko I, Arbeláez-Cortés E, Armbrecht I, Boekhout T, Cabra-García J, Domínguez-Haydar Y, Nates-Parra G, Gutiérrez-Lamus D, Higuera D, Isaacs-Cubides PJ, López-Quintero CA, Martinez E, Miranda-Esquivel D, Navarro-Iriarte LE, Visconti P (2016) Modelling and projecting the response of local assemblage composition to land use change across Colombia. Divers Distrib 22(11):1099–1111

Ernst R, Linsermair E, Rödel M-O (2006) Diversity erosion beyond the species level: Dramatic loss of functional diversity after selective logging in two tropical amphibian communities. Biol Conserv 133:143–155

Ernst R, Keller A, Landburg G, Grafe U, Linsenmair K, Rödel M, Dzioc F (2012) Common ancestry or environmental trait filters: cross-continental comparisons of trait–habitat relationships in tropical anuran amphibian assemblages. Glob Ecol Biogeogr 21:704–715

Etter A, van Wyngaarden W (2000) Patterns of landscape transformation in Colombia, with emphasis in the Andean region. AMBIO J Hum Environ 29(7): 432–439

Etter A, McAlpine C, Possingham H (2008) Historical patterns and drivers of landscape change in Colombia since 1500: a regionalized spatial approach. Ann Assoc Am Geogr 98(1):2–23

Ewers R, Didham R (2006) Confounding factors in the detection of species responses to habitat fragmentation. Biol Rev 81:117–142. https://doi.org/10.1017/S1464793105006949

Ewers R, Didham R (2008) Pervasive impact of large-scale edge effects on a beetle community. Proc Natl Acad Sci USA 105(14):5426–5429. https://doi.org/10.1073/pnas.0800460105

Eycott AE, Marzano M, Watts K (2011) Filling evidence gaps with expert opinion: the use of Delphi analysis in least-cost modelling of functional connectivity. Landsc Urban Plan 103:400–409. https://doi.org/10.1016/j.landurbplan.2011.08.014

Fahrig L (2003) Effects of habitat fragmentation on biodiversity. Annu Rev Ecol Evol Syst 34:487–515. https://doi.org/10.1146/annurev.ecolsys.34.011802.132419

Figueiredo L, Krauss J, Steffan-Dewenter I, Sarmento Cabral J (2019) Understanding extinction debts: spatio–temporal scales, mechanisms and a roadmap for future research. Ecography 42(12):1973–1990

Filippi E, Luiselli L (2006) Changes in community composition, habitats and abundance of snakes over 10 + years in a protected area in Italy: conservation implications. Herpetol J 16:29–36

Fischer J, Lindenmayer D (2000) An assessment of the published results of animal relocations. Biol Conserv 96:1–11. https://doi.org/10.1016/S0006-3207(00)00048-3

Fischer J, Lindenmayer DB (2007) Landscape modification and habitat fragmentation: a synthesis. Glob Ecol Biogeogr 16:265–280. https://doi.org/10.1111/j.1466-8238.2007.00287.x

Flynn DFB, Gogol-Prokurat M, Nogeire T, Molinari N, Richers BT, Lin BB et al (2009) Loss of functional diversity under land use intensification across multiple taxa. Ecology 12(1):22–33

Foley JA, DeFries R, Asner GP et al (2005) Global consequences of land use. Science 309:570–574

Franklin JF, Lindenmayer DB (2009) Importance of matrix habitats in maintaining biological diversity. Proc Natl Acad Sci 13, 106(2):349–350. https://doi.org/10.1073/pnas.0812016105

Galindo-Uribe DM, Hoyos-Hoyos JM, Isaacs-Cubides P, Corral-Gómez N, Urbina-Cardona N (2022) Classification and sensitivity of taxonomic and functional diversity indices of anurans in the Andean coffee cultural landscape. Ecol Indic 136:108650

García H, Corzo G, Isaacs P, Etter E (2014) Distribution and current status of the remnants of the tropical dry forest biome in Colombia: inputs for its management. In: Pizano C, García H (eds) The tropical dry forest in Colombia. Institute of Biological Resources Research Alexander von Humboldt (IAvH)., Bogotá D.C., Colombia, pp 229–251

Gardner T, Barlow J, Peres C (2007) Paradox, presumption and pitfalls in conservation biology: the importance of habitat change for amphibians and reptiles. Biol Conserv 138:166–179

Gibbons JW, Scott DE, Ryan TJ, Buhlmann KA, Tuberville TD, Metts BS, Greene JL, Mills T, Leiden Y, Poppy S, Winne CT (2000) The global decline of reptiles, deja-vu amphibians. Bioscience 50:653–667

Gitay H, Bastow-Wilson J, Lee WG (1996) Species redundancy: a redundant concept? J Ecol 84:121–124

Glor RE, Flecker AS, Benard MF et al (2001) Lizard diversity and agricultural disturbance in a Caribbean forest landscape. Biodivers Conserv 10:711–723. https://doi.org/10.1023/A:1016665011087

Gonthier DJ, Ennis K, Farina S, Yi H, Iverson A, Batár P, Rudolph J, Tscharntke T, Cardinale B, Perfect I (2014) Biodiversity conservation in agriculture requires a multi-scale approach. Proc R Soc B Biol Sci 281(1791):20141358

Granda-Rodriguez HD, Zarate-Tirado HA, Robledo-Buitrago D, Urbina-Cardona JN (2025) Edge effects on amphibians in transformed tropical dry forest landscapes: the relationship between functional and taxonomic diversity. Glob Ecol Conserv 59:e03553

Hansen MC et al (2013) High-resolution global maps of 21st-century forest cover change. Science 342:850–853. https://doi.org/10.1126/science.1244693

Hedges B, Powell R (2016) Mabuya mabouya (errata version published in 2017). The IUCN Red List of Threatened Species 2016: e.T47102893A115397074. https://doi.org/10.2305/IUCN.UK.2016-3.RLTS.T47102893A47102903.en

Hernández-Ordóñez O, Santos BA, Pyron RA, Arroyo-Rodríguez V, Urbina-Cardona JN, Martínez-Ramos M et al (2019) Species sorting and mass effect along forest succession: evidence from taxonomic, functional, and phylogenetic diversity of amphibian communities. Ecol Evol 9(9):5206–5218

Hobbs R, Higgs E, Harris JA (2009) Novel ecosystems: implications for conservation and restoration. Trends Ecol Evol 24(11):599–605. https://doi.org/10.1016/j.tree.2009.05.012

Hokit DG, Branch LC (2003) Habitat patch size affects demographics of the Florida Scrub Lizard (*Sceloporus woodi*). J Herpetol 37(2):257–265. https://doi.org/10.1670/0022-1511(2003)037[0257:HPSADO]2.0.CO;2

Holdridge LR (1982) Ecology based on life zones. Inter-American Institute of Cooperation for Agriculture, San José de Costa Rica

Hoskin CJ (2011) The invasion and potential impact of the Asian House Gecko (Hemidactylus frenatus) in Australia. Austral Ecol 36(3):240–251

Huey RB (1982) Temperature, physiology, and the ecology of reptiles. In: Gans C, Pough FH (eds) Biology of the Reptilia. Physiology C. Physiological ecology, vol 12. Academic Press, New York, pp 25–91

IDEAM (2018) Results of deforestation monitoring 2017. Institute of Hydrology, Meteorology and Environmental Studies-IDEAM and Ministry of Environment-MINAMBIENTE, Bogotá D.C.

Kanowski J, Reis T, Catterall C, Piper S (2006) Factors affecting the use of reforested sites by reptiles in cleared rainforest landscapes in tropical and subtropical Australia. Restor Ecol 14:67–76. https://doi.org/10.1111/j.1526-100X.2006.00106.x

Kjoss V, Litvaitis J (2001) Community structure of snakes in a human-dominated landscape. Biol Conserv 98:285–292. https://doi.org/10.1016/S0006-3207(00)00167-1

Kupfer A, Müller H, Antoniazzi M, Jared C, Greven H, Nussbaum R, Wilkinson M (2006) Parental investment by skin feeding in a caecilian amphibian. Nature 440:926–929. https://doi.org/10.1038/nature04403

Laliberté E, Legendre P (2010) A distance-based framework for measuring functional diversity from multiple traits. Ecology 91(1):299–305

Lapwong Y, Dejtaradol A, Webb JK (2021) Shifts in thermal tolerance of the invasive Asian house gecko (Hemidactylus frenatus) across native and introduced ranges. Biol Invasions 23(4):989–996

Laurance W, Sayer J, Cassman KG (2014) Agricultural expansion and its impacts on tropical nature. Trends

Ecol Evol 29(2):107–116. https://doi.org/10.1016/j.tree.2013.12.001

Lehtinen RM, Ramanamanjato J, Raveloarison JG (2003) Edge effects and extinction proneness in a herpetofauna from Madagascar. Biodivers Conserv 12:1357–1370

López-Bedoya PA, Cardona-Galvis EA, Urbina-Cardona JN, Edwards FA, Edwards DP (2022) Impacts of pastures and forestry plantations on herpetofauna: a global meta-analysis. J Appl Ecol 59(12):3038–3048

López-Camacho R, Rojas J (eds) (2019) The dry forest in the CAR Territory. Regional Autonomous Corporation of Cundinamarca CAR—Francisco José de Caldas District University, Bogotá, D.C., Colombia

Luck GW, Harrington R, Harrison PA, Kremen C, Berry PM, Bugter R et al (2010) Quantifying the contribution of organisms to the provision of ecosystem services. Bioscience 59:223–235

Luedtke JA, Chanson J, Neam K, Hobin L, Maciel AO, Catenazzi A et al (2023) Ongoing declines for the world's amphibians in the face of emerging threats. Nature 622(7982):308–314

Marroquín-Páramo JA, Urbina-Cardona JN, Benítez-Malvido J, Suazo-Ortuño I (2021) Cumulative effects of high intensity hurricanes on herpetofaunal assemblages along a tropical chronosequence. For Ecol Manag 417:118505

Martínez-Abraín A, Jiménez J (2016) Anthropogenic areas as incidental substitutes for original habitat. Conserv Biol 30:593–598

Martín-López B, González JA, Díaz S, Castro I, García-Llorente M (2007) Biodiversity and human well-being: the role of functional diversity [Biodiversity and human well-being: the role of functional diversity]. Ecosystems 16:69–80

Maschio GF, Santos-Costa MC, Prudente AL (2016) Road-kills of snakes in a tropical rainforest in the central Amazon basin, Brazil. S Am J Herpetol 11: 46–53

Mason NW, De Bello F (2013) Functional diversity: a tool for answering challenging ecological questions. J Veg Sci 24(5):777–780

Mason NWH, Mouillot D, Lee WG, Wilson JB (2005) Functional richness, functional evenness and functional divergence: the primary components of functional diversity. Oikos 111:112–118

Meiri S (2008) Evolution and ecology of lizard body sizes. Glob Ecol Biogeogr 17:724–734

Mendenhall C, Frishkoff L, Santos-Barrera G, Pacheco J, Mesfun E, Quijano F, Ehrlich P, Ceballos G, Daily G, Pringle R (2014) Countryside biogeography of Neotropical reptiles and amphibians

Mioanovic M (2021) Functional diversity changes in native and alien urban flora over three centuries. Biol Invasions 23:2337–2353

Moreira LFB, Maltchik L (2014) Does organic agriculture benefit anuran diversity in rice fields? Wetlands 34: 725–733

Morse W, Schedlbauer J, Sesnie S, Finegan B, Harvey C, Hollenhorst S, Kavanagh K, Stoian D, Wulfhorst JD (2009) Consequences of environmental service payments for forest retention and recruitment in a Costa Rican Biological Corridor. Ecol Soc 14:23

Mouchet MA, Villéger S, Mason NWH, Mouillot D (2010) Functional diversity measures: an overview of their redundancy and their ability to discriminate community assembly rules. Funct Ecol 24:867–876

Nori J, Lemes P, Urbina-Cardona N, Baldo D, Lescano J, Loyola R (2015) Amphibian conservation, land-use changes and protected areas: a global overview. Biol Conserv 191:367–374

Nowakowski AJ, Watling JI, Thompson ME, Brusch GA, Catenazzi A, Whitfield SM et al (2018) Thermal biology mediates responses of amphibians and reptiles to habitat modification. Ecol Lett 21:345–355. https://doi.org/10.1111/ele.12901

Oksanen J, Blanchet FG, Friendly M, Kindt R, Legendre P, McGlinn D, Minchin P, O'Hara RB, Simpson G, Solymos P, Stevens MHH, Szöcs E, Wagner H (2020) Vegan community ecology package version 2.5–7 November 2020

Olalla-Tárraga MÁ, Rodríguez MÁ (2007) Energy and interspecific body size patterns of amphibian faunas in Europe and North America: anurans follow Bergmann's rule, urodeles its converse. Glob Ecol Biogeogr 16:606–617. https://doi.org/10.1111/j.1466-8238.2007.00309.x

Ouellette MH, Durand S, Legendre P (2019) cascadeKM: K-means partitioning using a range of values of K

Petchey OL, Gaston KJ (2006) Functional diversity: back to basics and looking forward. Ecol Lett 9:741–758

Pfeifer M, Lefebvre V, Peres CA, Banks-Leite C, Wearn OR, Marsh CJ et al (2017) Global creation of forest edges has a near-universal impact on forest vertebrates. Nature 551:187–191

Phillips HRP, Halley JM, Urbina-Cardona JN, Purvis A (2018) The effect of fragment area on site-level biodiversity. Ecography 40:1220–1231

Pianka ER (1966) Convexity, desert lizards, and spatial heterogeneity. Ecology 47:1055–1056

Pickett STML, Cadenasso JMG, Boone CG, Groffman PM, Irwin E, Kaushal SS, Marshall V, McGrath BP, Nilon CH, Pouyat RV, Szlavecz K, Troy A, Warren P (2011) Urban ecological systems: scientific foundations and a decade of progress. J Environ Manag 92(3):331–362, ISSN 0301-4797. https://doi.org/10.1016/j.jenvman.2010.08.022

Portillo-Quintero CA, Sánchez-Azofeifa GA (2010) Extent and conservation of tropical dry forests in the Americas. Biol Conserv 143(1):144–155. https://doi.org/10.1016/j.biocon.2009.09.020

Prugh L, Hodges K, Sinclair A, Brashares J (2009) Effect of habitat area and isolation on fragmented animal populations. Proc Natl Acad Sci USA 105:20770–20775. https://doi.org/10.1073/pnas.0806080105

Purvis A, Cardillo M, Grenyer R, Collen B (2005) Correlates of extinction risk: phylogeny, biology, threat and scale. In: Purvis A, Gittleman JL, Brooks T (eds) Phylogeny and conservation. Cambridge University Press, Cambridge, pp 295–316

R Core Team (2014) R: a language and environment for statistical. R Foundation for Statistical Computing, Vienna, Austria

Ramírez-Arce DG, Ochoa-Ochoa LM, Lira-Noriega A, Martorell C (2025) Reptile diversity patterns under climate and land use change scenarios in a subtropical Montane Landscape in Mexico. J Biogeogr 52(1): 108–121

Ramos E, Meza-Joya FL (2018) Reptile road mortality in a fragmented landscape of the middle Magdalena Valley, Colombia. Herpetol Notes 11:81–91

Reading CJ, Luiselli LM, Akani GC, Bonnet X, Amori G, Ballouard JM, Filippi E, Naulleau G, Pearson D, Rugiero L (2010) Are snake populations in widespread decline? Biol Lett 6:777–780

Ribeiro J, Colli GR, Batista R, Soares R (2017a) Landscape and local correlates with anuran taxonomic, functional and phylogenetic diversity in rice crops. Landsc Ecol 32:1599–1612

Ribeiro J, Colli G, Caldwell J, Ferreira E, Batista R, Soares A (2017b) Evidence of neotropical anuran community disruption on rice crops: a multidimensional evaluation. Biodivers Conserv 26:3363–3383. https://doi.org/10.1007/s10531-017-1410-0

Ribeiro J, Colli GR, Soares A (2019) Landscape correlates of anuran functional connectivity in rice crops: a graph-theoretic approach. J Trop Ecol 35(3):118–131

Richardson M (2023) Threatened and recently extinct vertebrates of the world: a biogeographic approach. Cambridge University Press. Cambridge, UK

Ricotta C (2005) A note on functional diversity measures. Basic Appl Ecol 6(5):479–486

Ries L, Fletcher RJ, Battin J, Sisk TD (2004) Ecological responses to habitat edges: mechanisms, models, and variability explained. Ann Rev Ecol Syst 35:491–522

Rincón-Aranguri M, Urbina-Cardona N, Galeano SP, Bock BC, Páez VP (2019) Road kill of snakes on a highway in an Orinoco ecosystem: landscape factors and species traits related to their mortality. Trop Conserv Sci 12:1940082919830832

Rincón-Aranguri M, Toro-Cardona FA, Galeano SP, Roa-Fuentes L, Urbina-Cardona N (2023) Functional diversity of snakes is explained by the landscape composition at multiple areas of influence. Ecol Evol 13(7): e10352

Rivas CA, Navarro-Cerrillo RM (2024) Forest fragmentation and connectivity in South American dry forests. Biodivers Conserv 33(11):3015–3037

Roach NS, Urbina-Cardona N, Lacher TE Jr (2020) Land cover drives amphibian diversity across steep elevational gradients in an isolated neotropical mountain range: implications for community conservation. Glob Ecol Conserv 22:e00968

Rödder D, Solé M, Böhme W (2008) Predicting the potential distributions of two alien invasive Housegeckos (Gekkonidae: Hemidactylus frenatus, Hemidactylus mabouia). North-West J Zool 4(2):236

Rosas-Espinoza VC, Rodríguez-Zaragoza FA, Álvarez-Grzybowska E, Peña-Joya KE, Santiago-Pérez AL, Godoy-González AA, Huerta-Martínez FM (2024) Taxonomic and functional diversity of reptiles in a heterogeneous landscape of Jalisco State, West-Central Mexico. Diversity 16(7):394

Sala O, Chapin FS III, Armesto JJ, Berlow E, Bloomfield J, Dirzo RH, Huber-Sannwald E, Huenneke L, Jackson R, Kinzig AP, Leemans R, Lodge D, Mooney H, Oesterheld MIN, Poff N, Sykes M, Walker B, Walker M, Wall D (2000) Global biodiversity scenarios for the year 2100. Science 287(5459):1770–1774. https://doi.org/10.1126/science.287.5459.1770

Salgado-Negret B, Paz H (2015) Scaling from functional traits to population, community and ecosystem processes. In: Salgado-Negret B (ed) Functional ecology as an approach to the study, management and conservation of biodiversity: protocols and applications. Alexander von Humboldt Biological Resources Research Institute, Bogotá D.C., pp 13–35

Saunders DA, Hobbs RJ, Margules CR (1991) Biological consequences of ecosystem fragmentation: a review. Conserv Biol 5:18–32

Scherer-Lorenzen M (2009) Biodiversity and ecosystem functioning: basic principles. Biodiversity: structure and function. In: Encyclopedia of life support systems, vol 1, pp 68–88. Developed under the Auspices of the UNESCO, Eolss Publishers, Paris, France. https://www.eolss.net/sample-chapters/c09/E4-27-02-02.pdf

Schneider-Maunoury L, Lefebvre V, Ewers RM, Medina-Rangel GF, Peres CA, Robinson R, Somarriba E, Urbina-Cardona JN, Pfeifer M (2016) Abundance signals of amphibians and reptiles indicate strong edge effects in Neotropical fragmented forest landscapes. Biol Conserv 200:207–215

Silvera-Chimá EDCS, Vargas JCA, Urbina-Cardona N (2024) Edge effect on anurans of the Tropical Dry Forest, pastures and oil palm plantations, and its relationship with environmental variables in the Colombian Caribbean. Latin Am Herpetol J 7(1): e584–e149

Solar RR, Barlow J, Ferreira J, Berenguer E, Lees AC, Thomson JR, Gardner TA (2015) ¿Qué tan generalizada es la homogeneización biótica en paisajes de bosques tropicales modificados por el hombre? Ecol Lett 18:1108–1118

Srivastava DS, Vellend M (2005) Investigación de la función de la biodiversidad en los ecosistemas: ¿es relevante para la conservación? Annu Rev Ecol Evol Syst 36:267–294

Stuart SN, Chanson JS, Cox NA, Young BE, Rodrigues ASL, Fischman DL, Waller RW (2004) Estado y tendencias de las disminuciones y extinciones de anfibios en todo el mundo. Science 306:1783–1786

Tan WC, Herrel A, Rödder D (2023) A global analysis of habitat fragmentation research in reptiles and amphibians: what have we done so far? Biodivers Conserv 32(2):439–468

Thompson ME, Nowakowski AJ, Donnelly MA (2016) The importance of defining focal assemblages when evaluating amphibian and reptile responses to land use. Conserv Biol 30(2):249–258. https://doi.org/10.1111/cobi.12637

Tilman D, Knops J, Wedin D, Reich P, Ritchie M, Siemann E (1997) The influence of functional diversity and composition on ecosystem processes. Science 277:1300–1302

Tilman D, Isbell F, Cowles JM (2014) Biodiversity and ecosystem functioning. Ann Rev Ecol Evol Systemat 45:471–493. https://doi.org/10.1146/annurev-ecolsys-120213-091917

Tingley R, Hitchmough RA, Chapple DG (2013) Life-history traits and extrinsic threats determine extinction risk in New Zealand lizards. Biol Conserv 165:62–68

Tobar-Suárez C, Urbina-Cardona N, Villalobos F, Pineda E (2022) Amphibian species richness and endemism in tropical montane cloud forests across the Neotropics. Biodivers Conserv 31(1):295–313

Urbina-Cardona JN, Castro F (2011) Distribución del nicho ecológico actual y futuro de anfibios y reptiles invasores: Competencias en las Corporaciones Autónomas Regionales y de Desarrollo Sostenible en Colombia. In: Vargas-Ríos O, Reyes SP (eds) La restauración ecológica en la práctica. Memorias del I Congreso Colombiano de Restauración Ecológica & II Simposio Colombiano de Experiencias en Restauración Ecológica. Grupo de Restauración Ecológica, Universidad Nacional de Colombia—GREUNAL. ISBN 978-958-719-741-9, pp 180–188

Urbina-Cardona JN, Navas CA, González I, Gómez-Martínez MJ, Llano-Mejía J, Medina-Rangel GF, Blanco A (2014) Determinantes de la distribución de los anfibios en el Bosque Seco Tropical de Colombia: herramientas para su conservación. In: Pizano C, García H (eds) El Bosque Seco Tropical en Colombia. Instituto de Investigación de Recursos Biológicos Alexander von Humboldt (IAvH). Bogotá, D.C., Colombia, pp 163–189

Urbina-Cardona JN, Bernal EA, Giraldo-Echeverry N, Echeverry-Alcendra A (2015) El monitoreo de la herpetofauna en los procesos de restauración ecológica: indicadores y métodos. Capítulo 2.7. In: Aguilar-Garavito M, Ramírez W (eds) Monitoreo a procesos de restauración ecológica, aplicado a ecosistemas terrestres. Instituto de Investigación de Recursos Biológicos Alexander von Humboldt (IAvH). Bogotá D.C., Colombia. 250 p, pp 134–147

Urbina-Cardona N, Angulo A, Turner A, Cogalniceanu D, Wickramasinghe D, Suazo-Ortuño I, Watling J, Acevedo LD, Doherty-Bone T, Hernandez-Ordoñez O, Tarrant J (2024) Habitat loss: protection and management. Chapter 5. In: Wren S, Borzée A, Marcec-Greaves R, Angulo A (eds) IUCN SSC Amphibian Specialist Group Amphibian conservation action plan: A status review and roadmap for global amphibian conservation. IUCN SSC Occasional Paper, No 57. IUCN, Gland, pp 115–146. https://doi.org/10.2305/QWVH2717

Van Wilgen NJ, Richardson DM (2012) The roles of climate, phylogenetic relatedness, introduction effort, and reproductive traits in the establishment of non-native reptiles and amphibians. Conserv Biol 26(2):267–277. https://doi.org/10.1111/j.1523-1739.2011.01804.x

Vandermeer J, Perfecto I (2007) The agricultural matrix and a future paradigm for conservation. Conserv Biol 21:274–277

Vanessa J, Palomino-Cuéllar J, Nicolás, Urbina-Cardona (2025) Environmental filtering and limiting similarity shape the taxonomic and functional diversity of amazonian floodplain frogs across pasture–edge–interior gradients Global Ecology and Conservation 62e03817. https://doi.org/10.1016/j.gecco.2025.e03817

Vargas-Santamaría F, Flores-Ville Vetter D, Hansbauer MM, Végvári Z, Storch I (2011) Predictors of forest fragmentation sensitivity in Neotropical vertebrates: a quantitative review. Ecography 34:1–8

Vié J, Hilton-Taylor C, Stuart SN (2009) Wildlife in a changing world—an analysis of the 2008 IUCN Red List of Threatened Species. IUCN, Gland

Villéger S, Mason NWH, Mouillot D (2008) New multidimensional functional diversity indices for a multifacted framework in functional ecology. Ecology 89:2290–2301. https://doi.org/10.1890/07-1206.1

Violle C et al (2007) Let the concept of trait be functional! Oikos 116:882–892

Watling JI, Nowakowski AJ, Donnelly MA, Orrock JL (2011) Meta-analysis reveals the importance of matrix composition for animals in fragmented habitat. Glob Ecol Biogeogr 20:209–217. https://doi.org/10.1111/j.1466-8238.2010.00586.x

Weterings R, Vetter KC (2018) Invasive house geckos (Hemidactylus spp.): their current, potential and future distribution. Curr Zool 64(5):559–573

Wilcove DS, Rothstein D, Dubow J, Phillips A, Losos E (1998) Quantifying threats to imperiled species in the United States: assessing the relative importance of habitat destruction, alien species, pollution, overexploitation, and disease. Bioscience 48(8):607–615. https://doi.org/10.2307/1313420

Zabala-Forero F, Urbina-Cardona JN (2021) Taxonomic and functional diversity to landscape transformation: relationship of amphibian assemblages with changes in land use and coverage. Revista Mexicana de Biodiversidad 92:e923443.

Zabala-Forero F, Cortés-Gómez AM, Urbina-Cardona N (2025) How low-abundance amphibians shape functional diversity across tropical forest succession stages? Ecol Indic 171:113140

Biogeographic Patterns of Richness and Endemism in Liolaemidae: Identifying Hotspots for the Most Diverse Andean Reptile Radiation

Damien Esquerré and Alexander Skeels

1 Introduction

1.1 Liolaemidae: Diverse Patterns of Diversity

The vertebrate diversity of the Andes is in great part composed by lizards in the family Liolaemidae (Fig. 1), particularly by the hyper diverse genus *Liolaemus* (Skeels et al. 2023). This genus showcases an outstanding species richness (290 described species) and ecological diversity, and is emerging as one of the most important model systems to study diversification processes (Schulte et al. 2000; Harmon et al. 2003; Pincheira-Donoso et al. 2015; Esquerré et al. 2019a; Ocampo et al. 2024). Aside from thriving in most habitats across the southern half of South America, some *Liolaemus* inhabit extremely hostile environments, including mountains over 5000 m a.s.l., the freezing Tierra del Fuego at the southern end of the continent, and the hyper-arid Atacama Desert (Abdala et al. 2021a; Cerdeña et al. 2021; Jaksic 2022). They also display numerous transitions between viviparous and oviparous reproduction modes (Schulte et al. 2000; Esquerré et al. 2019a) and have evolved a variety of different phenotypes and habitat-use strategies (Espinoza et al. 2004; Tulli et al. 2016; Aguilar-Puntriano et al. 2018; Ocampo et al. 2024). Interestingly, the two other genera in the Liolaemidae display contrasting evolutionary patterns. *Phymaturus* is a fairly diverse group of 52 species, but unlike *Liolaemus*, their ecology, life history, and phenotype are highly conserved: all species are adapted to cold rocky environments and are viviparous, herbivorous, and share the basic morphology of a short snout, a stocky dorsoventrally flattened body, and a spiny tail (Díaz-Gómez 2009; Morando et al. 2013; Lobo et al. 2015; Marín et al. 2018; Abdala et al. 2021b). Most strikingly, the sister lineage of *Liolaemus* and *Phymaturus* is the monotypic *Ctenoblepharys*, and the single representative (*C. adspersa*) is a slender desert-adapted oviparous lizard from south-central Peru, representing over 40 million years without diversification (Esquerré et al. 2019a) (Fig. 1).

Morphology and molecular data both strongly support two main clades in *Liolaemus*, classified as the *Liolaemus* (sensu stricto) and the *Eulaemus* subgenera (Laurent 1983; Etheridge 1995; Schulte et al. 2000; Esquerré et al. 2019a). Because of the relative dominance of the *Liolaemus* and *Eulaemus* subgenera on the west and east sides of the Andes, respectively, they have been referred to as the Chilean and Argentinean groups (Laurent 1983). Interestingly, the *walkeri* clade, a group of high-altitude

D. Esquerré (✉)
Environmental Futures Research Centre, School of Science, University of Wollongong, Wollongong, NSW, Australia

A. Skeels
Research School of Biology, Australian National University, Canberra, ACT, Australia

J. N. Urbina-Cardona et al. (eds.), *Andean Herpetofauna*, Sustainable Development Goals Series,
https://doi.org/10.1007/978-3-032-00074-3_5

Fig. 1 Representative photographs of members of the three genera in Liolaemidae: the monotypic *Ctenoblepharys*, the cold-adapted herbivores *Phymaturus*, and the hyper-diverse *Liolaemus*. Species are from left to right then top to bottom: *Ctenoblepharys adspersa*, *Phymaturus* cf. *darwini*, *Liolaemus nigroviridis*, *L. incaicus*, *L. pictus*, *L. tenuis*, and *L. leopardinus*

species mostly found on the highlands of Peru, with strong phenotypic affinities to the *alticolor* clade within the *Liolaemus* subgenus (Portelli and Quinteros 2018; Quinteros et al. 2020), belongs to neither *Liolaemus* (sensu stricto) nor *Eulaemus*, and is inferred as the sister group to the clade formed by both subgenera by both mitochondrial and nuclear DNA (Esquerré et al. 2019a, 2022). This indicates three major clades in *Liolaemus*. *Phymaturus* is divided into two main clades. The *palluma* group is mostly restricted to the central portion of the Southern Andes, at high altitudes between Chile and Argentina, while the *patagonicus* group is mostly distributed across the Patagonian Steppe in northern Argentinean Patagonia (Morando et al. 2013; Esquerré et al. 2019a). Given the substantial diversity of these lizards, in particular *Liolaemus*, these major clades are further classified into a number of nested clades. Briefly, the *Eulaemus* group is subdivided into three major lineages (*lineomaculatus*, *boulengeri*, and *montanus* groups), while the *Liolaemus* subgenus is further subdivided into two lineages (*chilensis* and *nigromaculatus* sections), and each of these groups is further divided into a number of clades (Lobo and Quinteros 2005; Pincheira-Donoso et al. 2008b; Abdala and Quinteros 2014; Olave et al. 2015; Abdala et al. 2020; Morando et al. 2020; Esquerré et al. 2022).

1.2 The Role of the Andes in the Diversity of Liolaemidae

The Andes play a central role in the diversity and evolution of liolaemids, and the importance of this massive mountain chain on the biogeography and diversification of *Liolaemus* has long been acknowledged (Hellmich 1951; Cei 1979; Fuentes and Jaksic 1979; Young-Downey 1998). Considering the Patagonian Steppe and the Chilean Coastal Range as part of the same broad geological system as the Andes (Skeels et al. 2023), all *Phymaturus* and 259 of the 290 (89%) of *Liolaemus* species are found in Andean regions (Fig. 2). Schulte et al. (2000) was one of the first to analytically show that liolaemids likely originated in the Andes and that their diversification was underpinned by recurrent vicariance and dispersal between Andean regions (Schulte et al. 2000). Biogeographic reconstruction analyses highlight the Central Andes between central Chile and Argentina, or the Altiplano in the Andes of northern Chile and Argentina and southern Peru and Bolivia, as the most likely cradles for Liolaemidae, *Liolaemus*, and most of the main clades (Gómez and Lobo 2006; Díaz Gómez 2011; Portelli and Quinteros 2018; Esquerré et al. 2019a; Ruiz et al. 2020; Portelli et al. 2022). These alternative scenarios are likely due to different phylogenetic hypotheses used for biogeographic reconstruction. For instance, the sister clade to *Liolaemus* (sensu stricto) and *Eulaemus* being either the Peruvian Andes *walkeri* group or the mostly Patagonian *Phymaturus* has different implications for the geographic origins of *Liolaemus*. Some analyses also place the origins of the diversification of *Phymaturus* in either the central Southern Andes or Patagonia (Díaz Gómez 2011; Esquerré et al. 2019a).

In recent years, the development of new sophisticated tools to investigate diversification processes opened the doors for a number of studies to look at the role of the Andes in boosting diversification of South American organisms, such as plants, reptiles, amphibians, and birds (Lagomarsino et al. 2016; Quintero and Jetz 2018; Rangel et al. 2018; García-Rodríguez et al. 2021; Boschman and Condamine 2022). The role of Andean orogeny in the diversification rates of liolaemid lizards has also been investigated in recent years. Mechanistically, a correlation between speciation and mountain formation is likely caused by habitat fragmentation for cold-adapted organisms that become isolated across mountains, much like speciation across oceanic islands formed by volcanoes (Rowe et al. 2019). This idea can be qualitatively supported by the frequent discovery of new *Liolaemus* species restricted to narrow mountain formations and isolated from their closest relatives by lower elevation valleys (Chaparro et al. 2020; Avila et al. 2021). Even though *Liolaemus* split from *Phymaturus* likely sometime between 30 and 50 million years ago (Esquerré et al. 2019a), *Liolaemus* started experiencing accelerated speciation around 20 million years ago, which coincides with an intensified period of Andean uplift (Garzione et al. 2008; Boschman and Condamine 2022). Time-variable diversification models using historical Andean elevation data find high statistical support of changes in elevation driving speciation rates (Esquerré et al. 2019a), but some of the

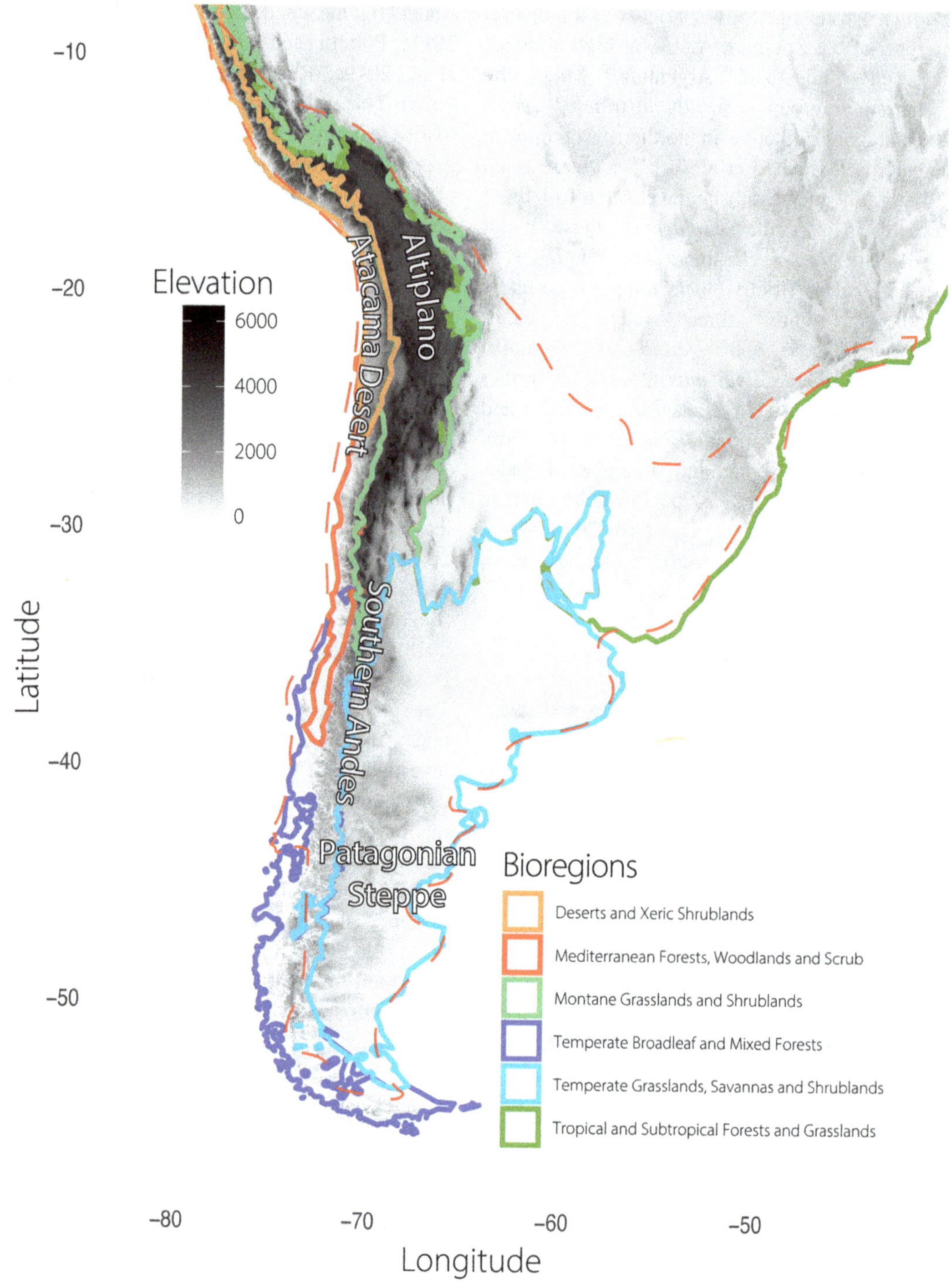

Fig. 2 Elevation and biome boundaries across southern South America. Red dashed line is an approximate description of the distribution Liolaemidae (for more details, see Fig. 4). Biome boundaries are modified from Olson et al. (2009) and define major vegetation types around the globe. Elevation data is taken from the digital elevation model of Farr and Kobrick (2000) at 1 km resolution

patterns are not supported if taking into account rate heterogeneity (Olave et al. 2020). In any case, the complex process of Andean uplift is oversimplified by a linear series of elevational estimates, because different Andean regions have experience uplift at different rates, and much of the southern Andean region that Liolaemidae occupy has not experienced major uplift in the past 30 Ma compared to more dynamic northern Andean regions. To address this, Skeels et al. (2023) implemented a diversification model using spatial reconstructions of the South American topographic and climatic landscape (Boschman 2021), finding that at the southern latitudes that the Liolaemidae occupy, mid-elevations (>2000 m) offer enough topographic complexity to promote speciation without the high extinction rates experienced at higher elevations. This topographic "Goldilocks Zone," in combination with moderate climates that reduce extinction risks in ectotherms, is partly responsible for the high diversity of liolaemids.

Some studies indicate that possibly mountain orogeny itself is not the driver of increased diversification, but that conditions found on mountains promote speciation. For instance, habitat heterogeneity that offers ecological opportunity might be a mechanism for increased speciation in mountains, as more ecological niches facilitate divergence of closely related species (Quintero and Jetz 2018; Boschman and Condamine 2022; Alencar et al. 2024). This can also be enhanced by the fact that viviparity, which is an adaptation to cold climates, might also promote diversification by accelerating postzygotic isolation between populations due to viviparity-driven conflict between mother and embryo (Zeh and Zeh 2008; Pyron and Burbrink 2014). Additionally, larger sizes, which tend to evolve in colder climates in some ectotherms (Cruz et al. 2005; Pincheira-Donoso et al. 2008a; Domínguez-Guerrero et al. 2024), might promote wider dispersal and therefore diversification of lineages which colonize new regions (Alencar et al. 2024). Finally, climate change, and in particular glacial cycles throughout the Pliocene and Pleistocene, has likely expanded, contracted, and fragmented species based on their thermal niches. These rearrangements of species geographic ranges might have also played a role in conjunction with the complex topography of the Andes to promote diversification via allopatric speciation (Fuentes and Jaksic 1979; Victoriano et al. 2008; Breitman et al. 2012; Cianferoni et al. 2013; Wallis et al. 2016; Esquerré et al. 2019b; Sánchez et al. 2023; Skeels et al. 2023).

The formation of the Andes dramatically transformed the climatic landscape of South America, driving the evolution of species and population-level diversity, and also indirectly shaping the ecological diversity of its inhabitants, including liolaemids. Besides shaping thermal biology (Azócar et al. 2013), cold mountain tops have provided strong selective pressures for the recurrent evolution of viviparous reproduction (Pincheira-Donoso et al. 2013, 2017; Cruz et al. 2014; Esquerré et al. 2019a; Zimin et al. 2022), larger sizes (Cruz et al. 2005; Pincheira-Donoso et al. 2008a; Domínguez-Guerrero et al. 2024), and herbivory (Espinoza et al. 2004; Ocampo et al. 2024). The rain shadow effect promoting the extreme aridity of the Atacama Desert (Houston and Hartley 2003) has resulted in liolaemids convergently evolving morphological adaptations to life on Earth's driest desert, including a toad-like head with a short snout and enlarged ciliary scales (Aguilar-Puntriano et al. 2018). These are just a few examples to illustrate the importance of this mountain range of South America's biodiversity.

We are now beginning to understand patterns of evolutionary diversification in Liolaemidae due to recent advances in phylogenetics. Recently uncovered phylogenetic relationships of Liolaemidae offer a fundamental framework to explore biogeography of the clade, to more deeply explore how the Andes has shaped the diversity and distribution of this lineage. Yet spatial patterns of richness, endemism, and evolutionary significant areas remain to be explored. The objective of this chapter is to use a comprehensive spatial database of Liolaemidae to uncover these patterns and identify areas of particular importance that capture hotspots of diversity and unique evolutionary history. Given the geographically structured distribution of the different clades of Liolaemidae along the Andes we predict several areas of unique evolutionary importance for the family.

2 Methods

2.1 Geographic Ranges

We obtained point occurrence records for 383 taxa of Liolaemidae expanding on the dataset used by Skeels et al. (2023), incorporating recent taxonomic changes and tissue collection records (Abdala et al. 2023; Ruiz et al. 2023; González-Gutiérrez 2024; Mella Ávila and Marambio-Alfaro 2024; Sánchez et al. 2024). We manually cleaned these records, removing clearly erroneous records and cross referencing each taxon's distribution using Abdala et al. (2021a, b). There are on average 23.3 observed occurrences per taxon with a range of 1–448. Occurrences are left skewed such that most species had few occurrence records, and 33 species were known from only a single record. The vast majority of this dataset, accessible from Skeels et al. (2023), is available on Zenodo (https://doi.org/10.5281/zenodo.14219557). From these occurrences, we obtained an estimate of each species geographic range using the rangeBuilder method (Rabosky et al. 2016). This approach builds polygons describing the outline of species geographic distributions using a combination of methods (alpha hulls, minimum convex hulls, and buffers) to estimate ranges with varying numbers of occurrence records. For species with more than five records we built alpha hulls using the "alphahull" package in R (Pateiro-López and Rodríguez-Casal 2010) by sequentially increasing the alpha parameter value until a contiguous species range was created that contained at least 99% of the occurrence records. For species with 3–5 occurrences, or for those species for whom an alpha hull did not converge given the fixed criteria, we estimated a minimum convex hull using the *st_convex_hull* function in the "sf" package (Pebesma 2018; Pebesma and Bivand 2023). For species with fewer than three records, we buffered each point with a 0.1° radius, or approximately 10 km at the equator, using the *st_buffer* function in "sf." The ocean was masked from each species estimated range and we turned these ranges into presence-absence data at 0.1° grid cell resolution for downstream analyses.

2.2 Phylogeny

We used the phylogeny of Esquerré et al. (2019a) as a starting point for our analysis which contained 258 Liolaemidae taxa. To have a complete representation of our dataset in the phylogeny, we imputed the missing taxa on the tree. For this, we assigned species that were missing in our base phylogeny to specific clades in the tree based on recent molecular or morphology-based phylogenies (Abdala et al. 2020; Quinteros et al. 2020; Morando et al. 2021; Esquerré et al. 2022; Sánchez et al. 2024). We then attached each missing species to a polytomy at the most recent common ancestor of all species in that clade in the phylogeny.

2.3 Spatial Diversity Patterns

We estimated major spatial biodiversity patterns from this matched spatial and phylogenetic dataset: species richness, weighted endemism, phylogenetic endemism, hotspots, and phyloregions (see explanations below).

We estimated species richness (SR) as the sum total number of species found in each 0.1° grid cell across South America. Next, to determine regions with a high proportion of range restricted species, we estimated weighted endemism (WE) following the approach of Crisp et al. (2001). WE is the sum total of species in each grid cell, weighted by the inverse of their geographic range size, such that widespread species contribute smaller values to the sum and small-ranged species contribute larger values. We additionally estimated phylogenetic endemism (PE) as a complementary spatial measure of range restricted lineages which incorporates phylogenetic branch lengths. We estimated PE following the method of Rosauer et al. (2009), where PE is the total phylogenetic branch length connecting species in a grid cell, after dividing each branch length by the total range size of its descendants so that branches with widespread descendants contribute less to the sum than branches with narrowly distributed descendants. We defined "hotspots" as regions with the greatest concentration of species richness, weighted endemism, or phylogenetic endemism (i.e., areas of conservation importance). We used a threshold of the top 10% of grid cells in the domain to define grid cells as a hotspot. We highlighted regions which are hotspots for single metrics (e.g., WE or PE) or multiple metrics (e.g., WE and PE; Fig. 3).

Finally, we performed a bioregionalization exercise to infer regions of endemism and unique evolutionary history across the Andes based on the composition of taxa and branches of the phylogeny following the method of Daru et al. (2017) using the R package "phyloregions" and associated protocol described in Daru et al. (2020), as follows: We estimated the degree of turnover of species and phylogenetic lineages (taxonomic β-diversity and phylogenetic β-diversity, respectively) to measure pairwise dissimilarity between grid cells. Two sites can have different compositions because one has a subset of taxa found in another (nestedness) or because they have unique compositions (replacement). We used the Simpson index of dissimilarity to measure β-diversity as this considers the replacement of taxa (rather than nestedness) which is better suited to investigate unique evolutionary arenas. We selected the best hierarchical clustering algorithm (UPGMA) of β-diversity values using the *select_linkage* function to cluster these into *k* spatial biogeographic units. We selected the optimal number of units by iterating the procedure over increasing values of *k* from 3 to 20 and finding the "elbow" at which increased values of *k* do not greatly increase explained variance (ev) using the function *optimal_phyloregion*. We found that either 14 or 7 phyloregions correspond to the "elbow" of explained variance for increasing values of *k*. We opted to use seven phyloregions as this is more easily interpreted.

3 Results and Discussion

3.1 Spatial Diversity Patterns

Liolaemidae diversity is concentrated in the Southern Andes in central and southern Chile and Argentina (Fig. 3). Richness tapers from this region northward toward the Atacama Desert and Altiplano, eastward toward the tropical grasslands and forests of eastern South America, and southward toward Patagonia. Patterns of weighted endemism (WE) and phylogenetic endemism (PE) differ from that of richness in having higher values extending northward along the Andean spine, indicating that more species and lineages with narrow distributions are found in these more northern ranges. Areas of concentrated richness and endemism are highlighted with hotspots (Fig. 3). We found that areas that are hotspots for richness alone are found in the south-central region, while regions that are hotspots for PE, WE, or both are found in the northern part of the distribution. Hotspots for all three metrics (SR, WE, PE) are concentrated in the central Southern Andes (areas discussed in more detail in Sect. 3.3).

3.2 Phyloregions

Phyloregion-1 is found in the northern part of the distribution of Liolaemidae and is driven by the presence of the deeply divergent *walkeri* group in *Liolaemus* (Fig. 4). Phyloregion-2 is distributed across much of the eastern and northern distribution and is driven by the two closely related and widespread *boulengeri* and *montanus* groups of the *Eulaemus* subgenus. Phyloregion-3 is disjunct from south to north and again in the east and is driven primarily by the occurrence of the *chilensis* section of the *Liolaemus* subgenus. Phyloregions 4, 5, and 6 are in the western Andes, Atacama, and Altiplano regions and are driven by the distribution of the *nigromaculatus* section of the *Liolaemus* subgenus and the *Phymaturus paluma* group and highlights that phyloregions can turnover in short geographic distances in this environmentally complex area.

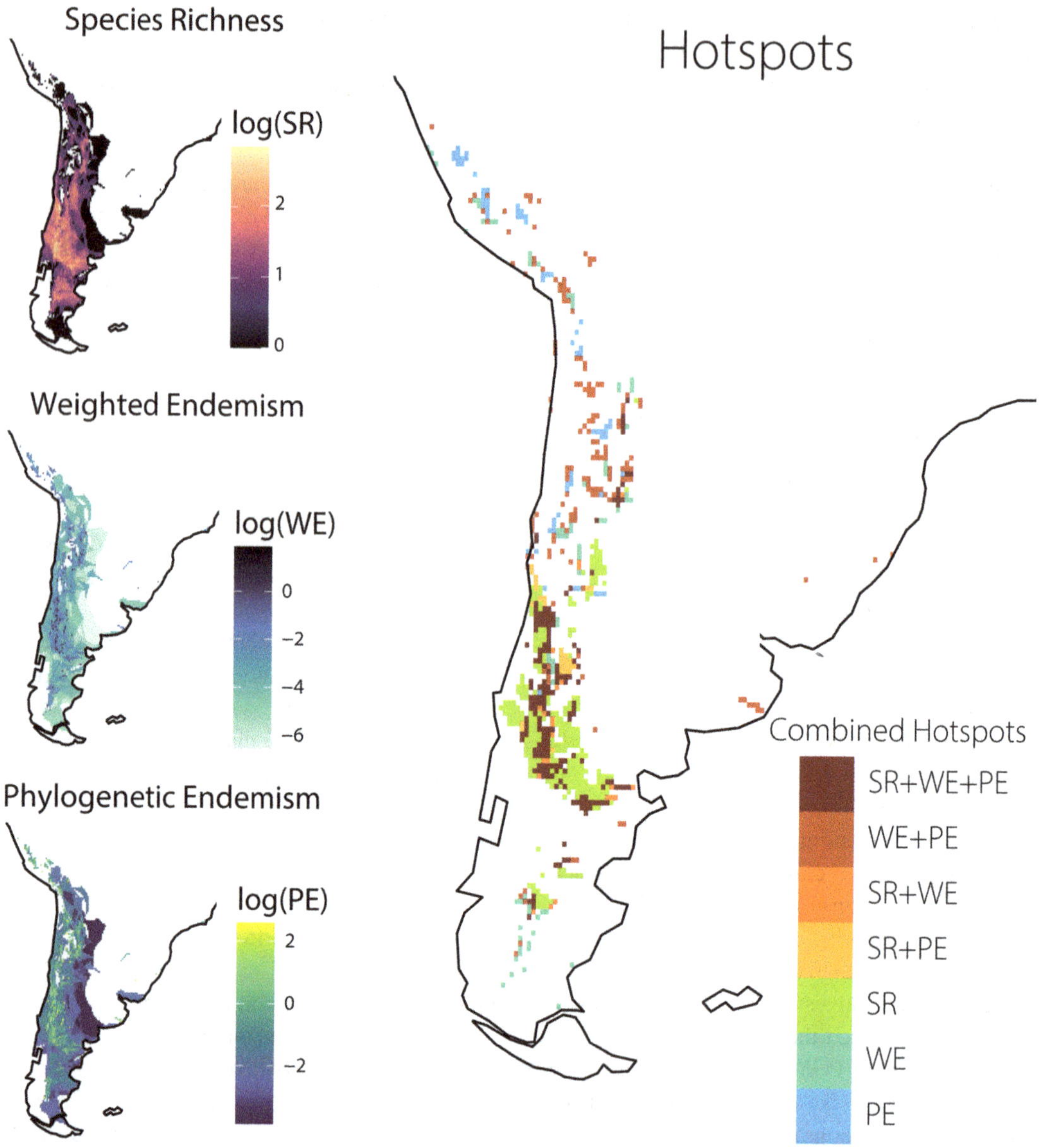

Fig. 3 Spatial distribution of species richness, weighted endemism, phylogenetic endemism, and cells containing the highest 10% of values for these three metrics (defined as hotspots) for Liolaemidae across 0.1° grid cells in South America

Phyloregion-7 is in the southern part of the distribution and is driven by the *lineomaculatus* section of the *Eulaemus* subgenus and the *Phymaturus patagonicus* group. The presence and absence of each taxon in the seven phyloregions is presented in Table S1 in Zenodo (https://doi.org/10.5281/zenodo.14219557).

3.3 Biogeographically Important Regions for Liolaemidae

Considering the results from the sections above and the general geography of southern South America, we can qualitatively identify five broad areas of biogeographic importance for Liolaemidae, going from north to south:

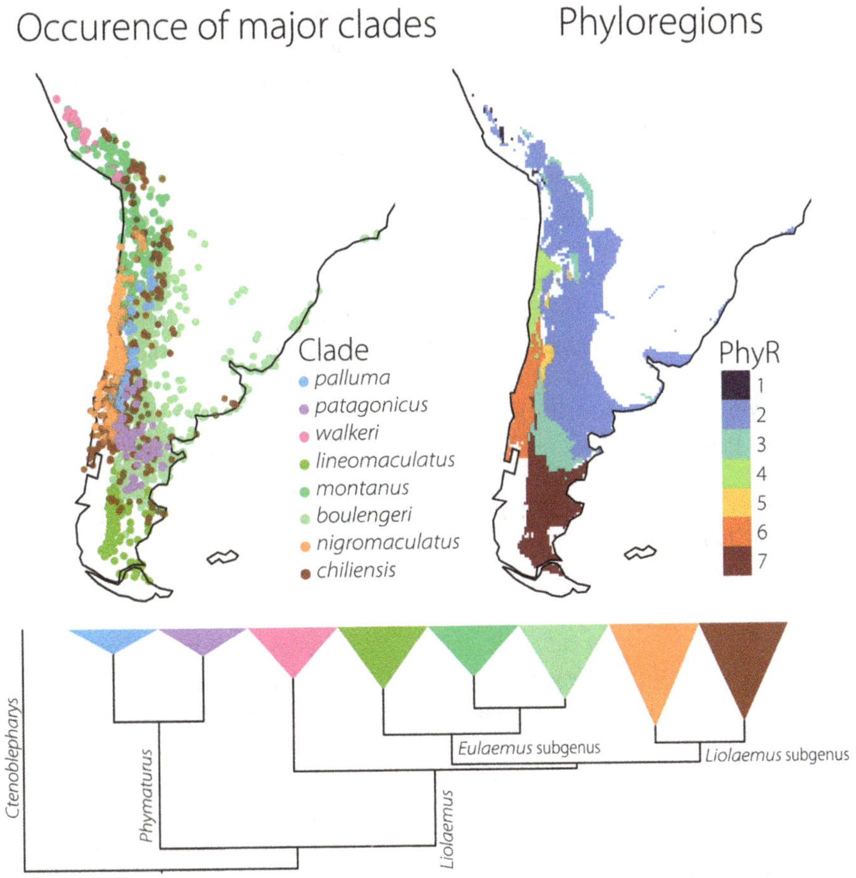

Fig. 4 The distribution of phylogenetic patterns in Liolaemidae. The left-hand panel shows the distribution of point occurrence records for infra-generic Liolaemidae clades. The right-hand panel shows the distribution of phyloregions—areas of endemism estimated using phylogenetic beta-diversity for the group. The bottom panel shows a schematic of the phylogenetic groups in Liolaemidae. (Based on Esquerré et al. 2019a, b) with the clades colored according to the occurrence records

(a) A northern Altiplano region along the Andes between central and south Peru, northern Chile, and western Bolivia, roughly between latitudes 13°S and 18°S. This region is characterized by relatively low richness, but high weighted and phylogenetic endemism (Fig. 3). This means that although there are not as many species here as in the south, there are many species with restricted ranges and that are phylogenetically distinct. For example, phyloregion-1 is entirely restricted to the highlands of this region, which is likely driven in great part by the presence of the *walkeri* clade (Fig. 4). This group of lizards is endemic to the highlands of Peru and adjacent parts of Chile (Aguilar et al. 2013), and morphologically resemble species of the *alticolor-bibronii* clade of the *Liolaemus* subgenus (Portelli and Quinteros 2018; Quinteros et al. 2020). However, molecular data place the *walkeri* clade as the sister clade to *Liolaemus* + *Eulaemus* subgenera, being the most basal *Liolaemus* clade (Esquerré et al. 2019a, 2022). This region also has representatives of the *Eulaemus* subgenus (*montanus* group) along with the *Liolaemus* (sensu stricto) subgenus (*alticolor-bibronii* clade from the chilensis section), and correspondingly, parts of phyloregion-2 and -3.

(b) The southern Altiplano and northern section of the Southern Andes, going from around the highlands of Potosí Department in southern Bolivia, Antofagasta Region in Chile, and Jujuy Province in Argentina, to the Andes between the Coquimbo Region in Chile and San Juan Province in Argentina, roughly from latitudes 19°S to 30°S. This segment of the Andes is characterized by high richness, high weighted endemism, and high phylogenetic endemism (Fig. 3), and combines parts of phyloregion-2, -3, -4 and -5 (Fig. 4). This region is rich in species of the *Eulaemus* subgenus, particularly the *montanus* and *boulengeri* (mostly *darwinii* clade) groups (Abdala et al. 2020). There is also high richness for the *Liolaemus* subgenus, especially the *alticolor-bibronii* and *capillitas* groups from the *chilensis* section, as well as some of the *nigroviridis* group, a group of montane species of the *nigromaculatus* section which is a major component of phyloregion-4. Phylogenetic diversity is further enhanced by a high prevalence of species of the *Phymaturus palluma* group, particularly in the Argentinean highlands of Catamarca, La Rioja, and San Juan.

(c) The central portion of the Southern Andes, starting roughly at latitude 33°S in the Valparaíso Region of Chile and the Mendoza Province of Argentina, south to about latitude 41°S, in Los Lagos Region of Chile and Rio Negro Province of Argentina. This region has the highest richness and endemism of *Liolaemus* (Fig. 3), containing mostly phyloregion-6 but also -5 and part of -3 (Fig. 4). This area is dominated by the *Liolaemus* subgenus, where most of the species of its two largest clades, the *elongatus-kriegi* and *gravenhorstii-bibronii* complexes (sensu Esquerré et al. 2022) are concentrated. On the western slopes (Chilean side) there are also a number of montane species from the *nigromaculatus* section. The *Eulaemus* subgenus has a relatively lower representation here, but there are several species of the *boulengeri* group on the eastern slopes (Argentinean side of the Andes). This area is also rich in species of *Phymaturus*, particularly the *palluma* section, but including some species of the *patagonicus* section at its southern end. This region is likely the most topographically rugged, a center of speciation and possibly the cradle of *Liolaemus* (Esquerré et al. 2019a). A recent study found that it might be the most diverse by providing the ideal conditions for high allopatric speciation rates and by lacking the higher extinction rates associated with higher elevations toward the north (Skeels et al. 2023).

(d) A northern Patagonia region in Argentina in the Chubut Province approximately from latitudes 41°S to 43°S, going east from the high Andes in the Chilean-Argentinean border. This area contains high species richness and endemism (Fig. 3) and is mostly represented by phyloregion-3 (Fig. 4). There is high diversity of the *Liolaemus* subgenus, mostly represented by the *elongatus-kriegi* complex and *bibronii* clade. The dominant group here is the *Eulaemus* subgenus represented mostly by the *boulengeri* group, particularly the *fitzingerii* and *boulengeri* clades. This is also the center of speciation for the *Phymaturus patagonicus* section.

(e) A southern Patagonia region in the Santa Cruz Province around latitudes 45°S to 48° S, which, like the region described above, features another topographically complex system. By far the dominant group here is the *lineomaculatus* section of the *Eulaemus* subgenus, especially toward Tierra del Fuego at the end of the continent. However, there are also species from the *boulengeri* group, *chilensis* section, and *Phymaturus patagonicus* section. Patagonia has long been considered a center of speciation for *Liolaemus* and *Phymaturus* (Cei 1979; Pereyra 1992; Morando et al. 2003; Lobo and Quinteros 2005; Díaz Gómez 2011;

Portelli and Quinteros 2018; Esquerré et al. 2019a). It might also be the place of origin of *Phymaturus*, but in the case of *Liolaemus*, their diversity there has more likely resulted from different dispersals from other parts of the distribution, with subsequent diversification (Morando et al. 2007; Breitman et al. 2011; Portelli and Quinteros 2018; Esquerré et al. 2019a). Studies in Patagonia highlight the importance of Quaternary glacial cycles in fragmenting, contracting, and expanding populations, which has resulted in complex and porous speciation processes in these lizards, marked by intermittent pulses of isolation and gene flow (Morando et al. 2003, 2004; Avila and Morando 2006; Olave et al. 2011; Breitman et al. 2012; Fontanella et al. 2012; Grummer et al. 2018; Sánchez et al. 2023, 2024). However, as some other studies suggest (Vera-Escalona et al. 2012; Esquerré et al. 2019b), it is very likely that similar processes have taken place all along the Andes.

Aside from these main areas which we highlight based on patterns of richness and endemism, there are other regions in South America that harbor unique assemblages of *Liolaemus*. The Atacama Desert is the most arid desert on the planet and contains a number of species adapted to its extreme conditions, most notably species of the *nigromaculatus* section and the *montanus* group. From the latter, we observe a few species that display a divergent slender morphology that has even resulted in their classification in different genera than *Liolaemus* in earlier studies before the phylogenetic era (Donoso-Barros 1971, 1972; Núñez and Yáñez 1984). Interestingly, recent molecular phylogenetics has revealed that these species in fact are all part of the *montanus* group but do not form a monophyletic group, having evolved their peculiar morphology, along with *Ctenoblepharys adspersa*, through convergent evolution (Aguilar-Puntriano et al. 2018). On the far eastern side of the range of *Liolaemus,* along the Atlantic coast of Argentina, Uruguay, and southern Brazil, we observe species of the *wiegmannii* group (part of the *Eulaemus* subgenus) (Etheridge 2000; Villamil et al. 2019). Even though this is a very low-diversity region for liolaemids, it is biogeographically important, particularly because some of these dune specialists are threatened by habitat destruction (Kacoliris et al. 2009; Rocha et al. 2009; Almeida-Santos et al. 2015).

3.4 Future Directions and Closing Remarks

Precise identification of regions of biogeographic importance relies on good taxonomy. Although our knowledge of the diversity of Liolaemidae has increased dramatically over the past two decades, the true diversity of this family is still far from being fully resolved. New species are still being described every year, and the rapid advance of genomic tools is also revealing that the established limits between many species need to be reassessed (Sánchez et al. 2023, 2024). Moreover, phylogenetics and species delimitation in this group is partly hindered by an unusually high prevalence of interspecific hybridization and introgression (Olave et al. 2018; Esquerré et al. 2022; Sánchez et al. 2024).

Another potential bias to our results is the geographically uneven spread of taxonomic efforts, which can lead to high richness estimates in well-studied areas where the diversity is well described and high-weighted endemism in understudied areas where species are known from single or few localities. For instance, areas like Patagonia and the central and southern Andes have been studied more intensely than the Altiplano of Peru and Bolivia, which might partly explain some of the observed patterns. Fortunately, in recent years, more systematic research has been established in the northern range of Liolaemidae, but there is still undescribed diversity, and many species are still known only from their type localities (Aguilar-Puntriano and Salazar 2021; Nori et al. 2021; Quiroz et al. 2021; Valladares-Faúndez et al. 2021). Nevertheless, we believe the broad patterns identified here are biologically real and that future advances into the systematics and taxonomy of Liolaemidae will refine, rather than reshape, these results.

We find that all areas of importance in terms of richness and endemism are on Andean regions, highlighting the importance of this mountain chain for the diversity of Liolaemidae, and more broadly reinforcing the importance of mountains as centers of biodiversity. Reaching and sampling a lot of the Andean areas where liolaemids live is logistically and financially challenging, sometimes requiring high levels of fitness and mountaineering skills to perform multi-day expeditions to remote areas. To add to this, many of these areas are used for mining or hydroelectric power generation and access is highly restricted. There are likely many populations and even species to be discovered in these inaccessible areas, and thus a full picture of the biodiversity of the Andes lies in a relatively distant future. Nonetheless, knowledge on the biogeography and evolution of this region is moving faster than ever, emphasizing the importance of the Andes for the biodiversity of South America.

References

Abdala CS, Quinteros AS (2014) Los últimos 30 años de estudios de la familia de lagartijas más diversa de Argentina. Actualización taxonómica y sistemática de Liolaemidae. Cuadernos de Herpetología 28:55–82

Abdala CS, Quinteros AS, Semhan RV, Arroyo ALB, Schulte J, Paz MM, Ruiz-Monachesi MR, Laspiur A, Aguilar-Kirigin AJ, Gutiérrez-Poblete R, Faundez PV, Valdés J, Portelli S, Cruz RS, Aparicio J, Garcia N, Langsroth R (2020) Unravelling interspecific relationships among highland lizards: first phylogenetic hypothesis using total evidence of the Liolaemus montanus group (Iguania: Liolaemidae). Zool J Linnean Soc 189:349–377

Abdala CS, Laspiur A, Scrocchi G, Semhan RV, Lobo F, Valladares P (eds) (2021a) Liolaemidae: Sistemática, distribución e historia natural de una de las familias de vertebrados más diversas del cono sur de Sudamérica, vol 1. RIL Editores – Universidad de Tarapacá, Santiago

Abdala CS, Laspiur A, Scrocchi G, Semhan RV, Lobo F, Valladares P (eds) (2021b) Liolaemidae: Sistemática, distribución e historia natural de una de las familias de vertebrados más diversas del cono sur de Sudamérica, vol 2. RIL Editores – Universidad de Tarapacá, Santiago

Abdala CS, Chafrat PA, Chaparro JC, Procheret IE, Valdes J, Lannutti V, Perez L, Quinteros S (2023) A new species of Liolaemus (Iguania: Liolaemidae) from the hot deserts of northern Patagonia, Argentina. Eur J Taxon 890:136–164

Aguilar C, Wood PL, Cusi JC, Guzman A, Huari F, Lundberg M, Mortensen E, Ramirez C, Robles D, Suarez J, Ticona A, Vargas VJ, Venegas PJ, Sites JW (2013) Integrative taxonomy and preliminary assessment of species limits in the Liolaemus walkeri complex (Squamata, Liolaemidae) with descriptions of three new species from Peru. ZooKeys 364:47–91

Aguilar-Puntriano C, Salazar EB (2021) Comentarios y adiciones sobre especies recientemente descritas del grupo Liolaemus montanus procedentes de Perú. Cuadernos de Herpetología 35:91–100

Aguilar-Puntriano C, Avila LJ, la Riva ID, Johnson L, Morando M, Troncoso-Palacios J, Wood PL Jr, Sites JW Jr (2018) The shadow of the past: convergence of young and old South American desert lizards as measured by head shape traits. Ecol Evol 67:14–11

Alencar L, Schwery O, Gade M, Domínguez-Guerrero S, Tarimo E, Bodensteiner B, Uyeda J, Muñoz M (2024) Opportunity begets opportunity to drive macroevolutionary dynamics of a diverse lizard radiation. EcoEvoRxiv. https://doi.org/10.32942/x2cs4g

Almeida-Santos P, Militão CM, Nogueira-Costa P, Menezes VA, Rocha CFD (2015) Thermal ecology of five remaining populations of an endangered lizard (Liolaemus lutzae) in different Restinga habitats in Brazil. J Coast Conserv 19:1–9

Avila LJ, Morando M, Sites JW Jr (2006) Congeneric phylogeography: hypothesizing species limits and evolutionary processes in Patagonian lizards of the Liolaemus boulengeri group (Squamata: Liolaemini). Biol J Linn Soc 89:241–275

Avila LJ, Vrdoljak JE, Medina CD, Massini JG, Pérez CHF, Sites JWJ, Morando M (2021) A new species of Liolaemus (Reptilia: Squamata) of the Liolaemus capillitas clade (Squamata, Liolaemini, Liolaemus elongatus kriegi group) from sierra de Velasco, La Rioja Province, Argentina. Zootaxa 4903:194–216

Azócar DLM, Vanhooydonck B, Bonino MF, Perotti MG, Abdala CS, Schulte JA, Cruz FB (2013) Chasing the Patagonian sun: comparative thermal biology of Liolaemus lizards. Oecologia 171:773–788

Boschman LM (2021) Andean mountain building since the late cretaceous: a paleoelevation reconstruction. Earth-Sci Rev 220:103640

Boschman LM, Condamine FL (2022) Mountain radiations are not only rapid and recent: ancient diversification of South American frog and lizard families related to Paleogene Andean orogeny and Cenozoic climate variations. Glob Planet Chang 208:103704

Breitman MF, Avila LJ, Sites JW Jr, Morando M (2011) Lizards from the end of the world: phylogenetic relationships of the Liolaemus lineomaculatus section (Squamata: Iguania: Liolaemini). Mol Phylogenet Evol 59:364–376

Breitman MF, Avila LJ, Sites JW, Morando M (2012) How lizards survived blizzards: phylogeography of the Liolaemus lineomaculatus group (Liolaemidae) reveals multiple breaks and refugia in southern Patagonia and their concordance with other codistributed taxa. Mol Ecol 21:6068–6085

Cei JM (1979) In: Duellman WE (ed) The Patagonian herpetofauna, pp 309–339. The Museum of Natural History, The University of Kansas, Lawrence, Kansas

Cerdeña J, Farfán J, Quiroz AJ, Cerdeña J (2021) A high mountain lizard from Peru: the world's highest-altitude reptile. Herpetozoa 34:61–65

Chaparro JC, Quiroz AJ, Mamani L, Gutiérrez RC, Condori P, la Riva ID, Herrera-Juárez G, Cerdeña J, Arapa LP, Abdala CS (2020) An endemic new species of Andean lizard of the genus Liolaemus from southern Peru (Iguania: Liolaemidae) and its phylogenetic position. Amphib Reptile Conserv 14:47–63

Cianferoni F, Yáñez RP, Palma RE, Garín CF, Torres-Pérez F (2013) Deep divergences within Liolaemus nigroviridis (Squamata, Liolaemidae) lineages associated with sky islands in Central Chile. Zootaxa 3619:59–69

Crisp MD, Laffan S, Linder HP, Monro A (2001) Endemism in the Australian flora. J Biogeogr 28:183–198

Cruz FB, Fitzgerald LA, Espinoza RE, Schulte JA 2nd (2005) The importance of phylogenetic scale in tests of Bergmann's and Rapoport's rules: lessons from a clade of South American lizards. J Evol Biol 18:1559–1574

Cruz FB, Azócar DLM, Bonino MF, Schulte JA II, Abdala CS, Perotti MG (2014) Clima, distribución geográfica y viviparismo en especies de Liolaemus (Reptilia: Squamata): cuando las hipótesis se ponen a prueba. Ecosistemas 23:37–45

Daru BH, Elliott TL, Park DS, Davies TJ (2017) Understanding the processes underpinning patterns of phylogenetic regionalization. Trends Ecol Evol 32:845–860

Daru BH, Karunarathne P, Schliep K (2020) Phyloregion: R package for biogeographical regionalization and macroecology. Methods Ecol Evol 11:1483–1491

Díaz Gómez JM (2011) Estimating ancestral ranges: testing methods with a clade of Neotropical lizards (Iguania: Liolaemidae). PLoS One 6:e26412

Díaz-Gómez JM (2009) Historical biogeography of Phymaturus (Iguania: Liolaemidae) from Andean and patagonian South America. Zool Scr 38:1–7

Domínguez-Guerrero S, Esqu D, Burress E, Maciel-Mata CA, Alencar L, Munoz M (2024) Viviparity imparts a macroevolutionary signature of ecological opportunity in Liolaemus lizards. Res Sq. https://doi.org/10.21203/rs.3.rs-3293672/v1

Donoso-Barros R (1971) The genera Ctenoblepharis and Phrynosaura. Herpetol Rev 3:85

Donoso-Barros R (1972) Contribución al conocimiento del género Ctenoblepharis Tschudi y Phrynosaura Werner (Sauria, Iguanidae). Bol Soc Biol Concepc 44:129–134

Espinoza RE, Wiens JJ, Tracy CR (2004) Recurrent evolution of herbivory in small, cold-climate lizards: breaking the ecophysiological rules of reptilian herbivory. Proc Natl Acad Sci 101:16819–16824

Esquerré D, Brennan IG, Catullo RA, Torres-Pérez F, Keogh JS (2019a) How mountains shape biodiversity: the role of the Andes in biogeography, diversification, and reproductive biology in South America's most species-rich lizard radiation (Squamata: Liolaemidae). Evolution 73:214–230

Esquerré D, Ramírez-Alvarez D, Pavón-Vázquez CJ, Troncoso-Palacios J, Garín CF, Keogh JS, Leaché AD (2019b) Speciation across mountains: phylogenomics, species delimitation and taxonomy of the Liolaemus leopardinus clade (Squamata, Liolaemidae). Mol Phylogenet Evol 139:106524

Esquerré D, Keogh JS, Demangel D, Morando M, Avila LJ, Sites JW, Ferri-Yáñez F, Leaché AD (2022) Rapid radiation and rampant reticulation: phylogenomics of South American Liolaemus lizards. Syst Biol 71:286–300

Etheridge R (1995) Redescription of Ctenoblepharys adspersa Tschudi, 1845, and the taxonomy of Liolaeminae (Reptilia: Squamata: Tropiduridae). Am Mus Novit 3142:1–34

Etheridge R (2000) A review of lizards of the Liolaemus wiegmannii group (Squamata, Iguania, Tropiduridae), and a history of morphological change in the sand-dwelling species. Herpetol Monogr 14:293–352

Farr TG, Kobrick M (2000) Shuttle radar topography mission produces a wealth of data. Eos Trans Am Geophys Union 81:583–585

Fontanella FM, Feltrin N, Avila LJ, Sites JW Jr, Morando M (2012) Early stages of divergence: phylogeography, climate modeling, and morphological differentiation in the South American lizard Liolaemus petrophilus (Squamata: Liolaemidae). Ecol Evol 2:792–808

Fuentes ER, Jaksic FM (1979) Lizards and rodents: an explanation for their relative species diversity in Chile. Arch Biol Med Exp 12:179–190

García-Rodríguez A, Martínez PA, Oliveira BF, Velasco JA, Pyron RA, Costa GC (2021) Amphibian speciation rates support a general role of mountains as biodiversity pumps. Am Nat 198:E68–E79

Garzione CN, Hoke GD, Libarkin JC, Withers S, MacFadden B, Eiler J, Ghosh P, Mulch A (2008) Rise of the Andes. Science 320:1304–1307

Gómez JMD, Lobo F (2006) Historical biogeography of a clade of Liolaemus (Iguania: Liolaemidae) based on ancestral areas and dispersal-vicariance analysis (DIVA). Pap Avulsos Zool 46:261–274

González-Gutiérrez PA (2024) Corrección de la localidad tipo de Liolaemus rosenmanni Núñez y Navarro, 1992 (Squamata: Liolaemidae) y esclarecimiento de su real distribución geográfica en el norte de Chile. Cuad Herpetol 38:65–69

Grummer JA, Morando MM, Avila LJ, Sites JW Jr, Leaché AD (2018) Phylogenomic evidence for a recent and rapid radiation of lizards in the Patagonian Liolaemus fitzingerii species group. Mol Phylogenet Evol 125:243–254

Harmon LJ, Schulte JA, Larson A, Losos JB (2003) Tempo and mode of evolutionary radiation in iguanian lizards. Science 301:961–964

Hellmich WC (1951) On ecotypic and autotypic characters, a contribution to the knowledge of the evolution of the genus Liolaemus (Iguanidae). Evolution 5:359

Houston J, Hartley AJ (2003) The central Andean west-slope rainshadow and its potential contribution to the origin of hyper-aridity in the Atacama Desert. Int J Climatol 23:1453–1464

Jaksic FM (2022) Historical account and current ecological knowledge of the southernmost lizard in the world, Liolaemus magellanicus (Squamata: Liolaemidae). Rev Chil Hist Nat 95:7

Kacoliris FP, Williams JD, Arcaute C (2009) Home range size and overlap in Liolaemus multimaculatus (Squamata: Liolamidae) in pampean coastal dunes of Argentina. S Am J Herpetol 4:229

Lagomarsino LP, Condamine FL, Antonelli A, Mulch A, Davis CC (2016) The abiotic and biotic drivers of rapid diversification in Andean bellflowers (Campanulaceae). New Phytol 210:1430–1442

Laurent RF (1983) Contribución al conocimiento de la estructura taxonómica del género Liolaemus Wiegmann (Iguanidae). Boletín de la Asociación Herpetológica Argentina 1:16–18

Lobo F, Quinteros S (2005) A morphology-based phylogeny of Phymaturus (Iguania: Liolaemidae) with the description of four new species from Argentina. Pap Avulsos Zool 45:143–177

Lobo F, Barrasso DA, Hibbard T, Basso NG (2015) On the evolution and diversification of an Andean clade of reptiles: combining morphology and DNA sequences of the palluma group (Liolaemidae: Phymaturus). Zool J Linnean Soc 176:648–673

Marín AG, Olave M, Avila LJ, Sites JW Jr, Morando M (2018) Evidence of body size and shape stasis driven by selection in Patagonian lizards of the Phymaturus patagonicus clade (Squamata: Liolaemini). Mol Phylogenet Evol 129:226–241

Mella Ávila J, Marambio-Alfaro Y (2024) Reptiles of the Atacama region, northern Chile: localities, and representation in provinces, ecogeographic landscapes, vegetation formations and protected areas. Gayana 87: 108–122

Morando M, Avila LJ, Sites JW (2003) Sampling strategies for delimiting species: genes, individuals, and populations in the Liolaemus elongatus-kriegi complex (Squamata: Liolaemidae) in Andean-Patagonian South America. Syst Biol 52:159–185

Morando M, Avila LJ, Baker J, Sites JW Jr (2004) Phylogeny and phylogeography of the Liolaemus darwini complex (Squamata: Liolaemidae): evidence for introgression and incomplete lineage sorting. Evolution 58: 842–861

Morando M, Avila LJ, Turner CR, Sites JW Jr (2007) Molecular evidence for a species complex in the patagonian lizard Liolaemus bibronii and phylogeography of the closely related Liolaemus gracilis (Squamata: Liolaemini). Mol Phylogenet Evol 43:952–973

Morando M, Avila LJ, Perez CHF, Hawkins MA, Sites JW Jr (2013) A molecular phylogeny of the lizard genus Phymaturus (Squamata, Liolaemini): implications for species diversity and historical biogeography of southern South America. Mol Phylogenet Evol 66:694–714

Morando M, Olave M, Avila LJ, Sites JW Jr, Leaché AD (2020) Phylogenomic data resolve higher-level relationships within South American Liolaemus lizards. Mol Phylogenet Evol 147:106781

Morando M, Olave M, Sites JW Jr, Avila LJ (2021) Molecular phylogeny of the "salty" Liolaemus species (Reptilia: Squamata): anomalus group, endemic of the Argentina Monte Desert. Cuadernos de Herpetología 35:79–89

Nori J, Semhan R, Abdala CS, Rojas-Soto O (2021) Filling Linnean shortfalls increases endemicity patterns: conservation and biogeographical implications for the extreme case of Liolaemus (Liolaemidae, Squamata) species. Zool J Linn Soc 194:592–600

Núñez H, Yáñez J (1984) Ctenoblepharis erroneus nov. sp. de Iguanidae para la zona norte de Chile. Boletín del Museo Nacional de Historia Natural de Chile 40: 91–95

Ocampo M, Pincheira-Donoso D, Rios RS (2024) Patterns of morphological diversification are influenced by dietary evolution in a highly species-rich lizard radiation. Front Ecol Evol 12:1361799

Olave M, Martinez LE, Avila LJ, Sites JW, Morando M (2011) Evidence of hybridization in the Argentinean lizards Liolaemus gracilis and Liolaemus bibronii (Iguania: Liolaemini): an integrative approach based on genes and morphology. Mol Phylogenet Evol 61: 381–391

Olave M, Avila LJ, Sites JW Jr, Morando M (2015) Model-based approach to test hard polytomies in the Eulaemus clade of the most diverse South American lizard genus Liolaemus (Liolaemini, Squamata). Zool J Linnean Soc 174:169–184

Olave M, Avila LJ, Sites JW, Morando M (2018) Hybridization could be a common phenomenon within the highly diverse lizard genus Liolaemus. J Evol Biol 31:893–903

Olave M, Avila LJ, Sites JW, Morando M (2020) How important is it to consider lineage diversification heterogeneity in macroevolutionary studies? Lessons from the lizard family Liolaemidae. J Biogeogr 1:6–12

Olson DM, Dinerstein E, Wikramanayake ED, Burgess ND, Powell GVN, Underwood EC, D'amico JA, Itoua I, Strand HE, Morrison JC, Loucks CJ, Allnutt TF, Ricketts TH, Kura Y, Lamoreux JF, Wettengel WW, Hedao P, Kassem KR (2009) Terrestrial ecoregions of the world: a new map of life on earth. Bioscience 51:933–938

Pateiro-López B, Rodríguez-Casal A (2010) Generalizing the convex hull of a sample: the R package alphahull. J Stat Softw 34:1–28

Pebesma E (2018) Simple features for R: standardized support for spatial vector data. R J 10:439

Pebesma E, Bivand R (2023) Spatial data science, with applications in R. Chapman and Hall/CRC

Pereyra EA (1992) Sistematica y relaciones evolutivas de las especies de Phymaturus Gravenhorst, 1838 (Sauria-Liolaeminae). Thesis of Magister, Universidad de Chile

Pincheira-Donoso D, Hodgson DJ, Tregenza T (2008a) The evolution of body size under environmental gradients in ectotherms: why should Bergmann's rule apply to lizards? BMC Evol Biol 8:68

Pincheira-Donoso D, Scolaro JA, Sura P (2008b) A monographic catalogue on the systematics and phylogeny of the south American iguanian lizard family Liolaemidae (Squamata, Iguania). Zootaxa 1800:1–85

Pincheira-Donoso D, Tregenza T, Witt MJ, Hodgson DJ (2013) The evolution of viviparity opens opportunities for lizard radiation but drives it into a climatic cul-de-sac. Glob Ecol Biogeogr 22:857–867

Pincheira-Donoso D, Harvey LP, Ruta M (2015) What defines an adaptive radiation? Macroevolutionary diversification dynamics of an exceptionally species-rich continental lizard radiation. BMC Evol Biol 15:1–13

Pincheira-Donoso D, Jara M, Reaney A, Roa RG, Córdoba MS, Hodgson DJ (2017) Hypoxia and hypothermia as rival agents of selection driving the evolution of viviparity in lizards. Glob Ecol Biogeogr 26:1238–1246

Portelli SN, Quinteros AS (2018) Phylogeny, time divergence, and historical biogeography of the South American Liolaemus alticolor-bibronii group (Iguania: Liolaemidae). PeerJ 6:e4404–e4430

Portelli SN, Abdala CS, Schulte J, Gmez JMD, Fernndez LD, Quinteros AS (2022) Historical biogeographic reconstruction of the South American Liolaemus boulengeri group (Iguania: Liolaemidae). S Am J Herpetol 25:41–56

Pyron RA, Burbrink FT (2014) Early origin of viviparity and multiple reversions to oviparity in squamate reptiles. Ecol Lett 17:13–21

Quintero I, Jetz W (2018) Global elevational diversity and diversification of birds. Nature 10:e1001292

Quinteros AS, Ruiz-Monachesi M, Abdala CS (2020) Solving the Liolaemus bibronii puzzle, an integrative taxonomy approach: redescription of L. bibronii and description of three new species (Iguania: Liolaemidae). Zool J Linnean Soc 189:315–348

Quiroz AJ, Huamaní-Valderrama L, Gutiérrez RC, Aguilar-Kirigin AJ, López-Tejeda E, Lazo-Rivera A, Huanca-Mamani W, Valladares-Faúndez P, Morrone JJ, Cerdeña J, Chaparro JC, Abdala CS (2021) An endemic and endangered new species of the lizard Liolaemus montanus group from southwestern Peru (Iguania: Liolaemidae), with a key for the species of the L. reichei clade. Zool Stud 60:e23

Rabosky ARD, Cox CL, Rabosky DL, Title PO, Holmes IA, Feldman A, McGuire JA (2016) Coral snakes predict the evolution of mimicry across New World snakes. Nat Commun 7:11484

Rangel TF, Edwards NR, Holden PB, Diniz-Filho JAF, Gosling WD, Coelho MTP, Cassemiro FAS, Rahbek C, Colwell RK (2018) Modeling the ecology and evolution of biodiversity: biogeographical cradles, museums, and graves. Science 361:eaar5452

Rocha CFD, Siqueira CD, Ariani CV (2009) The endemic and threatened lizard Liolaemus lutzae (Squamata: Liolaemidae): current geographic distribution and areas of occurrence with estimated population densities. Zoologia 26:454–460

Rosauer D, Laffan SF, Crisp MD, Donnellan SC, Cook LG (2009) Phylogenetic endemism: a new approach for identifying geographical concentrations of evolutionary history. Mol Ecol 18:4061–4072

Rowe KC, Achmadi AS, Fabre P, Schenk JJ, Steppan SJ, Esselstyn JA (2019) Oceanic islands of Wallacea as a source for dispersal and diversification of murine rodents. J Biogeogr 46:2752–2768

Ruiz MS, Portelli SN, Hibbard TN, Quinteros AS (2020) Discovering the biogeographic history using predefined areas and explicit geographical data in the South American Liolaemus elongatus group (Iguania: Liolaemidae). Herpetol J 30:53–68

Ruiz S, Chafrat P, Quipildor M, Valdecantos S, Lobo F (2023) Description of a new species of the Liolaemus elongatus group (Squamata: Iguania) through integrative taxonomy. Cuad Herpetol 37:105–129

Sánchez KI, Huesa EGD, Breitman MF, Avila LJ, Sites JW Jr, Morando M (2023) Complex patterns of diversification in the gray zone of speciation: model-based approaches applied to Patagonian Liolaemid lizards (Squamata: Liolaemus kingii clade). Syst Biol 72. https://doi.org/10.1093/sysbio/syad019

Sánchez KI, Recknagel H, Elmer KR, Avila LJ, Morando M (2024) Tracing evolutionary trajectories in the presence of gene flow in South American temperate lizards (Squamata: Liolaemus kingii group). Evolution 78: 716–733

Schulte JA, Macey JR, Espinoza RE, Larson A (2000) Phylogenetic relationships in the iguanid lizard genus Liolaemus: multiple origins of viviparous reproduction and evidence for recurring Andean vicariance and dispersal. Biol J Linn Soc 69:75–102

Skeels A, Esquerré D, Lipsky D, Pellissier L, Boschman LM (2023) Elevational Goldilocks zone underlies the exceptional diversity of a large lizard radiation (Liolaemus; Liolaemidae). Evolution 77:2672–2686

Tulli MJ, Cruz FB, Kohlsdorf T, Abdala V (2016) When a general morphology allows many habitat uses. Integr Zool 11:483–499

Valladares-Faúndez P, León PF, Chipana CJ, Guzmán MN, Ignacio-Apaza J, Musaja CC, Langstroth R, Aguilar-Kirigin A, Gutierrez RC, Abdala CS (2021) A new lizard of the Liolaemus montanus group that inhabits the hyperarid desert of southern Peru. Amphib Reptile Conserv 15:10–22

Vera-Escalona I, D'Elía G, Gouin N, Fontanella FM, Muñoz-Mendoza C, Sites JW, Victoriano PF (2012) Lizards on ice: evidence for multiple refugia in Liolaemus pictus (Liolaemidae) during the last glacial maximum in the southern Andean beech forests. PLoS One 7:e48358

Victoriano PF, Ortiz JC, Benavides E, Adams BJ, Sites JW Jr (2008) Comparative phylogeography of codistributed species of Chilean Liolaemus (Squamata: Tropiduridae) from the central-southern Andean range. Mol Ecol 17:2397–2416

Villamil J, Avila LJ, Morando M, Sites JW Jr, Leaché AD, Maneyro R, Camargo A (2019) Coalescent-based species delimitation in the sand lizards of the Liolaemus wiegmannii complex (Squamata: Liolaemidae). Mol Phylogenet Evol 138:89–101

Wallis GP, Waters JM, Upton P, Craw D (2016) Transverse alpine speciation driven by glaciation. Trends Ecol Evol 31:916–926

Young-Downey AR (1998) Phylogenetic studies on Liolaemus (Suaria: Tropiduridae): an interpretation based on molecular data and a biochemical test of biogeographic hypothesis. University of Miami

Zeh JA, Zeh DW (2008) Viviparity-driven conflict: more to speciation than meets the fly. Ann N Y Acad Sci 1133:126–148

Zimin A, Zimin SV, Shine R, Avila L, Bauer A, Böhm M, Brown R, Barki G, Caetano GHO, Herrera FC, Chapple DG, Chirio L, Colli GR, Doan TM, Glaw F, Grismer LL, Itescu Y, Kraus F, LeBreton M, Martins M, Morando M, Murali G, Nagy ZT, Novosolov M, Oliver P, Passos P, Pauwels OSG, Pincheira-Donoso D, Ribeiro-Junior MA, Shea G, Tingley R, Torres-Carvajal O, Trape J, Uetz P, Wagner P, Roll U, Meiri S (2022) A global analysis of viviparity in squamates highlights its prevalence in cold climates. Glob Ecol Biogeogr 31. https://doi.org/10.1111/geb.13598

The Climatic Niche Evolution of South American Squamates (Reptilia: Squamata)

Julián A. Velasco, Juan D. Vásquez-Restrepo, and Jhan C. Salazar

1 Introduction

Understanding the role of ecological and evolutionary processes in the assembly, generation, and maintenance of mountain biodiversity is a central topic in macroecology and biogeography (Hoorn et al. 2018; Rahbek et al. 2019a, b). Recent studies have made progress by integrating geophysical framework to explain how mountains have accumulated biodiversity (Hoorn et al. 2018; Dolby et al. 2022; Dolby 2021). At least two hypotheses—the "diversification rate" and the "time-for-speciation" effect—have been proposed to explain how regional mountain biotas were assembled through time (Hutter et al. 2017). Both hypotheses suggest that the regional assembly process is a combination of high rates of in situ diversification and repeated colonization events from adjacent lowland areas (Hoorn et al. 2018; Goldberg et al. 2011; Perrigo et al. 2020). However, the impact of these historical processes affecting mountain diversity beyond taxonomic facets (e.g., niche trait disparity) remains underexplored in the literature (Moore and Donoghue 2009; Cerezer et al. 2023).

The uplift of South America's Andean region had driven extensive accumulation of old and new lineages of plants and animals (Hughes and Eastwood 2006; Weir 2006; Santos et al. 2009; Hutter et al. 2017). As a result of complex past geological processes, the Andes are recognized as one of the most biodiverse regions in the world (Hutter et al. 2017; Rahbek et al. 2019a, b; García-Rodríguez et al. 2021; Pérez-Escobar et al. 2022). Increasing evidence suggests that exceptional lineage diversification, often identified as adaptive radiations, was triggered by the Andean uplift in the last 15 mya in several groups (Hughes and Eastwood 2006; Weir 2006; Santos et al. 2009; Hutter et al. 2017; Tribble 2024). These rapid radiations are associated with changes in topography and past climatic change promoted different modes of speciation and niche evolution (Wiens 2004; Hua and Wiens 2013; Hutter et al. 2017). For instance, Hutter et al. (2017) found evidence of increasing species diversification and elevational shifts for several frog lineages across South America and extensively the Andes. Similarly, Meseguer et al. (2022) found that many Neotropical taxa exhibit a macroevolutionary signature of expansion over time with increasing diversification rates likely triggered by past climatic changes.

J. A. Velasco (✉)
Instituto de Ciencias de la Atmósfera y Cambio Climático, Universidad Nacional Autónoma de México, Mexico city, Mexico
e-mail: javelasco@atmosfera.unam.mx

J. D. Vásquez-Restrepo
Department of Ecology and Evolutionary Biology, Princeton University, Princeton, USA
e-mail: dvasquez@princeton.edu

J. C. Salazar
Department of Biology and Biomedical Sciences, Washington University, St. Louis, MO, USA
e-mail: jhancsalazar@wustl.edu

J. N. Urbina-Cardona et al. (eds.), *Andean Herpetofauna*, Sustainable Development Goals Series,
https://doi.org/10.1007/978-3-032-00074-3_6

Based on these previous studies, it is possible to hypothesize how species might either track ecological niche requirements or adapt to new regimes as mountains emerge and create new ecological spaces. For instance, as the Andean mountains rose lineages that colonize higher elevations would need to tolerate colder temperatures through local adaptation or phenotypic plasticity. Accordingly, independent colonization of mountains by unrelated lineages can be associated with repeated climatic niche evolution, particularly toward lower thermal limits. Similarly, lineages moving from Andean mountains to lowlands would need to tolerate warmer temperatures, again through local adaptation or phenotypic plasticity. In other words, the recurrent biogeographic shifts across elevations, whether upward or downward, exhibited by different lineages during its evolutionary history (Santos et al. 2009; Antonelli et al. 2018) could be coupled with repeated climatic niche shifts. However, it is still unknown how frequent these biogeographic transitions are for many clades and whether they are consistently associated with evolutionary shifts in climatic niches.

The general association between biogeographic movements and trait evolution has been explored in several studies (Moore and Donoghue 2009; Hutter et al. 2013). A decoupling between biogeographic movements and climatic niche shifts would align with a scenario of phylogenetic niche conservatism (PNC; (Wiens and Graham 2005)). However, the formal tests for the PNC hypothesis have varied, and some studies continue to use metrics of phylogenetic signal to evaluate PNC scenarios across different clades or regions (Saban et al. 2023). It is important to clarify that different evolutionary processes can result in similar values of phylogenetic signal, and there is no direct evidence that high phylogenetic signal necessarily implies evolutionary conservatism (Losos 2008; Revell et al. 2008; Münkemüller et al. 2015). Furthermore, the invariance of a trait within clades (i.e., the lack of phylogenetic signal) may be interpreted as evidence of strong conservatism and support a strict PNC scenario (Losos 2008; Wiens 2008; Münkemüller et al. 2015).

Some earlier studies provided evidence for a weak niche conservatism as species colonized highlands in the Andes region. For instance, Hutter et al. (2013) found an association between changes in elevational distributions and rates of climatic niche evolution for Andean glassfrogs, suggesting the absence of PNC in this clade. However, these studies did not examine whether recurrent biogeographic transitions between regions (lowlands vs. highlands) were coupled with shifts in climatic niche evolution. Therefore, it can be expected that colonization of the Andean highlands might be associated with a shift in rates of climatic niche evolution. Species may need to evolve quickly to colder environments as they move to higher elevations. This potential coupling between elevational transitions and shifts in evolutionary rates of climatic niche supports a niche divergence scenario, where species colonize regions with different environmental conditions, leading to trait niche divergence as an expected outcome. Investigating these biogeographic and niche shift associations may provide valuable insights into how unrelated taxa evolved (convergent) traits to tolerate colder or warmer temperatures throughout their biogeographic history, which often involved recurrent past movements across elevations.

In this chapter, we explore the potential association between biogeographical transitions (from lowlands to highlands and vice versa) and shifts in tempo and mode of climatic niche evolution across diverse clades of Squamate reptiles in South America. We compiled an extensive dataset that includes phylogenetic, geographical, and climatic information for 870 species. We classified species as mountain-dwelling or non-mountain-dwelling using a recent global mountain inventory based on topographic roughness. Additionally, we quantified the number of transitions into and out of the Andean mountains using stochastic character mapping reconstructions and implemented evolutionary models to test for differences in tempo and mode of climatic niche evolution between lowland and mountain species across the phylogeny.

2 Materials and Methods

2.1 Data Compilation

We obtained distributional information for species occurring in South America from the International Union for Conservation of Nature range maps (IUCN, https://www.iucnredlist.org) and the Global Biodiversity Information Facility (GBIF, https://www.gbif.org). Due to the lack of clear and unambiguous definition in the literature regarding what a mountain is, we used the most recent inventory of the world's mountains (Snethlage et al. 2022) to classify species as Andean mountain-dwelling (hereafter referred to as mountain) and non-Andean-mountain-dwelling (hereafter referred to as lowland). Specifically, we adopted an arbitrary criterion of at least a 15% overlap between the species' range and the Andean mountain polygons to consider a species as an Andean mountain dwelling. In total, we compiled a list of 1020 species based on range maps.

We obtained occurrence records for species from the Global Biodiversity Information Facility (GBIF; doi: 10.15468/dl.ut6jzr) using the R package rgbif v.3.7.9 (Chamberlain et al. 2017). We then cleaned the records using functions from the R package CoordinateCleaner (Zizka et al. 2019). Specifically, we identified and removed common errors, such as geographic outliers, zero or identical coordinates, and records near institutions or centroids to ensure higher-quality occurrence data (Zizka et al. 2019). Subsequently, we filtered out occurrence records using species range maps from IUCN retaining a total of 261,918 records for 870 species occurring in South America. We did not conduct spatial thinning filtering processes because we are not using species distribution models, which are very sensitive to bias sampling and uneven spatial distribution. Instead, we estimated elevational ranges based on the entire occurrence dataset, including outlier occurrence records, to capture distributional limits more accurately. Furthermore, for each occurrence record, we extracted information of elevation and three bioclimatic variables: annual mean temperature (AMT), the maximum temperature of the warmest month (maxTWM), and the minimum temperature of the coldest month (minTCM), available in the Chelsa database (Brun et al. 2022). The spatial resolution of raster layers was approximately 1 km^2 in each grid cell. These bioclimatic variables will serve as proxies for thermal preferences of each species (Qu and Wiens 2020). Accordingly, we averaged the values of each bioclimatic variable for each species. Finally, we calculated each species' elevational and latitudinal midpoints as the difference between maximum and minimum values, divided by two according to standard macroecological protocols (Rohde et al. 1993).

2.2 Phylogenetic Comparative Analyses

We estimated the ancestral character states for a binary character representing mountain (1) vs. lowland (0) for South American squamate species, and the number of transitions between each character state, using a stochastic character mapping approach. This was conducted with a pruned Squamata phylogenetic tree for South American species. Stochastic character mapping was implemented using the make.simmap function under an equal-rates model in the R package phytools v.2.1.1 (Revell 2024). To incorporate topological uncertainty in the reconstructions (Rangel et al. 2015), we used 100 trees from the posterior distribution derived from the phylogeny presented by Tonini et al. (2016). Next, we fitted several evolutionary models to evaluate the tempo and mode of climatic niche evolution across the South American species tree. Specifically, we applied climatic niche data to a set of Brownian-Motion (BM) and Ornstein-Uhlenbeck (OU) models (Beaulieu et al. 2012) available in the R package OUwie v.2.11 (Beaulieu and O'Meara 2023) to assess differences in evolutionary rates (σ^2) and the rate of adaptation to an optimal state (α) between character states (i.e., lowland vs. mountain). A key difference between the BM and OU models is that the BM model describes purely random trait evolution, while the OU model incorporates both random fluctuations and a deterministic tendency toward a preferred trait value (Blomberg et al. 2020), making it more suitable for modeling scenarios involving stabilizing selection. We compared all models using the Akaike Information Criterion weights (Burnham and Anderson 2002).

3 Results

The entire assemblage of lizards and snakes occurring in the mountain regions in South America comprises various clades across the entire Squamate phylogeny (Fig. 1) with some particular exceptional radiations within certain genera, such as *Anolis*, *Liolaemus*, and *Atractus* (Fig. 2). These three clades exhibit a considerable species richness in both lowland and highland regions, suggesting that they have evolved repeatedly in warm and cold environments.

The three bioclimatic variables examined display similar distributions across elevations and latitudes (Fig. 3). The lowest annual mean temperature (AMT) and minimum temperature of the coldest month (minTCM) values were found in species occurring in southern latitudes (greater than 40° in latitudinal midpoint). Notably we did not observe a clear trend of low or high values for these bioclimatic variables across the elevational distribution of species (Fig. 3). When we analyzed the variation of these variables by discriminating between species assigned to mountains or lowlands, we found substantial variation both within and between families (Fig. 4). Each bioclimatic variable exhibits a different distribution pattern across mountains and lowlands among different families. Clades with a greater number of species in mountain regions, such as Liolaemidae and Anolidae, showed lower values for all variables in mountain areas compared to lowland areas (Fig. 4).

Across the squamate radiation in South America, transitions from mountain to lowland areas were more frequent than the reverse (Fig. 5). We estimated an average of 300 transitions from mountains to lowlands, contrasting with an average of 260 transitions from lowlands to mountains. These repeated shifts between lowland and mountain environments suggest that squamate lineages can adapt quickly to both cold and warm environments.

The three climatic variables examined demonstrated a better fit to models that allow for variation in parameters across different states (lowland vs. mountain; Table 1). For AMT and minTCM, the model with the highest Akaike weight was a Brownian motion model that allows the rate to vary across the tree for lowland and highland states. In contrast, for the maximum temperature of the warmest month (maxTWM), the models with higher Akaike weights were an Ornstein-Uhlenbeck model (OUM) with different means and a single α and σ^2 across states, as well as an Ornstein-Uhlenbeck model (OUMV) with different means and σ^2 across states. Finally, we found a higher evolutionary rate for AMT and minTCM in the mountain lineages compared to lowland lineages. However, there was considerable variation and overlap in both cases when accounting for topological uncertainty (Fig. 6). In contrast, we did not observe any differences in evolutionary rates for maxTWM.

4 Discussion

Our analyses of biogeographic transitions and climatic niche evolution in South American lizards and snakes provide compelling evidence for recurrent climatic niche evolution associated with highland colonization. Notably, the differences in tempo of climatic niche evolution between highland and lowland lineages indicate that thermal adaptations are linked to movements across biogeographic areas. In addition, the high frequency of biogeographic movements at high elevations appears to be accompanied by a repeated evolution of the minimum temperature of the coldest month, serving as a proxy for cold tolerance. These findings suggested that movement across biogeographic areas, particularly the colonization of mountain regions like the Andes, not only acts as a catalyst for an increased of species diversity (Moore and Donoghue 2009; Lagomarsino et al. 2016) but can also drive rapid morphological changes (Cerezer et al. 2023) and ecological niche shifts (this study).

The uplift of the Andean region has served as a catalyst for species diversification across several clades, influencing shifts in the tempo of trait

Fig. 1 Phylogenetic position of South American lizard and snake species inhabiting mountain regions (blue color) in South America across the entire Squamate tree

evolution, including body size (Cerezer et al. 2023). According to our results, Andean uplift likely also promoted climatic niche evolution, particularly along the thermal niche axis. Notably, we observed the fastest rates of niche evolution for the minimum temperature of the coldest month—a key indicator of cold tolerance—in mountain lineages. This supports our hypothesis that there is an association between highland colonization and the evolution of thermal tolerances toward cold temperatures (Muñoz et al. 2014). In particular, Muñoz et al. (2014) demonstrated that cold tolerance evolved more rapidly than heat tolerance in *Anolis* lizards along an elevation gradient in Hispaniola. Diurnal *Anolis* that inhabit in colder environments may be unable to engage in thermoregulatory behavior at night, exposing them to selection pressures for colder tolerance thereby increasing the lability of this trait. We hypothesize that diurnal species in highlands environments may exhibit varying efficiency of degrees of thermoregulatory behaviors to

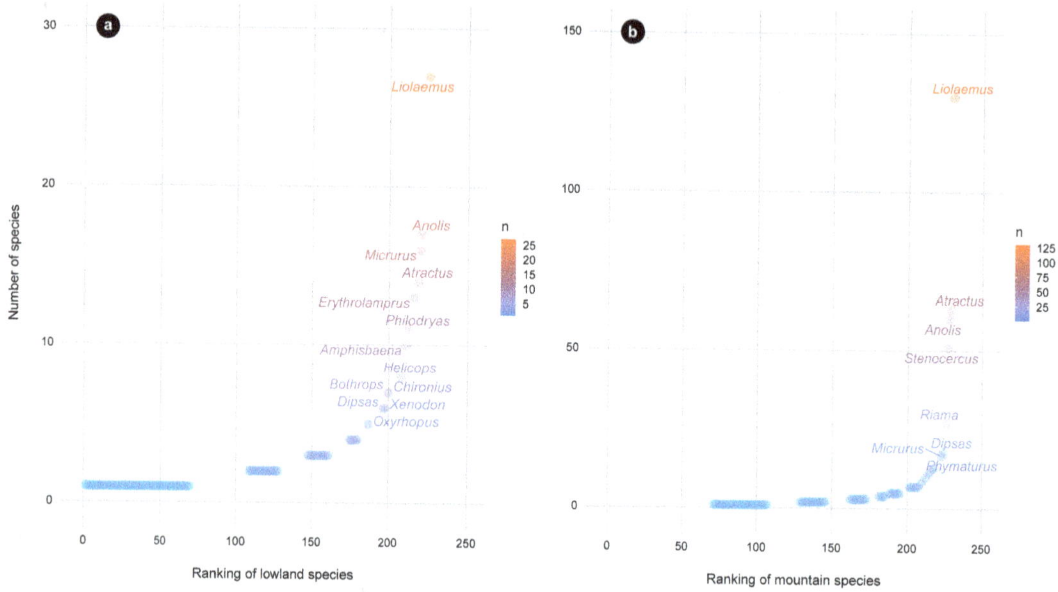

Fig. 2 Ranking of clades according to species richness inhabiting lowland (**a**) and mountain (**b**) regions in South America. This figure is color-blind friendly

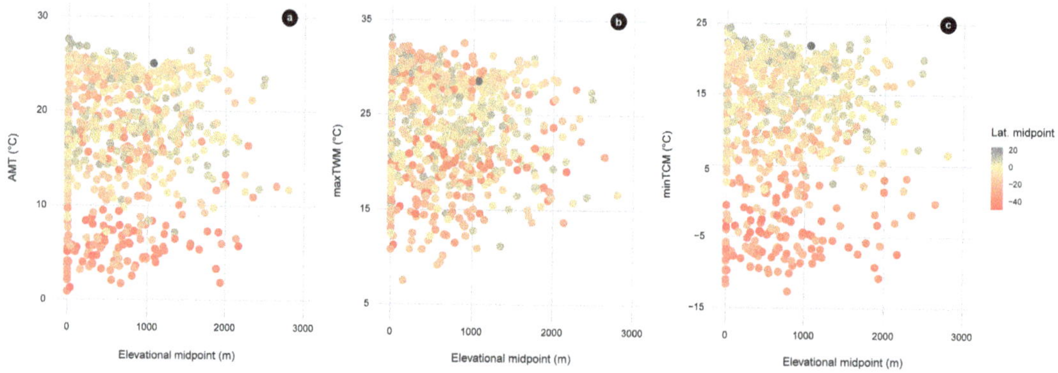

Fig. 3 Scatterplots of three bioclimatic variables for each species examined in this chapter across the elevational and latitudinal midpoints. Annual mean temperature (**a**), maximum temperature of the warmest month (**b**), minimum temperature of the coldest month (**c**). This figure is color-blind friendly

maintain stable body temperatures. In contrast, many Squamate species, particularly snakes, tend to be nocturnal, which may lead to a non-strict thermoconformer behavior facilitated by reduced thermal complexity at night (Dubiner et al. 2024). It is known that diurnal species exhibit higher body temperatures than nocturnal species (Meiri et al. 2013; Moreira et al. 2021). However, specific variation in thermoregulatory behaviors and thermal physiological traits across Squamate clades, particularly across elevational gradients, remains largely unexplored. Previous research have reported a considerable variation in behavioral thermoregulation among 69 Amazonian lizard species (Diele-Viegas et al. 2018). These findings highlight the challenges of categorizing species into discrete thermoregulatory strategy, especially for lowland species (Diele-Viegas et al. 2018). More research is needed to understand how physiological traits

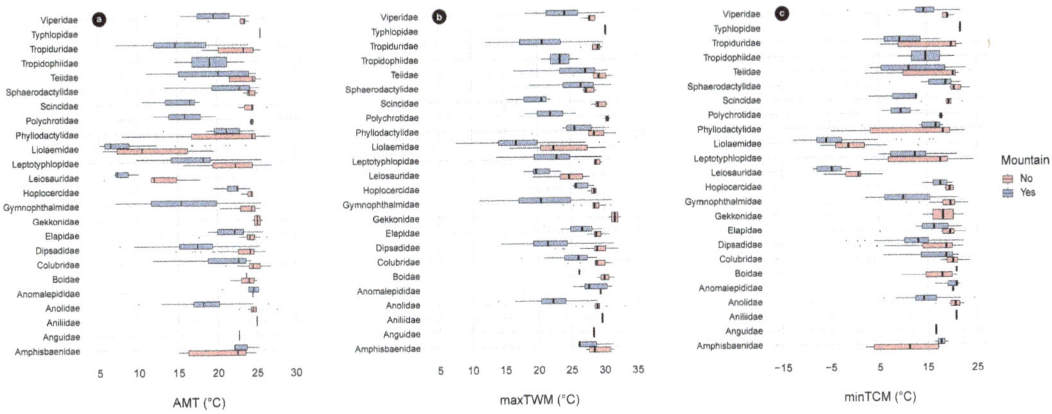

Fig. 4 Boxplots of bioclimatic variables across Squamate families assigned to mountain (0) or lowlands (1). Annual mean temperature (**a**), maximum temperature of the warmest month (**b**), minimum temperature of the coldest month (**c**). This figure is color-blind friendly

vary across day-night activity and elevations, providing further insights into the relationship between highland colonization and the evolution of cold tolerance in Squamates.

The maximum temperature of the warmest month, which serves as a proxy of heat tolerance, did not exhibit differences in evolutionary rates between lowland and mountain lineages. This finding supports previous research indicating that heat tolerance is less labile or exhibits stronger stasis than cold tolerance (Bennett et al. 2021). The absence of a correlation between maximum temperature and elevations across latitudes (Fig. 3) suggests that lizards and snakes are likely operating near their upper thermal limits, regardless of their geographical location. These patterns indicate a strong niche conservatism concerning upper thermal limits, a trend observed in both marine and terrestrial taxa across various latitudes (Sunday et al. 2011). Recent studies have shown that lizards can adapt to increasing temperatures, particularly in urban environments (Campbell-Staton et al. 2020; Campbell-Staton et al. 2021). For example, *Anolis cristatellus*, a common lizard in Puerto Rico that inhabits lowland areas, has demonstrated the ability to adapt to higher temperatures in urban environments multiple times due to genetic changes in a protein synthesis gene linked to enhance resilience under thermal stress (Campbell-Staton et al. 2020). It would be worthwhile to investigate whether squamate species have evolved upper thermal tolerances in regions that have experienced significant increases in maximum temperatures since the pre-industrial period. As projected future temperatures escalate under various climate change scenarios, it is expected that these species may be at greater risk. However, the spatial variability of these projected maximum temperatures changes and how species may be exposed based on their thermal tolerances remains largely unknown. Some studies using realized climatic niche variables, as applied here, have indicated that thermal exposure is expected to be higher in tropical regions, particularly for reptiles (Trisos et al. 2020).

The bioclimatic variables employed in this study have been widely used to infer the ecological niche (Hutchinson 1957; Holt 2009) and to predict the species geographical distribution (Peterson et al. 2011). These variables effectively capture the distributional limits of taxa and correlate well with thermal physiology traits (Qu and Wiens 2020). For example, Rolland et al. (2018) found that rates of climatic niche evolution are higher in endotherms than in ectotherms, suggesting that a narrow climatic tolerance may constrain evolutionary pace. Similarly, Qu and Wiens (2020) demonstrated that rates of climatic niche evolution derived from bioclimatic

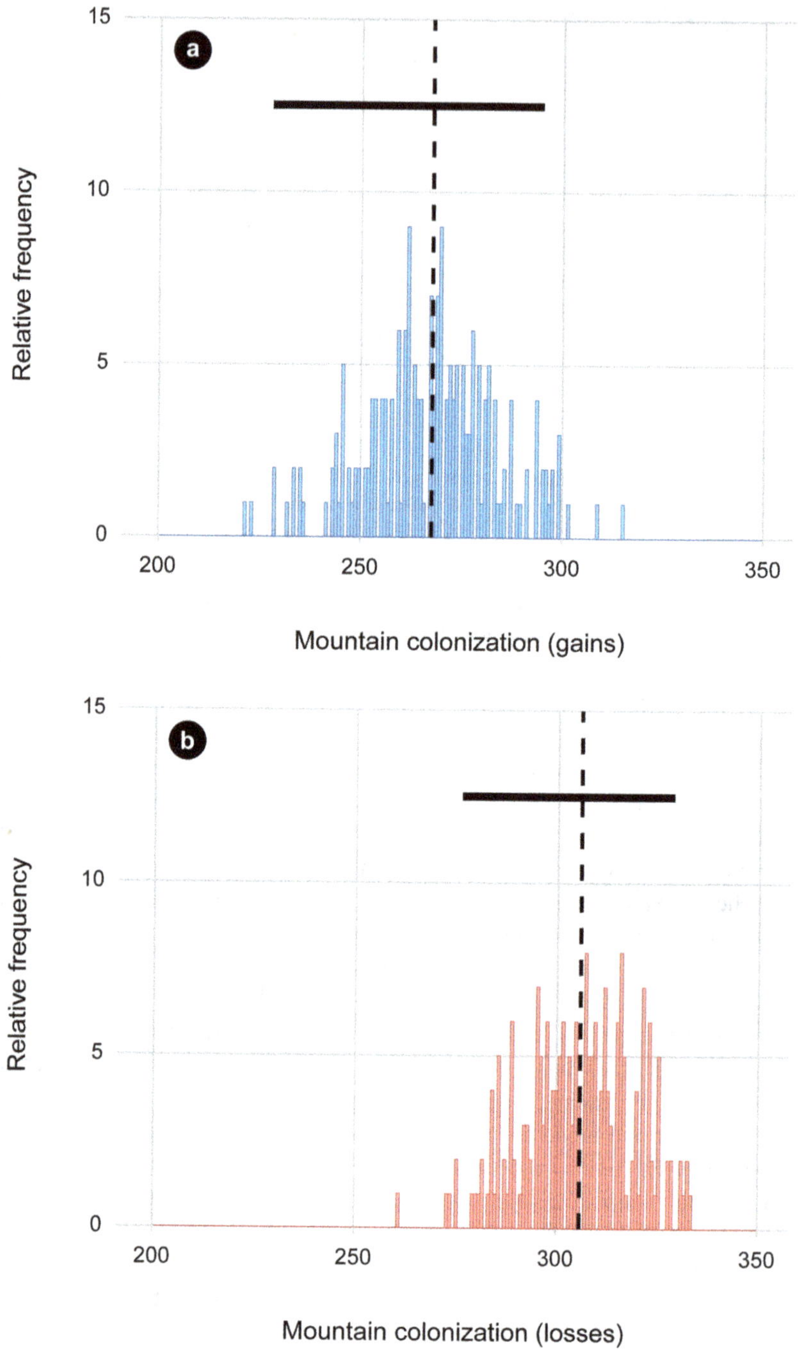

Fig. 5 Evolutionary transitions into (**a**) and out (**b**) of the mountain regions in South America across the Squamate tree. This figure is color-blind friendly

variables are generally higher than those derived from thermal physiological traits (e.g., CTmin and CTmax); however, both estimates show good correlation across different taxa. Using these proxy variables allows us to gain insights into potential pathways for species in cold or warm environments through a combination of spatial and phylogenetic comparative methods. This approach is particularly relevant given the absence of thermal physiology data for many species, including squamates (Bennett et al. 2018). However, we advocate for further research

Table 1 Evolutionary model selection for three bioclimatic variables fitted using a Brownian motion and Ornstein-Uhlenbeck model set

Variable	Model	AICc	Δ AICc	Akaike weights
AMT	BM1	4562.89	4.92	0.03
	BMS	*4557.97*	*0.00*	*0.35*
	OU	4564.91	6.93	0.01
	OUM	4564.08	6.11	0.02
	OUMV	*4558.44*	*0.47*	*0.28*
	OUMA	4558.65	1.67	0.15
	OUMVA	4559.58	1.61	0.16
maxTWM	BM1	4361.59	18.19	0.00
	BMS	4362.81	19.41	0.00
	OU	4362.98	19.59	0.00
	OUM	*4343.46*	*0.07*	*0.40*
	OUMV	*4343.39*	*0.00*	*0.41*
	OUMA	4345.49	2.09	0.14
	OUMVA	4347.48	4.02	0.05
minTCM	BM1	4965.81	15.32	0.39
	BMS	*4950.49*	*0.00*	*0.84*
	OU	4967.83	17.33	0.00
	OUM	4969.84	19.35	0.00
	OUMV	4954.54	4.05	0.11
	OUMA	4960.29	9.80	0.01
	OUMVA	4956.24	5.75	0.05

BM1 single-rate Brownian motion, *BMS* variable-rate Brownian motion, *OU* Ornstein-Uhlenbeck with a single optimum, *OUM* Ornstein-Uhlenbeck model with different state means but with a single α and σ^2, *OUMV* Ornstein-Uhlenbeck with different state means and σ^2, *OUMA* Ornstein-Uhlenbeck with different state means and α, *OUMVA* Ornstein-Uhlenbeck with different state means, α, and σ^2

on thermal physiology across elevation gradients in squamates to validate these findings, especially considering the limitations of using coarse-grain climatic information.

There is currently no consensus on the definition of mountain areas worldwide, leading to conflicting results in various studies. In our research, we classify species using a spatially explicit approach based on the percentage of overlap between species' ranges and a mountain inventory derived from terrain features in a DEM-layer (Snethlage et al. 2022). This methodology has been succesfully applied to delineate mountain amphibian assemblages worldwide (García-Rodríguez et al. 2021) and offers the advantage of defining a mountain areas based on topographical complexity rather than solely on elevational criteria. For instance, Frishkoff et al. (2022) used a threshold of 700 meter elevation to classify mountain species in Hispaniola. Additional studies are needed to assess how different criteria for defining mountain regions can impact estimates of biogeographic movements and climatic niche evolution.

Our results reinforce the idea that biogeographic transitions can drive changes in phenotypes, particularly in climatic niche traits, across various Squamate lineages. We propose that the relationship between biogeographic movements and phenotypic evolution is facilitated by ecological opportunity, which is a prerequisite for adaptive radiation phenomena (Losos et al. 2010; Stroud and Losos 2016). Ecological opportunity appears when new areas are available for species colonization (Stroud and Losos 2016). As mountain uplift, these newly accessible areas tend to be colder and often contain a depauperate biota, including limited vegetation. Consequently, lineages may successfully colonize these areas if they already have the

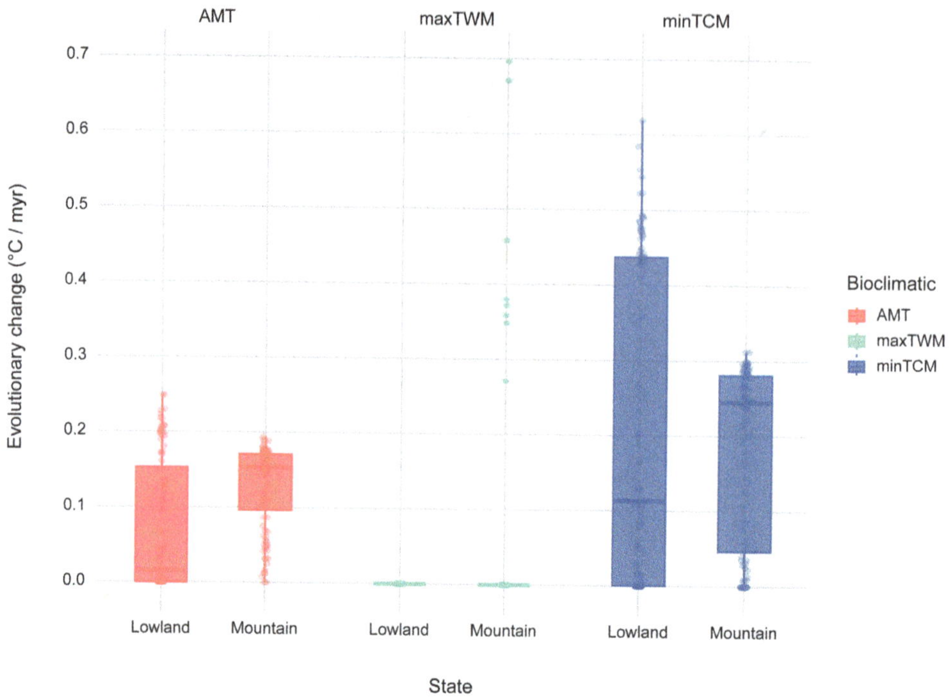

Fig. 6 Rates of climatic niche evolution using three proxy climate variables estimated with evolutionary models with variable parameters across states (lowland vs. highland). Rate values are estimated in temperature changes (°C) value per million years. This figure is color-blind friendly

adaptations or evolve traits that enable them to tolerate cold temperatures. Further research is needed to determine whether different clades occupying mountain regions in South America have experienced similar evolutionary pathways regarding climatic niche evolution and cold tolerance.

Acknowledgments JAV is grateful to the Dirección General de Asuntos del Personal Académico DGAPA (UNAM-PAPIIT IA206523) for the funding provided to complete this project. The authors are grateful to Carla Piantoni and an anonymous reviewer for beneficial corrections and suggestions in the text.

Conflicts of Interest This study was funded by a DGAPA-PAPIIT project IA206523 assigned to Julián A. Velasco. The authors have no conflicts of interest to declare relevant to this chapter's content.

Data Availability All datasets and R codes required to reproduce the analyses and figures are publicly available at https://doi.org/10.6084/m9.figshare.26039287.v3.

References

Antonelli A, Zizka A, Carvalho FA, Scharn R, Bacon CD, Silvestro D, Condamine FL (2018) Amazonia is the primary source of neotropical biodiversity. Proc Natl Acad Sci 115(23):6034–6039

Beaulieu JM, O'Meara BC (2023) OUwie: Analysis of Evolutionary Rates in an OU Framework. https://github.com/thej022214/OUwie.

Beaulieu JM, Jhwueng D-C, Boettiger C, O'Meara BC (2012) modeling stabilizing selection: expanding the ornstein–uhlenbeck model of adaptive evolution. Evolution 66(8):2369–2383

Bennett JM, Sunday JM, Calosi P, Villalobos F, Martínez B, Molina-Venegas R, Araújo MB et al (2021) The evolution of critical thermal limits of life on earth. Nat Commun 12(1):1198

Bennett PC, Clusella-Trullas S, Martínez B, Sunday J, Algar AC, Araújo MB et al (2018) GlobTherm, a global database on thermal tolerances for aquatic and terrestrial organisms. Sci Data 5(1):1–7

Blomberg SP, Rathnayake SI, Moreau CM (2020) Beyond Brownian motion and the Ornstein-uhlenbeck process: stochastic diffusion models for the evolution of quantitative characters. Am Nat 195(2):145–165

Brun P, Zimmermann NE, Hari C, Pellissier Loïc, Karger DN (2022) Global climate-related predictors at kilometer resolution for the past and future. Earth Syst Sci Data 14(12):5573–5603

Burnham D, Anderson K (2002) Model selection and multimodel inference: a practical information-theoretic approach. Springer, New York, NY

Campbell-Staton SC, Velotta JP, Winchell KM (2021) Selection on adaptive and maladaptive gene expression plasticity during thermal adaptation to urban heat islands. Nat Commun 12(1):6195

Campbell-Staton SC, Winchell KM, Rochette NC, Fredette J, Maayan I, Schweizer RM, Catchen J (2020) Parallel selection on thermal physiology facilitates repeated adaptation of city lizards to urban heat islands. Nat Ecol Evol 4(4):652–658

Cerezer FO, Dambros CS, Coelho MTP, Cassemiro FAS, Barreto E, Albert JS, Wüest RO, Graham CH (2023) Accelerated body size evolution in upland environments is correlated with recent speciation in South American freshwater fishes. Nat Commun 14(1):6070

Chamberlain, Scott, Karthik Ram, Vijay Barve, Dan Mcglinn, and Maintainer Scott Chamberlain. (2017). Package 'Rgbif'. Interface to the Global Biodiversity Information Facility 'API 5 (0.9).

Diele-Viegas LM, Vitt LJ, Sinervo B, Colli GR, Werneck FP, Miles DB, Magnusson WE et al (2018) Thermal physiology of Amazonian lizards (Reptilia: Squamata). *PLoS One* 13(3):e0192834

Dolby GA (2021) Towards a unified framework to study causality in earth–life systems. Mol Ecol 30(22): 5628–5642

Dolby GA, Bennett SEK, Dorsey RJ, Stokes MF, Riddle BR, Lira-Noriega A, Munguia-Vega A, Wilder BT (2022) integrating earth–life systems: a geogenomic approach. Trends Ecol Evol 37(4):371–384

Dubiner S, Aguilar Rocío, Anderson RO, Arenas Moreno DM, Avila LJ, Boada-Viteri E, Castillo M et al (2024) A global analysis of field body temperatures of active squamates in relation to climate and behaviour. Glob Ecol Biogeogr 33(4):e13808

Frishkoff LO, Lertzman-Lepofsky G, Luke D, Mahler. (2022) Evolutionary opportunity and the limits of community similarity in replicate radiations of island lizards. Ecol Lett 25(11):2384–2396

García-Rodríguez A, Velasco JA, Villalobos F, Parra-Olea G (2021) Effects of evolutionary time, speciation rates and local abiotic conditions on the origin and maintenance of amphibian montane diversity. Glob Ecol Biogeogr 30(3):674–684

Goldberg EE, Lancaster LT, Ree RH (2011) Phylogenetic inference of reciprocal effects between geographic range evolution and diversification. Syst Biol 60(4): 451–465

Holt (2009) Bringing the Hutchinsonian Niche into the 21st century: ecological and evolutionary perspectives. Proc Natl Acad Sci 106(supplement_2):19659–19665

Hoorn C, Perrigo A, Antonelli A (2018) Mountains, climate and biodiversity. Wiley

Hua X, Wiens JJ (2013) How does climate influence speciation? Am Nat 182(1):1–12

Hughes C, Eastwood R (2006) Island radiation on a continental scale: exceptional rates of plant diversification after uplift of the Andes. Proc Natl Acad Sci 103(27): 10334–10339

Hutchinson GE (1957) Concluding remarks. In: Cold spring harbor symposia on quantitative biology, vol 22. Cold Spring Harbor Laboratory Press, pp 415–427

Hutter CR, Guayasamin JM, Wiens JJ (2013) Explaining andean megadiversity: the evolutionary and ecological causes of glassfrog elevational richness patterns. Ecol Lett 16(9):1135–1144

Hutter CR, Lambert SM, Wiens JJ (2017) Rapid diversification and time explain amphibian richness at different scales in the tropical Andes, earth's most biodiverse hotspot. Am Nat 190(6):828–843

Lagomarsino LP, Condamine FL, Antonelli A, Mulch A, Davis CC (2016) The abiotic and biotic drivers of rapid diversification in a Ndean Bellflowers (Campanulaceae). New Phytol 210(4):1430–1442

Losos JB (2008) Phylogenetic niche conservatism, phylogenetic signal and the relationship between phylogenetic relatedness and ecological similarity among species. Ecol Lett 11(10):995–1003

Losos JB, Luke Mahler D et al (2010) Adaptive radiation: the interaction of ecological opportunity, adaptation, and speciation. Evol Since Darwin: The First 150:381–420

Meiri S, Bauer AM, Chirio L, Colli GR, Das I, Doan TM, Feldman A et al (2013) Are lizards feeling the heat? a tale of ecology and evolution under two temperatures. Glob Ecol Biogeogr 22(7):834–845

Meseguer AS, Michel A, Fabre P-H, Pérez OA, Escobar GC, Riina R, Antonelli A, Antoine P-O, Delsuc F, Condamine FL (2022) Diversification dynamics in the neotropics through time, clades, and biogeographic regions. elife 11:e74503

Moore BR, Donoghue MJ (2009) A Bayesian approach for evaluating the impact of historical events on rates of diversification. Proc Natl Acad Sci 106(11): 4307–4312

Moreira MO, Yan-Fu Q, Wiens JJ (2021) Large-scale evolution of body temperatures in land vertebrates. Evol Lett 5(5):484–494

Münkemüller T, Boucher FC, Thuiller W, Lavergne S (2015) Phylogenetic niche conservatism–common pitfalls and ways forward. Funct Ecol 29(5):627–639

Munoz MM, Stimola MA, Algar AC, Conover A, Rodriguez AJ, Landestoy MA, Bakken GS, Losos JB (2014) Evolutionary stasis and lability in thermal physiology in a group of tropical lizards. Proc R Soc B Biol Sci 281(1778):20132433

Pérez-Escobar OA, Zizka A, Bermúdez MA, Meseguer AS, Condamine FL, Hoorn C, Hooghiemstra H et al (2022) The Andes through time: evolution and distribution of Andean Floras. Trends Plant Sci 27(4): 364–378

Perrigo A, Hoorn C, Antonelli A (2020) Why mountains matter for biodiversity. J Biogeogr 47(2):315–325

Peterson AT, Soberón J, Pearson RG, Anderson RP, Martínez-Meyer E, Nakamura M, Araújo MB (2011) ecological niches and geographic distributions (MPB-49). In: Ecological niches and geographic distributions (MPB-49). Princeton University Press

Qu Y-F, Wiens JJ (2020) Higher temperatures lower rates of physiological and niche evolution. Proc R Soc B 287(1931):20200823

Rahbek C, Borregaard MK, Antonelli A, Colwell RK, Holt BG, Nogues-Bravo D, Rasmussen CMØ et al (2019a) building mountain biodiversity: geological and evolutionary processes. Science 365(6458): 1114–1119

Rahbek C, Borregaard MK, Colwell RK, Dalsgaard BO, Holt BG, Morueta-Holme N, Nogues-Bravo D, Whittaker RJ, Fjeldså J (2019b) Humboldt's enigma: what causes global patterns of mountain biodiversity? Science 365(6458):1108–1113

Rangel TF, Colwell RK, Graves GR, Fučíková K, Rahbek C, Diniz-Filho JAF (2015) Phylogenetic uncertainty revisited: implications for ecological analyses. *Evolution* 69(5):1301–1312

Revell LJ (2024) Phytools 2.0: an updated r ecosystem for phylogenetic comparative methods (and other things). *Peer J* 12:e16505

Revell LJ, Harmon LJ, Collar DC (2008) Phylogenetic signal, evolutionary process, and rate. *Syst Biol* 57(4): 591–601

Rohde K, Heap M, Heap D (1993) Rapoport's rule does not apply to marine teleosts and cannot explain latitudinal gradients in species richness. *Am Nat* 142(1): 1–16

Rolland J, Silvestro D, Schluter D, Guisan A, Broennimann O, Salamin N (2018) The impact of endothermy on the climatic niche evolution and the distribution of vertebrate diversity. Nat Ecol Evol 2(3):459–464

Saban KE, Yan-Fu Q, Wiens JJ (2023) Niche conservatism over deep timescales and diverse niche axes in land vertebrates. Glob Ecol Biogeogr 32(9): 1522–1534

Santos JC, Coloma LA, Summers K, Caldwell JP, Ree R, Cannatella DC (2009) Amazonian amphibian diversity is primarily derived from late Miocene Andean Lineages. *PLoS Biol* 7(3):e1000056

Snethlage MA, Geschke J, Ranipeta A, Jetz W, Yoccoz NG, Körner C, Spehn EM, Fischer M, Urbach D (2022) A hierarchical inventory of the world's mountains for global comparative mountain science. *Sci Data* 9(1):149

Stroud JT, Losos JB (2016) Ecological opportunity and adaptive radiation. Annu Rev Ecol Evol Syst 47:507–532

Sunday JM, Bates AE, Dulvy NK (2011) Global analysis of thermal tolerance and latitude in ectotherms. Proc R Soc B Biol Sci 278(1713):1823–1830

Tonini JF, Riva KH, Beard RB, Ferreira WJ, Alexander R, Pyron. (2016) Fully-sampled phylogenies of squamates reveal evolutionary patterns in threat status. Biol Conserv 204:23–31

Tribble CM, Alzate-Guarín F, Gándara E, Vartoumian A, Burleigh JG, Zenil-Ferguson R, Specht CD, Rothfels CJ (2024) the rapid radiation of bomarea (Alstroemeriaceae: Liliales), driven by the rise of the Andes. Evolution 78(2):221–236

Trisos CH, Merow C, Pigot AL (2020) The projected timing of abrupt ecological disruption from climate change. Nature 580(7804):496–501

Weir JT (2006) Divergent timing and patterns of species accumulation in lowland and highland neotropical birds. *Evolution* 60(4):842–855

Wiens JJ (2004) Speciation and ecology revisited: phylogenetic niche conservatism and the origin of species. *Evolution* 58(1):193–197

Wiens JJ (2008) Commentary on losos (2008): Niche conservatism déjà Vu. *Ecol Lett* 11(10):1004

Wiens JJ, Graham CH (2005) Niche conservatism: integrating evolution, ecology, and conservation biology. Annu Rev Ecol Evol Syst 36:519–539

Zizka A, Silvestro D, Andermann T, Azevedo J, Ritter CD, Edler D, Farooq H et al (2019) CoordinateCleaner: standardized cleaning of occurrence records from biological collection databases. Methods Ecol Evol 10(5):744–751

Part II

Evolutionary Histories and Life Strategies

The Shared Evolutionary History of South America's Mountain Biotas: Genetic Evidence from Amphibians and Reptiles

Ivan Prates, Antoine Fouquet, Miguel Trefaut Rodrigues, and Philippe J. R. Kok

1 Introduction

The mountains of South America (Fig. 1) are some of Earth's most iconic natural features. The Andean chain, spanning nearly 9000 km across seven countries and reaching up to 6000 m a.s.l., includes habitats as diverse as glaciated peaks, alpine grasslands, forests, and deserts (Graham 2009; Pérez-Escobar et al. 2022). It is the continent's major topographic feature, with an area around 1.5 million km^2 above 1000 m in its tropical section. However, two other regions harbor significant mountains. The first is the Guiana Shield's Pantepui region, which encompasses the *tepuis*, precipitous table-top mountains that tower dramatically from the surrounding lowland forests and savannas (Mayr and Phelps 1967). The second is Brazil's Atlantic Forest, whose mountains harbor unique ecosystems like the rocky *campo rupestre* savannas and native *Araucaria* pine forests (Ab'Sáber 2003). Unlike the Andes, Pantepui and the Atlantic Forest mountains do not exceed 3000 m and are discontinuous, forming "archipelagos" of uplands (1000–2000 m) and highlands (2000–3000 m). Moreover, the area corresponding to distinct elevations (Table 1) differs markedly between Pantepui (uplands: 83,000 km^2; highlands: 7300 km^2) and the Atlantic Forest (uplands: 110,000 km^2; highlands: 220 km^2), particularly between 1500 and 2000 m (Pantepui: 28,300 km^2; Atlantic Forest: 4800 km^2). Despite these differences, all three regions stand out by having remarkably high biological diversity, harboring some of the world's most species-rich biotas and high levels of endemism (Jenkins et al. 2013).

This high diversity is illustrated by amphibians and squamate reptiles. South America's montane herpetofauna is particularly diverse at intermediate elevations in tropical latitudes, with species richness peaking around 1500 m, where temperatures are mild and rainfall is high (Wiens et al. 2007; Hutter et al. 2013; Meza-Joya and Torres 2016; Brown et al. 2020). This band of high richness also corresponds to that of the highest endemism. For instance, despite achieving much lower maximum elevations, the Atlantic Forest has an almost equivalent number of endemic genera and species than the Andes, reflecting the large extent between 1000 and 1500 m in both regions (Table 1). Specifically, 33 genera (containing 384 upland or highland species) of amphibians are endemic to the

I. Prates (✉)
Department of Biology, Lund University, Lund, Sweden

A. Fouquet
Université Paul Sabatier, Centre de Recherche sur la Biodiversité et l'Environnement, Toulouse, France

M. T. Rodrigues
Instituto de Biociências, Departamento de Zoologia, Universidade de São Paulo, São Paulo, SP, Brazil

P. J. R. Kok
Department of Ecology and Vertebrate Zoology, University of Lodz, Łódź, Łódźkie, Poland

J. N. Urbina-Cardona et al. (eds.), *Andean Herpetofauna*, Sustainable Development Goals Series,
https://doi.org/10.1007/978-3-032-00074-3_7

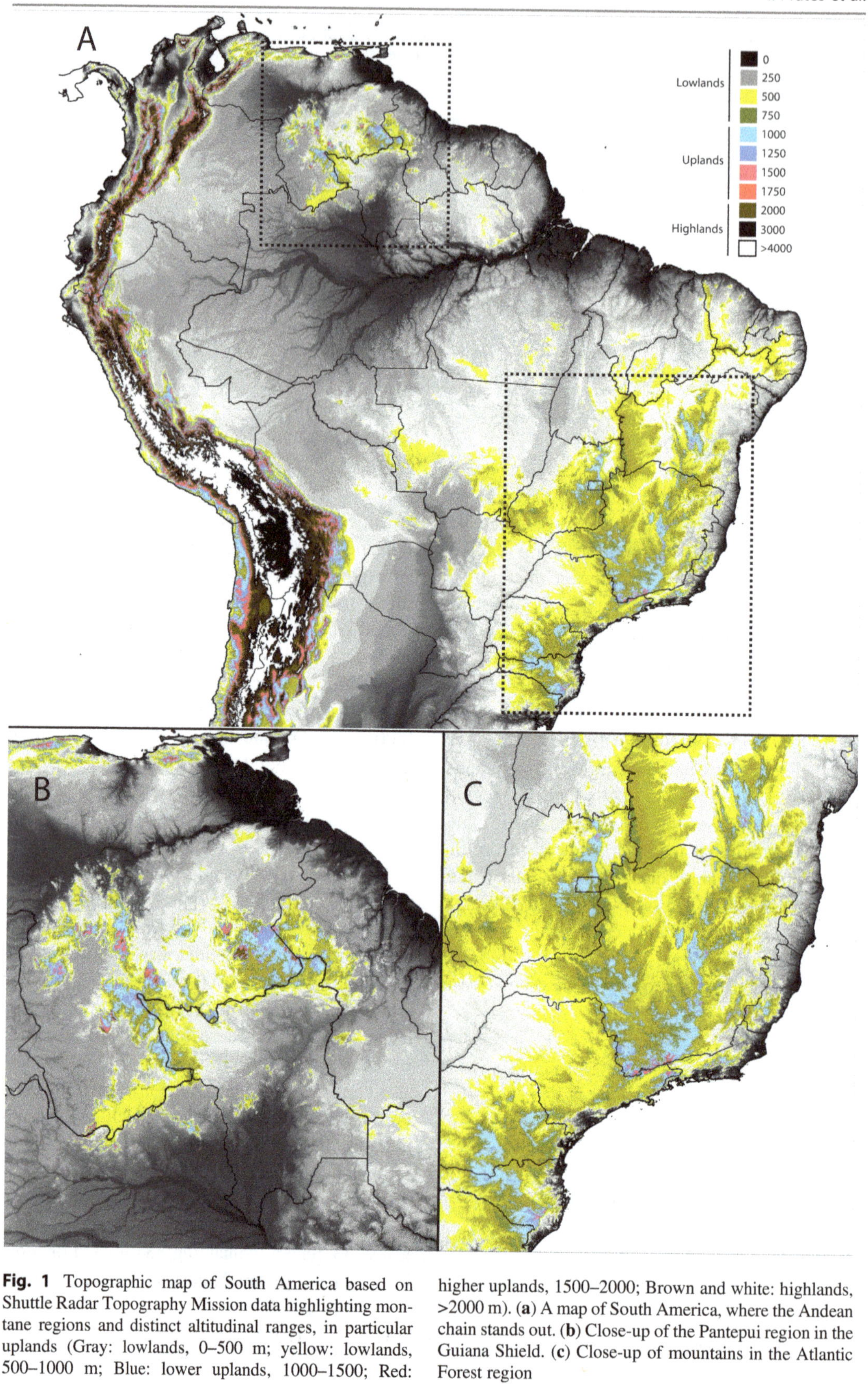

Fig. 1 Topographic map of South America based on Shuttle Radar Topography Mission data highlighting montane regions and distinct altitudinal ranges, in particular uplands (Gray: lowlands, 0–500 m; yellow: lowlands, 500–1000 m; Blue: lower uplands, 1000–1500; Red: higher uplands, 1500–2000; Brown and white: highlands, >2000 m). (**a**) A map of South America, where the Andean chain stands out. (**b**) Close-up of the Pantepui region in the Guiana Shield. (**c**) Close-up of mountains in the Atlantic Forest region

Table 1 Area corresponding to mid- and high-elevational bands in the Guiana Shield, Atlantic Forest, and Andean regions

Altitudinal band (m)	Area for Pantepui (km^2)	Area for the Atlantic Forest (km^2)	Area for the Andes (km^2)
1000–1500	54,268	102,000	270,000
1500–2000	28,308	4774	199,000
1000–2000	82,576	107,000	469,000
2000–3000	7259	219	336,000
3000–4000	0	0	456,000
4000–5000	0	0	327,000
>5000	0	0	21,600

Data from McCain and Grytnes (2010), Meza-Joya and Torres (2016), and García-Rodríguez et al. (2021)

Atlantic Forest, compared to 41 genera (with 518 upland or highland species) in the Andes. By contrast, levels of endemism are lower in Pantepui, with 12 endemic genera (comprising 55 upland or highland species), possibly reflecting the prevalence of highlands relative to uplands. The historical landscape and climatic processes underlying the configuration and extent of present-day alpine habitats have likely played a major role in the diversification of regional herpetofaunas (Box 1).

Box 1: Brief overview of environmental changes in South America over the Cenozoic

The biotic history of the Neotropics has been greatly affected by geological and climatic processes happening during the Cenozoic (66 Ma to the present), following the breakup of the western Gondwana paleocontinent and the Cretaceous–Paleogene boundary (Fig. 2). During its long period of "splendid isolation," South America has undergone radical geological and climatic transformations. By the end of the Cretaceous, a multitude of lineages underwent mass extinction worldwide, and the first closed-canopy forests dominated by angiosperms were formed (Jaramillo 2023). Early in the Paleogene (~66–23 Ma), a large portion of northern South America was covered by continuous tropical rainforests, indicating year-round warm and humid conditions. Transitions between the Eocene and Oligocene (~34 Ma) and the Paleogene and Neogene (~23 Ma), which involved brutal global cooling and the onset of growth of the polar ice caps, marked the shift from a "greenhouse" Earth to an "icehouse" one (Westerhold et al. 2020). Because the continent remained roughly at the same latitude as temperatures decreased throughout the Cenozoic, its extent under typically tropical conditions also decreased. Substantial changes occurred in floras and faunas around the world, reflecting the shift from warm and wet climatic regimes to colder and drier ones (Hutchinson et al. 2021), notably in Amazonia (Antoine et al. 2021). This shift also led to the formation of the Dry Diagonal, which separated Amazonia from the Atlantic Forest.

The central Andes (which extends south into southern Peru) and western cordillera (Ecuador and Colombia) progressively reached high elevations during the Paleogene, but the rest of the Andes rose only later, during the Neogene (from 23 Ma onward) (Boschman 2021). Most of the currently dominant South American clades date back to the Neogene (Hoorn et al. 2010; Boschman and Condamine 2022), including many anuran genera (Santos et al. 2009; Kok et al. 2017, 2018; Réjaud et al. 2020; Moraes et al. 2022; Fouquet et al. 2022a, b, 2024a; Ortiz et al. 2023). During this period, Pantepui and Atlantic Forest topography remained relatively stable, although weathering dramatically dissected Pantepui summits (Kok 2013). The central and Northern Andes rapidly rose as the Neogene advanced, with dramatic consequences to the hydrology of Amazonia, notably the formation of mega-wetlands in the west (Pebas system) and, eventually, the onset of a continental Amazon drainage system flowing east into the Atlantic Ocean some 9 Ma (Hoorn et al. 2017). The orogeny of the Andes also dramatically modified regional climates, blocking the easterly rainfall

(continued)

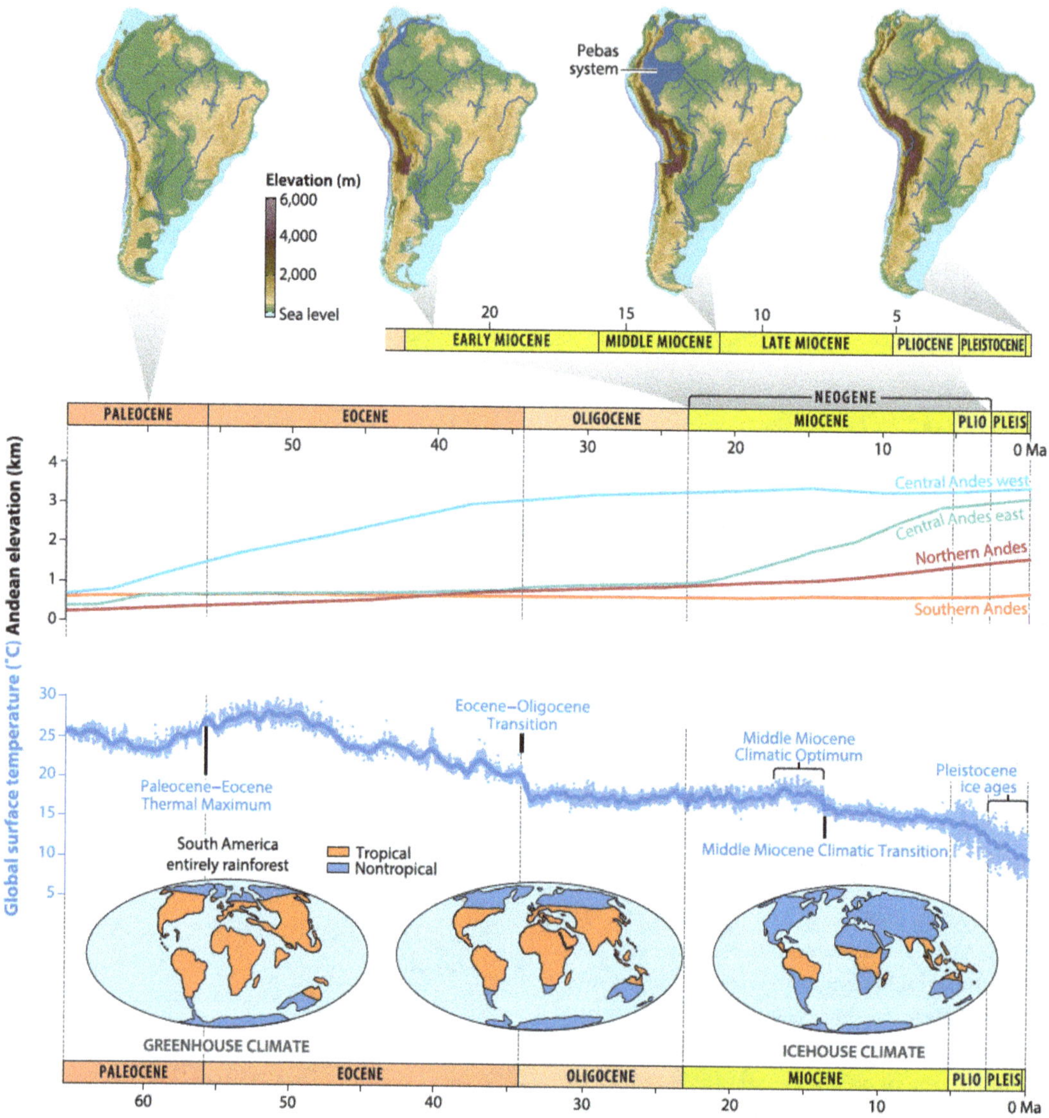

Fig. 2 Overview of Cenozoic geological and climatic evolution in South America, modified from Hoorn et al. (2023) (original licensed under a Creative Commons Attribution 4.0 International License). Andean uplift and river evolution maps (top row) from Boschman (2021) and Boschman et al. (2021); Andean uplift curves of the Northern (13°N–14°S), Central (14°S–34°S), and Southern (34°S–56°S) Andes based on Boschman et al. (2021); global surface temperature curve based on oxygen isotope ratio data of benthic foraminifera shells from Westerhold et al. (2020)

(Garreaud 2009). Global cooling intensified during the Miocene, 15 Ma onward. Finally, the Pleistocene (last 2.6 Ma) has seen rapid and pronounced climatic oscillations, which alternately favored or prevented biotic exchanges between Amazonia and the Atlantic Forest (Ledo and Colli 2017) and vertical dispersal between uplands and lowlands across the continent (Lötters et al. 2010). These events have likely transformed faunas at regional and continental scales, being widely invoked to explain patterns of dispersal, divergence, and extinction inferred from genetic analyses of extant amphibians and squamate reptiles and other clades.

Uncovering the factors behind species diversification and geographic distributions within the Andes, Pantepui, and the Atlantic Forest is an active research field. In each of these regions, paleoenvironmental and genetic analyses of animal and plant clades have inferred how climate change and mountain uplift and erosion over the last millions of years (Box 1) have led to vicariance and speciation on opposite sides of peaks and valleys (Kok et al. 2017; Rahbek et al. 2019; Recoder et al. 2020); how climate-driven altitudinal shifts in habitat distributions have, at certain times, fostered population connectivity, and at others, divergence on isolated summits (Rull 2004b); how species have persisted on stable summits or via altitudinal range changes as habitats shifted upslope or downslope (Rull 2004a; Carnaval et al. 2009); and the role of spatial environmental heterogeneity in limiting geographic distributions and promoting adaptive divergence (Fouquet et al. 2012b; Carnaval et al. 2014). By focusing on clades that have diversified within mountain chains, these studies have provided insights into the processes that generated present-day patterns of endemism in each region. Among these clades, terrestrial, non-volant, ectothermic tetrapods have emerged as informative models of diversification processes acting at both broad and narrow spatial and temporal scales.

At broader spatial scales, biogeographic investigations have revealed an intriguing pattern: many species and clades occurring on summits and slopes are more closely related to lineages inhabiting montane regions hundreds to thousands of kilometers away. The Andes, Pantepui, and Atlantic Forest are separated by varied lowland habitats, including Amazonian rainforests and the savannas and shrublands of the Caatinga, Cerrado, and Chaco (together referred to as South America's Dry Diagonal). Nevertheless, floristic and faunistic similarity suggests a shared history between those mountains. Many bird and plant taxa regarded as single species occur in both the Andes and Atlantic Forest or Pantepui highlands (Mayr and Phelps 1967; Huber 1988; Désamoré et al. 2014), which would imply conspecific populations separated by vast tracts of unsuitable lowland habitats. At deeper evolutionary scales, phylogenetic investigations have uncovered sister relationships between clades occurring on those mountains but absent from the intervening lowlands or only marginally occupying them (Leite et al. 2015). Some of these clades show contrasting species richness, for instance, with many species in the Andes but few in the Atlantic Forest, as is the case of subclades within Holoadeninae (Motta et al. 2021), Bufonidae (Moraes et al. 2022), Craugastoridae (Heinicke et al. 2015), and Hemiphractidae anurans (Castroviejo-Fisher et al. 2015). Why and how shared or closely related lineages occur in distant mountains is poorly understood. Importantly, such disjunctions might have resulted from biogeographic processes fundamentally different from those explaining most species formation and accumulation within regions. Among them, potentially important are jump dispersal, sky island hopping, broad-scale extinctions and lineage turnover in the lowlands, and repeated dispersals and adaptation across highland and lowland settings (see below).

This contribution seeks to help characterize these knowledge gaps by reviewing historical biogeographic investigations of amphibians and reptiles. These diverse, largely dispersal-limited organisms have provided crucial insights into how climate and landscape shifts through time have affected species and habitats. We start by summarizing major hypotheses invoked to explain biotic similarity across the Andes, Pantepui, and Atlantic Forest. To assess whether these hypotheses explain diversity patterns in herpetofauna, we review the genetic evidence of past dispersal, vicariance, and extinctions. We conclude by outlining outstanding questions and challenges to uncover the shared herpetofaunal history of South America's strikingly diverse mountains.

2 The Shared History of Montane Biotas: Biogeographic Hypotheses

Several hypotheses have been proposed to explain a pattern of shared species and clades between disjunct geographic regions in the Neotropics (Fig. 3). An influential idea posits that this sharing results from the historical establishment of intermittent corridors of suitable habitat (Baker et al. 2020; Fig. 3a). Pulses of habitat connectivity provide opportunities for organisms to colonize new areas, where they encounter and interact with new environments and species and sometimes initiate evolutionary radiations (Petren et al. 2005; Hughes and Eastwood 2006). In South America, habitat connections have been widely invoked to explain discontinuous distributions in rainforest species. For instance, genetic analyses of lizards, birds, and small mammals support pulses of species exchange between Amazonia and Atlantic Forest, particularly during presumably wetter and warmer phases of the Quaternary (Costa 2003; Prates et al. 2016a, 2018; Ledo and Colli 2017). From the perspective of cool mountain habitats (e.g., those above 2000 m), however, it is unclear how such warmer periods would have created suitable corridors allowing species now restricted to uplands and highlands to disperse between distant parts of the continent.

Besides former climate-mediated corridors of suitable habitat, other mechanisms have been proposed to explain species and clade sharing across montane regions. One hypothesis posits that summit species would have (perhaps repeatedly) adapted to lowlands settings, thus dispersing and colonizing other summits (Fig. 3b). This hypothesis has been invoked to explain disjunct ranges in Pantepui birds (Mayr and Phelps 1967) and plants (Huber 1988). However, this scenario might imply at least some populations being able to persist in the lowlands. Another hypothesis proposes that species restricted to highlands would have expanded their ranges via sequential episodes of jump dispersal between isolated mountaintops (Mayr and Phelps 1967; Fig. 3c). This "sky island hopping" model might explain the disjunct distributions of certain bird, plant, and insect taxa, but it is unclear if it applies to terrestrial organisms with limited dispersal capacities, such as amphibians and reptiles (Kok et al. 2012; Salerno et al. 2012). Finally, an alternative model proposes that such disjunct distributions result from the extinction of widespread ancestral species and clades, which would have spanned the lowland areas separating distant mountains (Mayr and Phelps 1967; Fig. 3d). This idea appears to have received relatively less attention than hypotheses based on dispersal, potentially because of considerable difficulties in inferring extinction and ancestral ranges based on data from extant taxa alone.

Below, we discuss the evidence for each of these hypotheses for amphibians and reptiles. We group empirical examples by the spatial and temporal scale of diversification—from across regions in deep time to within regions at recent timescales.

3 Diversification Across Montane Regions

Increased access to genetic samples has allowed phylogenetic and biogeographical investigations of South American amphibian and reptile clades at a continental scale. An emerging pattern from these studies is that of sister or close relationships between lineages restricted to distant mountain ranges while being only distantly related to lowland lineages. This is the case of several pairs of sister groups distributed, on the one hand, in the Andes and, one the other hand, in Pantepui, such as the Bufonidae toad clades *Atelopus* (primarily Andean) and *Oreophrynella* (Pantepui) (Kok et al. 2018; Moraes et al. 2022) and the Aromobatidae frog clades *Rheobates* (Andes) and *Anomaloglossus* (Pantepui) (Vacher et al. 2024); *Atelopus* and *Anomaloglossus* have subsequently diversified in the lowlands. Other studies have reported close relationships between clades from the Andes and Atlantic Forest, as is the case of the Bufonidae toads *Osornophryne* (Andes) and *Frostius* (Atlantic Forest) (Moraes et al.

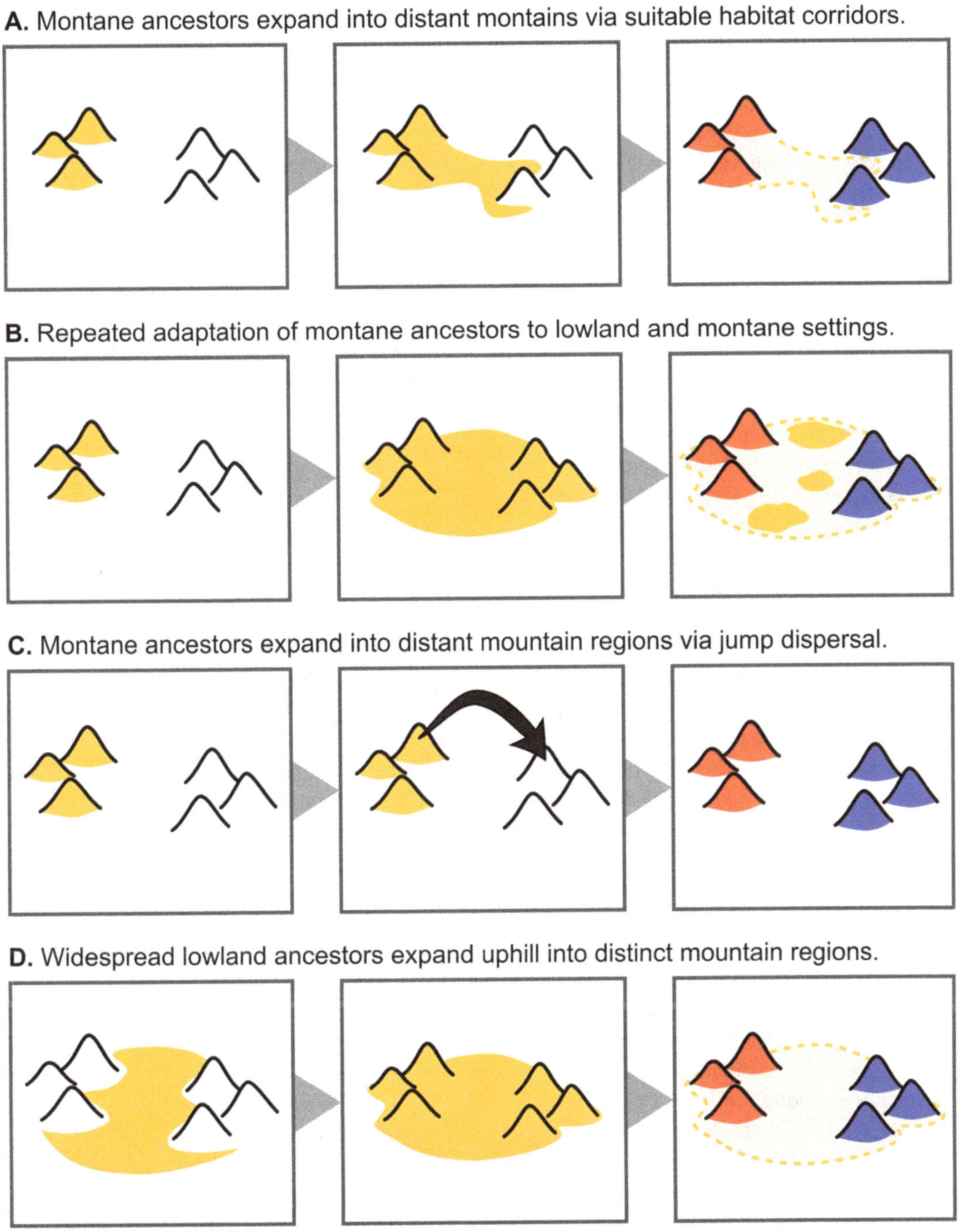

Fig. 3 Biogeographic hypotheses invoked to explain shared faunas and close phylogenetic relationships across distant mountain systems in South America. (**a**) Historical establishment of intermittent corridors of suitable habitat provides opportunities for organisms to colonize new areas. (**b**) Summit species would have adapted to lowlands settings, thus dispersing and colonizing other summits. (**c**) Species restricted to highlands would have expanded their ranges via sequential episodes of jump dispersal between isolated mountaintops. (**d**) Widespread ancestral species or clades formerly spanning the lowland areas separating distant mountains would go extinct following dispersal uphill, resulting in disjunct ranges on mountains

2022), genera within Holoadeninae frogs (Motta et al. 2021), and early diverging lineages within Hemiphractidae *Gastrotheca* frogs (Castroviejo-Fisher et al. 2015). Other studies have found close relationships between Pantepui and Atlantic Forest clades, as is the case of the frog families Caligophrynidae (Pantepui) and Brachycephalidae (Atlantic Forest) (Fouquet et al. 2024b).

Beyond pairs of clades, several studies have found close relationships between montane species from all three regions—the Andes, Pantepui, and Atlantic Forest. The Dactyloidae lizards *Anolis nasofrontalis*, *Anolis neglectus*, and *Anolis pseudotigrinus*, known from one or few montane sites in central Atlantic Forest, form a clade with species that have narrow ranges in the Peruvian or Ecuadorian Andes or on Venezuelan *tepuis*. This clade shares a common ancestor dating back to around 27 million years ago, and it diverged from clades including lowland Amazonian and Dry Diagonal species some 45 million years ago (Prates et al. 2017, 2020). In anurans, several radiations show early divergences between clades primarily restricted to distinct regions. This is the case of Cophomantini frogs (*Hyloscirtus* in the Andes; *Myersiohyla* and *Nesorohyla* in Pantepui; *Aplastodiscus* and *Bokermannohyla* in the Atlantic Forest) (Portik et al. 2023), Bufonidae toads (*Atelopus*, *Nannophryne* and *Osornophryne* in the Andes; *Oreophrynella* in Pantepui; *Dendrophryniscus*, *Melanophryniscus*, and *Frostius* in the Atlantic Forest) (Moraes et al. 2022), and Hemiphractidae frogs (*Cryptobatrachus*, *Flectonotus*, and *Gastrotheca* in the Andes and the Atlantic Forest; *Stefania* in Pantepui; *Fritziana* in the Atlantic Forest) (Castroviejo-Fisher et al. 2015). Divergences between these clades are generally old, predating the Neogene (23 Ma onward). In some cases, species from primarily montane clades occur in the adjacent lowlands, reflecting more recent downslope dispersals (Teixeira et al. 2012; Prates et al. 2015; Kok et al. 2017, 2018; Moraes et al. 2022; Vacher et al. 2024).

Two interrelated processes have been invoked to explain such clade disjunctions. The first is a history of downhill movement followed by dispersal into distant mountains (Rull 2004a). Alternatively, this pattern might reflect previously widespread ancestors that persisted solely at higher elevations due to unfavorable climates, habitats, or biotic interactions in the lowlands, thus fostering divergence in isolated mountains (Mayr and Phelps 1967; Pie et al. 2018). In both scenarios, extinction of ancestral species from the lowlands would have led to the disjunct clade ranges we see today. These hypotheses predict ancestors that were adapted to environmental conditions that were presumably widespread but are now limited to mountain summits or slopes, thus explaining range restriction in extant montane species. Among these presumed conditions are much lower temperatures. Notably, the transitions between the Eocene and Oligocene (~34 Ma) and the Paleogene and Neogene (~23 Ma) were characterized by pronounced cooling at a global scale, which intensified during the Miocene (~15 Ma onward). These conditions might have allowed habitats or organisms that are now typically alpine to have expanded and dispersed. However, the geographic extent of these expansions, and the factors leading to subsequent extinctions and lineage turnover in the lowlands, are unclear. Remarkably, some lowland species nested in clades composed primarily of montane lineages might represent instances of incomplete extinction in the lowlands, as is the case of Eleutherodactylidae *Adelophryne* and Strabomantidae *Oreobates* frogs (Teixeira et al. 2012; Fouquet et al. 2012a).

The hypotheses above rely on niche conservatism at the clade level, with little to no evolution of ecological tolerances across species. Conserved physiologies overall seem more common in amphibians and squamate reptiles than in other tetrapods (Rolland et al. 2018), a mechanism invoked to explain range restriction in montane herpetofauna in other parts of the globe, such as salamanders in North America (Kozak and Wiens 2006). However, there is evidence that ecological tolerances do not always constrain the distributions of mountain species. Experiments of thermal physiology in the Pantepui lizard *Riolama grandis* (Gymnophthalmidae) suggest that maximum yearly temperatures in the lowlands should not restrict this species to summits (Recoder et al. 2020). The same scenario was found for the lizards *Caparaonia itaiquara* and *Colobodactylus dalcyanus* (Gymnophthalmidae) and *Aspronema dorsivittatum* (Scincidae) from montane Atlantic Forest (Strangas et al. 2019). These findings suggest that range restriction to mountains might instead result from ecological factors like biotic interactions or microhabitat availability, but further investigation is necessary.

4 Diversification on Summits

Processes similar to those invoked to explain disjunct clade distributions across montane regions might also explain geographic separation between closely related species within a given mountain chain. In the Atlantic Forest, the sister lizard species *A. neglectus* and *A. nasofrontalis* occur at montane sites ca. 400 km apart (Prates et al. 2020). A similar configuration exists in many Pantepui clades. For instance, species of *Oreophrynella* toads, *Stefania* frogs, and *Riolama* lizards are restricted to distant summits and slopes (Kok et al. 2017, 2018; Recoder et al. 2020). In the case of neighboring but isolated mountains, divergences between extant species are more recent than those across distant parts of the continent, concentrated in the Pliocene and Pleistocene (5.3 Ma onward) (Kok et al. 2012). For instance, in anurans *Melanophryniscus* (Bufonidae) and *Brachycephalus* (Brachycephalidae) limited to southern Atlantic Forest peaks, most divergences between extant species are concentrated in the last five million years (Firkowski et al. 2016; Pie et al. 2018). In sister *Anolis* lizard species from Atlantic Forest slopes, divergences similarly date back to around five million years ago (Prates et al. 2020). On the other hand, divergences between more distant sets of mountains can be ancient (>20 My old), as reported for species on Pantepui and Atlantic Forest summits separated by lowlands (Firkowski et al. 2016; Kok et al. 2017, 2018).

Similar to the case of shared lineages across mountain chains, within-chain divergences might be explained by downhill and uphill dispersal followed by extinction in the intervening lowlands (Fig. 3). In this case, altitudinal movements presumably reflect the well-described shifts in environmental gradients and habitat distributions during climatic oscillations in the Quaternary (Baker et al. 2020). This model has been invoked to explain species ranges in Atlantic Forest *Anolis* (Prates et al. 2020) and the Pantepui anurans *Oreophrynella* and *Tepuihyla* (Kok et al. 2012; Salerno et al. 2012). However, the likelihood of active dispersal among *tepui* summits in the toad genus *Oreophrynella* has recently been challenged by limited spatial movements and enhanced resilience in *Oreophrynella quelchii* (Kok et al. 2024). A scenario of cold-adapted lowland ancestors forced into mountains by warming climates has been proposed to explain range disjunctions among dozens of mountaintop anuran species (*Melanophryniscus* and *Brachycephalus*) in southern Atlantic Forest (Firkowski et al. 2016). By contrast, isolation on summits in Pantepui endemics would result from the erosion of a formerly continuous plateau of ancient sedimentary rock. This model has been invoked to explain range disjunctions and older divergences within lizard and frog clades, while species divergences might reflect more recent range shifts under fluctuating climates (Kok et al. 2017, 2018; Recoder et al. 2020). In the examples above, extinction resulting from changes in climate or topography has led to fragmented clade distributions across summits and slopes. This regional-scale scenario mirrors the presumed role of extinction in explaining mountain clade disjunctions at continental scales (see *Diversification across Montane Regions*, above).

5 Diversification Along Altitudinal Gradients

Finally, biogeographic investigations have also found that South America's mountains have frequently exchanged species and lineages with adjacent regions. In many frog clades, the Andes have both provided lineages to Amazonia and received from it; following colonizations, many such lineages have diversified (Santos et al. 2009; Lötters et al. 2010; Castroviejo-Fisher et al. 2014; Antonelli et al. 2018; Guillory et al. 2020; Muell et al. 2022). In *Leposoma* lizards (Gymnophthalmidae) of the Atlantic Forest, early diverging lineages are restricted to mid-elevations, while relatively more derived lineages occupy the lowlands (Pellegrino et al. 2011; Rodrigues et al. 2013). Similar events have been inferred between Pantepui and Amazonia (Fouquet et al. 2021; Vacher et al. 2024) and between the Atlantic Forest region and Amazonia or the Dry Diagonal (Fouquet et al. 2012a, 2014). Certain clades have undergone multiple biome transitions, colonizing environments increasingly more disparate from the alpine sources. This is the case of *Ameerega* frogs (Dendrobatidae), which have expanded from mid-elevations along the Andes into Amazonia and ultimately colonized and diversified in the Cerrado (Santos et al. 2009; Guillory et al. 2020). Lineage exchange along environmental and altitudinal gradients is now seen as a crucial process behind present-day biodiversity patterns across the continent.

Evidence of frequent habitat transitions poses fundamental questions about the role of adaptation in diversification and how spatial ecological gradients might explain among-species variation in geographic range size (Prates et al. 2022). For instance, it is unclear why certain lineages were able to expand into lowland habitats, disperse, and often diversify further, while others have remained restricted to higher elevations. These differences might merely reflect priority effects, whereby the order of species arrival into regional communities determines the outcome of ecological interactions and thus the probability of species establishment (Reijenga et al. 2021). Alternatively, these differences might reflect physiological or life history traits that pre-adapt organisms for colonizations, or varying capacities for ecophysiological or behavioral adaptation in the face of new environmental regimes (Navas 2002, 2006). Future investigations of how organismal traits influence historical biogeography would benefit from comparisons of ecophysiological tolerances across clades, complementing within-clade efforts (Kohlsdorf et al. 2001; Prates and Navas 2009; Titon Jr et al. 2010).

6 Challenges and Future Directions

Compared to studies of evolutionary diversification within South America's mountain chains, few investigations have focused on the processes underlying shared or closely related lineages across distant mountain regions. Nevertheless, biogeographic investigations of amphibians and reptiles seem to provide little support to hypotheses invoking localized habitat corridors, jump dispersal, sky island hopping, or repeated colonizations and adaptation across highlands and lowlands (Fig. 3)—processes likely important in other organisms, particularly those with higher vagility. Instead, the emerging picture from herpetofauna is more consistent with ancient, broadly distributed ancestors that have gone extinct across vast lowland areas while persisting on summits and slopes. These expansive extinctions might be associated with the Eocene–Oligocene and Paleogene–Neogene transitions (Antoine et al. 2021; Jaramillo 2023). Among the surviving lineages, some appear to have produced few species, while others have radiated extensively, notably in the Andes, likely because more upland and highland areas are available (Table 1). During the Neogene, lineage turnover and subsequent diversification in the lowlands would have contributed to present-day transitions in species composition along altitudinal gradients, as suggested for birds (Weir 2006).

The general picture described above highlights several open questions and outstanding challenges for future research. One challenge is inferring and characterizing how extinction has contributed to present-day spatial patterns of diversity. Clade disjunctions across mountain regions suggest that many of today's upland and highland species are the survivors of ancient lineages that were widespread and potentially diverse. Had these lineages persisted, genetic analyses might reveal a scenario of independent upward dispersals rather than close relationships between geographically distant clades. Importantly, phylogenetic clustering of extant mountain lineages has often been interpreted as indicative of "connections," a term that—sometimes vaguely or implicitly—appears to implicate the historical establishment of suitable habitat corridors across regions, as in studies of lizards, birds, and small mammals (Prates et al. 2017, 2020; Ledo and Colli 2017). This situation likely results from applying verbal biogeographical models—better suited for recent events like Quaternary, glaciation-driven changes in rainforest distributions—to deep-time contexts (Prates et al. 2016b, 2018).

Differentiating ancient dispersal from long-term persistence is difficult using only the phylogenetic patterns of extant species. This challenge can be addressed using tools that infer a clade's palaeodiversity dynamics, enabling tests for shifts in diversification rate at specific phylogenetic branching points (Mazet et al. 2023). Such approaches could provide insights into diversification rates at different time windows between montane clades and those that have diversified in the lowlands, particularly from the Neogene onward. However, they would also require a leap forward in our knowledge of species boundaries and comprehensive genetic datasets to provide reliable time-calibrated phylogenetic trees.

Besides analytical developments, crucial insights could be acquired from the paleontological record, particularly in Amazonia. Fossils could reveal whether montane lineages formerly occupied the lowlands, similar to palynological studies supporting downward expansion of highland plant taxa in the Pantepui at recent (Pleistocene) timescales (Rull et al. 2019). However, amphibian and reptile fossils predating the Neogene are scarce, and existing records mostly involve large, well-ossified forms from southern South America (Barcelos and dos Santos 2023). Moreover, it might be difficult to confidently infer the phylogenetic placement of fossils, as illustrated by the first Eocene–Miocene specimens of Brachycephaloidea—the most species-rich Neotropical frog clade (Jansen et al. 2023). While highlighting the need for building a paleontological foundation for South America's herpetofauna, these examples also illustrate the promise of existing and new collections to reveal fossils from key clades.

Besides scarce information on extinct species, vast knowledge gaps remain about present-day patterns of biodiversity in South America's mountains. Limited knowledge of species diversity, distributions, phylogenetic relationships, and the timing of divergences strongly constrains inferences of historical biogeography and the drivers of diversification (Vacher et al. 2020). The increasing availability of DNA sequence data has strengthened investigations of the history of clades and habitats by providing evidence of dispersal, speciation, and indirectly, extinction. In turn, improved geographic and taxonomic sampling has relied on groundwork efforts to survey biodiversity, particularly in remote and poorly known regions. Herpetologists working in the Neotropics continue to encounter new species, relictual populations, and taxa thought to be extinct, some of which have provided transformative insights into the composition and evolution of regional faunas (Teixeira et al. 2012; Prates et al. 2017, 2020; Moraes and Pavan 2018). A bewildering example is that of two recently described family-level lineages of highland Pantepui frogs, which diverged from their closest living relatives more than 45 million years ago (Fouquet et al. 2024b). Efforts to document mountain biodiversity are particularly pressing in the face of human-driven global change, as rising temperatures force montane habitats and organisms to move upslope, shrinking species ranges (Colwell et al. 2008; Sinervo et al. 2010; Courtois et al. 2016), whereas lower temperatures and higher rainfall on mountains increase the deposition of persistent airborne pollutants (Daly and Wania 2005). Expeditions and biodiversity inventories will continue to lay the foundation of conservation assessments and investigations of the shared history of South America's herpetofaunas.

Acknowledgments We thank our many students and colleagues who have supported our work in South America's mountains. In particular, we thank participants and facilitators of the Neblina and Imeri expeditions, including Paulo Muzy, Eloisa Cabral, General Sinclair James Mayer, the Brazilian Army, the Yanomami people, Instituto Chico Mendes de Conservação da Biodiversidade, and Fundação Nacional dos Povos Indígenas. IP acknowledges support from the European Research Council (ERC-Stg 948126). AF benefited from an "Investissement d'Avenir" grant managed by the Agence Nationale de la Recherche (CEBA rANR-10-LABX-25-01). MTR received funding from the Fundação de Amparo à Pesquisa do Estado de São Paulo (FAPESP 2011/50146-6 and 2024/05630-7) and Conselho Nacional de Desenvolvimento Científico e Tecnológico (CNPq 302864/2020-2). PJRK was supported by the Fonds voor Wetenschappelijk Onderzoek (FWO12A7614N and 12A7617N), the Polish National Science Center (NCN 2020/39/D/NZ8/02399/EXILE) and Marie Skłodowska-Curie Actions (MSCA 101022238/HOSTILE).

Competing Interests The authors have no conflicts of interest to declare.

References

Ab'Sáber AN (2003) Os domínios de natureza no Brasil: potencialidades paisagísticas. Atelie Editorial, São Paulo

Antoine P-O, Yans J, Castillo AA et al (2021) Biotic community and landscape changes around the Eocene–Oligocene transition at Shapaja, Peruvian Amazonia: regional or global drivers? Glob Planet Change 202: 103512

Antonelli A, Zizka A, Carvalho FA et al (2018) Amazonia is the primary source of Neotropical biodiversity. Proc Natl Acad Sci USA 115:6034–6039

Baker PA, Fritz SC, Dick CW et al (2020) Beyond Refugia: new insights on quaternary climate variation and the evolution of biotic diversity in tropical South America. In: Rull V, Carnaval AC (eds) Neotropical diversification: patterns and processes. Springer, pp 51–70

Barcelos LA, dos Santos RO (2023) The lissamphibian fossil record of South America. Paleobiodivers Paleoenviron 103:341–405

Boschman LM (2021) Andean mountain building since the late cretaceous: a paleoelevation reconstruction. Earth-Sci Rev 220:103640

Boschman LM, Condamine FL (2022) Mountain radiations are not only rapid and recent: ancient diversification of South American frog and lizard families related to Paleogene Andean orogeny and Cenozoic climate variations. Glob Planet Change 208:103704

Boschman LM, Cassemiro FAS, Carraro L et al (2021) South American freshwater fish diversity shaped by Andean uplift since the Late Cretaceous. bioRxiv. 2021.05.14.444133

Brown JL, Paz A, Reginato M et al (2020) Seeing the forest through many trees: multi-taxon patterns of phylogenetic diversity in the Atlantic Forest hotspot. Divers Distrib 26:1160–1176

Carnaval AC, Hickerson MJ, Haddad CFB et al (2009) Stability predicts genetic diversity in the Brazilian Atlantic forest hotspot. Science 323:785–789

Carnaval AC, Waltari E, Rodrigues MT et al (2014) Prediction of phylogeographic endemism in an environmentally complex biome. Proc Biol Sci 281:20141461

Castroviejo-Fisher S, Guayasamin JM, Gonzalez-Voyer A, Vilà C (2014) Neotropical diversification seen through glassfrogs. J Biogeogr 41:66–80

Castroviejo-Fisher S, Padial JM, De La Riva I et al (2015) Phylogenetic systematics of egg-brooding frogs (Anura: Hemiphractidae) and the evolution of direct development. Zootaxa 4004:1–75

Colwell RK, Brehm G, Cardelús CL et al (2008) Global warming, elevational range shifts, and lowland biotic attrition in the wet tropics. Science 322:258–261

Costa LP (2003) The historical bridge between the Amazon and the Atlantic Forest of Brazil: a study of molecular phylogeography with small mammals. J Biogeogr 30:71–86

Courtois EA, Michel E, Martinez Q et al (2016) Taking the lead on climate change: modelling and monitoring the fate of an Amazonian frog. Oryx 50:450–459

Daly GL, Wania F (2005) Organic contaminants in mountains. Environ Sci Technol 39:385–398

Désamoré A, Vanderpoorten A, Laenen B et al (2014) Biogeography of the Lost World (Pantepui region, northeastern South America): insights from bryophytes. Phytotaxa 9:254–265

Firkowski CR, Bornschein MR, Ribeiro LF, Pie MR (2016) Species delimitation, phylogeny and evolutionary demography of co-distributed, montane frogs in the southern Brazilian Atlantic Forest. Mol Phylogenet Evol 100:345–360

Fouquet A, Loebmann D, Castroviejo-Fisher S et al (2012a) From Amazonia to the Atlantic forest: molecular phylogeny of Phyzelaphryninae frogs reveals unexpected diversity and a striking biogeographic pattern emphasizing conservation challenges. Mol Phylogenet Evol 65:547–561

Fouquet A, Noonan BP, Rodrigues MT et al (2012b) Multiple quaternary refugia in the eastern Guiana Shield revealed by comparative phylogeography of 12 frog species. Syst Biol 61:461–489

Fouquet A, Santana Cassini C, Fernando Baptista Haddad C et al (2014) Species delimitation, patterns of diversification and historical biogeography of the Neotropical frog genus *Adenomera* (Anura, Leptodactylidae). J Biogeogr 41:855–870

Fouquet A, Leblanc K, Framit M et al (2021) Species diversity and biogeography of an ancient frog clade from the Guiana Shield (Anura: Microhylidae: *Adelastes*, *Otophryne*, *Synapturanus*) exhibiting spectacular phenotypic diversification. Biol J Linn Soc Lond 132:233–256

Fouquet A, Cornuault J, Rodrigues MT et al (2022a) Diversity, biogeography, and reproductive evolution in the genus *Pipa* (Amphibia: Anura: Pipidae). Mol Phylogenet Evol 170:107442

Fouquet A, Réjaud A, Rodrigues MT et al (2022b) Diversification of the *Pristimantis conspicillatus* group (Anura: Craugastoridae) within distinct neotropical areas throughout the Neogene. Syst Biodivers 20:1–16

Fouquet A, Ferrão M, Rodrigues MT et al (2024a) Integrative species delimitation and biogeography of the *Rhinella margaritifera* species group (Amphibia, Anura, Bufonidae) suggest an intense diversification throughout Amazonia during the last 10 million years. Syst Biodivers 22:2291086

Fouquet A, Kok PJR, Recoder RS et al (2024b) Relicts in the mist: two new frog families, genera and species highlight the role of Pantepui as a biodiversity museum throughout the Cenozoic. Mol Phylogenet Evol 191: 107971

García-Rodríguez A, Velasco JA, Villalobos F, Parra-Olea G (2021) Effects of evolutionary time, speciation rates and local abiotic conditions on the origin and maintenance of amphibian montane diversity. Glob Ecol Biogeogr 30:674–684

Garreaud R (2009) The Andes climate and weather. Adv Geosci 22:3–11

Graham A (2009) The Andes: a geological overview from a biological perspective. Ann Mo Bot Gard 96:371–385

Guillory WX, French CM, Twomey EM et al (2020) Phylogenetic relationships and systematics of the Amazonian poison frog genus *Ameerega* using ultraconserved genomic elements. Mol Phylogenet Evol 142:106638

Heinicke MP, Barrio-Amorós CL, Hedges SB (2015) Molecular and morphological data support recognition of a new genus of New World direct-developing frog (Anura: Terrarana) from an under-sampled region of South America. Zootaxa 3986:151–172

Hoorn C, Wesselingh FP, ter Steege H et al (2010) Amazonia through time: Andean uplift, climate change, landscape evolution, and biodiversity. Science 330:927–931

Hoorn C, Bogotá-A GR, Romero-Baez M et al (2017) The Amazon at sea: onset and stages of the Amazon River from a marine record, with special reference to Neogene plant turnover in the drainage basin. Glob Planet Change 153:51–65

Hoorn C, Lohmann LG, Boschman LM, Condamine FL (2023) Neogene history of the Amazonian flora: a perspective based on geological, palynological, and molecular phylogenetic data. Annu Rev Earth Planet Sci 51:419–446

Huber O (1988) Guayana highlands versus Guayana lowlands, a reappraisal. Taxon 37:595–614

Hughes C, Eastwood R (2006) Island radiation on a continental scale: exceptional rates of plant diversification after uplift of the Andes. Proc Natl Acad Sci USA 103: 10334–10339

Hutchinson DK, Coxall HK, Lunt DJ et al (2021) The Eocene–Oligocene transition: a review of marine and terrestrial proxy data, models and model–data comparisons. Clim Past 17:269–315

Hutter CR, Guayasamin JM, Wiens JJ (2013) Explaining Andean megadiversity: the evolutionary and ecological causes of glassfrog elevational richness patterns. Ecol Lett 16:1135–1144

Jansen O, Gómez RO, Fouquet A et al (2023) First Eocene–Miocene anuran fossils from Peruvian Amazonia: insights into neotropical frog evolution and diversity. Pap Palaeontol 9:e1542

Jaramillo C (2023) The evolution of extant South American tropical biomes. New Phytol 239:477–493

Jenkins CN, Pimm SL, Joppa LN (2013) Global patterns of terrestrial vertebrate diversity and conservation. Proc Natl Acad Sci USA 110:E2602–E2610

Kohlsdorf T, Garland T Jr, Navas CA (2001) Limb and tail lengths in relation to substrate usage in *Tropidurus* lizards. J Morphol 248:151–164

Kok PJR (2013) Islands in the sky: species diversity, evolutionary history, and patterns of endemism of the Pantepui Herpetofauna. PhD Dissertation, Leiden University

Kok PJR, Macculloch RD, Means DB et al (2012) Low genetic diversity in tepui summit vertebrates. Curr Biol 22:R589–R590

Kok PJR, Russo VG, Ratz S et al (2017) Evolution in the South American "Lost World": insights from multilocus phylogeography of stefanias (Anura, Hemiphractidae, *Stefania*). J Biogeogr 44:170–181

Kok PJR, Ratz S, MacCulloch RD et al (2018) Historical biogeography of the palaeoendemic toad genus *Oreophrynella* (Amphibia: Bufonidae) sheds a new light on the origin of the Pantepui endemic terrestrial biota. J Biogeogr 45:26–36

Kok PJR, Broholm TL, van Doorn L et al (2024) Lost in space and time: robust demography and enhanced resilience buffer adverse environmental effects in a highly isolated and sedentary pre-Pleistocene landscape vertebrate. BMC Ecol Evol 24:125

Kozak KH, Wiens JJ (2006) Does niche conservatism promote speciation? A case study in North American salamanders. Evolution 60:2604–2621

Ledo RMD, Colli GR (2017) The historical connections between the Amazon and the Atlantic Forest revisited. J Biogeogr 44:2551–2563

Leite YLR, Kok PJR, Weksler M (2015) Evolutionary affinities of the "Lost World" mouse suggest a late Pliocene connection between the Guiana and Brazilian shields. J Biogeogr 42:706–715

Lötters S, van der Meijden A, Rödder D et al (2010) Reinforcing and expanding the predictions of the disturbance vicariance hypothesis in Amazonian harlequin frogs: a molecular phylogenetic and climate envelope modelling approach. Biodivers Conserv 19: 2125–2146

Mayr E, Phelps WH (1967) The origin of the bird fauna of the south Venezuelan highlands. Bull Am Mus Nat Hist 136:269–328

Mazet N, Morlon H, Fabre P-H, Condamine FL (2023) Estimating clade-specific diversification rates and palaeodiversity dynamics from reconstructed phylogenies. Methods Ecol Evol 14:2575–2591

McCain CM, Grytnes JA (2010) Elevational gradients in species richness. In: Encyclopedia of life sciences (ELS). Wiley, pp 1–10

Meza-Joya FL, Torres M (2016) Spatial diversity patterns of *Pristimantis* frogs in the Tropical Andes. Ecol Evol 6:1901–1913

Moraes D, Pavan D (2018) On the occurrence of *Hemiphractus scutatus* (Spix, 1824) (Anura: Hemiphractidae) in eastern Amazonia. Amphib Reptile Conserv 12(1):5–14

Moraes LJCL, Werneck FP, Réjaud A et al (2022) Diversification of tiny toads (Bufonidae: *Amazophrynella*) sheds light on ancient landscape dynamism in Amazonia. Biol J Linn Soc Lond 136:75–91

Motta AP, Taucce PPG, Haddad CFB, Canedo C (2021) A new terraranan genus from the Brazilian Atlantic Forest with comments on the systematics of Brachycephaloidea (Amphibia: Anura). J Zool Syst Evol Res 59:663–679

Muell MR, Chávez G, Prates I et al (2022) Phylogenomic analysis of evolutionary relationships in *Ranitomeya* poison frogs (Family Dendrobatidae) using ultraconserved elements. Mol Phylogenet Evol 168: 107389

Navas CA (2002) Herpetological diversity along Andean elevational gradients: links with physiological ecology and evolutionary physiology. Comp Biochem Physiol A Mol Integr Physiol 133:469–485

Navas CA (2006) Patterns of distribution of anurans in high Andean tropical elevations: insights from integrating biogeography and evolutionary physiology. Integr Comp Biol 46:82–91

Ortiz DA, Hoskin CJ, Werneck FP et al (2023) Historical biogeography highlights the role of Miocene landscape changes on the diversification of a clade of Amazonian tree frogs. Org Divers Evol 23:395–414

Pellegrino KCM, Rodrigues MT, James Harris D et al (2011) Molecular phylogeny, biogeography and insights into the origin of parthenogenesis in the Neotropical genus *Leposoma* (Squamata: Gymnophthalmidae): ancient links between the Atlantic Forest and Amazonia. Mol Phylogenet Evol 61: 446–459

Pérez-Escobar OA, Zizka A, Bermúdez MA et al (2022) The Andes through time: evolution and distribution of Andean floras. Trends Plant Sci 27:364–378

Petren K, Grant PR, Grant BR, Keller LF (2005) Comparative landscape genetics and the adaptive radiation of Darwin's finches: the role of peripheral isolation. Mol Ecol 14:2943–2957

Pie MR, Faircloth BC, Ribeiro LF et al (2018) Phylogenomics of montane frogs of the Brazilian Atlantic Forest is consistent with isolation in sky islands followed by climatic stability. Biol J Linn Soc Lond 125:72–82

Portik DM, Streicher JW, Wiens JJ (2023) Frog phylogeny: a time-calibrated, species-level tree based on hundreds of loci and 5,242 species. Mol Phylogenet Evol 188:107907

Prates I, Navas CA (2009) Cutaneous resistance to evaporative water loss in Brazilian *Rhinella* (Anura: Bufonidae) from contrasting environments. Copeia 2009:618–622

Prates I, Rodrigues MT, Melo-Sampaio PR, Carnaval AC (2015) Phylogenetic relationships of Amazonian anole lizards (*Dactyloa*): taxonomic implications, new insights about phenotypic evolution and the timing of diversification. Mol Phylogenet Evol 82 (Pt A):258–268

Prates I, Rivera D, Rodrigues MT (2016a) A mid-Pleistocene rainforest corridor enabled synchronous invasions of the Atlantic Forest by Amazonian anole lizards. Molecular 25:5174–5186

Prates I, Xue AT, Brown JL et al (2016b) Inferring responses to climate dynamics from historical demography in neotropical forest lizards. Proc Natl Acad Sci USA 113:7978–7985

Prates I, Melo-Sampaio PR, de Oliveira Drummond L et al (2017) Biogeographic links between southern Atlantic Forest and western South America: rediscovery, re-description, and phylogenetic relationships of two rare montane anole lizards from Brazil. Mol Phylogenet Evol 113:49–58

Prates I, Penna A, Rodrigues MT, Carnaval AC (2018) Local adaptation in mainland anole lizards: integrating population history and genome–environment associations. Ecol Evol 8:11932–11944

Prates I, Melo-Sampaio PR, de Queiroz K et al (2020) Discovery of a new species of *Anolis* lizards from Brazil and its implications for the historical biogeography of montane Atlantic Forest endemics. Amphib-Reptil 41:87–103

Prates I, Singhal S, Marchán-Rivadeneira MR et al (2022) Genetic and ecogeographic controls on species cohesion in Australia's most diverse lizard radiation. Am Nat 199:E57–E75

Rahbek C, Borregaard MK, Antonelli A et al (2019) Building mountain biodiversity: geological and evolutionary processes. Science 365:1114–1119

Recoder R, Prates I, Marques-Souza S et al (2020) Lizards from the Lost World: two new species and evolutionary relationships of the Pantepui highland *Riolama* (Gymnophthalmidae). Zool J Linnean Soc 190:271–297

Reijenga BR, Murrell DJ, Pigot AL (2021) Priority effects and the macroevolutionary dynamics of biodiversity. Ecol Lett 24:1455–1466

Réjaud A, Rodrigues MT, Crawford AJ et al (2020) Historical biogeography identifies a possible role of Miocene wetlands in the diversification of the Amazonian rocket frogs (Aromobatidae: *Allobates*). J Biogeogr 47: 2472–2482

Rodrigues MT, Teixeira M Jr, Recoder RS et al (2013) A new species of *Leposoma* (Squamata: Gymnophthalmidae) with four fingers from the Atlantic Forest central corridor in Bahia, Brazil. Zootaxa 3635:459–475

Rolland J, Silvestro D, Schluter D et al (2018) The impact of endothermy on the climatic niche evolution and the distribution of vertebrate diversity. Nat Ecol Evol 2: 459–464

Rull V (2004a) An evaluation of the Lost World and Vertical Displacement hypotheses in the Chimantá massif, Venezuelan Guayana. Glob Ecol Biogeogr 13:141–148

Rull V (2004b) Is the "Lost World" really lost? Palaeoecological insights into the origin of the peculiar flora of the Guayana Highlands. Naturwissenschaften 91:139–142

Rull V, Montoya E, Nogué S et al (2019) Climatic and ecological history of Pantepui and surrounding areas. In: Rull V, Vegas-Vilarrúbia T, Huber O, Señaris C (eds) Biodiversity of Pantepui. Academic Press, pp 33–54

Salerno PE, Ron SR, Señaris JC et al (2012) Ancient tepui summits harbor young rather than old lineages of endemic frogs. Evolution 66:3000–3013

Santos JC, Coloma LA, Summers K et al (2009) Amazonian amphibian diversity is primarily derived from late Miocene Andean lineages. PLoS Biol 7:e56

Sinervo B, Méndez-de-la-Cruz F, Miles DB et al (2010) Erosion of lizard diversity by climate change and altered thermal niches. Science 328:894–899

Strangas ML, Navas CA, Rodrigues MT, Carnaval AC (2019) Thermophysiology, microclimates, and species distributions of lizards in the mountains of the Brazilian Atlantic Forest. Ecography 42:354–364

Teixeira M, Amaro RC, Recoder R et al (2012) A relict new species of *Oreobates* (Anura, Strabomantidae) from the Seasonally Dry Tropical Forests of Minas Gerais, Brazil, and its implication to the biogeography of the genus and that of South American Dry Forests. Zootaxa 3158:37–52

Titon B Jr, Navas CA, Jim J, Gomes FR (2010) Water balance and locomotor performance in three species of neotropical toads that differ in geographical distribution. Comp Biochem Physiol A Mol Integr Physiol 156:129–135

Vacher J-P, Chave J, Ficetola FG et al (2020) Large-scale DNA-based survey of frogs in Amazonia suggests a vast underestimation of species richness and endemism. J Biogeogr 47:1781–1791

Vacher J-P, Kok PJR, Rodrigues MT et al (2024) Diversification of the terrestrial frog genus *Anomaloglossus* (Anura, Aromobatidae) in the Guiana shield proceeded from highlands to lowlands, with successive loss and reacquisition of endotrophy. Mol Phylogenet Evol 192: 108008

Weir JT (2006) Divergent timing and patterns of species accumulation in lowland and highland neotropical birds. Evolution 60:842–855

Westerhold T, Marwan N, Drury AJ et al (2020) An astronomically dated record of Earth's climate and its predictability over the last 66 million years. Science 369:1383–1387

Wiens JJ, Parra-Olea G, García-París M, Wake DB (2007) Phylogenetic history underlies elevational biodiversity patterns in tropical salamanders. Proc Biol Sci 274: 919–928

Anurans of the Patagonian Andes: A Little-Known Variety of Reproductive and Developmental Strategies

Carmen A. Úbeda, Helen Díaz-Páez, and Felipe E. Rabanal

1 Introduction

Amphibians constitute the group of tetrapod vertebrates that presents the greatest variety of reproductive modes and strategies (Duellman and Trueb 1986). In turn, within amphibians, anurans (frogs and toads) constitute the order with the greatest diversification. This wide variety displayed by anurans has been studied mostly for assemblages of tropical forests of the planet, especially from the neotropics (e.g., Crump 1974, 2015; Haddad and Prado 2005). On the contrary, little is known about these aspects in complete faunas of amphibians from temperate and cold climates.

In this chapter, we analyze the reproductive strategies and larval ecological diversity of the anurans inhabiting the Patagonian Andes, a wide territory that extends in the southernmost territories of Chile and Argentina. This sector is inhabited by a varied fauna of anuran that presents a high degree of endemism, at the level of species, genus, and even family, and unique reproductive peculiarities (Formas 1979). In addition, it contains the southernmost anurans on the planet.

C. A. Úbeda (✉)
Universidad Nacional del Comahue, Bariloche, Argentina (now retired)

H. Díaz-Páez
Universidad de Concepción, Ciencias Básicas, Escuela de Educación, Los Ángeles, Chile

F. E. Rabanal
Proyecto Anfibios de Chile, Valdivia, Chile

2 Modes and Reproductive Strategies in Anurans and Ecological Diversity of Larvae

Amphibians present varied life cycles, which has allowed them to inhabit under different environmental conditions. Over two-thirds (68.1%) of the species of anurans with known reproductive modes have a complex and biphasic life cycle (Liedtke et al. 2022), with an aquatic phase that involves oviposition in water with external fertilization, the hatching of an aquatic larva (tadpole) that grows and develops until its metamorphosis, and a terrestrial or semi-aquatic phase in which the juvenile grows and matures to become a reproductive adult. However, there are multiple variants to this generalized ancestral scheme that make anurans unique among vertebrates, by adopting various forms of reproduction, which include from the place of oviposition and the type of egg and larva to the patterns of development of these (Wells 2007). In their evolution, some species would have migrated from the aquatic environment in search of reducing the predation of eggs and larvae and the risk of desiccation for the most vulnerable stages, resulting in the diversification of terrestrial strategies

J. N. Urbina-Cardona et al. (eds.), *Andean Herpetofauna*, Sustainable Development Goals Series,
https://doi.org/10.1007/978-3-032-00074-3_8

(Duellman and Trueb 1986). For several authors, terrestrial reproductive modes would be the trigger of the evolution of parental care, a behavior that is key to ensuring the survival of eggs and larvae in conditions of desiccation, and moisture-limiting conditions (Wells 2007).

The concept of reproductive mode in amphibians combines oviposition site, characteristics of eggs and clutches, trophism of the larva, type of larval development, and type of parental care, if it exists (Salthe 1969; Salthe and Duellman 1973; Salthe and Mecham 1974; Duellman and Trueb 1986; Duellman 1989). Since the nineteenth century, many classifications of reproductive modes in amphibians have been proposed (e.g., Boulenger 1886; Salthe and Duellman 1973; Crump 1974, 1982, 2015; Salthe and Mecham 1974; Duellman 1989). A significant contribution was made by Duellman and Trueb (1986), who recognized 29 reproductive modes for anurans based on a series of biological and environmental traits, such as the site of oviposition (land, water, or retention in the oviducts), the type of egg and clutch structure, the pattern of embryonic and larval development, the type of larval feeding or its absence, and parental care when it exists. Over time, the number of modes increased as new cases were discovered (e.g., 39 modes, Haddad and Prado 2005). The number continued to increase until Nunes-de-Almeida et al. (2021) redefined the concept of reproductive mode by incorporating some parameters, mainly physical, and removing other purely behavioral aspects, such as parental care. With this redefinition of the concept, they managed to approximately double the number of reproductive modes in amphibians (Nunes-de-Almeida et al. 2021).

Another frequently used concept is the reproductive strategy, which Wilbur (1977) defined as "a set of tactics that has been selected as the adaptations that have contributed, on the average, the greatest number of offspring to recent generations." An example of reproductive strategies are the well-known "r" and "K" strategies, which were discussed by Wilbur (1977).

The eggs and clutch structures of amphibians present a great variety, which Altig and McDiarmid (2007) described and classified by grouping them into independent eggs, three-dimensional arrangements, floating, froth nests (foam and bubbles), and linear arrangements.

On the other hand, anuran larvae present an extraordinary diversity of shapes, sizes, use of microhabitat, and behavior. The body shape and the fin's configuration vary across habitats and the morphological diversity of all the components of the oral apparatus is extraordinary (Altig and McDiarmid 1999). In an attempt to interpret larval morphology in the context of its ecology, Orton (1953) recognized different types of anuran larvae, derived from a generalized form (a typical larva of lentic waters), adapted to different ecological conditions or different forms of life, and highlighted that these types appeared repeatedly and independently through convergent evolution. Later, Altig and Johnston (1989) presented a grouping of anuran larvae into ecomorphological guilds, which are based on a heterogeneous set of characters such as developmental mode, microhabitat, feeding behavior, and external morphology. They recognized two major forms of development: endotrophic larvae, which obtain energy for development exclusively from parental sources (usually the yolk), and exotrophic larvae, which obtain their energy orally by consuming many types of food, as free-swimming larvae, and defined several types within each group. This classification scheme was revised by McDiarmid and Altig (1999), who regrouped the exotrophic larvae into a smaller number of types. More recently, Vera Candioti et al. (2023) classified all the anuran larvae of the world into seven ecomorphological guilds resulting from summarizing the ecomorphological guilds of McDiarmid and Altig (1999).

On a regional scale, there is a bibliographic background about reproductive modes and strategies and larvae of some of the groups of frogs or certain anurofaunas of Chile and Argentina (e.g., Formas 1981; Díaz and Valencia 1985; Lavilla 1988; Lavilla and Rougés 1992; Soto et al. 2008; Grosso et al. 2022).

3 The Patagonian Andes

The study area extends from the 36°S parallel southward along the Andes. In the Strait of Magellan, at 53°S, the mountain range is interrupted, to reappear on the Island of Tierra del Fuego, where this mountain range adopts a west–east orientation (Chioza 1976).

With a mountainous relief, the landscape ranges from high peaks, with hills that exceed 3000 m, to deep low-altitude valleys, with a west–east orientation, some of which cross the Andes transversely. The region is crossed by an extensive hydrographic network that drains its waters to the west into the Pacific Ocean or to the east into the Atlantic. There is a great variety of lotic and lentic, permanent and temporary water bodies, in addition to a series of related wetlands, such as peat bogs and *mallines*, which are fundamental for the varied fauna of anurans that inhabit them (Úbeda 1998).

In general terms, in the Patagonian Andes, the temperature decreases with the increase in latitude and altitude, with temperate to cold climates. The precipitation of the Andean region is produced by the effect of air currents coming from the Pacific Ocean, which, when colliding with the mountain range, rise, cool, and condense, depositing their moisture in the form of copious rain and snowfall. The rain shadow effect exerted by the Andes, which stands perpendicular to these moisture-laden winds, determines the existence of a marked gradient of decreasing precipitation in a west–east direction (Paruelo et al. 1998), which gives rise to one of the most marked vegetation gradients in the world (Coronato et al. 2017).

In the wide latitudinal, longitudinal, and altitudinal range of the Patagonian Andes, with complex topographies and under different climates, various macroenvironments develop, with a great variety of types of vegetation, from rainy temperate forests in the west, passing through to the north and east by transition forests, up to arid environments with xerophytic vegetation to the north and east of the Patagonian mountain range.

The austral temperate forests represent a biogeographic island separated from other forests on the continent by arid formations (Armesto et al. 1995). *Rainforests* dominated by species of *Nothofagus*, like the so-called "Valdivian rainforest", occur in areas of higher precipitation, particularly on the slopes and valleys of the western sector of the mountain range and in a narrow neighboring strip on the eastern side of the range (Fig. 1a). These are cold forests with several tree strata, dense and diverse understory, and a large number of climbing plants and epiphytes. In the southern region the rainforest is structurally less complex and alternate with peatlands (Fig. 1b).

Transition and ecotonal forests are found towards the east, with decreasing rainfall and a water deficit during the summer (Fig. 1c). These are drier, and present a distinctive composition and structure. In the northern sector, there are forests of *Araucaria,* pure or mixed with deciduous *Nothofagus* species, and forests of Patagonian cypress (*Austrocedrus chilensis*). In the southern region towards the steppe, the transition forests are mainly constituted by deciduous forests of *N. pumilio* and of *N. antarctica* (Fig. 1d).

As precipitation decreases even further to the east, in the eastern strip of the mountain range extend the *arid environments*, with different types of shrubby and herbaceous steppes that continue in the extensive extra-Andean Patagonian territory (Fig. 1e).

The ecosystems of the *high mountain environments* or high Andean, along with other similar systems around the world, make up the alpine biome, which includes all ecosystems that are found above the upper limit of the forests or tree line. These ecosystems, which extend to the line of permanent snow, are characterized by very harsh and restrictive ecological conditions for the development of vegetation and fauna and are among the most unfavorable terrestrial systems on the planet (Körner 2001). These have an alpine-type cold climate, with marked seasonality, with long winters and, for certain latitudes, several months of snow and ice coverage. In addition, they are subject to high insolation, with high UV radiation, and the partial pressure

Fig. 1 Macroenvironments of anuran species of the Patagonian Andes. (**a**) Valdivian rainforest. (**b**) Peatbog in the southern rainforest. (**c**) Montane transitional forest of *Nothofagus pumilio*. (**d**) Ecotonal transitional forest with Patagonian steppe. (**e**) Patagonian steppe and wetland (*mallín*) in the foreground. (**f**) High Andean landscape with oligotrophic melting lake

of oxygen and temperature decrease with altitude (Körner 2001).

Environmental heterogeneity, given by the interaction between the altitude, precipitation, orientation, and slope (which condition exposure to the sun and wind and the accumulation of snow) and the type of soil, determines the existence of various conditions for life, which allow the development of different plant communities (Ferreyra et al. 1998). The semi-deserts and high mountain steppes are the most widespread types of vegetation, on non-existent soils or with little water retention capacity (Fig. 1f) (Ferreyra et al. 1998). In sectors with a generally flat or flat–concave relief and gentle slope, on poorly drained soils, wetlands develop (*mallines* or *vegas*), evergreen, dense, and wet meadows.

In the high Andean landscape there is a variety of water bodies, the most conspicuous being the oligotrophic lakes. Fed by rainwater and meltwater (Fig. 1f), they are hyposaline, slightly acidic, and extremely transparent, with low attenuation of UV radiation, and freeze superficially for 6–8 months a year (Zagarese et al. 2000). These lakes typically drain through emissary streams with rocky bottoms and crystalline waters. Other streams flow along the slopes and end in wetlands or lakes, or continue their course toward their baseline. Within the wetlands, the microrelief determines the existence of a variety of shallow aquatic microenvironments, both lotic and lentic, permanent or temporary (Úbeda 2021).

4 Anurofauna of the Patagonian Andes

4.1 Exploring the Origins of the Patagonian Anurans

The evolution of the various reproductive modes of amphibians is determined by the various environmental pressures to which they are exposed, thus, the climatic characteristics of the habitat of the species assume great relevance (Wells 2007; Crump 2015). That is why, for adequate understanding of the diversity of the reproductive modes present in the anurans that inhabit current Patagonia, it is necessary to know their origins.

The complex geological history of South America is marked by orogenic processes, marine transgressions and regressions, glacial advances and retreats, and volcanism, all processes that have affected the anurofaunas present in Patagonia (Hoorn et al. 2010; Barcelos and Dos Santos 2023). The paleoclimatic conditions of Patagonia, with subtropical characteristics during the Paleocene, gave way during the Oligocene to a more arid and cold climate, marking a notable climatic change. This process had a high biological significance, which turned Patagonia from being an extraordinary center of evolution and probably also of dispersion, into a territory that became increasingly arid, especially east of the Andes range (Suazo and Gómez 2022).

Fossil records show that Patagonia had a large presence of anurans from the Pipidae family, currently extinct in its Patagonian distribution, and Calyptocephalellidae, which is currently present with the genera *Calyptocephalella* and *Telmatobufo*, which are the representatives of the oldest lineage of the current fauna (Barcelos and Dos Santos 2023).

Glacial processes affected the prevailing climate in Patagonia, causing the retraction of forests. In this scenario, the anurofauna was exposed to environmental variations, with populations isolated in forest refuges where species had to adapt to the new conditions provided by the environment (Scolaro et al. 2003). Under this dynamic of environmental changes, the different representatives of the anuran families that make up the current anurofauna of Chilean–Argentine Patagonia (Alsodidae, Batrachylidae, Bufonidae, Rhinodermatidae, Calyptocephalellidae and Leptodactylidae) were consolidated.

The existence of a temperate Patagonia would be associated with the predominance of the generalized aquatic reproductive mode (Modes 1 and 2 of Duellman and Trueb 1986 and Haddad and Prado 2005) in current Patagonian anurans that is typical of temperate environments (Duellman and Trueb 1986; Wells 2007; Nunes-de-Almeida et al. 2021). The climatic changes imposed by the Pleistocene glaciations make it necessary to analyze the selective pressures that these environmental variations would have imposed on the reproductive and developmental aspects of Patagonian anurofauna and their relationship with the emergence of terrestrial reproductive modes.

4.2 The Current Anurofauna of the Patagonian Andes

Groups of amphibians belonging to different evolutionary lineages inhabit the extensive Patagonian territory, many of them native to the region. Particularly, in the Patagonian Andes, south of the 36°S parallel, there is a varied fauna of anurans, rich in endemisms. Based on current taxonomic knowledge (Díaz-Páez et al. 2011; Blotto et al. 2013; Barrasso and Basso 2019; Pereyra et al. 2021; Correa 2022), this fauna includes 28 species of native anurans, grouped into 12 genera and six families (Table 1). However, owing to the refinement and emergence of new tools at the service of taxonomy, the aforementioned panorama may vary in the near future as currently, some taxa are in the process of revision in order to clarify the identity of some species.

As occurs in other parts of the globe, in Patagonia the specific richness decreases with the increase in latitude (Formas 1979; Úbeda 1998). Therefore, at the southern end of the Patagonian Andes, at the latitude of 52°S, only four species of anurans (Batrachyla antartandica, Nannophryne variegata, Chaltenobatrachus grandisonae and Pleurodema bufoninum) persist, with the last two species extending beyond 53°S, to become the southernmost species on the planet (Ortiz 2015; Cisternas-Medina et al. 2019; Correa 2022). The Island of Tierra del Fuego, which includes the southermost portion of the Andes Range, has no known amphibians species to date.

Anurans in the Patagonian Andes show a differential distribution within the large macroenvironments of the region, according to their ecological requirements and life strategies (Table 1).

This anurofauna inhabits a wide variety of environments and climates, and deploys very different life strategies. To illustrate the modes and reproductive strategies of these anurans, Tables 2 and 3 summarize the characteristics of their reproduction, and embryonic and larval development. The compiled information includes our own field data and a wide bibliography.

Most of the species covered in this study (60.27%) exhibit a generalized reproductive mode, that is, aquatic eggs and exotrophic aquatic larvae without parental care, and are included in the genera, *Telmatobufo, Atelognathus, Chaltenobatrachus, Hylorina, Alsodes, Pleurodema, Nannophryne*, and *Rhinella* (Tables 2 and 3). This generalized reproductive mode is typical of unpredictable environments, in which the population size can vary from year to year depending on environmental conditions (Duellman 1989). On the contrary, the species of three genera, *Batrachyla*, *Eupsophus*, and *Rhinoderma*, deploy specialized reproductive modes, with different degrees of terrestriality, which are described later in the framework of their respective environments (Sect. 4.2.1.1).

All species of the Patagonian Andes are ovuliparous, whether they present aquatic or terrestrial amplexus and oviposition. For most of the species where the amplexus is known, it is of the axillary type (Table 2), a type considered derived and more advantageous than the inguinal (Wells 2007). Some species also present an inguinal amplexus.

The vast majority of the species covered in this study have a seasonal pattern of reproduction and embryonic development, in the Austral spring–summer period, in line with what is expected for species of anurans from temperate climates of the world that present a marked seasonality (Duellman and Trueb 1986; Duellman 1989). Species of the genus *Batrachyla* have a different timing in their reproduction and embryonic development, concentrated in the Austral autumn, which we develop in Sect. 4.2.1.1.

Regarding larvae, from a morphological point of view, the majority of species have a generalized type larvae (sensu Orton 1953; Vera Candioti et al. 2023). From a trophic point of view, the majority of species have exotrophic larvae or active feeders (Table 3 and Fig. 2). Species of *Eupsophus* and *Rhinoderma* have endotrophic larvae, that is, they develop at the expense of the

Table 1 Anurans of the Patagonian Andes and their occurrence by macroenvironment. The macroenvironments are described in Sect. 3

Family	Species	Forests		Arid environments	High mountain environments
		Rainforests	Transitional and Ecotonal forests		
Calyptocephalellidae	*Calyptocephalella gayi*	X			
	Telmatobufo australis	X	X		
	Telmatobufo venustus	X	X		
Rhinodermatidae	*Rhinoderma darwinii*	X			
Batrachylidae	*Atelognathus nitoi*		X	X	
	Atelognathus patagonicus			X	X
	Batrachyla antartandica	X			
	Batrachyla fitzroya	X			
	Batrachyla leptopus	X			
	Batrachyla nibaldoi	X			
	Batrachyla taeniata	X			
	Chaltenobatrachus grandisonae	X			
	Hylorina sylvatica	X			
Alsodidae	*Alsodes australis*	X			
	Alsodes coppingeri	X			
	Alsodes gargola	X	X		X
	Alsodes igneus	X	X		
	Alsodes neuquensis		X	X	X
	Alsodes verrucosus	X	X		
	Eupsophus calcaratus	X			
	Eupsophus emiliopugini	X			
	Eupsophus roseus	X			
Leptodactylidae	*Pleurodema bufoninum*		X	X	X
	Pleurodema thaul	X	X		X
Bufonidae	*Nannophryne variegata*	X	X		X
	Rhinella papillosa	X	X	X	X
	Rhinella rubropunctata	X	X		
	Rhinella spinulosa	X	X	X	X

egg yolk or another parental source, without actively feeding. Among the exotrophic larvae, there are larvae with different morphological adaptations and habits, which can be classified as lentic–benthic, lentic–nectonic, and lotic–benthic (Table 3), following the concept of ecomorphological guilds proposed by Altig and Johnston (1989). These will be discussed in the

Table 2 Reproductive characteristics, oviposition, and embryonic development of anurans of the Patagonian Andes

Species	Max. adult size (mm)	Breeding season	Amplexus	Oocyte diameter (mm)	Yolk amount	Ovum pigmentation	Clutch size (number of eggs/ clutch)	Clutch structure	Oviposition sites and amplexus	Clutch and amplexus microhabitats	Clutch parental care	Duration of embryonic period	References
Calyptocephalella gayi	320	Austral spring–summer (August–January/ February)	Axillary	2.7–3.1	Intermediate	Strongly pigmented in animal hemisphere	1000–10,000	Initially globular, it expands in contact with water (Fig. 4b)	Aquatic	Shallow areas with abundant aquatic vegetation	During first days of embryonic development	4–9 days	Oliver Schneider (1930), Jorquera and Izquierdo (1964), Formas (1979), Hermosilla and Acuña (2004), Vallejos and Pizarro (2004), Vélez et al. 2014
Telmatobufo australis	75	NA	NA	8.4 ± 0.43	Abundant	Scarce or absent	NA, Presumably small	NA Presumably grouped eggs	Aquatic	Lotic waters, eggs are presumably adhered to the bottom of rocks	NA	NA	Rabanal and Núñez (2008), Formas et al. (2001)
Telmatobufo venustus	70	NA	NA	4.7 ± 0.24	Abundant	Scarce or absent	NA Presumably small	NA Presumably grouped eggs	Aquatic	Lotic waters, eggs are presumably adhered to the bottom of rocks	NA	NA	Rabanal and Núñez (2008), Formas et al. (2001)
Rhinoderma darwinii	30	Throughout the year, with predominance between September and March	Loose and repeated	3–4	Abundant	Absent	Typically 4–24 eggs/an exceptional single record of 34 eggs	Cluster	Terrestrial	Humid ground, with abundant surrounding vegetation	In captivity, probably attended by male	19–25 days	Wilhelm (1927), Wilhelm (1932), Pflaumer (1935), Jorquera

													et al. (1972, 1981), Busse (2004), Cabeza-Alfaro et al. (2021)
Atelognathus nitoi	45	Austral early spring–early summer (October–January)	Axillary	1.4–2.3	42.6 ± 8.8%	Intense in animal hemisphere	50–297	Spherical to ovoid mass attached to submerged small branches or stems (Fig. 4d)	Aquatic	Shallow waters in permanent and temporary vegetated lagoons	Absent	NA	Úbeda et al. (1999), Meriggio et al. (2004), Úbeda (2006), Grosso et al. (2022), Úbeda (personal observation)
Atelognathus patagonicus	50	Austral spring	NA	NA	NA	Intense	NA	Mass	Aquatic	Clusters attached to aquatic vegetation of steppe lagoons and high Andean lakes	NA	NA	Cei (1965), Cei and Roig (1968), Úbeda (2021)
Batrachyla antartandica	45	Austral summer–early autumn (December–May)	Inguinal/ axillary	2.5–4	58.4 ± 9%	Pigmented in the animal hemisphere and ashy in the vegetal hemisphere	45–74	Clusters of independent adhered eggs (Fig. 3b)	Aquatic and alternatively in phytotelmata	Eggs attached to cavities in *Sphagnum* peat bogs, floodable ground in wetlands, and phytotelmata walls	Present	From 10 days in water, no data on the ground	Barrio (1967a), Formas and Pugín (1971), Formas (1976), Díaz et al. (1987), Rabanal et al. (2021), Grosso et al. (2022), Úbeda (personal observation)

(continued)

Table 2 (continued)

Species	Max. adult size (mm)	Breeding season	Amplexus	Oocyte diameter (mm)	Yolk amount	Ovum pigmentation	Clutch size (number of eggs/ clutch)	Clutch structure	Oviposition sites and amplexus	Clutch and amplexus microhabitats	Clutch parental care	Duration of embryonic period	References
Batrachyla fitzroya	35.2	Presumably austral summer–early autumn	NA	NA	Abundant	Intense in the animal hemisphere	NA	Cluster of eggs	Terrestrial	Humid floodable ground, under refuges, with abundant surrounding vegetation	NA	NA	Úbeda (personal observation)
Batrachyla leptopus	40	Austral summer–autumn (January–April)	Inguinal/ axillary	3–4	60.4 ± 5.1%	Intense in animal hemisphere	60–220	Clusters of eggs scattered	Terrestrial	Humid floodable ground, under refuges	Present	From 16 days in water to 66 days on ground	Busse (1971), Formas (1976), Díaz et al. (1987), Úbeda and Nuñez (2006), Jara (2022), Grosso et al. (2022)
Batrachyla nibaldoi	42	Austral summer (January–February)	Axillary (Fig. 3a)	cCa. 2.7	Abundant	Intense in the animal pole	30	Groups of stacked eggs (Fig. 3a)	Terrestrial	Floodable ground in peat bogs and wetlands	Present	NA	Rabanal (2010, 2011), Rabanal (personal observation)
Batrachyla taeniata	46	Austral late summer–autumn (FebruaryMay)	Axillary	1.6–3	Abundant	Intense in the animal pole	22–399	Groups of stacked eggs (Fig. 4a)	Terrestrial	Floodable ground, under refuges such as logs, barks, and vegetation	Present	From 6 weeks in water to 64 days on ground	Formas (1976), Jorquera and Molinari (1979), Díaz et al. (1987), Úbeda and Nuñez (2006), Jara et al. (2021), Grosso et al. (2022)

Chaltenobatrachus grandisonae	46	Austral spring (October–December)	Axillary	NA	Moderate	Intense in the animal hemisphere	14–30	Submerged mass of eggs adhered to small branches, *Sphagnum*, or rocks (Fig. 4e)	Aquatic	Shallow lagoons, vegetated pools, and cavities in peat bogs	Absent	NA	Basso et al. (2011), Cisternas et al. (2013), Díaz-Páez (personal observation)
Hylorina sylvatica	66	Austral spring–early summer (October–January)	Axillary	1.71–3.06	51.5 ± 5.2%	Scarce	400–600	Single eggs scattered or in groups of 2 to 4, fallen on the bottom or adhered to aquatic vegetation	Aquatic	Shore areas with aquatic vegetation in lagoons	Absent	10–16 days	Barrio (1967b), Formas and Pugín (1978b), Formas (1981), Díaz (1986), Charrier (2019), Jara (2022), Úbeda et al. (2022), Grosso et al. (2022)
Alsodes australis	56.8	Presumably austral spring	NA	NA	NA	NA	NA	NA	Aquatic	Presumably under rocks in mountain streams and high-altitude lakes	NA	NA	Formas et al. (1997)
Alsodes coppingeri	58.7	Presumably austral spring/summer	NA	NA	NA	Presumably absent	NA	NA	Aquatic	Under rocks in mountain streams and high-altitude lakes	NA	NA	Formas et al. (2008), Barrasso et al. (2022)

(continued)

Table 2 (continued)

Species	Max. adult size (mm)	Breeding season	Amplexus	Oocyte diameter (mm)	Yolk amount	Ovum pigmentation	Clutch size (number of eggs/ clutch)	Clutch structure	Oviposition sites and amplexus	Clutch and amplexus microhabitats	Clutch parental care	Duration of embryonic period	References
Alsodes gargola	70	Austral summer (December–January)	NA	NA	NA	NA	30–40	Clusters of eggs	Aquatic	Under submerged rocks or in underwater cavities on the edges of mountain streams and high-altitude permanent pools and lakes	NA	NA	Gallardo (1970), Úbeda (2021)
Alsodes igneus	66.7	Presumably austral spring	NA	$\bar{x} = 3.92$ (intra-ovarian)	NA	Animal pole black and vegetal pole creamy–yellowish	NA	NA	Aquatic	Presumably under rocks in mountain streams	NA	NA	Cuevas and Formas (2005)
Alsodes neuquensis	56	NA	NA	NA	NA	NA	NA	NA	Aquatic	Presumably under submerged rocks or in underwater cavities on the edges of mountain streams and high-altitude permanent pools and lakes	NA	NA	Cei (1976, 1980), Úbeda (2021)
Alsodes verrucosus	55	NA	NA	NA	NA	NA	NA	Presumably cluster of eggs	Aquatic	Presumably under submerged rocks in mountain streams	NA	NA	Rabanal and Núñez (2008)

Eupsophus calcaratus	41	Austral winter-spring (June–December)	NA	NA	Abundant	Absent	NA	Cluster of eggs, partially chained (Fig. 3c)	Terrestrial	Cavities and tunnels with water in the forest ground and peat bogs	Present	NA	Márquez et al. (2005), Rabanal and Nuñez (2008), Rabanal (personal observation)
Eupsophus emiliopugini	64	Austral spring (October–November)	NA	5.78–7.2	Abundant	Absent	88–166	Cluster of eggs	Terrestrial	Water-filled hollows under moss on the bank of streams and peat bogs	Present	NA	Formas and Vera (1980), Formas (1989b), Úbeda (1998), Úbeda and Nuñez (2006), Rabanal and Núñez (2008), Vera Candioti et al. (2011), Rabanal (personal observation)
Eupsophus roseus	50	Austral spring (September–December)	NA	2.8–4.0	Abundant	Absent	53–206	Cluster of eggs, partially chained	Terrestrial	Cavities and tunnels with water in the forest ground	Present	NA	Formas and Vera (1980), Úbeda and Nuñez (2006), Vera Candioti et al. (2005)
Pleurodema bufoninum	55	Austral spring (September–December depending on altitude)	Inguinal	1.26–2.94	Moderate	Intense in the animal hemisphere	513–2913	Strands of irregular diameter	Aquatic	Shallow and sunny waters (small ponds, and backwaters in temporary or permanent streams)	Absent	1–2 days	Weigandt et al. (2004), Jara et al. (2007)

(continued)

Table 2 (continued)

Species	Max. adult size (mm)	Breeding season	Amplexus	Oocyte diameter (mm)	Yolk amount	Ovum pigmentation	Clutch size (number of eggs/ clutch)	Clutch structure	Oviposition sites and amplexus	Clutch and amplexus microhabitats	Clutch parental care	Duration of embryonic period	References
Pleurodema thaul	66	Austral late winter–early summer (July–January)	Inguinal	1.4–2.48	Moderate	Intense in the animal hemisphere	213–>1000	Strands that can be coiled around in the submerged vegetation, adopting a globular shape (Fig. 4f)	Aquatic	Bottom or over vegetation of shallow and sunny waters (ponds and small lagoons typically vegetated)	Absent	4–8 days	Díaz (1986), Díaz-Páez and Ortiz (2001), Izquierdo and Pereda (1964), Formas (1979), Jara and Perotti (2009), Moncada and Úbeda (2020)
Nannophryne variegata	57	Austral spring (September–December)	Axillary	Small	Scarce	Very intense	34–450	Strings, the eggs are laid in a double spiral string	Aquatic	Shallow waters, peat bogs, wetlands, calm streams, and backwaters	Absent	Short	Hock (1967), Formas and Pugín (1978a), Cisternas et al. (2013), Úbeda (1998, 2021)
Rhinella papillosa	100	Austral spring, explosive or prolonged breeding	Axillary	1.56 ± 0.9	Scarce	Dark brown, almost black, in a the animal hemisphere toward creamy pale gray in the vegetal pole	3250–7100	Long strings of very uniform diameter (Fig. 4c)	Aquatic	Shallow waters (small ponds, and backwaters in streams)	Absent	4 days	Jara et al. (2006, 2007), Sympson et al. (2006)

Rhinella rubropunctata	65	Austral spring	NA	NA	NA	NA Presumably intense	NA Presumably numerous	Presumably strings	Aquatic	Shallow waters on the shores of lakes and reservoirs, inundation lagoons in alluvial plains	Absent	NA	Veloso (2006), Formas and Pugín (1978b), Úbeda (personal observation)
Rhinella spinulosa	95	Austral late winter–early summer, explosive or prolonged breeding	Axillary	1.5–2.8	Scarce	Very intense in the animal hemisphere	Hundreds to thousands	Long strings of uniform diameter	Aquatic	Shallow waters (ponds, small lagoons, and backwaters in streams)	Absent	NA	Cei (1961), Soto et al. (2008), Formas (1979) (as *Bufo arunco*), Díaz-Páez (personal observation)

For species whose distribution exceeds the study area, only the data corresponding to the populations of Patagonian Andes are provided. Species are ordered by family according to Table 1

NA data not available

Table 3 Characteristics of the larvae and larval development of the anurans of the Patagonian Andes

Species	Larval pigmentation	Type of larva (according to trophism and habit)	Larval ecomorphological type (sensu Altig and Johnston 1989)	Larval environment	Larval habitat	Parental care of larvae	Maximum larval size (mm)	Duration of larval period	Froglet size (mm)	References
Calyptocephalella gayi	Intense	Exotrophic, free living	Lentic–benthic (Fig. 2b)	Aquatic	Lentic waters (vegetated lagoons)	Absent	200	Up to 4 years	22–67	Díaz (1986), Hermosilla and Acuña (2004), Vallejos and Pizarro (2004), Díaz-Páez (personal observation), Rabanal (personal observation)
Telmatobufo australis	Intense	Exotrophic, free living	Lotic–benthic suctorial (Fig. 2e)	Aquatic	Rushing mountain streams	Absent	73.3 ± 4.99	NA	35.6	Formas (1972, 1979), Catchpole and Medina (2018), Formas et al. (2001)
Telmatobufo venustus	Intense	Exotrophic, free living	Lotic–benthic suctorial	Aquatic	Rushing mountain streams	Absent	107.4	NA	41	Díaz et al. (1983), Díaz-Páez and Alveal (2021)
Rhinoderma darwinii	Intense	Endotrophic (yolk and nutritional supplies secreted by the male's vocal sac)	Endotrophic exoviviparous	Male vocal sac	Inner vocal sac	Neomelia (inside the vocal sac) (Fig. 3d)	18	52–53 days	12	Jorquera et al. (1972, 1981), Busse (2004), Lavilla (1987), Formas (2013)
Atelognathus nitoi	Intense	Exotrophic, free living	Lentic–benthic	Aquatic	Permanent and temporary lagoons	Absent	75	Dual strategy with seasonal larvae (5 months) and annual larvae with one overwintering (12–13 months)	17–23	Basso and Úbeda (1997), Úbeda and Basso (2003), Úbeda et al. (1999), Meriggio et al. (2004), Úbeda (2006), Úbeda (personal observation)

Atelognathus patagonicus	Intense	Exotrophic, free living	Lentic–benthic	Aquatic	Typically permanent lagoons and high Andean lakes	Absent	77	Dual strategy with seasonal larvae (5 months) and annual larvae with one overwintering	21–24	Cei (1965), Cuello et al. (2014, 2017), Úbeda (2021)
Batrachyla antartandica	Intense	Exotrophic, free living	Lentic–benthic	Aquatic	Shallow waters in lagoons and peat bogs; alternatively phytotelmata	Absent; possibly present in phytotelmata	50	7–12 months with one overwintering	13–16	Formas (1976), Formas and Pugín (1971), Úbeda (1998), Rabanal et al. (2021), Úbeda (personal observation)
Batrachyla fitzroya	NA	NA	NA	Aquatic	NA	NA	NA	NA	NA	Úbeda (personal observation)
Batrachyla leptopus	Intense	Exotrophic, free living	Lentic–benthic	Aquatic	Temporary and permanent shallow and vegetated lagoons, small pools in the forest	Absent	50.39	7–9 months with one overwintering	11.61–15.23	Busse (1971), Jara (2022)
Batrachyla nibaldoi	Intense	Exotrophic, free living	Lentic–benthic	Aquatic	Temporary bodies of water, shallow lagoons, and small pools in peat bogs	Absent	32	7 months, with one overwintering	14	Rabanal (2011)
Batrachyla taeniata	Intense	Exotrophic, free living	Lentic–benthic (Fig. 2c)	Aquatic	Vegetated lagoons, temporary bodies of water, shallow lagoons	Absent	49.15	6–9 months with one overwintering	12–14	Formas (1976), Sallaberry et al. (1981), Díaz et al. (1987), Jara et al. (2021), Jara (2022)
Chaltenobatrachus grandisonae	Intense	Exotrophic, free living	Lentic–benthic	Aquatic	Temporary and permanent small shallow ponds, vegetated with	Absent	56	Dual, seasonal larvae (10–12 weeks) or annual larvae	15–18	Úbeda (1998), Basso et al. (2011), Cisternas et al. (2013)

(continued)

Table 3 (continued)

Species	Larval pigmentation	Type of larva (according to trophism and habit)	Larval ecomorphological type (sensu Altig and Johnston 1989)	Larval environment	Larval habitat	Parental care of larvae	Maximum larval size (mm)	Duration of larval period	Froglet size (mm)	References
					organic matter in the bottom, small ponds in peat bogs			with one overwintering		
Hylorina sylvatica	Moderate	Exotrophic, free living	From lentic–nectonic to lentic-benthic (Fig. 2d)	Aquatic	Typically permanent lentic and vegetated water bodies	Absent	84	12–14 months	21–28.5	Formas and Pugín (1978b), Úbeda (1998), Cárdenas-Rojas et al. (2007), Jara (2022), Úbeda et al. (2022)
Alsodes australis	Intense	Exotrophic, free living	Lotic–benthic	Aquatic	Mountain streams	Absent	66.5 ± 5.2	Presumably very long (years)	NA	Formas et al. (1997)
Alsodes coppingeri	Intense	Exotrophic, free living	Lotic–benthic	Aquatic	Mountain streams	Absent	66.5	Presumably very long (years)	NA	Formas et al. (2008), Barrasso et al. (2022)
Alsodes gargola	Intense	Exotrophic, free living	Lotic–benthic/ lentic–benthic	Aquatic	Backwaters and slow water sites in streams, high Andean lakes	Absent	95	4 years	20–26	Gallardo (1970), Úbeda (1998), Pillado et al. (2000), Logares and Úbeda (2006), Úbeda and Casanovas (2012), Úbeda (2021)
Alsodes igneus	Intense	Exotrophic, free living	Lotic–benthic	Aquatic	Mountain streams	Absent	61	Presumably very long (years)	NA	Cuevas and Formas (2005)
Alsodes neuquensis	Intense	Exotrophic, free living	Lotic–benthic/ lentic–benthic	Aquatic	Backwaters and slow water sites in streams, high Andean lakes	Absent	80	Very long	22–26	Cei (1970, 1976), Pillado et al. (2000), Barrasso et al. (2016), Úbeda (2021)

Alsodes verrucosus	Intense	Exotrophic, free living	Lotic–benthic	Aquatic	Mountain streams	NA	60	Presumably very long	NA	Formas and Brieva (2004)
Eupsophus calcaratus	Absent	Endotrophic	Nidicolous	Aquatic	Cavities and tunnels with water in the forest ground and peat bogs	Present	22.1	At least 50 days	6.5–7.1	Formas (1989a)
Eupsophus emiliopugini	Absent	Endotrophic (Fig. 2f)	Nidicolous	Aquatic	Water-filled hollows under moss on the bank of streams and peat bogs	Present	24.7	64 days	8.2–10.6	Formas and Pugín (1978b), Formas (1989b), Vera Candioti et al. (2011), Rabanal (personal observation)
Eupsophus roseus	Absent	Endotrophic	Nidicolous	Aquatic	Cavities and tunnels with water in the forest ground	Present	25	NA, at least 58 days	8	Formas and Pugín (1978a, b), Díaz (1986), Vera Candioti et al. (2005), Rabanal (personal observation)
Pleurodema bufoninum	Intense	Exotrophic, free living	Lentic–benthic with morphological phenotypic plasticity	Aquatic	Lentic and shallow waters (small ponds and backwaters in creeks)	Absent	64	30–45 days	13–17	Weigandt et al. 2004, Perotti et al. (2011), Úbeda (2021)
Pleurodema thaul	Intense	Exotrophic, free living	Lentic with benthic and nektonic characters and morphological phenotypic plasticity	Aquatic	Temporary water bodies, shallow and vegetated lagoons	Absent	60	45–90 days	12–22.4	Úbeda (1998), Jara and Perotti (2009), Perotti et al. (2011), Barrasso et al. (2012), Moncada and Úbeda (2020), Jara (2022)
Nannophryne variegata	Black uniform	Exotrophic, free living	Lentic–benthic (Fig. 2a)	Aquatic	Lentic, sunny, and shallow waters (small ponds, peat	Absent	41	Short (spring–summer)	7–10.1	Donoso-Barros (1975), Formas and Pugín (1978a), Úbeda (1998),

(continued)

Table 3 (continued)

Species	Larval pigmentation	Type of larva (according to trophism and habit)	Larval ecomorphological type (sensu Altig and Johnston 1989)	Larval environment	Larval habitat	Parental care of larvae	Maximum larval size (mm)	Duration of larval period	Froglet size (mm)	References
					bogs, and backwaters in creeks)					Cisternas et al. (2013), Úbeda (2021)
Rhinella papillosa	Intense and uniform brown	Exotrophic, free living	Lentic–benthic	Aquatic	Lentic, sunny, and shallow waters (small ponds and backwaters in creeks and streams)	Absent	37	7–10 weeks (spring)	10–15	Jara et al. (2006, 2007), Sympson et al. (2006), Vera Candioti et al. (2020), Úbeda (2021)
Rhinella rubropunctata	Dark and uniform	Exotrophic, free living	Lentic–benthic	Aquatic	Lentic and shallow waters (small ponds and temporary lagoons)	Absent	27	Spring–early summer	8.15–14	Formas and Pugín (1978b), Veloso (2006), Úbeda (personal observation)
Rhinella spinulosa	Intense and uniform brown	Exotrophic, free living	Lentic–benthic	Aquatic	Lentic, sunny, and shallow waters (small ponds and backwaters in creeks and streams)	Absent	35	3–4 months (spring–summer)	10–13	Donoso Barros (1975), Vera Candioti et al. (2020), Díaz-Páez (personal observation)

For species whose distribution exceeds the study area, only the data corresponding to the populations of the Patagonian Andes are provided. Species are ordered by family according to Table 1

NA data not available

Fig. 2 Representative larvae of different ecomorphological types (sensu Altig and Johnston 1989) of the anurans of the Patagonian Andes. (**a–e**) Exotrophic larvae. (**a**) *Nannophryne variegata* (lentic–benthic). (**b**) *Calyptocephalella gayi* (lentic–benthic). (**c**) *Batrachyla taeniata* (lentic–benthic). (**d**) Early larva of *Hylorina sylvatica* (lentic–nectonic). (**e**) *Telmatobufo australis* (lotic–benthic suctorial). (**f**) Endotrophic nidicolous larva of *Eupsophus emiliopugini*

context of their respective environmental scenarios (Sects. 4.2.1, 4.2.2, and 4.2.3).

The high proportion of species (50%) that present prolonged larval periods overwintering as such is notable in this fauna of the Patagonia Andes (Table 3). This type of strategy (larval overwintering) seems to be common and successful among the Patagonian anurans of temperate and cold climates and is present in six genera, *Atelognathus*, *Batrachyla*, *Hylorina*, optionally in *Chaltenobatrachus* (Díaz 1986; Úbeda et al. 1999, 2022; Úbeda 2006; Basso et al. 2011; Cuello et al. 2014; Jara et al. 2021), and, very markedly, in *Calyptocephalella* and *Alsodes*, with larvae that spend more than one winter in the water body (Cei 1962; Díaz 1986; Díaz and Valencia 1985; Úbeda and Casanovas 2012). Although there are no data, it is likely that *Telmatobufo* species also have this characteristic, given the size that the larvae reach and the low temperatures of the waters that they inhabit.

The life strategies of the species groups in the context of the different macroenvironments that they inhabit and their prevailing conditions are presented and analyzed below.

4.2.1 Anurans of the Forests

The greatest richness of anurans in the Patagonian Andes is associated to forests, especially to rainforests, which are mostly found in the western strip of the Patagonian Andean range (Table 1).

Almost two thirds of the anurans of the forests as a whole (17 species) have the generalized reproductive mode. However, in the rainforests there are a significant number of species, belonging to the genera *Batrachyla*, *Eupsophus*, and

Rhinoderma (Table 1) that present specialized reproductive modes, with different degrees of terrestriality (Table 2), favored by environmental conditions of high rainfall and great ambient humidity, especially in the lower strata of these forests.

The larvae of the anurans of these forests present different ecomorphological types (sensu Altig and Johnston 1989). Several species have lentic–benthic type larvae with depressed bodies, low fins, and dorsal eyes, which develop either in very shallow water bodies, such as the bufonids *Rhinella* and *Nannophryne* (Fig. 2a) or in deeper ones, such as *Calyptocephalella*, *Batrachyla*, and *Atelognathus* (Fig. 2b, c). Other species have lentic–nectonic type larvae such as the early larvae of *Hylorina sylvatica,* with high fins and lateral eyes (Fig. 2d). On the other hand, the six present species of *Alsodes* have larvae of benthic habit and as they inhabit streams in sectors of lower current, they present some morphological characteristics that allow them to face a moderate current, such as the muscular tail, the low and subparallel fins, the large oral disk with a certain capacity for adherence to the substrate, although structurally little modified, preserving the basic configuration of the labial denticles (FD = 2/3) and a wide rostral gap in the marginal papillae (e.g., Pillado et al. 2000; Echeverría et al. 2001; Formas and Brieva 2004; Cuevas and Formas 2005; Barrasso et al. 2016). The most extreme case of larval adaptation corresponds to the species of the genus *Telmatobufo* (Fig. 2e), which present notable morphological adaptations to lotic habitats with strong currents (Table 3), which are detailed in Section 4.2.1.1. Some species, such as *Hylorina sylvatica* and *Calyptocephalella gayi*, can change larval habit throughout their development, a situation that is developed further ahead.

Next, we discuss the composition and characteristics of the frogs of the rainforests separately from those of the transition forests found in the east.

4.2.1.1 Anurans of the Rainforest

The fauna of these forests is the richest in the Patagonian Andes (24 species, Table 1) and presents the greatest diversification of life history strategies. The variety of microenvironments present in the lower strata is very suitable for the establishment and reproduction of several species of anurans. High rainfall and dense canopy cover favor fairly stable conditions in the understory, with low light and constant ambient humidity, which attenuate the maximum temperatures and maintain a moist soil, in parts flooded. To these characteristics is added the existence of temporary and permanent ponds and mountain streams. In this complex environmental scenario we can find nine species contained in three endemic genera, *Eupsophus, Batrachyla*, and *Rhinoderma*, which present specialized reproductive modes and strategies, discussed below (Fig. 3).

All known species of the genus *Batrachyla* inhabit these rainforests and present a specialized reproductive mode that includes terrestriality in oviposition and embryonic development (Table 2). Reproduction occurs in the summer and early austral autumn, unlike the temporal pattern of the other species of these forests, which are spring and summer breeders. Amplexus and oviposition occur on floodable moist soil under shelters, such as vegetation, barks, and fallen logs, where embryonic development also occurs under parental care (Fig. 3a) (Úbeda and Nuñez 2006; Rabanal et al. 2021). One of the species, *Batrachyla antartandica*, also presents an alternative reproductive mode, involving the use of flooded or water-filled cavities in fallen or standing logs (phytotelmata) as a site for mating call, reproduction, development, and metamorphosis (Fig. 3b), being the only species of the austral temperate forests that presents the possibility of a complete arboreal life cycle (Rabanal et al. 2021). The species of *Batrachyla* have yolk-rich eggs, with a mechanically firm and elastic gelatinous capsule, with hygrophilic properties, useful in the water regulation of embryos against changes in environmental and soil humidity (Capurro 1958; Cei and Capurro 1958; Busse 1971). Embryonic development on land is prolonged and can extend even further and enter a final stage of intracapsular resistance (Fig. 4a) in which organogenesis continues and the hind limb sketches appear (Cei and Capurro 1958; Busse 1971; Jorquera

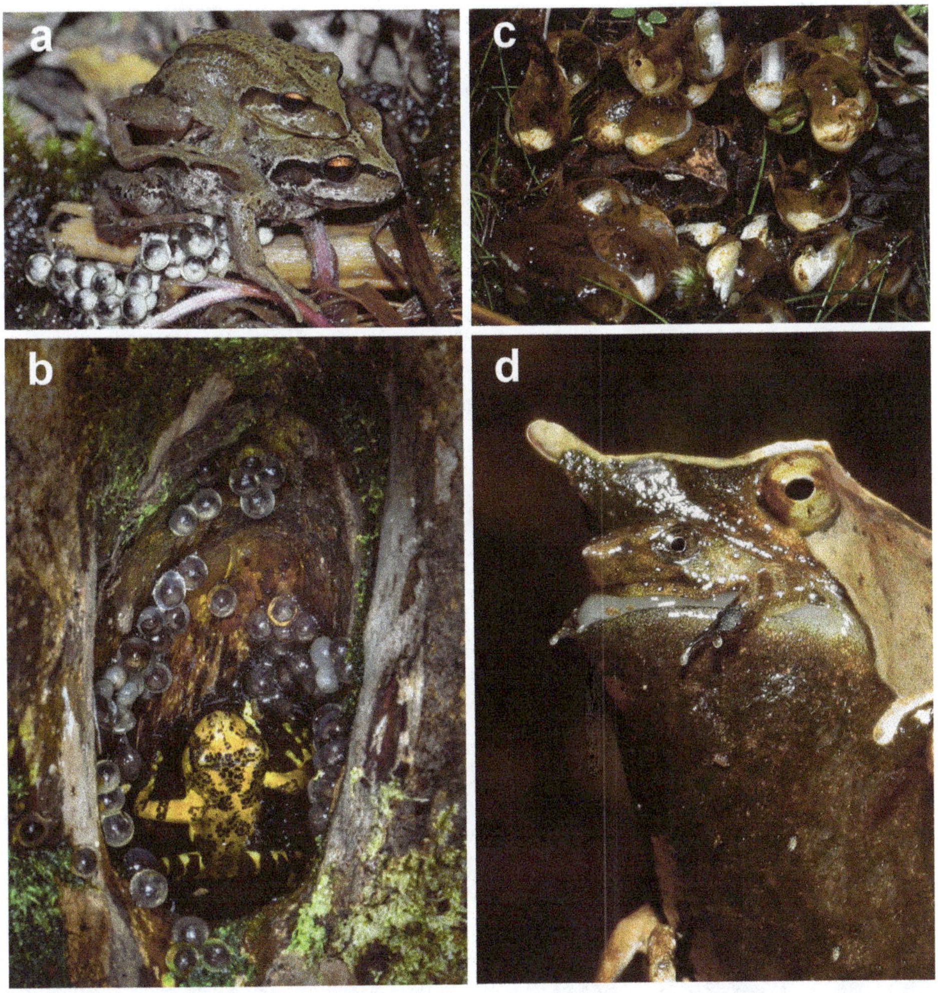

Fig. 3 Specialized reproductive modes of some of the anuran species of the Patagonian Andes. (**a**) Terrestrial axillary amplexus and clutch of *Batrachyla nibaldoi*. (**b**) Clutch and parental care in *Batrachyla antartandica* in phytotelma. (**c**) Parental care carried out by the male *Eupsophus calcaratus* on its clutch. (**d**) Neomelia and expulsion of a froglet from the vocal sac of the male *Rhinoderma darwinii*

and Molinari 1979; Jara et al. 2021; Rabanal et al. 2021), waiting for autumn rains that flood the site and trigger the hatching of the exotrophic larvae that start their free life in lentic waters (Fig. 2c). If the rains are delayed, the larvae can hatch and resist a few more days on moist soil (Cei and Capurro 1958; Busse 1971; Jorquera and Molinari 1979). The larval period is prolonged and the larvae overwinter to metamorphose in the spring or summer, depending on the species (Table 3). This also occurs in *B. antartandica* within the previously described phytotelmata, where cohorts of several reproductive events can coexist, even from two consecutive seasons (Rabanal et al. 2021). The partially terrestrial reproductive mode and the temporality in the embryonic and larval development, with larvae that reach their maximum size in the spring, allows the species of *Batrachyla* to reduce competition for resources with other frogs that reproduce in spring and evade predation by aquatic insects that colonize the environment in this season (Jara et al. 2021).

Fig. 4 Types of clutches of some representative anurans of the Patagonian Andes. (**a**) Clutch of *Batrachyla taeniata* with embryos in the intracapsular resistance stage. (**b**) Floating mass of *Calyptocephalella gayi*. Note the dry foam in contact with the air. (**c**) Egg strings of *Rhinella papillosa*. (**d**) Egg mass of *Atelognathus nitoi*. (**e**) Egg mass of *Chaltenobatrachus grandisonae*. (**f**) Clutch of *Pleurodema thaul* in its two possible states: strand and globular

On the other hand, the five species of the genus *Eupsophus* that inhabit the Patagonian Andes are restricted to these rainforests (Formas 1980, 1989b). Reproduction begins in spring. Males emit their nuptial vocalizations from small cavities and tunnels in the ground or in stream banks (Penna and Veloso 1990) (Table 2). Amplexus and oviposition occur in these microenvironments, partially flooded, with some water replacement, dark and communicated with the outside, often under fallen logs or large rocks. The eggs are relatively large (2.8–7.2 mm, Formas and Vera 1980; Vera Candioti et al. 2011) owing to the large amount of yolk they contain, and their color varies from white to yellow. The male stays with the eggs, providing at list the simplest form of parental care (Úbeda and Nuñez 2006; Nuñez and Úbeda 2009; Nuñez et al. 2012) (Fig. 3c). The larvae are endotrophic and nidicolous (sensu Altig and Johnston 1989);

they do not present pigmentation during the early stages of development, highlighting the protruding abdomen and yellowish-white color due to the high yolk content (Fig. 2f). In relation to the type of endotrophic nutrition, these larvae present a reduction of the keratinized structures of the oral apparatus (Formas 1989b; Vera Candioti et al. 2005, 2011; Nuñez and Úbeda 2009). The larvae remain in the flooded cavities, under parental care, until they complete the metamorphosis in a relatively short time (around 60 days), exclusively at the expense of the yolk, that is, without active feeding. The metamorphs, with some remaining yolk in the abdomen, are small, pigmented, and reach the surface of the ground to start their new terrestrial stage of active feeding as inhabitants of the forest floor in humid, cool, and shady sites. The specialized and nidicolous reproductive mode, the low number of eggs, and the parental care of eggs and larvae configure a typical "K" strategy in the sense of Wilbur (1977).

Finally, the most astonishing example of specialized reproductive mode and larval development, unique among amphibians, that has evolved in these rainforests, is that of Darwin's frog, *Rhinoderma darwinii,* which completes its entire life cycle independently of the aquatic environment. It presents a type of reproduction completely different from the generalized pattern of anurans, with parental care taken to its maximum expression. In spring–summer, the male attracts the females with a high-pitch nuptial vocalization, similar to the chirping of some birds (Wilhelm 1927; Pflaumer 1935). During a loose amplexus on the moist ground, the female typically deposits between 4 and 24 independent, whitish, and large eggs, about 3–4 mm in diameter (Jorquera et al. 1972; Busse 2004; Cabeza-Alfaro et al. 2021), although Wilhelm (1932) reported the laying of 34 eggs, which seems to be an exceptional case. Observations under laboratory conditions suggest that, once spawning is complete, the male would remain in the vicinity of the clutch site, occasionally returning to visit. However, these data, considered a presumed early custody of eggs, should be taken with caution, as it may be influenced by the conditions of confinement. Nevertheless, the male's return and attention to egg-laying does occur toward the end of embryonic development, which is well documented. After about 19–25 days, the male, stimulated by the movement of the embryos, introduces them into his mouth where they slide into the vocal sac (Jorquera et al. 1981; Busse 2004). Recently, Cabeza-Alfaro et al. (2021) pointed out that males actively help in the hatching of the larvae, breaking the egg membranes with their mouths, and then introducing them into the vocal sac. There, the larvae develop until they complete metamorphosis at the expense of the egg yolk and of nutritive substances that the male secretes from the epithelium of the vocal sac (Jorquera et al. 1972; Goicoechea et al. 1986). The maximum number of larvae reported inside the vocal sac is 22 (Wilhelm 1927), although they typically tend to harbor between five and eight at the same time (Cabeza-Alfaro et al. 2021). After about 50 days, through strong spasmodic contractions of the body, the male expels the tiny fully metamorphosed froglets (Fig. 3d). All these characteristics configure a typical "K" reproductive strategy, in the sense of Wilbur (1977).

The rest of the species of the rainforests present aquatic eggs and larvae, with different types of clutch structures (independent eggs, strings, strands, and masses), types of larvae, and larval periods of different duration, even in sympatry under the same general environmental conditions (Tables 2 and 3). Within this group of species, some endemic examples that illustrate this variety are provided below.

Calyptocephalella gayi, known as the Chilean giant frog, is a species adapted for life in vegetated lentic waters, such as lakes and lagoons, although it can also be found in backwaters of low-flow rivers. Reproductive activity in Patagonian populations occurs between the months of August and January (Table 2) (Cei 1962; Hermosilla and Acuña 2004), a period in which nuptial vocalizations occur, with low pitch tones, audible at great distances, emitted by semi-submerged and hidden males in shallow areas with abundant aquatic vegetation, forming powerful choruses (Penna and Veloso 1990). During axillary amplexus, the female deposits between 1000 and 10,000 eggs (Formas 1979) wrapped in a floating gelatinous mass, which expands upon contact with water, with the male cooperating in the formation of bubbles during amplexus (Acuña et al. 2014). Then, upon contact with the air, it dries and adheres strongly to the surrounding vegetation (Fig. 4b). Field and captivity observations demonstrated that males provide parental care during embryonic development, usually positioning themselves under the egg mass or next to it, presumably protecting the eggs and oxygenating them through water currents or vortices under the egg mass (Hermosilla and Acuña 2004). The larvae of this species are the largest among the Patagonian species and are among the largest in the world, being able to reach 20 cm in length and in some cases taking up to 4 years to complete metamorphosis (Vélez et al. 2014). The larvae correspond to the lentic–benthic ecomorphological type (Table 3), with an oval body, dorsolateral eyes, and a tail ending in a point, whose height does not exceed the maximum height of the body, and has a dorsal fin that does not extend over the body (Fig. 2b). However, field and captivity observations show that the larva occupies the entire water column, being observable both at the bottom and mid-water with slow swimming, similar to what happens with larvae of the lentic–nectonic type. In the early stages of development, the larvae feed on the jelly that forms the matrix of the egg clutch (Vallejos and Pizarro 2004), to later feed on algae, zooplankton, and detritus from plant or animal matter (Acuña et al. 2014). These large larvae can transform into very large metamorphs (Table 3). Being larvae that hatch in large quantities ("r" strategy), which reach a large size and therefore biomass, and that occupy the entire water column, the important ecological role that they play in the ecosystems that they inhabit can be inferred.

Hylorina sylvatica also has an aquatic reproduction in lentic water bodies, such as vegetated lakes and lagoons (Table 2). Between the months of October and January, intense male choruses occur, vocalizing from the vegetation in the shallowest areas (Úbeda et al. 2022), with a low pitch and powerful sounds that resemble the pounding of dry logs together. These choruses are audible at a great distance (Barrio 1967b). The axillary amplexus occurs in the water, while the couple are swimming at the surface (Úbeda et al. 2022). The eggs, which have moderate pigmentation of the animal pole, are independent and can fall to the bottom or adhere to the aquatic vegetation (Úbeda et al. 2022). Around 480 to 600 are laid (Barrio 1967b; Charrier 2019). After 10 to 16 days, small larvae hatch and during the early stages present a morphology associated with a lentic–nectonic habit (Fig. 2d), to later, as growth and development progress, transform into lentic–benthic larvae (Úbeda et al. 2022) (Table 3). The larval period is prolonged, just over a year, with larvae that overwinter (sometimes under a layer of surface ice) and continue growing in the spring to metamorphose in the summer (Formas and Pugín 1978b; Úbeda et al. 2022). Throughout this prolonged larval period the larvae can reach up to 84 mm (Úbeda 1998) and transform into metamorphs between 24 and 28.5 mm (Table 3).

On the other hand, in the species of the endemic genera *Alsodes* and *Telmatobufo* reproduction takes place in cold mountain streams. The known egg clutches for some species of *Alsodes* consist of groups of a few whitish eggs, rich in yolk, which are deposited in water under rocks and in cavities of the margins of mountain streams, high Andean lakes, and high Andean pools with water replacement (Gallardo 1970; Corbalán et al. 2014). The development of the larvae (described in Sect. 4.2.1) can last several years, as has been proven for *Alsodes gargola* and *A. pehuenche*, with larvae that can go through four winters until completing metamorphosis (Corbalán et al. 2014; Úbeda and Casanovas 2012; Úbeda 2021). This characteristic would be expected for the rest of the congeneric species that inhabit Patagonian environments.

In the case of *Telmatobufo*, of the four species currently recognized, only the egg clutch of *Telmatobufo bullocki* is known, described by Moreno Puig (2014). Although this is not a species of the Patagonian Andes, as it is endemic to the Cordillera de Nahuelbuta in Chile, it is possible to hypothesize that the rest of the species of this genus could present similarities, as they share most of the known traits of their natural history. Thus, it is estimated that the eggs of the Andean representatives, *T. venustus* and *T. australis*, are likely to be laid on the rocky bed of streams of clear, cold, and well-oxygenated water, hidden and adhered to submerged rocks, in areas of moderate current. It is estimated that the number of eggs per clutch should be around 200, being nonpigmented and of large size (Table 2). The most conspicuous adaptation of these species is related to the larvae (Fig. 2e), which present notable morphological adaptations to mountain streams (Formas 1972). They are of the lotic–benthic suctorial type (Table 3), with an extreme modification of the oral disk. This has transformed into a large suctorial disk, sucker-like, that allows them to attach to the rocks, while they scrape the periphyton and the attached organic matter to feed and, at the same time, resist being swept downstream. In addition to the characteristics of the oral disk, the larvae present a depressed and robust body, dorsal eyes, and a strongly developed caudal musculature, with low paddle-shaped fins inserted into the last third of the tail (Formas 1972, 1988; Díaz et al. 1983) (Fig. 2e). These larvae reach a significant size (Table 3), and their growth and development at low temperatures would suggest that, like the larvae of *Alsodes*, they could take more than a year to complete metamorphosis.

4.2.1.2 Anurans of the Transition and Ecotonal Forests

The anuran fauna of the transition forests, with 13 species, is poorer than that of the rainforests (Table 1) and presents a lesser variety of modes and reproductive strategies and of reproductive habitats, for example, temporary ponds and permanent lagoons and streams.

All species have a generalized reproductive mode and seasonal reproduction. Most lay their pigmented eggs exposed to the sun, except for species of *Alsodes* and *Telmatobufo*, which hide them under rocks or in cavities on the banks of permanent streams (Table 2).

All species have exotrophic and pigmented aquatic larvae, with a predominance of the lentic–benthic ecomorphological type (sensu Altig and Johnston 1989), except for the larvae of the species of *Alsodes* and *Telmatobufo* of the lotic-benthic type, as mentioned in Sects. 4.2.1 and 4.2.1.1, as well as in Table 3. Short larval periods predominate during the most productive time of the water bodies, that is, the austral spring–summer season. The longest larval period of this fauna in these transition forests is represented by the species of *Alsodes*, with a duration of 4 years in *Alsodes gargola* (Úbeda and Casanovas 2012; Úbeda 2021). *Atelognathus nitoi*, which reproduces in lagoons during the spring, presents a double larval development strategy, with seasonal larvae and others that overwinter to metamorphose at the beginning of the following austral summer (Úbeda et al. 1999). This versatility also allows them to inhabit temporary lagoons with a hydroperiod of at least 5–6 months.

4.2.2 Anurans of Arid Environments

In the structurally simpler environmental scenario of the arid environments of the Patagonian Andes, the anuran fauna is comparatively poor, with the smallest number of species (six species) (Table 1). Some of them are highly adapted to the rigorous environmental conditions of the Patagonian steppes and others, of semi-aquatic habit, are associated with lagoons and permanent streams. All have generalized reproductive modes and seasonal reproduction, in different reproductive habitats (Table 2). Some reproduce in ponds and small lentic water bodies, often temporary, and have short larval periods during the spring–summer period, such as *Rhinella spinulosa*, *R. papillosa*, and *Pleurodema bufoninum* (Tables 2 and 3). On the contrary, *Atelognathus patagonicus* and *A. nitoi* reproduce in permanent and temporary lagoons and have a double strategy with seasonal larvae and larvae that overwinter, whereas *Alsodes neuquensis* reproduces in permanent streams and presents a very long larval period (Úbeda 2021). Two of the strategies present in these arid environments are discussed below.

Rhinella papillosa is a toad well adapted to the arid environments of the Patagonian steppe, although it also inhabits open and sunny environments in different Andean Patagonian forests and high-altitude wetlands (Úbeda 2021). Owing to its reproductive characteristics, *Rhinella papillosa* is a good example of an "r" reproductive strategy (sensu Wilbur 1977). In early spring, males and females congregate to reproduce in small, shallow, and sunny bodies of water, typically temporary, and in shallow areas with little or no current in rivers and streams; no mating call is known in this species. During a prolonged axillary amplexus, females release thousands of small (1.5 ± 0.9 mm, Sympson et al. 2006) and intensely pigmented eggs arranged in long hyaline gelatinous strings of regular diameter (Fig. 4c) (Sympson et al. 2006). After a short embryonic period of about 4 days, the small larvae hatch at an early stage (18–19, sensu Gosner 1960), very melanized, which initially lie recumbent and with little mobility under the remains of the strings, to develop quickly, reaching stage 26, and actively feed during the 1st week (Sympson et al. 2006). The larvae, with low fins, present a lentic–benthic ecomorphological type and have a uniform dark-brown color. On sunny days, in the afternoon they form dense aggregations in the shallowest and warmest areas of the bodies of water that they inhabit, while they feed, swim, or rest on the bottom (Jara et al. 2006). This behavior contributes to accelerating the growth of the larvae. The larval period demands 7–10 weeks, which are concentrated during the most productive period of the aquatic habitats, and can exhibit phenotypic plasticity in the face of the desiccation of the water bodies (Perotti et al. 2011). The larvae reach a size of 37 mm, to then metamorphose quite synchronously, generating large amounts of small toadlets of 11–15 mm, which begin their terrestrial stage. In the laboratory, it was found that the larvae can withstand high densities and complete metamorphosis with remarkable synchronism (Jara et al. 2007). Although in steppe environments the species behaves as an explosive breeder, some populations of forest environments, with a higher frequency of rainfall during the spring, exhibit a prolonged breeding pattern (Sympson et al. 2006).

A very different strategy is presented by the semi-aquatic frog *Atelognathus patagonicus*, which inhabits permanent and temporary endorheic lagoons located in arid environments of northwestern Patagonia and in high Andean lakes (Cei and Roig 1968; Úbeda 2021). This species shows remarkable plasticity both in the larval stage and in the post-metamorphic stage. Reproduction is seasonal during the spring and early summer (Table 2). Cei (1965) briefly described the egg clutches as "gelatinous clutches of intensely pigmented eggs attached to submerged stalks of aquatic plants." The larvae are pigmented, have a lentic–benthic ecomorphological type, and live mainly among dense aquatic vegetation (Cuello et al. 2017). In permanent lagoons, the population presents a dual larval development strategy, in which some larvae from early ovipositions have a short larval period and metamorphose in the same growing season in which they hatched, whereas other larvae from later ovipositions have a prolonged larval period, overwinter, and metamorphose in the following spring. As a result, overwintering larvae reach a larger size than seasonal larvae (Cuello et al. 2014). In temporary lagoons, the desiccation of the water body toward the end of summer, along with the deterioration of limnological conditions, induces the acceleration of larval development, resulting in a short larval period. This remarkable plasticity in the larval development strategy allows the species to inhabit environments as diverse as temporary or permanent lagoons with variable hydroperiods over the years (resulting from multi-year climatic cycles) and to overcome periods of extraordinary droughts in these typically permanent lagoons. Post-metamorphic individuals exhibit a remarkable phenotypic plasticity, presenting two reversible morphotypes, one aquatic and one terrestrial, according to the environmental conditions of the lagoons (Cuello et al. 2008).

4.2.3 High Mountain Anurans

Anurans that managed to colonize and establish themselves in the high Andean environments of Patagonia, above the ecological line of the "tree line," deserve a special mention. These environments have a low specific richness, with eight species (Table 1), which constitute a particular subset of the lower altitude anurofauna (Úbeda 2021).

Of the eight species present, *Pleurodema bufoninum*, *P. thaul*, *Nannophryne variegata*, *Rhinella papillosa*, and *R. spinulosa* have a clearly terrestrial habit and deploy an "r" reproductive strategy sensu Wilbur (1977). The other three species (*Alsodes gargola, A. neuquensis*, and *Atelognathus patagonicus*) have semi-aquatic habits.

In the vastness of the high Andean landscape, individuals do not distribute homogeneously, but appear aggregated in certain habitat patches, such as wetlands that constitute the sources of streams, low-order streams, and ultraoligotrophic lakes, with few anuran species inhabiting in sympatry (Úbeda 2021). Within these patches, larval populations can reach significant permanent biomasses, with hundreds of large larvae in certain microenvironments, such as small pools at the sources of streams (Úbeda 2021). Environments such as rocky outcrops, scree, and lakes at the foot of glaciers are not inhabited by anurans, regardless of altitude.

The species of these rigorous high mountain ecosystems share many characteristics (Tables 2 and 3). First, all have the generalized reproductive mode, with aquatic ovipositions and exotrophic aquatic larvae, without parental care. Within this mode, most have eggs and larvae in lentic waters; only the species of *Alsodes* reproduce in streams, in sites with little current, in addition to oligotrophic lakes. Second, in this high mountain anurofauna, the specialized reproductive modes present in the forests at lower altitude are not represented.

In addition, all species have a seasonal breeding pattern, within the bounded growing season characteristic of these high mountain systems, either from early spring or later in the summer. Most species (six out of eight) place their clutches exposed to the sun, in accordance with the intense melanization of their eggs. The photoprotective effect that melanin has against ultraviolet radiation is widely documented (e.g., Perotti and Diéguez 2006). On the contrary, the species of *Alsodes* that oviposits the eggs hidden under rocks or in cracks, have depigmented eggs.

All species have larvae of the generalized central type (sensu Orton 1953), none present extreme adaptations to a certain type of habitat. They have a lentic–benthic ecomorphological type (sensu Altig and Johnston 1989), except for the two species of *Alsodes* that have a lotic–benthic larva, with adaptations to a moderate current (Pillado et al. 2000; Úbeda 2021). The benthic habit is associated with the consumption of periphyton and detritus, as has been proven for the two high Andean species of *Alsodes* (Baffico and Úbeda 2006).

In addition, all high Andean species have pigmented larvae, in association with their sun-exposed habitats; the maximum pigmentation is presented by the two species of bufonids (Fig. 2a), with intense dorsal and ventral melanization. These species inhabit very shallow waters, only 1–2 cm deep. Melanization, in addition to protecting from UVB radiation, promotes the heating of the larvae and therefore accelerates their growth and development.

The majority of species (five out of eight) complete their larval development within the same growing season, when the bodies of water are at their most productive in terms of their primary production. As mentioned above (Sects. 4.2.1.2 and 4.2.2), *Atelognathus patagonicus* has a dual strategy (with seasonal larvae and others that overwinter until the following spring), whereas the two species of *Alsodes* have a very prolonged larval period, having being verified to be 4 years and a bit more in a high Andean population of *Alsodes gargola* (Úbeda and Casanovas 2012; Úbeda 2021).

It is noteworthy that sympatric species facing the same general environmental conditions in most cases segregate into different microenvironments and deploy very different strategies of reproduction and larval development, which are equally successful (Úbeda 2021). For example, *Nannophryne variegata* begins to reproduce in early spring, just after the thaw, in shallow ponds fed with surface runoff water, lays strings with numerous eggs, have relatively small larvae (Fig 2a), and a larval period bounded to the growing season, which ends with small metamorphs (Formas and Pugín 1978a; Úbeda 2021). On the contrary, the sympatric species *Alsodes gargola* reproduces in permanent bodies of water, such as streams and high Andean lakes (Fig. 1f), places its few eggs under rocks and in cavities in the margins, and has a very long larval period through which the larvae reach a significant size and transform into metamorphs comparatively larger than those of *N. variegata* (Úbeda 2021).

In addition to reproductive and developmental aspects, the life strategies of high Andean species are complemented by characteristics of their thermal physiology. The existing information, although fragmentary, is sufficient to notice the wide range of activity temperatures that some of these species present. It has been reported that different stages of these anurans respond to mechanical stimuli, are active, and even exhibit reproductive activity at very low temperatures (Hock 1967; Logares and Úbeda 2006; Navas et al. 2010; Bonino et al. 2020).

In species that have long larval periods (*Alsodes gargola*, *Alsodes. neuquensis*, and *Atelognathus patagonicus*), the entire larval population and a significant part of the population of juveniles and adults (of aquatic habit) overwinter in permanent lakes frozen on the surface, where the larvae can even continue feeding, as was proven for the larvae of *A. gargola* (Baffico and Úbeda 2006). In contarst, during the short summer there is a large daily thermal amplitude, with nocturnal frosts and exceptional maximums, the smaller and shallower water bodies being the ones in which these changes are most noticeable (Navas et al. 2010; Bonino et al. 2020; Úbeda 2021).

In relation to the thermal variability of their habitats, in larvae of *R. papillosa*, *P. bufoninum*, *P. thaul*, and *A. gargola*, a physiological eurythermy was proven with a wide range of thermal tolerance (Navas et al. 2010; Bonino et al. 2020). All have a high critical thermal maximum, higher (36–38 °C) in *R. papillosa*, *P. bufoninum*, and *P. thaul*, which inhabit shallow and small environments, than in *A. gargola* (32–33 °C), which inhabits permanent water bodies with greater thermal stability (Navas et al. 2010; Bonino et al. 2020). All have a critical thermal minimum close to 0 °C; additionally, in *R. papillosa*, *P. bufoninum*, and *A. gargola* a certain capacity for resistance to freezing temperatures was demonstrated, as they are able to resist a few hours confined within the ice (Navas et al. 2010).

5 Final Considerations

Amphibians, being both aquatic and terrestrial organisms, increase their possibilities of using different environments and sites for embryonic and larval development and, therefore, present a greater diversification of strategies than other tetrapod vertebrates (Nunes-de-Almeida et al. 2021).

The large macroenvironments present in the Patagonian Andes harbor different sets of anurans that display a wide variety of reproductive strategies, development patterns, and larvae, a variety that is remarkable for the fauna of an austral territory with a temperate to cold climate. Nunes-de-Almeida et al. (2021) found that on a global scale the generalized reproductive modes (Modes 1 and 2 of Duellman and Trueb 1986 and Haddad and Prado 2005), i.e., aquatic eggs and exotrophic aquatic larvae, without parental care, are not the most common in anurans. However, in the anurofauna of the southern Andes, our results show that the majority of the species (64.28%) do present these modes, with Mode 1 (eggs and larvae in lentic waters) predominating over Mode 2 (eggs and larvae in lotic waters) of Duellman and Trueb 1986. A subset of nine species belonging to three endemic genera (*Batrachyla*, *Eupsophus* and *Rhinoderma*), which only inhabit the rainforests, present specialized reproductive modes and parental care. It is highly likely that the prevailing conditions in these forests (high precipitation and high environmental humidity) favored the evolution of these reproductive modes. However, no species presents direct development. Although in *Rhinoderma darwinii* small fully metamorphosed froglets emerge from the male's mouth, the larval stage have not been eliminated from the life cycle but occur within the vocal sac.

As analyzed in this work, the majority of species (24 out of 28 species) have exotrophic larvae and within these, generalized larvae predominate, in line with the global trend of 69% of anuran species (Vera Candioti et al. 2023). Only the species of *Telmatobufo* have specialized lotic larvae, a type present in 12% of the world's anurans whose larvae have been described (Vera Candioti et al. 2023). Within the exotrophic larvae of the Patagonian Andes the relatively high proportion of species with prolonged larval periods, which undergo one or more winters in the aquatic environment, is remarkable.

On the other hand, endotrophic larvae, less common among anurans worldwide (Vera Candioti et al. 2023), in the fauna of the Patagonian Andes, are represented by the nidicolous larvae of the three species of *Eupsophus* and the larva with neomelia in the vocal sac of the male *Rhinoderma darwinii.* These anurans, inhabitants of the rainforests, complete their larval development in a comparatively short span.

The comparative analysis of the reproductive characteristics and the embryonic and larval development of the species of the Patagonian Andes reveals the disparity of knowledge of them, with a majority of well-studied species and others with little or no information. Therefore, with the available knowledge, the majority of the species included in this work could be classified using the methods proposed by Duellman and Trueb (1986), Haddad and Prado (2005), Nunes-de-Almeida et al. (2021), and Liedtke et al. (2022) for reproductive modes, or for the types of larvae of Altig and Johnston (1989), McDiarmid and Altig (1999), and Vera Candioti et al. (2023). The use of classification methods that aspire to be applicable to world fauna requires detailed information of various reproductive aspects of the species, and its application entails a certain degree of difficulty, or even results in two categories for the same species.

In light of the above, it is evident how little has been disseminated globally about the state of the art on the reproductive aspects of the anurofauna of southern South America, considering that the most recent authors did not include in their classifications several of the species of the Patagonian Andes, presumably owing to lack of access to the necessary information.

With this contribution on the anurans of the Patagonian Andes, we hope to provide an overview of the state of knowledge regarding reproductive and developmental aspects, which will be a guide and stimulate future studies that provide new information and contribute toward a better understanding of the biology and ecology of this very particular fauna.

Acknowledgments The field, laboratory, and office tasks that contributed to this work were carried out with funding and equipment from the National University of Comahue (Argentina), the University of Concepción (Chile), and the Austral University of Chile (Chile).

Conflicts of Interest The authors have no conflicts of interest to declare that are relevant to the content of this chapter. All photographs belong to the authors.

References

Acuña OPL, Pulgar R, Vélez RCM et al (2014) Aspectos nutricionales. In: Vélez Rodríguez CM (ed) Manejo en cautiverio de la rana grande chilena *Calyptocephalella gayi* (Duméril and Bibron, 1841). Ediciones Universidad Santo Tomás, Santiago, pp 33–61

Altig R, Johnston GF (1989) Guilds of anuran larvae: relationships among developmental modes, morphologies, and habitats. Herpetol Monogr 3:81–109. https://doi.org/10.2307/1466987

Altig R, McDiarmid RW (1999) Body plan: development and morphology. In: McDiarmid RW, Altig R (ed) Tadpoles: the biology of anuran larvae. The University of Chicago Press, Chicago/London, pp 24–54

Altig R, Mcdiarmid RW (2007) Morphological diversity and evolution of egg and clutch structure in amphibians. Herpetol Monogr 21:1–32. https://doi.org/10.1655/06-005.1

Armesto JJ, León-Lobos P, Kalin-Arroyo M (1995) Los bosques templados del sur de Chile y Argentina: una isla biogeográfica. In: Armesto JJ, Villagrán C, Kalin-Arroyo M (eds) Ecología de los bosques nativos de Chile. Editorial Universitaria, Santiago, Chile, pp 23–28

Baffico GD, Úbeda CA (2006) Larval diet of the frog *Alsodes gargola* (Leptodactylidae: Telmatobiinae) and some ecological considerations on its role in alpine and mountain aquatic environments in Patagonia. Amphibia-Reptilia 27(2):161–168. https://doi.org/10.1163/156853806777239986

Barcelos LA, dos Santos RO (2023) The Lissamphibian fossil record of South America. Palaeobiodivers Palaeoenviron 103:341–405. https://doi.org/10.1007/s12549-022-00536-0

Barrasso DA, Basso NG (2019) Low genetic divergence but many names in the endemic Patagonian frogs of the genus *Atelognathus* (Anura, Batrachylidae): a molecular genetic and morphological perspective. J Zool Syst Evol Res 57:383–399. https://doi.org/10.1111/jzs.12259

Barrasso DA, Alcalde L, Martinazzo LB et al (2012) External morphology, chondrocranium, cranial muscles, and buccopharyngeal features of tadpoles of *Pleurodema thaul* (Anura: Leiuperidae): a comparison with *P. bufoninum*. Herpetologica 68(1):48–59. https://doi.org/10.2307/41406818

Barrasso DA, Alcalde L, Blotto BL et al (2016) Description of the tadpole of *Alsodes neuquensis* Cei, 1976 and comparison with the sibling species *A. gargola* Gallardo, 1970 (Amphibia, Anura, Alsodidae). Herpetol J 26:21–31. https://doi.org/10.1643/CH-18-138

Barrasso DA, Úbeda CA, Cotichelli L et al (2022) On the presence of *Alsodes coppingeri* (Anura, Alsodidae) in Argentina, with comments on other southern Alsodes. Neotrop Biodiver 8(1):21–30. https://doi.org/10.1080/23766808.2022.2029321

Barrio A (1967a) *Batrachyla antartandica* n. sp. (Anura, Leptodactylidae) descripción y estudio comparativo con la especie genotípica, *B. leptopus* Bell. Physis (Buenos Aires) 27(74):101–109

Barrio A (1967b) Observaciones etoecológicas sobre *Hylorina sylvatica* Bell (Anura, Leptodactylidae). Physis (Buenos Aires) 27(74):153–157

Basso NG, Úbeda CA (1997) The tadpole of *Atelognathus nitoi* (Leptodactylidae: Telmatobiinae). Alytes 15(3):121–126

Basso NG, Úbeda CA, Bunge MM et al (2011) A new genus of neobatrachian frog from southern Patagonian forests, Argentina and Chile. Zootaxa 3002:31–44. https://doi.org/10.11646/zootaxa.3002.1.3

Blotto BL, Nuñez JJ, Basso NG et al (2013) Phylogenetic relationships of a Patagonian frog radiation, the *Alsodes + Eupsophus* clade (Anura: Alsodidae), with comments on the supposed paraphyly of *Eupsophus*. Cladistics 29(2):113–131. https://doi.org/10.1111/j.1096-0031.2012.00417.x

Bonino MF, Cruz FB, Perotti MG (2020) Does temperature at local scale explain thermal biology patterns of temperate tadpoles? J Therm Biol 94 article:102744. https://doi.org/10.1016/j.jtherbio.2020.102744

Boulenger GA (1886) Remarks in connection with the preceding note. Ann Mag Nat Hist 17:463–464

Busse K (1971) Desarrollo de *Batrachyla leptopus* Bell con observaciones sobre su ecología y comportamiento (Amphibia, Leptodactylidae). Inv Zool Chilenas 15:5–64

Busse K (2004) Biología de la reproducción del sapito de Darwin (*Rhinoderma darwinii)* y su cría en cautividad. In: Iriarte A, Tala C, González B, Zapata B, González G, Maino M (eds) Cría en cautividad de fauna chilena, Ediciones SAG-Parque Metropolitano-Fac. Veterinaria, U. de Chile, Santiago, pp 137–146

Cabeza-Alfaro O, Valenzuela-Sánchez A, Alvarado-Rybak M et al (2021) First case of male alloparental care in amphibians: tadpole stealing in Darwin's frogs. Endang Species Res 45:331–335. https://doi.org/10.3354/esr01138

Capurro SLF (1958) Nota sobre la ovipostura y desarrollo de *Eupsophus taeniatus* (Girard). Inv Zool Chilenas 4:208

Cárdenas-Rojas DR, Rabanal F, Formas JR (2007) The tadpole of *Hylorina sylvatica (Anura: Cyclorhamphidae)* in southern Chile. Zootaxa 1580:51–62. https://doi.org/10.11646/zootaxa.1580.1.4

Catchpole S, Medina M (2018) Hábitat de las larvas de *Telmatobufo australis* Formas 1972 (Amphibia, Anura, Calyptocephalellidae): características fisicoquímicas, hidrodinámicas y morfológicas del medio fluvial. Bol Chil Herpetol 5:26–28

Cei JM (1961) *Bufo arunco* (Molina) y las formas chilenas de *Bufo spinulosus*. Inv Zool Chilenas 7:59–81

Cei JM (1962) Batracios de Chile. Ediciones de la Universidad de Chile, Santiago de Chile

Cei JM (1965) The tadpole of *Batrachophrynus patagonicus* Gallardo. Herpetologica 20:242–245

Cei JM (1970) Fluctuaciones biocenóticas y relictos herpetológicos de la planicie de Lonco-Luan (Neuquén). Acta zool lilloana 23:193–200

Cei JM (1976) Remarks on some neotropical amphibians of the genus *Alsodes* from southern Argentina (Anura Leptodactylidae). Atti Soc ital Sci nat Museo civ Stor nat Milano 117(3–4):159–164

Cei JM (1980) Amphibians of Argentina. Monitore zool ital (NS) Monogr 2

Cei JM, Capurro L (1958) Biología y desarrollo de *Eupsophus taeniatus* Girard. Inv Zool Chilenas 4:159–182

Cei JM, Roig VG (1968) Telmatobiinos de las lagunas basálticas de Neuquén (Anura, Leptodactylidae). Physis 27:265–284

Charrier A (2019) Guía de campo Anfibios de los Bosques de la zona centro sur y Patagonia de, Chile edn. Corporación Chilena de la Madera, Chile

Chiozza E (1976) El país de los Argentinos: la Patagonia, la Antártida Argentina y el Nordeste. Centro Editor de América Latina, Buenos Aires

Cisternas J, Correa C, Velásquez N et al (2013) Reproductive features of *Chaltenobatrachus grandisonae* (Anura: Batrachylidae) within a protected area in Patagonia, Chile. Rev Chil Hist Nat 86:365–368. https://doi.org/10.4067/S0716-078X2013000300013

Cisternas-Medina I, Ortiz JC, Úbeda C et al (2019) Distribución geográfica del sapito de rayas o sapo variegado *Nannophryne variegata* Günther, 1870, nuevas localidades y comentarios sobre su hábitat en Chile y Argentina. Gayana 83(1):33–45. https://doi.org/10.4067/S0717-65382019000100033

Corbalán V, Debandi G, Martínez F et al (2014) Prolonged larval development in the Critically Endangered Pehuenche's frog *Alsodes pehuenche*: implications for conservation. Amphibia-Reptilia 35:283–292. https://doi.org/10.1163/15685381-00002951

Coronato A, Mazzoni E, Vázquez M et al (2017) Patagonia: una síntesis de su geografía física. Ediciones Universidad Nacional de la Patagonia Austral, Río Gallegos

Correa C (2022) Lista viva de las especies de anfibios de Chile (Versión 2022.1). Ediciones de la Asociación Red Chilena de Herpetología, RECH. Available at www.herpetologiadechile.cl

Crump ML (1974) Reproductive strategies in a tropical Anuran community. Misc publ Univ Kans 61:1–68

Crump ML (1982) Amphibian reproductive ecology on the community level. In: Scott NJ Jr (ed) Herpetological Communities U.S. Fish and Wildlife Research report 13, Washington D.C, pp 21–36

Crump ML (2015) Anuran reproductive modes: evolving perspectives. J Herpetol 49(1):1–16. https://doi.org/10.1670/14-097

Cuello ME, Úbeda CA, Bello MT (2008) Relationship between morphotypes of *Atelognathus patagonicus* (Anura, Neobatrachia) and environmental conditions: evidence and possible explanation. Phyllomedusa 7(1): 35–44. https://doi.org/10.11606/issn.2316-9079.v7i1p35-44

Cuello ME, Úbeda CA, Bello MT et al (2014) Plastic patterns in larval development of Endangered endemic *Atelognathus patagonicus*: implications for conservation strategies. Endang Species Res 23:83–92. https://doi.org/10.3354/esr00550

Cuello ME, Úbeda CA, Bello MT (2017) Habitat associations for the endangered frog *Atelognathus patagonicus* within the aquatic environment: key microhabitats for conservation. Herpetol Conserv Biol 12(2):410–421

Cuevas CC, Formas JR (2005) A new frog of the genus *Alsodes* (Leptodactylidae) from the Tolhuaca National Park, Andes Range, southern Chile. Amphibia-Reptilia 26:39–48. https://doi.org/10.1163/1568538053693288

Díaz N (1986) Biosistemática de los Leptodactylidae chilenos. An Mus Hist Nat Valparaiso Chile 17:65–85

Díaz NF, Valencia J (1985) Larval morphology and phenetic relationships of the Chilean *Alsodes*, *Telmatobius*, *Caudiverbera* and *Insuetophrynus* (Anura: Leptodactylidae). Copeia 1985(1):175–181. https://doi.org/10.2307/1444807

Díaz N, Sallaberry M, Nuñez H (1983) The tadpole of *Telmatobufo venustus* (Anura: Leptodactylidae) with a consideration of generic relationships. Herpetologica 39(2):111–113. https://doi.org/10.4067/S0716-078X2006000400003

Díaz NF, Sallaberry M, Valencia J (1987) Microhabitat and reproductive traits in populations of the frog *Batrachyla taeniata*. J Herpetol 21(4):317–323. https://doi.org/10.2307/1563973

Díaz-Páez H, Alveal N (2021) A new geographic record of the endangered *Telmatobufo venustus* (Amphibia: Calyptocephalellidae) in the Biobío Region, Chile. Rev Chil Hist Nat 94(1):3. https://doi.org/10.1186/s40693-021-00100-8

Díaz-Páez H, Ortiz JC (2001) The reproductive cycle of *Pleurodema thaul* (Anura, Leptodactylidae) in central Chile. Amphibia-Reptilia 22:431–445. https://doi.org/10.1163/15685380152770390

Díaz-Páez H, Vidal MA, Ortiz JC et al (2011) Taxonomic identity of the Patagonian frog *Atelognathus jeinimenensis* (Anura: Neobatrachia) as revealed by molecular and morphometric evidence. Zootaxa 2880: 20–30. https://doi.org/10.11646/zootaxa.2880.1.2

Donoso-Barros R (1975) Dos larvas de *Bufo* chilenos no conocidas. Not Mens Mus Nac Hist Nat 19(230–231): 6–7

Duellman WE (1989) Alternative life-history styles in anuran amphibians: evolutionary and ecological implications. In: Bruton MN (ed) Alternative life-history styles of animals. Kluwer Academic Publishers, Dordrecht, pp 101–126

Duellman WE, Trueb L (1986) Biology of amphibians. McGraw-Hill, New York

Echeverría DD, Alonso CA, Pillado MS et al (2001) Microanatomía del aparato bucal y cavidad bucofaríngea de la larva de *Alsodes gargola* Gallardo, 1970 (Leptodactylidae, Telmatobiinae). Alytes 18(3–4):187–200

Ferreyra M, Cingolani A, Ezcurra C et al (1998) High-Andean vegetation and environmental gradients in northwestern Patagonia, Argentina. J Veg Sci 9:307–316. https://doi.org/10.2307/3237095

Formas JR (1972) A second species of Chilean frog genus *Telmatobufo* (Anura: Leptodactylidae). J Herpetol 6(1):1–3. https://doi.org/10.2307/1563086

Formas JR (1976) Descriptions of *Batrachyla* (Amphibia, Anura, Leptodactylidae) tadpoles. J Herpetol 10(3): 221–225. https://doi.org/10.2307/1562983

Formas JR (1979) La herpetofauna de los bosques temperados de Sudamérica. In: Duellman WE (ed) The South American herpetofauna: its origin, evolution, and dispersal, University of Kansas, Museum of Natural History, Monograph 7, Lawrence, Kansas, USA, pp 341–369

Formas JR (1980) The chromosomes of *E. calcaratus* and the karyological evolution of the genus *Eupsophus*

(Anura: Leptodactylidae). Experientia 36:1163–1164. https://doi.org/10.1007/BF01976101

Formas JR (1981) Adaptaciones larvarias de los anuros del bosque temperado austral de Sudamérica. Medio Ambiente 5(1–2):15–21

Formas JR (1988) The tadpole of *Telmatobufo bullocki* (Anura: Leptodactylidae). Herpetologica 44(4): 458–460

Formas JR (1989a) The tadpole of *Eupsophus calcaratus* in Southern Chile. J Herpetol 23(2):195–197. https://doi.org/10.2307/1564035

Formas JR (1989b) A new species of *Eupsophus* (Amphibia: Anura: Leptodactylidae) from Southern Chile. Proc Biol Soc Wash 102(3):568–576

Formas JR (2013) External morphology, chondrocranium, hyobranchial skeleton, and external and internal oral features of *Rhinoderma rufum* (Anura, Rhinodermatidae). Zootaxa 3641(4):395–400. https://doi.org/10.11646/zootaxa.3641.4.6

Formas JR, Brieva L (2004) The tadpoles of *Alsodes vanzolinii* and *A. verrucosus* (Anura: Leptodactylidae) with descriptions of their internal oral and chondrocranial morphology. Amphibia-Reptilia 25: 151–164. https://doi.org/10.1163/1568538041231238

Formas JR, Pugín E (1971) Reproducción y desarrollo de: *Batrachyla antartandica* (Barrio) (Anura, Leptodactylidae). Bol Mus Nac Hist Nat Chile 32: 201–213. https://doi.org/10.54830/bmnhn.v32.1971.559

Formas JR, Pugín E (1978a) Tadpoles of *Eupsophus roseus* and *Bufo variegatus* (Amphibia, Anura) in Southern Chile. J Herpetol 12(2):243–246. https://doi.org/10.2307/1563413

Formas JR, Pugín E (1978b) Tadpoles of *Hylorina sylvatica*, *Eupsophus vittatus*, and *Bufo rubropunctatus* in Southern Chile. Herpetologica 34(4):355–358

Formas JR, Vera MA (1980) Reproductive patterns of *Eupsophus roseus* and *E. vittatus*. J Herpetol 14:11–14. https://doi.org/10.2307/1563869

Formas JR, Úbeda C, Cuevas C et al (1997) *Alsodes australis*, a new species of Leptodactylid frog from the temperate *Nothofagus* forest of Southern Chile and Argentina. Stud Neotrop Fauna Environ 32:200–211. https://doi.org/10.1080/01650521.1997.11432423

Formas JR, Núñez JJ, Brieva LM (2001) Osteología, taxonomía y relaciones filogenéticas de las ranas del género *Telmatobufo* (Leptodactylidae). Rev Chil Hist Nat 74(2):365–387. https://doi.org/10.4067/S0716-078X2001000200013

Formas JR, Núñez J, Cuevas C (2008) Identidad de la rana austral chilena *Eupsophus coppingeri* (Amphibia, Anura, Neobatrachia): evidencias morfológicas, cromosómicas y moleculares. Rev Chil Hist Nat 81: 3–20. https://doi.org/10.4067/S0716-078X2008000100001

Gallardo JM (1970) A propósito de los Telmatobiinae (Anura, Leptodactylidae) patagónicos. Neotropica 16(50):73–85

Goicoechea O, Garrido O, Jorquera B (1986) Evidence for a trophic paternal-larval relationship in the frog *Rhinoderma darwinii*. J Herpetol 20:168–178. https://doi.org/10.2307/1563941

Gosner KL (1960) A simplified table for staging anuran embryos and larvae with notes on identification. Herpetologica 16:183–190

Grosso J, Barrasso DA, Baldo D et al (2022) Developmental diversity in an endemic family of Patagonian frogs (Anura: Batrachylidae). Biol J Linn Soc 136:240–263. https://doi.org/10.1093/biolinnean/blac032

Haddad CFB, Prado CPA (2005) Reproductive modes in frogs and their unexpected diversity in the Atlantic forest of Brazil. Bioscience 55(3):207–217

Hermosilla I, Acuña PL (2004) Crianza en cautividad de la rana gigante chilena *Caudiverbera caudiverbera*. In: Iriarte A, Tala C, González B, Zapata B, González G, Maino M (eds) Cría en cautividad de fauna chilena Ediciones SAG-Parque Metropolitano-Fac. Veterinaria, U. de Chile, Santiago, pp 197–122

Hock RJ (1967) Temperature effect on breeding of the toad, *Bufo variegatus*, in southern Chile. Copeia 1967(1):227–230. https://doi.org/10.2307/1442205

Hoorn C, Wesselingh FP, ter Steege H et al (2010) Amazonia through time: Andean uplift, climate change, landscape evolution, and biodiversity. Science 330(6006):927–931. https://doi.org/10.1126/science.1194585. PMID: 21071659

Izquierdo L, Pereda J (1964) Influencia de la temperatura ambiental sobre el desarrollo de *Pleurodema bibroni*. Arch Biol Med Exp 1:141–151

Jara FG (2022) Reproductive biology of the emerald frog *Hylorina sylvatica* (Anura: Batrachylidae) in Northwest Patagonia, Argentina. Biota Neotropica 22(4): e20221355. https://doi.org/10.1590/1676-0611-bn-2022-1355

Jara FG, Perotti MG (2009) La rana de cuatro ojos en la Laguna Fantasma de Bariloche. Desde la Patagonia, difundiendo saberes 6:10–15

Jara FG, Lozada M, Úbeda CA (2006) *Bufo spinulosus papillosus* (NCN). Tadpole behavior. Herpetol Rev 37(2):201

Jara FG, Úbeda CA, Lozada M (2007) Effect of larval density on development and metamorphosis in two anuran species (Bufonidae and Leptodactylidae) of the Patagonian steppe. Cuad Herpetol 21(1):53–58

Jara FG, Úbeda C, Moncada M et al (2021) Natural history traits of the terrestrial breeding frog *Batrachyla taeniata* (Anura: Batrachylidae) in wet meadows of Patagonia. Stud Neotrop Fauna Environm 56. https://doi.org/10.1080/01650521.2021.2005394

Jorquera B, Izquierdo L (1964) Tabla de desarrollo normal de *Calyptocephalella gayi* (Rana chilena). Biologica 36:43–53

Jorquera B, Molinari E (1979) Estados de desarrollo normal de *Batrachyla taeniata* (Girard). Medio Ambiente 4(1):89–102

Jorquera B, Pugín E, Goicoechea O (1972) Tabla de desarrollo normal de *Rhinoderma darwini*. Arch Med Vet 4(2):5–19

Jorquera B, Pugín E, Garrido O et al (1981) Procedimiento de desarrollo en dos especies del Género *Rhinoderma*. Medio Ambiente 5:58–71

Körner C (2001) Alpine ecosystems. In: Levin SA (ed) Encyclopedia of biodiversity, vol 1. Academic Press, San Diego, pp 133–144

Lavilla EO (1987) La Larva de *Rhinoderma darwinii* D. y B. (Anura: Rhinodermatidae). Acta zool lilloana 39(1): 81–88

Lavilla EO (1988) Lower Telmatobiinae (Anura: Leptodactylidae): generic diagnoses based on larval characters. Occ Pap Mus nat Hist Univ Kansas 124: 1–19

Lavilla EO, Rougés M (1992) Reproducción y desarrollo de anuros argentinos. Asociación Herpetológica Argentina, Serie de divulgación 5, 65 pp

Liedtke HC, Wiens JJ, Gomez-Mestre I (2022) The evolution of reproductive modes and life cycles in amphibians. Nat Commun 13:7039. https://doi.org/10.1038/s41467-022-34474-4

Logares RE, Úbeda CA (2006) First insights into the overwintering biology of *Alsodes gargola* frogs and tadpoles inhabiting harsh Andean Patagonian alpine environments. Amphibia-Reptilia 27(2):263–267. https://doi.org/10.1163/156853806777239913

McDiarmid RW, Altig R (1999) Research, material and techniques. In: McDiarmid RW, Altig R (eds) Tadpoles: the biology of anuran larvae. The University of Chicago Press, Chicago/London, pp 7–23

Márquez R, Penna M, Marques P et al (2005) Diverse types of advertisement calls in the frogs *Eupsophus calcaratus* and *E. roseus* (Leptodactylidae): a quantitative comparison. Herpetol J 15(4):257–263

Meriggio V, Veloso A, Young S et al (2004) *Atelognathus jeinimenensis* n.sp. de Leptodactylidae para el Sur de Chile. Boletín del Museo Nacional de Historia Natural (Chile) 53:99–123. https://doi.org/10.54830/bmnhn.v53.2004.301

Moncada M, Úbeda C (2020) Patrón de reproducción y ovipostura de *Pleurodema thaul* (Schneider 1799) (Anura, Leptodactylidae) en poblaciones de la vertiente oriental de la Cordillera de los Andes. Bol Chil Herpetol 7:12–19

Moreno Puig V (2014) La ovipostura de *Telmatobufo bullocki* Schmidt, 1952 (Amphibia, Anura, Calyptocephalellidae). Bol Chil Herpetol 1:17

Navas CA, Úbeda CA, Logares R et al (2010) Thermal tolerances in tadpoles of three species of Patagonian anurans. S Am J Herpetol 5(2):89–96. https://doi.org/10.2994/057.005.0203

Nunes-de-Almeida CHL, Haddad CFB, Toledo LF (2021) A revised classification of the amphibian reproductive modes. Salamandra 57(3):413–427. Available at https://www.salamandra-journal.com/index.php/contents/2021-vol-57/2054-nunes-de-almeida-c-h-l-c-f-b-haddad-l-f-toledo-1/file

Nuñez JJ, Úbeda CA (2009) The tadpole of *Eupsophus nahuelbutensis* (Anura: Neobatrachia): external morphology, chondrocranium, and comments on its natural history. Zootaxa 2126:58–68. https://doi.org/10.11646/zootaxa.2126.1.3

Nuñez JJ, Rabanal FE, Formas JR (2012) Description of a new species of *Eupsophus* (Amphibia: Neobatrachia) from the Valdivian Coastal range, Southern Chile: an integrative taxonomic approach. Zootaxa 3305:53–68. https://doi.org/10.11646/zootaxa.3305.1.3

Oliver Schneider C (1930) Observaciones sobre batracios chilenos. Rev Chil Hist Nat 34:220–223. Available at https://rchn.biologiachile.cl/pdfs/1930/1/Oliver_1930b.pdf

Ortiz JC (2015) Anfibios de las turberas del extremo austral de Chile. In: Domínguez E, Vega-Valdés D (eds) Funciones y servicios ecosistémicos de las turberas en Magallanes. Colección de libros INIA N° 33. Instituto de Investigaciones Agropecuarias. Centro Regional de Investigación Kampenaike. Punta Arenas, Chile, pp 227–238

Orton GL (1953) The systematics of vertebrate larvae. Syst Zool 2:63–75. https://doi.org/10.2307/2411661

Paruelo JM, Beltrán A, Jobbágy E et al (1998) The climate of Patagonia: general patterns and controls on biotic processes. Ecol Austral 8:85–101

Penna M, Veloso A (1990) Vocal diversity in frogs of the South American temperate forest. J Herpetol 24(1): 23–33. https://doi.org/10.2307/1564285

Pereyra MO, Blotto BL, Baldo D et al (2021) Evolution in the Genus *Rhinella*: a total evidence phylogenetic analysis of neotropical true toads (Anura: Bufonidae). Bull Am Mus Nat Hist 447(1):1–156. https://doi.org/10.1206/0003-0090.447.1.1

Perotti MG, Diéguez MC (2006) Effect of UV-B exposure on eggs and embryos of Patagonian anurans and evidence of photoprotection. Chemosphere 65:2063–2070. https://doi.org/10.1016/j.chemosphere.2006.06.055

Perotti MG, Jara FG, Úbeda CA (2011) Adaptive plasticity of life-history traits to pond drying in three species of Patagonian anurans. Evol Ecol Res 13:415–429. Available at https://www.evolutionary-ecology.com/issues/v13/n04/iiar2611.pdf

Pflaumer C (1935) Observaciones biológicas acerca de la *Rhinoderma darwinii* D & B. Rev Chil Hist Nat 39(1): 28–30. Available at http://rchn.biologiachile.cl/pdfs/1935/1/Pflaumer_1935a.pdf

Pillado MS, Alonso CA, Úbeda CA (2000) La larva de *Alsodes gargola* Gallardo, 1970 (Leptodactylidae, Telmatobiinae). Alytes 18(1–2):62–72. Available at https://ia800304.us.archive.org/8/items/biostor-286962/biostor-286962.pdf

Rabanal FE (2010) Amphibia, Anura, Ceratophryidae, *Batrachyla nibaldoi* Formas, 1997: latitudinal extension in Patagonia, southern Chile, and distributional range actualization. Check List 6(2):287–288. https://doi.org/10.15560/6.2.287

Rabanal FE (2011) Nuevos Aportes al conocimiento de la historia natural de *Batrachyla nibaldoi* Formas, 1997 (Anura: Ceratophryidae). In: Resúmenes XII Congreso Argentino de Herpetología, Asociación Herpetológica Argentina, Bariloche, 19–22 de septiembre de 2011

Rabanal FE, Nuñez JJ (2008) Anfibios de los bosques templados de Chile. Primera Edición, Universidad Austral de Chile, Valdivia

Rabanal FE, Úbeda C, Tejo CF et al (2021) Tree-holes as alternative reproductive sites of *Batrachyla antartandica* Barrio, 1967 (Anura: Batrachylidae). South Am J Herpetol 20:17–25. https://doi.org/10.2994/SAJH-D-18-00064.1

Sallaberry M, Valencia J, Díaz N (1981) Distribución y ambientes de *Batrachyla taeniata* (Girard) en Chile. Bol Mus Nac Hist Nat Chile 38:61–67. https://doi.org/10.54830/bmnhn.v38.1981.459

Salthe SN (1969) Reproductive modes and the number and sizes of ova in the urodeles. Am Midl Nat 81(2):467–490. https://doi.org/10.2307/2423983

Salthe SN, Duellman WE (1973) Quantitative constraints associated with reproductive mode in anurans. In: Vial JL (ed) Evolutionary biology of the Anurans: contemporary research on major problems. University of Missouri Press, Columbia, pp 229–249

Salthe SN, Mecham JS (1974) Reproduction and courtship patterns. In: Lofts B (ed) Physiology of the Amphibia, vol 2. Academic Press, New York, pp 309–521

Scolaro JA, Videla F, Cei JM (2003) Algunos modelos de especiación geográfica que interpretan aspectos de la diversidad herpetólogica andino-patagónica. Historia Natural (Segunda Serie), Buenos Aires 2(9):73–83

Soto ER, Sallaberry M, Núñez JJ et al (2008) Desarrollo larvario y estrategias reproductivas en anfibios. In: Vidal Maldonado MA, Labra Lillo A (eds) Herpetología de Chile. Science Verlag, Santiago, Chile, pp 333–357

Suazo Lara F, Gómez RO (2022) In the shadow of dinosaurs: late cretaceous frogs are distinct components of a widespread tetrapod assemblage across Argentinean and Chilean Patagonia. Cretac Res 131:105085. https://doi.org/10.1016/j.cretres.2021.105085

Sympson V, Jara FG, Úbeda CA (2006) *Bufo spinulosus papillosus* (NCN). Reprod Herpetol Rev 37(2):200–201

Úbeda CA (1998) Batracofauna de los Bosques Templados Patagónicos: un enfoque ecobiogeográfico. Tesis Doctoral. Facultad de Ciencias Exactas y Naturales, Universidad Nacional de Buenos Aires, Buenos Aires

Úbeda C (2006) La rana del Challhuaco: Biología y conservación. Desde la Patagonia, difundiendo saberes 3(4):16–20

Úbeda C (2021) Estrategias reproductivas, hábitats y otros aspectos ecológicos de los anfibios altoandinos en la vertiente oriental de la Cordillera de los Andes. Bol Chil Herpetol 8:10–21

Úbeda CA, Basso NG (2003) The tadpole of *Atelognathus salai* Cei, 1984 (Leptodactylidae: Telmatobiinae). Amphibia-Reptilia 24(1):112–118

Úbeda CA, Casanovas P (2012) Desarrollo larval prolongado en *Alsodes*: *Alsodes gargola* como modelo de estudio (Anura, Alsodidae). In: Resúmenes III Coloquio de Anfibios y Reptiles, Lican Ray, Chile, 6–8 diciembre 2012

Úbeda CA, Nuñez JJ (2006) New parental care behaviours in two telmatobiine genera from temperate Patagonian forests: *Batrachyla* and *Eupsophus* (Anura: Leptodactylidae). Amphibia-Reptilia 27:441–444. https://doi.org/10.1163/156853806778190015

Úbeda C, Zagarese H, Diaz M et al (1999) First steps towards the conservation of the microendemic Patagonian frog *Atelognathus nitoi*. Oryx 33(1):59–66. https://doi.org/10.1046/j.1365-3008.1999.00035.x

Úbeda C, Moncada M, Jara F (2022) Fenología del ciclo de vida de la rana esmeralda *Hylorina sylvatica* Bell 1843 (Anura, Batrachylidae) en un humedal semipermanente del noroeste de la Patagonia Argentina. Bol Chil Herpetol 9:1–11

Vallejos P, Pizarro C (2004) Crianza de la rana chilena (*Caudiverbera caudiverbera*) (Linnaeus, 1758), bajo sistema de cultivo semi-extensivo. In: Iriarte A, Tala C, González B, Zapata B, González G, Maino M (eds) Cría en cautividad de fauna chilena Ediciones SAG-Parque Metropolitano-Fac. Veterinaria, U. de Chile, Santiago, p 123–136

Vélez RCM, Núñez JJ, Acuña-O P et al (2014) Aspectos generales. In: Vélez Rodríguez CM (ed) Manejo en cautiverio de la rana grande chilena *Calyptocephalella gayi* (Duméril and Bibron, 1841). Ediciones Universidad Santo Tomás, Santiago, pp 15–24

Veloso A (2006) Batracios de las cuencas hidrográficas de Chile: origen, diversidad y estado de conservación. In: Vila I (ed) Macrófitas y vertebrados de los sistemas límnicos de Chile, Editorial Universitaria, p 103–140

Vera Candioti MF, Úbeda C, Lavilla EO (2005) Morphology and metamorphosis of *Eupsophus calcaratus* tadpoles (Anura: Leptodactylidae). J Morphol 264(2):161–177. https://doi.org/10.1002/jmor.10320

Vera Candioti MF, Nuñez JJ, Úbeda C (2011) Development of the nidicolous tadpoles of *Eupsophus emiliopugini* (Anura: Cycloramphidae) until metamorphosis, with comments on systematic relationships of the species and its endotrophic developmental mode. Acta Zool 92:27–45. https://doi.org/10.1111/j.1463-6395.2010.00448.x

Vera Candioti F, Grosso J, Pereyra MO et al (2020) Larval anatomy of Andean Toads of the *Rhinella spinulosa* Group (Anura: Bufonidae). Herpetol Monogr 34(1):116–130. https://doi.org/10.1655/HERPMONOGRAPHS-D-20-00001_hmon-34-01-05_11

Vera Candioti FV, Baldo D, Grosjean S et al (2023) Global shortfalls of knowledge on anuran tadpoles. npj Biodiver 2:22. https://doi.org/10.1038/s44185-023-00027-1

Weigandt M, Úbeda CA, Diaz M (2004) The larva of *Pleurodema bufoninum* Bell, 1843, with comments on its biology and on the egg strings (Anura, Leptodactylidae). Amphibia-Reptilia 25(4):429–437

Wells KD (2007) The ecology and behavior of amphibians. The University of Chicago Press, Chicago and London

Wilbur HM (1977) Propagule size, number, and dispersion pattern in *Ambystoma* and *Asclepias*. Am Nat 111(977):43–68. https://doi.org/10.1086/283137

Wilhelm GO (1927) La *Rhinoderma darwinii* D. y B. Bol Soc Biol Concepción 1(1–2):11–19. Available at https://bolsocbiolconcepc.cl/pdfs/v01(1-2)_1927/02-Wilhelm.pdf

Wilhelm GO (1932) Nuevas observaciones acerca de la neomelia de la *Rhinoderma darwinii*. Rev Chil Hist Nat 36:166–170. Available at http://rchn.biologiachile.cl/pdfs/1932/1/Wilhelm_1932.pdf

Zagarese HE, Diaz M, Pedrozo F et al (2000) Mountain lakes in northwestern Patagonia. Verh Internat Verein Limnol 27:533–538. https://doi.org/10.1080/03680770.1998.11901290

Unveiling Patagonian Lizards: A Three-Decade Journey through Ecology, Evolution, and Adaptation in a Changing World

Carla Piantoni, Jorgelina Boretto, Marlin Medina, Erika Kubisch, Jimena B. Fernández, Facundo Cabezas-Cartes, Nadia Vicenzi, Alejandro Laspiur, and Nora R. Ibargüengoytía

1 Beginnings: Early Research and Emerging Questions

The first studies on herpetofauna of the Andes and Patagonia were initiated at the beginning of the last century by herpetologists like Lorenz Müller (1868–1953), K. G. Walter Hellmich (1906–1974), José Miguel Cei (1918–2007; Fig. 1a), Roberto Donoso-Barros (1922–1975), Richard E. Etheridge (1929–2019), and José Alejandro Scolaro (1946-; Fig. 1a). Their publications laid the foundation for herpetology in the region, focusing on describing new species and deciphering the taxonomic and phylogenetic relationships among them. Inspired by the wide range of opportunities to study the life history, ecology, and physiology traits of several yet unkown species in such a pristine landscape, Dr. Nora R. Ibargüengoytía (Fig. 1a and e) began her extensive career with projects in San Carlos de Bariloche, Río Negro, Argentina. Her initial field campaigns and articles focused on the reproductive biology of liolaemid lizards, but her research quickly expanded to include studies on growth, ecology, physiology, evolution, and conservation, covering most reptile species in the region.

Established in 1996, the laboratory "Ecophysiology and Life History of Reptiles" initially centered on the reproductive biology of *Liolaemus pictus* (Figs. 2d, e, g and 3e–g), a medium-sized viviparous lizard with a broad distribution in Northern Patagonia (Figs. 1b-i, 2, and 3). The first goal was to know the basic features such as the timing of the reproductive events, clutch size, frequency of reproduction, and annual reproductive output. We found that this viviparous species from northern habitats up to moderate altitudes in the Andean-Patagonian Forest performed biennial to triennial reproductive cycles (Table 1). These observations caught the attention of a reviewer Dr. Oliver Pearson, who worked for

C. Piantoni (✉)
School of Life Sciences, University of Hawai'i at Mānoa, Honolulu, USA

J. Boretto · E. Kubisch · J. B. Fernández · F. Cabezas-Cartes · N. R. Ibargüengoytía
Instituto de Investigaciones en Biodiversidad y Medioambiente, Consejo Nacional de Investigaciones Científicas y Técnicas, Universidad Nacional del Comahue, San Carlos de Bariloche, Argentina

M. Medina
Centro de Investigación Esquel de Montaña y Estepa Patagónica, Consejo Nacional de Investigaciones Científicas y Técnicas, Universidad Nacional de la Patagonia San Juan Bosco, Esquel, Argentina

N. Vicenzi
Instituto Argentino de Investigaciones en Zonas Áridas, Consejo Nacional de Investigaciones Científicas y Técnicas, Mendoza, Argentina

A. Laspiur
Instituto de Ciencias Básicas, Facultad de Filosofía Humanidades y Artes, Universidad Nacional de San Juan, San Juan, Argentina

J. N. Urbina-Cardona et al. (eds.), *Andean Herpetofauna*, Sustainable Development Goals Series,
https://doi.org/10.1007/978-3-032-00074-3_9

Fig. 1 Early photos of the lab, and fieldwork in Jacobacci, Province of Rio Negro, Argentina. (**a**) During the XV Congreso Argentino de Herpetología in 2000, held in San Carlos de Bariloche, Dr. Alejandro Escolaro and Dra. Nora Ibargüengoytía extend a warm welcome to José Miguel Cei to know the new lab. (**b**) Dr. Jorgelina Boretto and Dr. Marlin Medina collecting in Jacobacci. (**c**) *Liolaemus elongatus*. (**d**) *Phymaturus spectabilis*. (**e**) Dr. Ibargüengoytía in Jacobacci, one of the main field localities. (**f**) *L. elongatus*. (**g**) Gathering of thermal data. (**h**) Field work in Jacobacci. (**i**) General view of the steppe toward the Andes

Fig. 2 Environments and herpetofauna found at low altitudes in the San Carlos de Bariloche region. (**a**) Lake Moreno—893 m asl. (**b**) *Diplolaemus sexcinctus*. (**c**) Lake Nahuel Huapi—767 m asl. (**d**, **e**) *Liolaemus pictus*. (**f**) *Nothofagus* forest by Lake Gutierrez—770 m asl. (**g**) *L. pictus* typical environment at Lake Moreno

Fig. 3 High-altitude environments and herpetofauna in San Carlos de Bariloche, Rio Negro, Argentina. (**a**) Andes mountains at about 1700 m asl. (**b**) Typical environment at 1100 m asl. (**c**) Forested area in Cerro Challhuaco, about 1100 m asl. (**d**) *Nothofagus* forest in winter. (**e**) Newborns of *Liolaemus pictus*. (**f**, **g**) Specimens of *L. pictus* in their typical environment. (**h**) General view of the lake and mountains area around Lake Nahuel Huapi

many years in Patagonia and developed a deep affection for these lands. Among the few studies on the reproductive biology of lioaemids at that time, Oliver described an annual cycle for *Liolaemus multiformis* which lives at 4000 m asl in Titicaca Lake, Peru. Therefore, he was skeptical about regarding Patagonia as a harsher environment compared to the high elevations in Peru where *L. multiformis* was found basking on snowy surfaces (Pearson 1954). After running the data collected by the lab, he confirmed that the analyses were conducted correctly and happily accepted this initial report for prolonged cycles (biennial to triennial cycle) in South America. In contrast to *L. pictus*, which needs at least 2 years, one activity season to perform vitellogenesis and another for pregnancy (Ibargüengoytía and Cussac 1996), *L. multiformis* reproduces annually. The question that arose was why congeners living at 4000 m asl in harsh environments in Peru reproduce more frequently than those in Patagonia. The response likely hinges on the more continental climate that *L. multiformis* occupies in Peru, resulting in a longer and hotter activity season, compared to the temperate oceanic climate in Patagonia (Ibargüengoytía 2004).

After this initial confirmation of prolonged reproductive cycles in lizards from Patagonia, the subsequent objective was to determine whether viviparity accounted for these extended cycles and to assess whether oviparous species exhibited annual reproductive patterns. These questions were explored in the dissertation of Dr. Marlin Medina (2010); Fig. 1b). We anticipated that the environment would not be the sole factor influencing the reproductive cycle; rather, the ability for thermoregulation would also play a relevant role in coping with this adverse climate. Thus, subsequent studies integrated reproductive research with field body temperatures and the efficacy of thermoregulation (Ibargüengoytía and Cussac 1998, 1999, 2002; Medina et al. 2009, 2011, 2012; Medina and Ibargüengoytía 2010).

2 Reproductive Patterns in the Patagonia Andes and Steppe

The comparisons of reproductive cycles and thermoregulation capabilities between oviparous (*Liolaemus* cf. *bibronii* and *Liolaemus boulengeri*) and viviparous (*Liolaemus lineomaculatus*) lizards from various latitudes and elevations demonstrated that these species compensate for environment harshness through effective thermoregulation. They all exhibit synchrony in the embryonic developmental period, which occurs during the warmest part of the activity season, irrespective of the reproductive mode, whether it occurs in nests or the female uteri (Table 1). In the viviparous *L. lineomaculatus*, there seems to be no clear benefit or selective advantage from the mother providing warmer or more stable incubation temperatures through active thermoregulation (as proposed by the cold climate or temperature stability hypotheses; Shine 2004) compared to oviparous species living at similar climatic conditions. The oviposition in *L.* cf. *bibronii* and *L. boulengeri* takes place around the beginning of summer, ensuring that the colder temperatures are experienced towards the end of the developmental period (Table 1). This timing is expected to have relatively little effect on the hatching date, as the developmental rates of the embryos exhibit low thermal dependence during this time (Shine 2002). The viviparous *L. lineomaculatus* is distributed within higher latitudes (from 37° to 52°S), compared to *L.* cf. *bibronii* and *L. boulengeri*, which inhabit Patagonia within the ranges from 31° to 50°S. These syntopic oviparous and viviparous liolaemids manage proximate constraints, such as an extreme thermal environment or limited food availability, due to

Table 1 Summary including details on the reproductive frequency in females, litter size, mean annual reproductive output, and reproductive cycle and sperm storage of various lizard species from the Argentinian Andes and Patagonia steppe

Species	Publication Year	Mode of reproduction	Locality of study (latitude, longitude and altitude)	Reproductive frequency in female	Litter/ clutch size	Mean annual reproductive output	Male reproductive cycle	Sperm storage	Investigators (authors)
Liolaemus pictus	1996	Viviparous	41° 10' S 71° 25' W, 711 m a. s.l.	Biennial–Triennial	3–6	1.9	Continuous	Yes	Ibargüengoytía and Cussac
Liolaemus elongatus	1998, 1999	Viviparous	39–41° S 70.5–71.6° W; 500–1800 m a.s.l.	Annual–Biennial	3–7	3.75	Seasonal	No	Ibargüengoytía and Cussac
Phymaturus tenebrosus	2004	Viviparous	40.5–41.6° S 70.5–71.5° W, 575–1230 m a.s.l.	Biennial	2	1	Postnuptial	Yes	Ibargüengoytía
Phymaturus antofagastensis	2006	Viviparous	27°02′00′′S; 68°04′11′′W), 4200 m a.s.l.	Biennial	2	1	Continuous	Yes	Boretto and Ibargüengoytía
Phymaturus punae	2007	Viviparous	28°59′ to 29°02'S; 69°29′ to 69°05'W; 3100–4200 m a.s.l.	Biennial	1–2	0.75	Prenuptial		Boretto et al.
Homonota darwinii	2007	Oviparous	41–45° S 69–71° W, 800 m a. s.l.	Annual–Biennial	1	0.75	Postnuptial	Yes	Ibargüengoytía and Casalins
Phymaturus zapalensis	2009	Viviparous	39° 44' S 70° 22' W, 824–1312 m a.s.l.	Annual–biennial	1–2	1.12	Postnuptial	No	Boretto and Ibargüengoytía
Liolaemus bibronii	2010	Oviparous	43° S 70° W, 626 m a.s.l. and 46° S, 71° W, 263 m a.s.l.	Annual	2–4	3	Postnuptial	Yes	Medina and Ibargüengoytía
Liolaemus boulengeri	2010	Oviparous	43° S 70° W, 626 m a.s.l. and 46° S, 71° W; 263 m a.s.l.	Annual	2–5	3.5	Postnuptial	Yes	Medina and Ibargüengoytía
Phymaurus aguanegra	2010	Viviparous	30°23'S, 69°34′ W; 2900 m a.s.l.	Biennial	1–2	0.75	Continuous	Yes	Cabezas-Cartes et al.
Phymaturus spectabilis	2014	Viviparous	41° S 69° W, 983–1064 m a. s.l.	Annual–Biennial	1–3	1.5	Postnuptial	No	Boretto et al.
Liolaemus lineomaculatus	2010	Viviparous	43° S 70° W, 626 m a.s.l. and 46° S, 71° W, 263 m a.s.l.	Annual	2–4	3	Postnuptial	Yes	Medina and Ibargüengoytía
Liolaemus magellanicus	2015 2017	Viviparous	50° S 72° W, 80 m asl, and 51° S, 71° W, 133 m a.s.l.	Annual-Biennial	3–4	2.55	Continuous	Yes	Fernández et al.
Liolaemus sarmientoi	2015 2017	Viviparous	50° S 72° W, 80 m asl, and 51° S, 71° W, 133 m a.s.l.	Annual–Biennial	2–7	3.09	Continuous	Yes	Fernández et al.
Phymaturus williamsi	2018	Viviparous	32°31'S; 69°42'W; 2.878 m a.s.l.	Biennial	1.9	0.95	Postnuptial	Yes	Boretto et al.
Phymaturus extrilidus	2022	Viviparous	30° 56′ 59″ S, 69° 04′ 50″ W; 3100 m a.s.l.	Biennial	2–3	1.14	Prenuptial		Boretto et al.

low temperatures and prolonged drought periods, by strategically allocating energy to balance maintenance, storage, growth, and reproduction functions. As a result, there is significant energetic investment in reproduction, enabling annual reproductive cycles, similar clutch sizes, and reproductive outputs regardless of their reproductive mode. *Liolaemus lineomaculatus* shares the same high latitudes with other oviparous species such as *Diplolaemus darwinii* (45°–52°S), and the nocturnal *Homonota darwinii* (34°–52°S; Cei 1986, Scolaro 2005; Fig. 4b and d). However, beyond latitude 52° S, the harsh environments are inhabited solely by two viviparous liolaemids, *L iolaemus magellanicus*, and *L iolaemus sarmientoi*.

Further studies were conducted to examine reproductive cycles and thermoregulation effectiveness, and integrated results were used to assess whether development at nest temperatures could negatively affect pregnancy and offspring fitness in species inhabiting the southernmost range in Patagonia. Of particular interest were the questions regarding the origin of viviparity in squamate reptiles and its correlation with cold climates (Mell 1929; Blackburn 1982; Shine 1983, 1985), as well as the stability of body temperatures (Charland and Gregory 1990; *The maternal manipulation hypothesis*, Shine 1995, 2004). Specifically, the low nest temperatures associated with cold-climate habitats are believed to be detrimental to the developing embryos in oviparous lizards (Andrews et al. 1997; Shine 2002).

3 Unveiling Reproductive Strategies in the Lizards from the "End of the World"

Thus, in 2015, the next inquiry was whether the southernmost lizards, the viviparous *L. sarmientoi* and *L. magellanicus* (from 50° and 51°S, respectively), living in environments subject to extreme changes, possess characteristics that enable them to reproduce successfully (Fernández 2016). We found that females of these species reproduce annually or biennially, probably influenced by the severity of energetic constraints imposed by the cold-temperate, harsh, and rapidly changing weather of Austral South America (Fernández et al. 2015). However, males exhibit a continuous reproductive cycle with sperm storage over winter, likely in response to the low availability of receptive females. This strategy provides a selective advantage by ensuring reproductive males are available as soon as females become receptive (Fernández et al. 2017a). Furthermore, pregnant females of *L. sarmientoi* prefer more stable body temperatures (T_b) compared to non-pregnant females. This behavior suggests a strategy to avoid high and variable temperatures that could accelerate embryo development and potentially impair offspring performance and survival. Precise thermoregulation by pregnant females enhances fitness-related offspring traits such as shorter gestation periods, improved locomotor performance, and better body condition. Conversely, exposure of pregnant females to low and changing temperatures resembling their natural environment (emulating nest temperatures of oviparous species) led to impaired embryonic development and reduced offspring performance and survival rate (Fernández et al. 2017b).

Other questions that surfaced while performing these studies included: Do females nourish embryos through placentation, or do they primarily rely on vitellogenesis? If so, how does this impact their ability to replenish energy reserves for initiating a hibernation period after parturition?

4 Placentation and Neonate Nourishment in the Genus *Liolaemus*

Studies on placentation in liolaemids were conducted by undergraduate researcher Melisa Crocco. This line of investigation examined the role of placental function in facilitating the survival of lizards in harsh environments. The findings were analyzed within a framework of the evolutionary transition from oviparity (egg-laying) to viviparity (live-bearing), drawing

Fig. 4 Field survey in Pilcaniyeu region, Province of Rio Negro, Argentina, after Puyehue-Cordón Caulle volcano eruption in 2011. (**a**) Rock outcrops covered in ash. (**b**) *Homonota darwinii* found in groups under the rocks. (**c**) Specimens of *Liolaemus elongatus*. (**d**) Specimen of *H. darwinii* partially covered with ash. (**e**) General view of the steppe covered in ash. (**f**) Male and females of *Phymaturus tenebrosus*. (**g**) The steppe landscape after the eruption, with the ground and air both blanketed by ash

comparisons with analogous processes observed in other reptiles and mammals. This research not only advanced knowledge of the survival strategies employed by liolaemid but also offered valuable insights into broader evolutionary patterns and reproductive adaptations across taxa.

Specifically, Melisa Crocco et al. (2008) demonstrated that *Liolaemus elongatus* (Figs. 1c and f, and 4c) is primarily a lecithotrophic species with placental structures specialized for oxygen uptake and inorganic nutrient transport. During development, there is a progression of the placenta, more specifically, from an allantoplacenta and an omphaloplacenta, and a secondary cleft opens in the main vitelline mass, above the first yolk cleft (at embryonic stage 42). This cleft separates the vitelline mass into two compartments, one embryonic that would be used for feeding the embryo during development and the other abembryonic that would remain after birth as an external yolk sac contained into extraembryonic membranes (Crocco et al. 2008). This hypothesis was reinforced by the observation that *L. elongatus* can bury eggs in an advanced stage of development with a transparent membrane, yet an external yolk reservoir remains. One possible explanation is that the newborn may rely on the external yolk supply until weather conditions become suitable for clutching. The scope of research on physiological and behavioral reproductive responses then expanded to include the viviparous genus *Phymaturus*, which occupies a wide range of elevations and latitudes under severe climatic conditions.

5 To the North, High in the Andes

The genus *Phymaturus* comprises 54 recognized species divided into two clades: the *Phymaturus palluma* group (23 species; Fig. 5c and d), distributed along the Andes in Argentina and Chile, and the *Phymaturus patagonicus* group (31 species), exclusively found in the steppe environments of Argentine Patagonia (Lobo and Barrasso 2021; Lobo et al. 2022). These lizards constitute an appealing model for ecological research due to their specialist saxicolous habitat, herbivorous diet, and viviparous reproductive strategy (Cei 1986).

The initial study on reproductive cycles in the genus *Phymaturus* was conducted by Ibargüengoytía in 2004. The research focused on different populations of *Phymaturus tenebrosus* from Neuquén and Río Negro Provinces (Fig. 4f) and demonstrated this species to be strictly biennial in its female reproduction (Table 1). In *P. tenebrosus*, vitellogenesis and pregnancy occur in alternating years, resulting in the simultaneous presence of vitellogenic and pregnant females throughout the entire activity season (Ibargüengoytía 2004). The most extended process in this species is vitellogenesis, which lasts 1 year, while gestation lasts approximately 4 months starting in mid-spring. After parturition at the end of summer, females enter a period of reproductively quiescent lasting approximately 8 months from autumn until the next activity season, when vitellogenesis resumes in spring (Ibargüengoytía 2004).

On the other hand, the Patagonian species *Phymaturus zapalensis* and *Phymaturus spectabilis* (Fig. 1d) exhibit plasticity in the timing and duration of female reproductive cycles (Boretto and Ibargüengoytía 2009; Boretto et al. 2014a). Studies of the reproductive biology of these species indicate that in environments with short, cold activity seasons and limited feeding opportunities, females must restock their energetic reserves over subsequent seasons to grow and store energy for the next reproductive cycle (Table 1). In contrast, under more favorable conditions, the female reproductive cycle can be completed in a single year. Variations in environmental constraints and diet among *Phymaturus* species may account for the different patterns of female reproduction observed. We found two life history patterns. *Phymaturus* species endemic to high elevations in the Andes, characterized by higher thermal amplitude and lower precipitations, typically exhibit an herbivorous diet and strict biennial female reproductive cycles. Conversely, species endemic to low altitudes and high latitudes in the Patagonian steppe of Argentina, where there is lower thermal

Fig. 5 Fieldwork and specimens from the Provinces of San Juan, Mendoza, and Catamarca, Argentina. (**a**) Environment of *Phymaturus aguanegra* a 4780 m asl at paso de Agua Negra, San Juan Province. (**b**) *P. aguanegra* from Agua Negra, San Juan. (**c**) *Phymaturus palluma* from Mendoza Province. (**d**) Typical environment of *P. palluma* at 2500 m asl, in the Aconcagua Provincial Park, Mendoza. (**e**) *Phymaturus antofagastensis* from Catamarca Province. (**f**) Typical environment of *P. antofagastensis* at 4500 m asl in Paso San Francisco in Catamarca

amplitude and higher precipitation, often have an omnivorous-insectivorous diet and display plasticity in their annual biennial reproductive cycles.

The *P. palluma* species group is found at high altitudes in the Andes mountains, ranging from 2000 to 4500 m asl, where they experience high thermal amplitude typical of high-elevation environments. All species in this group exhibit a strictly biennial female reproductive cycle (Table 1). Females require at least one activity season to undergo either vitellogenesis or pregnancy (*P. vociferator*, Habit and Ortiz 1996; *P. antofagastensis*, Boretto and Ibargüengoytía 2006, Fig. 5e and f; *P. punae*, Boretto et al. 2007; *P.* cf. *palluma* = *P. aguanegra*, Cabezas-Cartes et al. 2010; Fig. 5a and b; *P. williamsi*, Castro et al. 2018; *P. extrilidus*, Pizarro et al. 2022). Females, due to their higher reproductive effort compared to males, face physiological constraints related to the timing of vitellogenesis, pregnancy, and parturition, as well as to energy acquisition from the environment (Boretto et al. 2007). Yolk deposition into ovarian oocytes and the various forms of placentation observed in viviparous lizards are essential components of the reproductive effort and parental care displayed throughout the female reproductive cycle (Boretto et al. 2007). Newborns of several species of *Phymaturus* exhibit not only large fat bodies but also intra-abdominal yolk (Boretto et al. 2007; Boretto and Ibargüengoytía 2009; Cabezas-Cartes et al. 2010; Boretto et al. 2014a). Females appear to produce more yolk than is consumed by the developing embryo, resulting in remnant intra-abdominal yolk in the newborn. The presence of intra-abdominal yolk and fat bodies may help prevent starvation of newborns in harsh environmental conditions and increase survivorship, particularly if births occur late in the activity season (Boretto et al. 2007, 2014a, Boretto and Ibargüengoytía 2009, Cabezas-Cartes et al. 2010). Studies on placentation of *Phymaturus* have not been performed yet but would be interesting to know if they experience a similar progression of different types of placentae and a yolk cleft that could explain the intra-abdominal reserves of yolk observed in newborns (Boretto et al. 2007).

Despite interspecific variations in the timing of vitellogenesis and gestation among the different liolaemids endemic to the Andes highlands and high latitudes in Patagonia, births consistently occur during mid to late summer across all species. The convergence in birthing timing underscores the influence of climate on female reproductive schedules and the availability of food and warmer temperatures necessary for digestion, assimilation, and fat storage—essential for females and newborns preparing for brumation (reviewed in Boretto et al. 2020). Favorable environmental conditions during parturition are expected to enhance neonatal survival, particularly in lizards with small litter sizes such as *Phymaturus* (reviewed in Boretto et al. 2020, Boretto and Ibargüengoytía 2021).

While researching female reproductive cycles, questions arose regarding the endocrine regulation of prolonged female cycles, whether they are strictly biennial or annual-biennial. Previous studies have indicated that in viviparous female lizards, high levels of progesterone during gestation inhibit follicular development and vitellogenesis, thereby limiting the reproductive cycle due to the physiological inability of females to undergo vitellogenesis while gravid (Duvall et al. 1982; Callard et al. 1992; Custodia-Lora and Callard 2002).

A preliminary study conducted on endemic species such as *P. antofagastensis* (Fig. 5e and f) and *P. aguanegra* (Fig. 5a and b), inhabiting cold environments at high altitudes in the Andes mountains, showed high progesterone levels and low or undetectable levels of estrogen in pregnant females (Boretto et al. 2010a; Fig. 5). Similarly, females of *P. zapalensis*, endemic to the Laguna Blanca National Park and its surroundings in Argentine Patagonia, exhibited elevated progesterone levels during early and mid-gestation stages compared to late pregnancy or the vitellogenic period (Boretto et al. 2010a). These findings are consistent with earlier studies which suggest that the endocrine control mechanism induces incompatible physiological states in some liolaemids. Coupled with the thermal and energetic constraints of cold environments, these mechanisms restrict female reproductive cycles.

The knowledge of the prolonged female reproductive cycles led us to inquire how males reproduce when only a portion of the female population is receptive. Questions arise regarding whether males or females store sperm in their reproductive tracts and how males cope with reproductive asynchrony.

6 Males Adjust to a Low Frequency of Reproductive Females in Diverse Ways

In males, spermatogenesis is known to have a relatively low energy cost and is inhibited at body temperatures (T_b) below 20 °C (Saint Girons 1985). Therefore, it would be expected that thermal constraints at high altitudes in the Andes Mountains and high latitudes in Argentine Patagonia would also influence male reproductive strategies. The prolonged female reproductive cycle has led to a male-biased operational sex ratio (OSR; proportion of reproductive males to receptive females) in several species including *P. antofagastensis* (0.65:0.35; Boretto and Ibargüengoytía 2006; Fig. 5e and f), *P. aguanegra* (1.75:1; Cabezas-Cartes et al. 2010; Fig. 5a and b), *L. pictus* (1:0.33–0.5; Figs. 2d, e, g and 3e–g), *L. elongatus* (1:0.5–1; Ibargüengoytía and Cussac 1999; Figs. 1c, f and 4c), *P. tenebrosus* (1.5:1; Ibargüengoytía 2004; Fig. 4f), *P. spectabilis* (65–35; Boretto et al. 2014a; Fig. 1d), and *P. zapalensis* (72–28; Boretto and Ibargüengoytía 2009) (Table 1). This skewed ratio could lead to increased intrasexual competition among males, and indeed, male reproductive strategies exhibit considerable diversity.

In this regard, we discovered that phenotypic plasticity in females may serve as a reproductive adaptation enabling species to navigate the environmental challenges posed by the cold climate in the harsh environments of southern South America. According to Ibargüengoytía and Cussac (1999), the existence of severe thermal constraints for vitellogenesis and pregnancy sets off a chain of causal events: Environmental conditions impact female cycles, which in turn influence male cycle and male dimorphic traits. The biennial reproductive cycles of females and short activity seasons prompt males to adapt creatively to the low percentage of receptive females in the population.

An example of males coping with the presence of only 35% of reproductive females in the population is *P. antofagastensis* (Fig. 5e and f), which inhabits high elevations in the Andes. Males of this species exhibit continuous and asynchronous spermatogenesis and store sperm in their reproductive tract, which reduces competition among males and enables fertilization at any time during the activity season (Boretto and Ibargüengoytía 2006). This reproductive strategy is advantageous in situations where there is a low frequency of receptive females or when there is phenotypic plasticity in the timing of female reproductive events. In *P. antofagastensis*, female ovulation can occur either in autumn or in spring, with low rates of over-winter embryonic development or vitellogenesis, respectively. Consequently, males exhibit two peaks of spermiation and stored spermatozoa accordingly (Boretto and Ibargüengoytía 2006). Instead, males of *P. extrilidus*, another endemic species from the Andes highlands, exhibit an annual prenuptial reproductive cycle, with spermatogenesis occurring from spring to early autumn, synchronized with the follicular development of the females (Pizarro et al. 2022). The prenuptial spermatogenic cycle has also been described in other *Phymaturus* species such as *P. vociferator* (Habit and Ortiz 1996) and *P. punae* (Boretto et al. 2007), whereas species like *P. antofagastensis* (Boretto and Ibargüengoytía 2006; Fig. 5e and f) and *P. aguanegra* (Cabezas-Cartes et al. 2010 (Fig. 5a and b) exhibit continuous and asynchronous spermatogenic cycles. Other *Phymaturus* species perform postnuptial spermatogenic cycles, relying on stored sperm in deferent ducts associated with spring mating behaviors. This is the male reproductive strategy observed in *P. williamsi*, endemic to the Andes highlands in San Juan, Argentina (Castro et al. 2018); and for all studied *Phymaturus* endemics from Patagonia, such as *P. tenebrosus* (Ibargüengoytía 2004; Fig. 4f), *P. zapalensis* (Boretto and Ibargüengoytía 2009), and *P. spectabilis* (Boretto et al. 2014a; (Fig. 1d).

In *Liolaemus*, males of *L. sarmientoi* (Fernández et al. 2017a) and *L. pictus* (Ibargüengoytía and Cussac 1999) address female reproductive challenges by being permanently reproductive with spermatozoa in testicles or epididymis, ensuring availability for mating throughout the entire activity season. In addition, males of *L. sarmientoi* exhibit a notable color polymorphism, with red individuals displaying more aggressive behavior during agonistic encounters compared to the yellow males (Fernández et al. 2018). Interestingly, this is also observed in tropidurids: Red males of *Tropidurus* aff. *oreadicus* in the northern areas of the Cerrado Biome in Brazil are considerably more aggressive than the grey males (personal observation, Piantoni 2012). This polymorphism in males is commonly associated with other behaviors attributed to various reproductive and physiological strategies in lizards (Sinervo and Lively 1996; Sinervo et al. 2000; Healey et al. 2007; Miles et al. 2007; Bastiaans et al. 2013; Bruinjé et al. 2019; Brock et al. 2020). For instance, red males also tend to exhibit higher T_b and larger head sizes, traits that likely influence their reproductive success (Fernández et al. 2018).

7 Sertoli and Leydig Cell Shifts of Testosterone Production During the Activity Season in Synchrony with Female Cycles

Given the variability observed in male reproductive strategies and their synchronization with female cycles, the next step was to investigate potential mechanisms of plasticity in male reproductive behaviors, specifically in courtship and copulation, and their relationship with gametogenesis. To achieve this, the functional state of the testicular compartments was characterized through ultrastructural studies of cellular organelles involved in testosterone production complemented with the studies of serum testosterone concentrations across the reproductive cycle of *Phymaturus* species (Fig. 6).

Sertoli cells, found exclusively in seminiferous tubules, synchronize and maintain spermatogenesis (Bardin et al. 1988; Dubois et al. 1988), while Leydig cells synthesize androgens that influence courtship and mating behaviors (Dubois et al. 1988; Ibargüengoytia et al. 1999). Based on these distinctions, there exists a degree of independence between spermatogenesis and mating events (Mahmoud et al. 1985; Dubois et al. 1988; Mahmoud and Licht 1997). Studies on *P. antofagastensis* (Fig. 5e and f) revealed adaptive mechanisms where steroid synthesis in these cells occurs asynchronously, allowing males to adjust their reproductive cycles to environmental and physiological constraints and to undergo spermatogenic cycle and mate at different periods of the activity season, synchronizing their reproductive timing with females (Boretto et al. 2010b). Similarly, *P. punae* shows temporal asynchrony in steroid activity between Sertoli and Leydig cells, crucial for initiating spermatogenesis and supporting the prenuptial spermatogenic cycle and mating periods, respectively (Boretto et al. 2014b). This asynchrony appears more pronounced in *Phymaturus* species with continuous and asynchronous male reproductive cycles (e.g., *P. antofagastensis*, Boretto et al. 2010b) or postnuptial cycles (e.g., species from *P. patagonicus* group, like *P. zapalensis*, Boretto et al. 2012) than in species with prenuptial cycles (e.g., *P. punae*, Boretto et al. 2014b), emphasizing the adaptive significance of hormonal regulation in these lizards' reproductive strategies.

These studies represent significant progress in understanding the physiological and behavioral adaptations of males to the female cycles, within the cold environments of the Andean Mountains and the Patagonian steppe of Argentina. They highlight the existence of cellular mechanisms that enable the development of the spermatogenic cycle independently of the mating period (Boretto et al. 2010b, 2012, 2014b).

8 Growth and Life Events in High Altitudes and Latitudes

The environmental factors influencing the reproductive cycles of lizards in the Andes also significantly impact other aspects of their life history, including lifespan, sexual maturity, and energetic trade-offs (Fig. 7). Lower temperatures

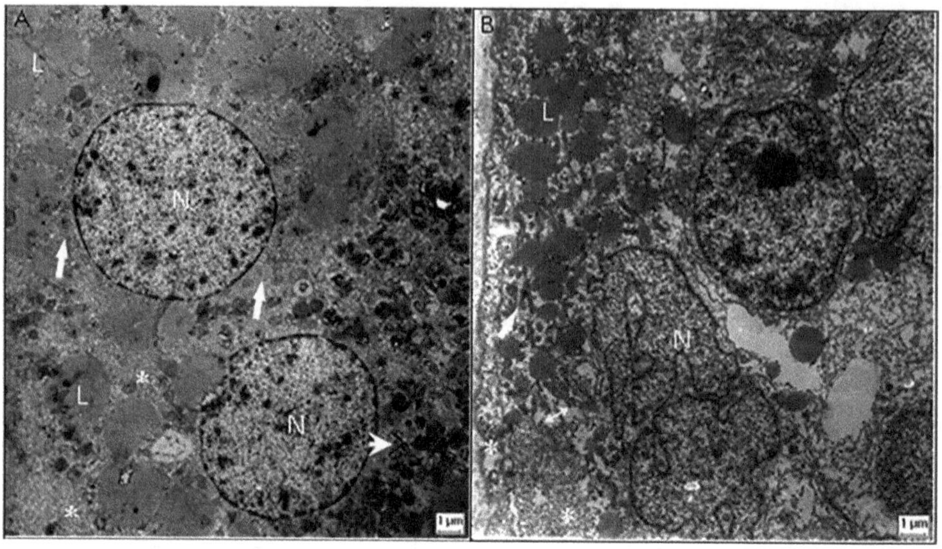

Fig. 6 Leydig and Sertoli cells of adult males of *Phymaturus antofagastensis* during the reproductive active season. (**a**, **b**) Sertoli cells with evidence of steroidogenic activity. N: cell nuclei. L: lipid droplets; *: smooth endoplasmic reticulum; arrows indicate the presence of mitochondria, arrowheads indicate the multilamellar bodies, and double-headed arrows indicate the presence of glycogen granules. Images also published by Boretto et al. 2010a, b

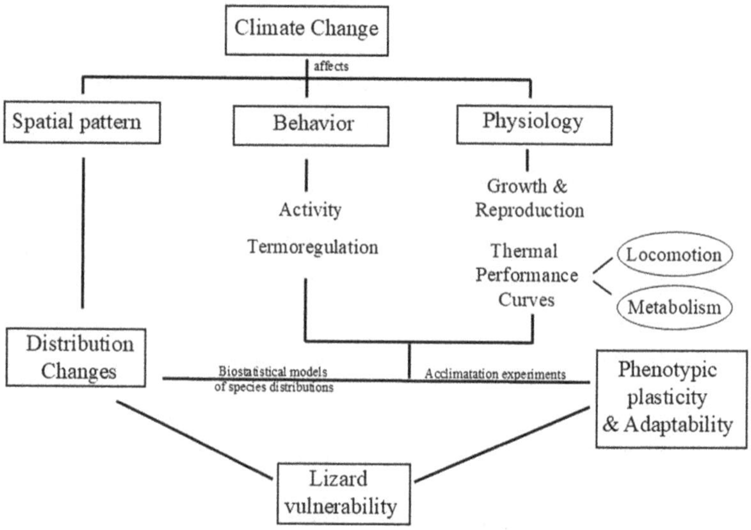

Fig. 7 Schematic representation of the effects of climate change on the life history traits of Andean lizards

throughout the year have been associated with slower growth rates and delayed sexual maturity. Furthermore, longevity has been seen to be influenced by the ambient temperatures (T_a) experienced by lizards over their lives. Given these considerations, several questions arise: How long do lizards live if they reproduce once every 2 years? At what age do they reach sexual maturity? And is there an energetic trade-off between growth and reproduction that leads to slower life cycles? Additional questions emerged throughout the course of the studies, including whether interspecific differences were associated with elevation and how various natural or anthropogenic disturbances influenced lizard populations. By understanding how lizards adapt to their environments and optimizing their reproductive success over their lifespan, we may be able to contribute to their conservation.

The investigation of these questions has been crucial in determining the annual reproductive output and life fecundity of several lizard species (Table 2), thereby contributing to their conservation status and vulnerability assessment (Abdala et al. 2012a, b). Since the early 2000s, the laboratory has employed skeletochronology as the most precise method available to address these inquiries (Piantoni et al. 2006a, b; Kubisch et al. 2012; Gutiérrez et al. 2013; Cabezas-Cartes et al. 2015). This technique has allowed for assessing the impact of chronic and acute natural disturbances on lizard populations and proposes a biological grounded conservation strategy. For instance, studies on reproductive biology and skeletochronology have provided precise estimates of average annual fecundity for numerous species, which are now integral to conservation assessments such as those conducted by Abdala et al. (2012a, b), in the "*Categorización del estado de conservación de las lagartijas y anfisbenas de la República Argentina*."

Initial studies using skeletochronology began in the early 2000s. Within Liolaemidae, analysis of growth marks found in femurs of *P. tenebrosus* (Fig. 4f) suggested that females reach sexual maturity at 7 years old, while males do so at 9 years old, with a maximum lifespan of 16 years (Piantoni et al. 2006a). This research proposed that severe environments in Patagonia play a significant role in shaping reproductive cycles, which collectively (i.e., delayed maturity and prolonged reproductive cycles conjunctively) result in low fecundity.

Years later, the technique enabled comparisons of growth patterns in populations of *L. pictus* (Figs. 2d, e, g and 3e–g) from low and high altitudes in the Andes near San Carlos de Bariloche, Argentina (Gutiérrez et al. 2013), revealing distinct environmental impacts, particularly from lower temperatures, on these life history traits. In 2006, Piantoni and collaborators analyzed growth and age in the southernmost gecko, *H. darwinii* (Fig. 4b and d), highlighting its k-selected traits of late maturity and long lifespan. These findings further supported the hypothesis that daytime burrow temperatures, akin to T_b registered in diurnal lizard species, meet the physiological growth needs of nocturnal species. This study was complemented by research on various aspects of *H. darwinii*, including thermophysiology (Aguilar and Cruz 2010), performance (Renner 1998; Ibargüengoytía et al. 2007), and diet (Kun et al. 2010).

However, the most intriguing findings regarding this species emerged in 2012 by Kubisch and collaborators. Using skeletochronology once again, museum specimens collected in Chubut Province (the "warm" site) in 1941 were compared with specimens from the same site collected in 2010. Additionally, the latter specimens were subsequently compared with specimens from Rio Negro Province (the "cold" site) collected in 1997–1998. This comparative analysis enabled examining temporal and geographical variations in life-history traits related to growth, specifically investigating whether these traits vary with environmental temperature. This study aimed to assess whether the plasticity in response to temperature variations, especially in the context of global warming, could potentially benefit species from colder regions rather than threaten their extinction. Interestingly, the results suggested that higher temperatures have a positive effect on growth, indicating that global warming might benefit species like *H. darwinii*, particularly populations in southern regions.

Studies utilizing skeletochronology continued in the group, integrating reproductive biology traits such as clutch size, reproductive frequency, and relative clutch mass with growth traits like longevity and age at sexual maturity. Consequently, life-history patterns were described for different lizard species of the Argentine Andes and Patagonian steppe. In *Phymaturus*, lizards exhibit "slow" life histories characterized by delayed sexual maturity and extended longevity (Table 2). Notably, distinct life-history patterns were observed between the two phylogenetic groups of the genus. Species of the *P. palluma* group (Fig. 5c and d) found in the Andes demonstrate a pattern of low reproductive investment

Table 2 Comparison of variables related to growth and reproduction in Andean-Patagonian lizards from Argentina, including their geographical location and elevation of the sampled site (m = meters above sea level), reproductive mode and diet, litter or clutch size, range of body sizes of adult individuals (SVL, mm), age of sexual maturity, maximum estimated age, adult life span in years (E), relative reproductive time (RT), and proportion of adult life (%)

Species	Locality	Reproductive mode and diet	Minimum and maximum adult size	Clutch/ litter size	Reproductive frequency of females	Age at sexual maturity	Maximum estimated age	E	RT	%
Homonota darwinii [a, b]	Ea. El Cóndor Pilcaniyeu (Río Negro) 922 m	Oviparous [c] Insectivorous [d]	♀: 38.0–47.0 ♂: 38.0–44.0	1	0–1	♀: 2 ♂: 2	♀: 7 ♂: 7	♀: 5 ♂: 5	♀: 3.5 ♂: 3.5	♀: 71.4 ♂: 71.4
Liolaemus pictus Low-altitude Site [e]	San Carlos de Bariloche (Río Negro) 771 m	Viviparous [f] Insectivorous	♀: 51.7–64.3 ♂: 49.8–63.2	3–6	0.3–0.5	♀: 4 ♂: 4	♀: 7 ♂: 9	♀: 3 ♂: 5	♀: 1.75 ♂: 2.25	♀: 42.8 ♂: 55.5
Liolaemus pictus High-altitude site [e]	Challhuaco mountain (Río Negro) 1769 m	Viviparous [f] Insectivorous	♀: 50.4–62.1 ♂: 51.4–62.0	3–6	0.3–0.5	♀: 3 ♂: 4	♀: 8 ♂: 8	♀: 5 ♂: 4	♀: 2.7 ♂: 2	♀: 62.5 ♂: 50.0
Phymaturus aguanegra [g]	Paso Agua Negra (San Juan) 2900 m	Viviparous [h] Herbivorous [i]	♀: 82.4–101.7 ♂: 83.4–104.5	2	2	♀: 7 ♂: 7	♀: 16 ♂: 16	♀: 9 ♂: 9	♀: 2.28 ♂: 2.28	♀: 56.2 ♂: 56.2
Phymaturus. spectabilis [j]	25 km S Ing. Jacobacci (Río Negro) 983–1064 m	Viviparous [k] Herbivorous [l]	♀: 83.1–89.7 ♂: 79.6–86.9	2	2	♀: 7 ♂: 6	♀: 12 ♂: 11	♀: 5 ♂: 5	♀: 1.71 ♂: 1.83	♀: 41.6 ♂: 45.4
Phymaturus tenebrosus [m]	Pilcaniyeu (Río Negro) 575–1230 m	Viviparous [n] Herbivorous [o]	♀: 88.0–109.0 ♂: 88.0–98.0	2	2	♀: 7 ♂: 9	♀: 16 ♂: 13	♀: 9 ♂: 4	♀: 2.28 ♂: 1.44	♀: 56.2 ♂: 30.7
Phymaturus zapalensis [g]	Near Laguna Blanca National Park, Zapala (Río Negro) 824–1312 m	Viviparous [p] Omnivorous [q]	♀: 79.02–96.6 ♂: 75.6–90.5	2	2	♀: 9 ♂: 8	♀: 14 ♂: 15	♀: 5 ♂: 7	♀: 1.55 ♂: 1.87	♀: 35.7 ♂: 46.6

(*a*) Piantoni et al. 2006b; (*b*) Kubisch et al. 2012; (*c*) Ibargüengoytía and Casalins 2007; (*d*) Kun et al. 2010; (*e*) Gutiérrez et al. 2013; (*f*) Ibargüengoytía and Cussac 1999; (*g*) Cabezas-Cartes et al. 2018; (*h*) Cabezas-Cartes et al. 2010; (*i*) Lobo et al. 2013; (*j*) Cabezas-Cartes et al. 2015; (*k*) Boretto et al. 2014a; (*l*) Scolaro 2006; (*m*) Piantoni et al. 2006b; (*n*) Ibargüengoytía 2004; (*o*) Abdala et al. 2012a; (*p*) Boretto and Ibargüengoytía 2009, (*q*) Boretto and Ibargüengoytía 2018 but longer reproductive spans. In contrast, species of the *P. patagonicus* group, inhabiting the Patagonian steppe, present shorter reproductive spans but higher reproductive frequency and effort (Boretto et al. 2018). Cabezas-Cartes et al. (2018) employed phylogenetic comparative analyses to demonstrate that longevity is primarily influenced by latitude (e.i., lizards from higher latitudes live longer), while mean annual temperature affects the age at sexual maturity (i.e., lizards from hotter climates mature earlier).

9 Lizard Sensitivity to Environmental Temperature Variation and Climate Change

The effects of acute and chronic disturbances, such as volcanic eruptions and climate change, highlight the vulnerability of lizard populations in the Andes and Patagonia, particularly those with slow life cycles and low reproductive rates, like the *Phymaturus* genus. These disturbances not only impact the ecology and reproduction of these species, as seen in the effects of ash deposition on locomotion and reproduction, but they also interact with the physiological sensitivity of reptiles to temperature variations. Since many physiological processes in reptiles are highly temperature-dependent, it is crucial to understand how these species respond to environmental temperatures that fall within or outside their optimal ranges (T_o). Given this common knowledge, we investigated the locomotion speeds and endurance of several Patagonian species across varying environmental temperatures. Studies included *Homonota darwinii* (Ibargüengoytía et al. 2007; Fig. 4b and d), *Liolaemus sarmientoi* (Fernández et al. 2011), *L. magellanicus* (Fernández et al. 2011), *L. pictus* (Kubisch et al. 2011; Figs. 2d, e, g and 3e–g), *L. elongatus* (Kubisch et al. 2023; Figs. 1c, f and 4c), *P. tenebrosus* (Cabezas-Cartes et al. 2019; Fig. 4f), *P. palluma* (Vicenzi et al. 2019b; Fig. 5c and d), *P. calcogaster* (Obregon et al. 2021), *P. zapalensis*, and *P. querque* (Cabezas-Cartes et al. 2023). Our findings revealed that most studied Patagonian species presently experience temperatures below their optimal range for maximum locomotor performance. However, some populations occasionally encounter temperature peaks that exceed their T_o-range. Therefore, the ongoing trend of global warming could potentially reduce running performance in these lizard populations.

The next step was to investigate whether these species exhibit phenotypic plasticity, allowing them to adjust their T_o for locomotion through acclimatization to different temperatures. Our results revealed that *L. pictus* demonstrated a limited response to acclimation (Kubisch et al. 2016). However, *L. elongatus* showed evidence of plasticity in locomotor performance when acclimated to higher temperatures, suggesting a potential adaptive advantage to mitigate the effects of rising environmental temperatures anticipated with climate change (Kubisch et al. 2023).

Several studies in the Patagonian steppe were conducted in collaboration with Alejandro Scolaro, who accompanied us on our explorations of Patagonia. The geological history of this region has produced several isolated basaltic plateaus that offer a variety of rocky microhabitats conducive to the development of the saxicolous genus *Phymaturus*, among other lizard species (Scolaro 2006).

This long-standing collaboration, spanning over three decades of fieldwork, led to the discovery of several new *Phymaturus* species. Notable, *Phymaturus ceii* (Scolaro and Ibargüengoytía 2007) was named in honor of Prof. Jose Miguel Cei; *Phymaturus manuelae* (Scolaro and Ibargüengoytía 2008) was named after Manuela Martínez, the daughter of Nora Ibargüengoytía, in recognition of the invaluable support provided by our families; and *Phymaturus sinervoi* (Scolaro et al. 2012) was named in tribute to Dr. Barry Sinervo. In 2009, a collaborative effort was initiated by Dr. Sinervo, Dr. Donald Miles, and Dr. Fausto Méndez de la Cruz along with other researchers from thirteen different countries, culminating in the publication of "Erosion of lizard diversity by climate change and altered thermal niches" in Science (Sinervo et al. 2010). This study modeled the extinction risks of 34 lizard families, and the projections were validated using extensive survey data and climate records from 26 researchers worldwide. The model, based on body temperature, incorporated lizard activity hours, reproductive data, and timing of breeding to assess key adaptations required to keep pace with global changes. The study also showed that rapid temperature changes have outpaced lizards ability to adapt, resulting in significant population declines. Since 1975, 12% of local lizard populations in Mexico have gone extinct. Extinctions have been observed in a variety of habitats, including tropical, temperate, rainforest, and desert environments. The study projects that local extinctions could reach 39% worldwide, with species extinctions potentially reaching 20% by 2080. This study was the starting point for the integration of the ecophysiological studies described above, including thermal physiology, reproduction, growth, performance, and the hours of restriction and activity (sensu Sinervo et al. 2010) to model the impact of climate change on lizard's populations in Patagonia and the high lands of the Andes.

In Mendoza, Argentina, we conducted comprehensive studies involving ecological, behavioral, and physiological aspects of *P. palluma* (Fig. 5c and d) to investigate the potential threat posed by increasing air temperature and reduced precipitation due to climate change. The research explored the thermal sensitivity of activity patterns, daily behavior, and locomotion in adult individuals of *P. palluma* (Vicenzi et al. 2017, 2019a, b), and the data allowed to predict the species vulnerability to climate change. Based on these predictions, we evaluated the magnitude of climate change exposure facing populations of *P. palluma* over the short and long term. The resulting assessment included estimating the risk of future extirpations and identifying critical conservation priorities (Vicenzi et al. 2017). To achieve this, we integrated physiological and ecological data from the species with historical climatic information records within its distribution area to model the current distribution. Subsequently, we then projected shifts in suitable habitats and identified potential future distributions under various climate change scenarios.

The model predicted a substantial reduction in the available habitat across the current distribution range of *P. palluma*. It also identified new potential distribution areas to the west and south in higher elevations of the Andes. However, due to the species' specialized habits, limited dispersal capability, and natural barriers, we consider the likelihood of expanding into these new sites to be minimal. Therefore, despite potential distribution shifts, *Phymaturus* species are unlikely to avoid population extirpations (Vicenzi et al. 2017). Similar models for three Patagonian species projected that by 2080, approximately 20% of populations of the nocturnal gecko *H. darwinii* (Fig. 4b and d), 15% of *L. pictus* (Figs. 2d, e, g and 3e, f, and g), and 26% of *L. elongatus* (Figs. 1c, f and 4c) could face extirpation.

However, these projections do not account for the species' ability to disperse to new habitable areas where they could potentially adapt. Many species do not possess the capacity to adjust to environmental change through phenotypic plasticity but could potentially move to a more favorable environment. For other species, this is impossible or very difficult. We found that *Liolaemus montanezi*, a microendemic lizard species found in the Andes, is another species at risk of extinction primarily due to its specialized habitat requirements and limited dispersal ability (Laspiur et al. 2021). The unique feature of its habitat seems to offer some thermal buffering against the immediate impacts of warming, particularly in the near-term. Within Leiosauridae, we began exploring the role of phenotypic plasticity in mediating resilience to climate change. We studied to what extent individuals of *Diplolaemus leopardinus*, a cold-adapted lizard, adjust various traits—such as metabolic rate, water loss, bite force, and preferred body temperature—to short-term temperature changes to understand the potential effects of warming on the species. Our studies revealed that recent thermal history plays a crucial role in shaping the response of these traits. Each trait adjusts independently; while traits like bite force enhance performance with warming, others exhibit less optimistic responses in the long term, such as the contraction of the activity window due to the selection of lower T_b. Additionally, low thermal plasticity in metabolism reduces net energy gain and threatens the survivorship and reproduction rates of the species (Vicenzi et al. 2020, 2021). Interestingly, we also observed that after a brief exposure to warm temperatures, such as during a heat wave, subsequent metabolic rates are compromised due to a significant increase in thermal sensitivity. This decoupled response likely leads to higher energetic maintenance costs and increased vulnerability to starvation in individuals. These studies emphasize the critical importance of comprehending the species' ecophysiology and its use of microhabitats in evaluating the risks posed by climate change and identifying necessary conservation measures.

Further studies of the Laboratory continue to use predictive models of climate change to assess the potential impact of global warming projected for 2050 and 2100 by the IPCC (2021) on the distribution of species we study and to predict possible extirpations and extinctions. The latest publication in Evolution (Ibargüengoytía et al. 2021) has also aimed to integrate the various research across genera in the Family Liolaemidae considering the effects of geographical factors (latitude, longitude, elevation), climatic variables (environmental temperatures and precipitation), and phylogenetic influences on ecophysiological variables. This study revealed two distinct patterns between the two main genera of the family Liolaemidae. The genus *Liolaemus* exhibits several evolutionary reversals in its thermal biology (T_p—preferred body temperature and T_b—field body temperature) and habits, which confer resilience to environmental changes and enable the exploitation of new niches. In contrast, the genus *Phymaturus* remains conservative in its ecological niche, retaining its saxicolous habit, viviparity, and feeding habits throughout evolution. However, the study also indicates that the expected temperature increases over the coming 50–80 years driven by climate warming and compounded by anthropogenic habitat reduction could lead to extirpations and extinctions, particularly among oviparous liolaemids. These species are endemic, have limited dispersal abilities, and currently inhabit warmer lowland environments where T_b are approaching and, in some cases, exceeds their T_p. Based on these investigations and the extensive data accumulated over more than three decades, we persist in exploring inquiries and endeavoring to unravel the diverse life history patterns and evolution processes revealed by our Andean species. Our ongoing research aims to deepen the understanding of how these complex interactions between phylogeny, ecology, and climate change will shape the future persistence, distribution, and evolutionary trajectories of Andean liolaemid lizards in an increasingly warming world.

Acknowledgments The studies were performed in stretched collaboration with Dr. Alejandro Scolaro (Centro Regional Patagonico, Madryn, Argentina), who shared his deep knowledge of Patagonia herpetofauna and accompanied us during most of the fieldwork. We also want to express our gratitude to Dr. Barry Sinervo (University of Santa Cruz, California), Dr. Donald Miles (Ohio University, Athens, USA), Dr. Fausto Méndez de la Cruz (Universidad Nacional Autónoma de México, Mexico), Dr. John D. Krenz (Minnesota State University, Mankato, USA), Dr. Prof. Carlos A. Navas (Universidade de São Paulo, Brazil), and to Dr. Víctor Cussac (INIBIOMA-UNCOMAHUE-CONICET, Argentina) for their friendships and professional contribution in the field, laboratory and writing manuscripts during the development of the Laboratory, and to Dr. Juan Santos (University of New York, USA), Dr. Xiang Ji and Dr. Yan-Fu Qu (Nanjing University, China) for their collaboration in several studies. Research was funded by the INIBIOMA-UNCOMAHUE-CONICET, IADIZA-CONICET, UNCUYO, and FONCYT.

Box 1. Effects of Puyehue-Cordon Caulle Ash Fall on Lizard Populations

Due to their slow life history traits and low reproductive frequency, many species in Patagonia, particularly the genus *Phymaturus*, were characterized as vulnerable. Consequently, the research in the Laboratory has focused extensively on understanding the life history and ecophysiology of these lizards to assess the impact of both acute and chronic natural disturbances, as well as anthropogenic, on their populations. For example, the effects of the 2011 volcanic eruption of Puyehue-Cordón Caulle complex on reptile populations in northern Patagonia (southern Rio Negro Province, Argentina; Fig. 4) were investigated. Specifically, following the eruption, research was conducted to assess the impact of ash accumulation on the physiological performance and reproductive responses of three *Phymaturus* lizard species. These studies revealed that the presence of ash in the substrate negatively affected the locomotion speed of some *Phymaturus* species strictly associated with rocky boulders, while species inhabiting more open habitats were unaffected (Cabezas-Cartes et al. 2014; Ibargüengoytía et al. 2016). Furthermore, the reproductive biology of *P. spectabilis* (Fig. 1d) following the ash fall showed a significant negative impact, with only 12% of females being

(continued)

reproductive during the activity season following the eruption. However, there was a notable increase in the number of reproductive females in the subsequent season, with 83% of females being reproductive, demonstrating great plasticity and rapid population recovery (Boretto et al. 2014c).

References

Abdala CS, Acosta LJ, Acosta JC, Álvarez BB, Arias F, Ávila LJ, Blanco MG, Bonino M, Boretto JM, Brancatelli G, Breitman MF, Cabrera MR, Cairo S, Corbalán VE, Hernando AB, Ibargüengoytía NR, Kacoliris FP, Laspiur A, Montero R, Morando M, Pelegrin N, Fulvio Pérez CH, Quinteros S, Semhan RV, Tedesco ME, Vega LE, Zalba SM (2012a) Categorización del estado de conservación de las lagartijas y anfisbenas de la República Argentina. Cuadernos de Herpetología 26(1):215–248

Abdala CS, Cruz FB, Semhan RV, Moreno-Azócar DL, Bonino MF, Paz MM (2012b) Taxonomic study and morphology based phylogeny of the patagonic clade *Liolaemus melanops* group (Iguania: Liolaemidae) with the description of three new taxa. Zootaxa 3163: 1–32

Aguilar R, Cruz F (2010) Refuge use in a Patagonian nocturnal lizard, *Homonota darwini*: the role of temperature. J Herpetol 44:236–241

Andrews RM, Qualls CP, Rose BR (1997) Effects of low temperature on embryonic development of *Sceloporus* lizards. Copeia 4:827–833

Bardin CW, Cheng C, Musto NA, Gunsalus GL (1988) The Sertoli Cell. In: Knobil E, Neill JD (eds) The physiology of reproduction, vol 1. Raven Press, New York, pp 933–976

Bastiaans E, Morinaga G, Gamaliel Castañeda Gaytán J, Marshall JC, Sinervo B (2013) Male aggression varies with throat color in 2 distinct population of the mesquite lizard. Behav Ecol 24(4):968–981

Blackburn DG (1982) Evolutionary origins of viviparity in the Reptilia. I Sauria Amphib-reptil 3:185–205. https://doi:10.1163/156853882X00419

Boretto JM, Ibargüengoytía NR (2006) Asynchronous spermatogenesis and biennial female cycle of the viviparous lizard *Phymaturus antofagastensis* (Liolaemidae): reproductive responses to high altitudes and temperate climate of Catamarca, Argentina. Amphibia-Reptilia 27:25–36

Boretto J, Ibargüengoytía NR (2009) *Phymaturus* of Patagonia Argentina: reproductive biology of *Phymaturus zapalensis* (Liolaemidae) and compared dimorphic traits in the genus. J Herpetol 43:96–104

Boretto JM, Ibargüengoytía NR (2018) *Phymaturus zapalensis* Diet. Herpetol Rev 49:745

Boretto JM, Ibargüengoytía NR (2021) La viviparidad en Liolaemidae: estado actual del conocimiento de la biología reproductiva y su importancia en la conservación de las especies. En: Abdala CS, Laspiur A, Scrocchi G, Semhan R, Lobo F, Valladares P (eds) Las Lagartijas de la Familia Liolaemidae: Sistemática, Distribución e Historia Natural de una de las familias de vertebrados más diversas del cono sur de Sudamérica 1. 306–310.Editorial RIL, Chile

Boretto JM, Ibargüengoytía NR, Acosta JC, Blanco GM, Villavicencio HJ, Marinero JA (2007) Reproductive biology and sexual dimorphism of a high-altitude population of the viviparous lizard *Phymaturus punae* from the Andes in Argentina. Amphibia-Reptilia 28: 1–7

Boretto JM, Jahn GA, Fornés MW, Ibargüengoytía NR (2010a) Variation in endocrine activity in relation to the reproductive cycles of viviparous lizards Phymaturus from Argentina. Paper presented at the VI Southern Connection Congress. Río Negro, San Carlos de Bariloche

Boretto JM, Ibargüengoytía NR, Jahn GA, Acosta JC, Vincenti AE, Fornés MW (2010b) Asynchronic steroid activity of Leydig and Sertoli cells related to spermatogenic and testosterone cycle in *Phymaturus antofagastensis*. Gen Comp Endocrinol 166:556–564

Boretto JM, Jahn GA, Fornés MW, Cussac VE, Ibargüengoytía NR (2012) How males synchronize their reproductive cycles with females to cope with seasonal climate: an endocrinal and ultrastructural study of *Phymaturus zapalensis* lizards (Liolaemidae). Herpetol J 22(1):33–42

Boretto JM, Cabezas-Cartes F, Tappari F, Méndez-De la Cruz F, Sinervo B, Scolaro JA, Ibargüengoytía NR (2014a) Reproductive biology of *Phymaturus spectabilis* (Liolaemidae): females skip reproduction in cold and harsh environments of Patagonia, Argentina. Herpetol Conserv Biol 9:170–180

Boretto JM, Cabezas-Cartes F, Kubisch EL, Sinervo B, Ibargüengoytía NR (2014b) Changes in female reproduction and body condition in an endemic lizard, *Phymaturus spectabilis*, following the Puyehue volcanic ashfall event. Herpetol Conserv Biol 9:181–191

Boretto JM, Cabezas Cartes F, Kubisch E, Sinervo B, Ibargüengoytía NR (2014c) Changes in female reproduction and body condition in an endemic lizard, *Phymaturus spectabilis*, following the Puyehue volcanic ashfall event. Herpetol Conserv Biol 9(1):181–191

Boretto JM, Cabezas-Cartes F, Ibargüengoytía NR (2018) Slow life histories in lizards living in the highlands of the Andes Mountains. J Comp Physiol B 188:491–503

Boretto JM, Fernández JB, Cabezas-Cartes F, Medina MS, Ibargüengoytía NR (2020) Reproductive biology of lizards from Patagonia Argentina: physiological and behavioral adaptations to cold and harsh environments. In: Morando M, Ávila LJ (eds) *Lizards of Patagonia: diversity, systematics, biogeography and biology of the reptiles at the end of the world*. ISBN: 978-3-030-42751-1. Springer, Cham

Brock KM, Baeckens S, Donihue CM, Martín J, Pafilis P, Edwards DL (2020) Trait differences among discrete morphs of a color polymorphic lizard, *Podarcis erhardii*. Peer J 8:e10284

Bruinjé AC, Moura MO, Maggi BS, São-Pedro VA, Pessoa DM, Costa GC (2019) Conspecifics of the striped lava lizard are able to distinguish sex and male colour morphs in apparently homogeneous dull dorsal colouration. Amphibia-Reptilia 40(2):149–162

Cabezas-Cartes F, Boretto JM, Acosta JC, Jahn G, Blanco G, Laspiur A, Ibargüengoytía NR (2010) Reproductive biology of *Phymaturus cf. palluma*: a vulnerable lizard from the highlands of the Andes, San Juan, Argentina. Herpetol Conserv Biol 5:430–440

Cabezas-Cartes F, Kubisch EL, Ibargüengoytía NR (2014) Consequences of volcanic ash deposition on the locomotor performance of the *Phymaturus spectabilis* lizard from Patagonia, Argentina. J Exp Zool A Ecol Genet Physiol 321(3):164–172

Cabezas-Cartes F, Boretto JM, Ibargüengoytía NR (2015) Age, growth and life-history parameters of an endemic vulnerable lizard from Patagonia, Argentina. Herpetol J 25(4):215–224

Cabezas-Cartes F, Boretto JM, Ibargüengoytía NR (2018) Effects of climate and latitude on age at maturity and longevity of lizards studied by skeletochronology. Integr Comp Biol 58(6):1086–1097

Cabezas-Cartes F, Fernández JB, Duran F, Kubisch EL (2019) Potential benefits from global warming to the thermal biology and locomotor performance of an endangered Patagonian lizard. Peer J 7:e7437. https://doi.org/10.7717/peerj.7437

Cabezas-Cartes F, Kubisch EL, Duran F, Boretto JM (2023) Comparative thermal sensitivity of locomotor performance and vulnerability to global warming of two sympatric *Phymaturus* lizards from cold environments of Patagonia (Argentina). Biol J Linn Soc 140(2):261–276

Callard IP, Fileti L, Pérez L, Sorbera L, Giannoukus G, Klosterman L, Tsang P, Cracken JMC (1992) Role of the corpus luteum and progesterone in the evolution of vertebrate viviparity. Am Zool 32:264–275

Castro SA, Boretto JM, Blanco GM, Acosta JC (2018) Adjustment of the reproductive activity of vulnerable lizard *Phymaturus williamsi* at high altitudes. Herpetol Conserv Biol 13(1):283–293

Cei J (1986) Reptiles del Centro-oeste y Sur de la Argentina. Herpetofauna de las Zonas Aridas y Semiáridas. In: Monografía IV Museo Regionale di Scienze Naturali, 1st edn, Torino

Charland MB, Gregory PT (1990) The influence of female reproductive status on thermoregulation in a viviparous, *Crotalus viridis*. Copeia 1990:1089–1098. https://doi.org/10.2307/1446493

Crocco M, Ibargüengoytía NR, Cussac V (2008) Contributions to the study of oviparity–viviparity transition: Placentary structures of *Liolaemus elongatus* (Squamata: Liolaemidae). J Morphol 269(7):865–874

Custodia-Lora N, Callard IP (2002) Progesterone and progesterone receptors in reptiles. Gen Comp Endocrinol 127(1):1–7

Dubois W, Pudney J, Callard IP (1988) The annual testicular cycle in the turtle, *Chrysemys picta*: a histochemical and electron microscopic study. Gen Comp Endocrinol 71:191–204

Duvall D, Guillette LJ Jr, Jones RE (1982) Environmental control of reptilian reproductive cycles. In: Gans C, Harvey Pough F (eds) Biology of the Reptilia. Physiology D. Academic, London, pp 201–231

Fernández JB, Smith JJR, Scolaro JA, Ibargüengoytía NR (2011) Performance and thermal sensitivity of the southernmost lizards in the world, *Liolaemus sarmientoi* and *Liolaemus magellanicus*. J Therm Biol 36:15–22

Fernández JB (2016) Ventajas de la viviparidad: Adaptaciones fisiológicas y sociales de los lagartos más australes del mundo *Liolaemus sarmientoi* y *Liolaemus magellanicus* al clima frío. PhD Dissertation. Centro Regional Universitario Bariloche, Universidad Nacional del Comahue. Directora: Nora R. Ibargüengoytía, Co-director: Barry R. Sinervo. from Patagonia, Argentina (Reptilia: Iguania: Liolaemidae). Zootaxa 1939:38–50

Fernández JB, Medina SM, Kubisch EL, Manero AA, Scolaro JA, Ibargüengoytía NR (2015) Female reproductive biology of the lizards *Liolaemus sarmientoi* and *L. magellanicus* from the southern end of the world. Herpetol J 25:101–108

Fernández JB, Kubisch EL, Ibargüengoytía NR (2017a) Viviparity advantages in the lizard *Liolaemus sarmientoi* from the end of the world. Evol Biol 44: 325–338. https://doi.org/10.1007/s11692-017-9410-2

Fernández JB, Medina SM, Kubisch EL, Scolaro JA, Ibargüengoytía NR (2017b) Reproductive strategies in males of the world's southernmost lizards. Int Zool 12:132–147

Fernández JB, Bastiaans E, Medina M, Méndez De la Cruz FR, Sinervo BR, Ibargüengoytía NR (2018) Behavioral and physiological polymorphism in males of the austral lizard *Liolaemus sarmientoi*. J Comp Physiol A 204:219–230

Gutiérrez JA, Piantoni C, Ibargüengoytía NR (2013) Altitudinal effects on life history parameters in populations of *Liolaemus pictus argentinus* (Sauria: Liolaemidae). Acta Herpetologica 8(1):1–9

Habit EM, Ortiz JC (1996) Ciclo reproductivo de *Phymaturus flagellifer* (Reptilia, Tropiduridae). Bol Soc Biol Concepción (Chile) 67:7–14

Healey M, Uller T, Olsson M (2007) Seeing red: morph-specific contest success and survival rates in a colour-polymorphic agamid lizard. Anim Behav 74(2): 337–341

Ibargüengoytía NR (2004) Prolonged cycles as a common reproductive pattern in viviparous lizards from Patagonia, Argentina: reproductive cycle of *Phymaturus patagonicus*. J Herpetol 38(1):73–79

Ibargüengoytía NR, Casalins LM (2007) Reproductive biology of the southernmost gecko *Homonota darwini*: convergent patterns among southern hemisphere reptiles living in harsh environments. J Herpetol 41: 72–80

Ibargüengoytía NR, Cussac VE (1996) Reproductive biology of the viviparous lizard, *Liolaemus pictus* (Tropiduridae): biennial female reproductive cycle? Herpetol J 6:137–143

Ibargüengoytía NR, Cussac VE (1998) Reproduction of the viviparous lizards *Liolaemus elongatus* in the highlands of Patagonia: plastic cycles in *Liolaemus* as a response to climate? Herpetol J 8:99–105

Ibargüengoytía NR, Cussac VE (1999) Male response to low frequency of female reproduction in viviparous *Liolaemus* (Tropiduridae). Herpetol J 9:111–117

Ibargüengoytía NR, Cussac VE (2002) Body temperatures of two viviparous *Liolaemus* lizard species, in Patagonian rain forest. Herpetol J 12:131–134

Ibargüengoytia NR, Pastor LM, Pallares J (1999) A light microscopy and ultrastructural study of the testes of tortoise *Testudo graeca* (Testudinae). J Submiscrosc Cytol Pathol 31:221–230

Ibargüengoytía NR, Renner M, Boretto J, Piantoni C, Cussac V (2007) Thermal effects on locomotion in the nocturnal gecko *Homonota darwini* (Gekkonidae). Amphibia-Reptilia 28:235–246

Ibargüengoytía NR, Cabezas-Cartes F, Boretto JM, Piantoni C, Kubisch EL, Fernández MS, Sinervo B (2016) Volcanic ash from Puyehue-Cordón Caulle eruptions affects running performance and body condition of *Phymaturus* lizards in Patagonia, Argentina. Biol J Linn Soc 118(4):842–851

Ibargüengoytía NR, Medina M, Laspiur A, Yan-Fu Q, Ramirez Peralta CA, Sinervo B, Miles DB (2021) Looking at the past to infer into the future: thermal traits track environmental change in Liolaemidae. Evolution 75:2348–2370. https://doi.org/10.1111/evo.14333

IPCC (2021) Intergovernmental panel on climate change. Climate change 2021: the physical science basis. In: Contribution of working group I to the fifth assessment of the intergovernmental panel on climate change. Cambridge University Press, Cambridge, U.K

Kubisch EL, Fernández JB, Ibargüengoytía NR (2011) Is locomotor performance optimised at preferred body temperature? A study of *Liolaemus pictus argentinus* from northern Patagonia. Argentina. J Therm Biol 36: 328–333

Kubisch E, Piantoni C, Williams J, Scolaro A, Navas CA, Ibargüengoytía NR (2012) Do higher temperatures increase growth in the nocturnal gecko *Homonota darwini* (Gekkota: Phyllodactylidae)? A skeletochronological assessment analyzed at temporal and geographic scales. J Herpetol 46(4):587–595

Kubisch EL, Fernández JB, Ibargüengoytía NR (2016) Vulnerability to climate warming of *Liolaemus pictus* (Squamata, Liolaemidae), a lizard from the cold temperate climate in Patagonia, Argentina. J Comp Physiol B 186:243–253

Kubisch EL, Fernández JB, Ibargüengoytía NR (2023) Thermophysiological plasticity could buffer the effects of global warming on a Patagonian lizard. J Exp Zool A 339(6):590–601

Kun ME, Piantoni C, Krenz JD, Ibargüengoytía NR (2010) Dietary analysis of *Homonota darwini* (Squamata: Gekkonidae) in northern Patagonia. Curr Zool 56(4): 406–410

Laspiur A, Santos JC, Medina SM, Pizarro JE, Sanabria EA, Sinervo B, Ibargüengoytía NR (2021) Vulnerability to climate change of a microendemic lizard species from the Central Andes. Sci Rep 11(1):11653

Lobo F, Barrasso DA (2021) Diversidad insospechada y explosión de estudios en las lagartijas saxícolas del género *Phymaturus*. In: Abdala CS, Laspiur A, Scrocchi G, Semhan R, Lobo F, Valladares P (eds) Las lagartijas de la Familia Liolaemidae: Sistemática, distribución e historia natural de una de las familias de vertebrados más diversas del cono sur de Sudamérica. Universidad de Tarapacá, Arica

Lobo F, Laspiur A, Acosta JC (2013) Description of new andean species of the genus *Phymaturus* (Iguania: Liolaemidae) from northwestern Argentina. Zootaxa 3683:117–132

Lobo F, Barrasso DA, Valdecantos S, Giraudo AR, Di Pietro DO, Basso NG (2022) Exploring the morphological diversity of Patagonian clades of *Phymaturus* (Iguania: Liolaemidae). Integrative study and the description of two new species. Cuad Herp 36:197–231

Mahmoud IY, Licht P (1997) Seasonal changes in gonadal activity and the effects of stress on reproductive hormones in the common snapping turtle, *Chelydra serpentina*. Gen Comp Endocrinol 107(3):359–372

Mahmoud IY, Cyrus RV, Bennett TM, Woller MJ, Montag DM (1985) Ultrastructural changes in testes of the snapping turtle, *Chelydra serpentina* in relation to plasma testosterone, Δ5-3β-hydroxysteroid dehydrogenase, and cholesterol. Gen Comp Endocrinol 57(3):454–464

Medina M (2010) Adaptaciones morfológicas y fisiológicas ligadas a la transición oviparidad-viviparidad en lagartos de climas fríos: reproducción y termofisiología. PhD Dissertation. Centro Regional Universitario Bariloche, Universidad Nacional del Comahue, San Carlos de Bariloche, Rio Negro

Medina M, Ibargüengoytía N (2010) How do viviparous and oviparous lizards reproduce in Patagonia? A comparative study of three species of *Liolaemus*. J Arid Environ 74(9):1024–1032. https://doi.org/10.1016/j.jaridenv.2010.02.004

Medina M, Gutierrez J, Scolaro A, Ibargüengoytía NR (2009) Thermal responses to environmental constraints in two populations of the oviparous lizard *Liolaemus bibronii* in Patagonia, Argentina. J Therm Biol 34:32–40

Medina M, Scolaro A, Mendez-De La Cruz F, Sinervo B, Ibargüengoytía NR (2011) Thermal relationships between body temperature and environment conditions set upper distributional limits on oviparous species. J Therm Biol 36:527–534

Medina M, Scolaro A, Méndez-De La Cruz F, Sinervo B, Miles DB, Ibargüengoytía NR (2012) Thermal biology of genus *Liolaemus*: a phylogenetic approach reveals advantages of the genus to survive climate change. J Therm Biol 37:579–586

Mell R (1929) Preliminary contributions to an ecology of east Asiatic reptiles, especially snakes. Lingnan Sci J 8: 187–197

Miles DB, Sinervo B, Hazard LC, Svensson EI, Costa D (2007) Relating endocrinology, physiology and behaviour using species with alternative mating strategies. Funct Ecol 21(4):653–665

Obregon RL, Scolaro JA, Ibargüengoytía NR, Medina M (2021) Thermal biology and locomotor performance in *Phymaturus calcogaster*: are Patagonian lizards vulnerable to climate change? Integr Zool 16:53–66

Pearson OP (1954) Habits of the lizard *Liolaemus multiformis multiformis* at high altitudes in southern Peru. Copeia 2:111–116

Piantoni C, Ibargüengoytía NR, Cussac VE (2006a) Age and growth of the Patagonian lizard *Phymaturus patagonicus*. Amphibia-Reptilia 27:385–392

Piantoni C, Ibargüengoytía NR, Cussac VE (2006b) Growth and age of the southernmost nocturnal gecko *Homonota darwini*: a skeletochronological assessment. Amphibia-Reptilia 27:393–400

Pizarro JE, Laspiur A, Acosta JC, Blanco GM, Boretto JM (2022) High reproductive effort in a vulnerable lizard from high altitudes in Argentina: reproductive biology and sexual dimorphism in *Phymaturus extrilidus*. An Acad Bras Cienc 94(4):e20210179

Renner ML (1998) Ecofisiología del lagarto nocturno Homonota darwini (Gekkonidae), actividad muscular en relación con la temperatura. Centro Regional Universitario Bariloche, Universidad Nacional del Comahue, Licenciatura en Biología

Saint Girons H (1985) Comparative data on lepidosaurian reproduction and some time tables. In: Gans C (ed) Biology of the Reptilia 15. Wiley, New York, pp 35–58

Scolaro A (2005) Reptiles Patagónicos: Sur. Guía de Campo. Universidad Nacional de la Patagonia, Trelew

Scolaro JA (2006) Reptiles patagónicos norte: Guía de Campo, 1st edn. Universidad Nacional de la Patagonia, Comodoro Rivadavia

Scolaro JA, Ibargüengoytía NR (2007) A new species of *Phymaturus* from rocky outcrops in the central steppe of Rio Negro province, Patagonia, Argentina (Reptilia: Iguania: Liolaemidae). Zootaxa 1524:47–55

Scolaro JA, Ibargüengoytía NR (2008) A new fragment for the understanding of the puzzling evolutive process of the *Phymaturus* genus: a new species of the patagonicus group. Zootaxa 1939(1939):38–50

Scolaro JA, Méndez de la Cruz FR, Ibargüengoytía NR (2012) A new species of *Phymaturus* of the *patagonicus* clade (Squamata, Liolaemidae) from an isolated plateau of southwestern Rio Negro Province, Argentina. Zootaxa 3451:17–30

Shine R (1983) Reptilian viviparity in cold climates: testing the assumptions of an evolutionary hypothesis. Oecologia 57:397–405. https://doi.org/10.1007/BF00377186

Shine R (1985) The evolution of viviparity in reptiles: an ecological analysis. In: Gans C, Billet F (eds) Biology of Reptilia, vol 15. Wiley, New York, pp 605–695

Shine R (1995) A new hypothesis for the evolution of viviparity in reptiles. Am Nat 145:809–823. https://doi.org/10.1086/285769

Shine R (2002) An empirical test of the 'predictability' hypothesis for the evolution of viviparity in reptiles. J Evol Biol 15:553–560. https://doi.org/10.1046/j.1420-9101.2002.00420.x

Shine R (2004) Does viviparity evolve in cold climate reptiles because pregnant females maintain stable (not high) body temperatures? Evolution 58:1809–1818

Sinervo B, Lively CM (1996) The rock-paper-scissors game and the evolution of alternative male reproductive strategies. Nature 380:240–243

Sinervo B, Svensson E, Comendant T (2000) Density cycles and an offspring quantity and quality game driven by natural selection. Nature 406:985–988

Sinervo B, Méndez De La Cruz FR, Miles DB, Heulin B, Bastiaans E, Villagran-Santa Cruz M, Lara-Resendiz R, Martínez-Méndez M, Calderon-Espinosa M, Mesa-Lázaro R et al (2010) Erosion of lizard diversity by climate change and altered thermal niches. Science 328:894–899

Vicenzi NP, Corbalán V, Miles DB, Sinervo B, Ibargüengoytía NR (2017) Range increment or range detriment? Predicting potential changes in distribution caused by climate change for the endemic high-Andean lizard *Phymaturus palluma*. Biol Conserv 206:151–160

Vicenzi N, Ibargüengoytía NR, Corbalán V (2019a) Activity patterns and thermoregulatory behavior of the lizard *Phymaturus palluma* in Aconcagua Provincial Park, Argentine. Andes Herp Con Bio 14:337–348

Vicenzi N, Kubisch E, Ibargüengoytía NR, Corbalán V (2019b) Thermal sensitivity of performance of *Phymaturus palluma* (Liolaemidae) in the highlands of Aconcagua: vulnerability to global warming in the Andes. Amphibia-Reptilia 40:207–218

Vicenzi N, Laspiur A, Sassi PL, Massarelli R, Krenz J, Ibargüengoytía NR (2020) Impact of temperature on bite force and bite endurance in the leopard iguana (*Diplolaemus leopardinus*) in the Andes Mountains. J Exp Biol 223(12):jeb221382

Vicenzi N, Bacigalupe LD, Laspiur A, Ibargüengoytía NR, Sassi PL (2021) Could plasticity mediate highlands lizards' resilience to climate change? A case study of the leopard iguana (*Diplolaemus leopardinus*) in Central Andes of Argentina. J Exp Biol 224(14):jeb242647

Drivers of Physiological Responses and Range Distribution in the Context of Rapaport's Rule: The Case of East Andean Liolaemidae Lizard Species

Félix B. Cruz, Marcelo F. Bonino, Débora L. Moreno Azocar, Oscar A. Stellatelli, Facundo Luna, and Daniel Antenucci

1 Introduction

Variation in metabolic rate across species is an important trait in the context of animal behavior, physiology, ecology, and evolution (Kooijman 2000; Speakman et al. 2004; Brown et al. 2004; Angilletta 2009; Careau et al. 2010; White and Kearney 2013; Muñoz et al. 2014; Pettersen et al. 2018). As a physiological response, metabolic rate—a standard measurement of minimal energy expenditure at a specific ambient temperature (T_a)—may be linked to large-scale spatiotemporal variation in climate, latitude, elevation (Angilletta 2001; Cruz et al. 2011), and seasonality (Tsuji 1988), all considered extrinsic variables. Metabolic rate is also influenced by intrinsic factors of the individuals, such as body temperature (T_b), body size (Bennett 1988; Gillooly et al. 2001), internal organ relative size (Wiersma et al. 2012; Kozłowski et al. 2020), trophic level, and phylogenetic affinities (Bennet 1988). Moreover, physiological responses in organisms, particularly within ectotherms, may result from a combination of extrinsic and intrinsic variables (McNab 2002; Spicer and Gaston 1999). Regarding extrinsic variables, several climatic factors have been identified as affecting the metabolism in both snakes (Dupoué et al. 2017) and lizards (Plasman et al. 2020). Specifically, aridity has been shown to reduce metabolic rates, particularly within ectothermic species (Auer et al. 2020). Among intrinsic aspects, nutritional state and quality (e.g., Zanotto et al. 1997; Simandle et al. 2001), as well as reproductive state (e.g., Angilletta and Sears 2000; Finkler and Cullum 2002; Finkler 2006), can influence metabolic rates (Naya et al. 2008). The quality and type of food also impact metabolic rates, for example, carnivores are expected to have higher metabolic rates than herbivores, according to the food habits hypothesis (McNab 1986). Additionally, the digestive tract links energy availability and intake to other vital functions (Karasov 1990; Bozinovic 1993; Secor 2001). However, the digestive tract is energetically costly to maintain (Cant et al. 1996) and represents an important energy-saving mechanism (Dietz and Piersma 2007; Naya et al. 2008). Consequently, larger energetically expensive organs, such as the digestive tract, are expected to correlate with higher physiological

Supplementary Information: The online version contains supplementary material available at https://doi.org/10.1007/978-3-032-00074-3_10.

F. B. Cruz (✉) · M. F. Bonino · D. L. Moreno Azocar
Instituto de Investigaciones en Biodiversidad y Medioambiente (INIBIOMA), Universidad Nacional del Comahue - Consejo Nacional de Investigaciones Científicas y Técnicas, San Carlos de Bariloche, Río Negro, Argentina
e-mail: cruzf@comahue-conicet.gob.ar

O. A. Stellatelli · F. Luna · D. Antenucci
Instituto de Investigaciones Marinas y Costeras (IIyMC), Universidad Nacional de Mar del Plata - Consejo Nacional de Investigaciones Científicas y Técnicas, Mar del Plata, Buenos Aires, Argentina

J. N. Urbina-Cardona et al. (eds.), *Andean Herpetofauna*, Sustainable Development Goals Series,
https://doi.org/10.1007/978-3-032-00074-3_10

demands and capacities (Diamond 1998; Naya et al. 2008). Previous studies have shown a link between intestinal size and metabolic rate across species, revealing that herbivorous species with larger guts are more prevalent in cold, high-latitude environments, compared to insectivores with shorter digestive tracts (Luna et al. 2017).

As mentioned before, energetic metabolism responses may be explained by several hypotheses, such as the metabolic cold adaptation hypothesis and the latitudinal compensation hypothesis; these hypotheses link variations in physiological functions to climate, suggesting a potential influence of environmental temperature on metabolic rates (Scholander et al. 1953; Levinton and Monahan 1983). In general, it is expected that cold-climate species will show higher metabolic rates than related species from warmer climates as compensation for the lower availability of favorable environmental conditions (Scholander et al. 1953; Clarke 1993; Addo-Bediako et al. 2002). This scenario also applies to studying differences in metabolic rates associated with altitude, latitude, and harsh habitat conditions (Gaston 2003; Gaston et al. 2009; Bozinovic et al. 2011a). The climate variability hypothesis suggests that climatic fluctuations impact animals differently depending on their latitude. According to this hypothesis, individuals at higher latitudes, where climate is more variable, are expected to exhibit greater levels of metabolic flexibility. This increased flexibility allows them to thrive across a broader range of climatic conditions and geographic areas (Carter et al. 2023). These relationships may drive the macroecological pattern known as Rapoport's rule, which posits that species at higher latitudes, where conditions are harsher, tend to have broader geographic distribution compared to species at lower, less harsh latitudes. For example, temperate lizard species exhibit a more pronounced acclimation response at the whole-organism level compared to tropical species, as indicated by respiratory gas exchange rates (Sun et al. 2022). However, latitudinal patterns have not been universally consistent across studies, with numerous exceptions reported in the literature (e.g., Šizling et al. 2009; Tomašových et al. 2015). Climatic variables such as ambient temperature and precipitation may not follow a strict north/south pattern due to the complexity of elevation and orography. It is more likely that associations between climatic factors and ecological patterns will be supported when using direct climatic measurements (e.g., air temperature or its variability), even if these associations do not adhere to a strict geographic pattern (Blackburn and Ruggiero 2001; Beck et al. 2016).

An organism's ability to cope with new environmental conditions is constrained by its physiology (Wiens et al. 2010). However, the driving force of evolution, through the natural selection process on key traits, may enable the exploitation of new resources in this novel environment (Rabosky 2014; Stroud and Losos 2016). To understand the mechanisms allowing organisms to adapt to new environments or expand their ranges, it is essential to identify the factors that facilitate or constrain niche expansion. For example, climate may either limit or facilitate the expansion of a species distribution, particularly in ectotherms (Dillon et al. 2010; Araújo et al. 2013; Pie et al. 2021a). Therefore, the physiological responses of an organism may be associated with its geographic range as well as its intrinsic characteristics (Agosta et al. 2013; Pie et al. 2021b). Species with plastic traits, such as a broad thermal tolerance range (TTR)—the range of temperatures within which the species can maintain metabolic activity—may be better to adjust to changes and expand their geographic ranges or tolerate variations in their current habitats (Bonino et al. 2015). The pattern of geographic variation in TTR reflects differences in metabolic rates among populations; lizards from colder, more thermally variable environments typically have lower standard oxygen consumption rates and greater cold tolerances (Campbell-Staton et al. 2018). Another aspect to consider is body size; larger species are expected to show a broader geographic range according to Bergmann's rule (Bergmann 1847; Addo-Bediako et al. 2000; Ashton 2002), which could be explained by their thermal inertia or resistance to starvation (Arnett and Gotelli 1999). Another aspect to consider is the relationship between diet and activity, known as the activity hypothesis (Garland 1983; Muñoz-García and Williams 2005). This hypothesis posits that carnivore species, due to their higher trophic level, require larger home ranges and, consequently, larger geographic ranges compared to herbivore species. Furthermore, plant communities can influence the distribution of animal species by shaping the physical structure of the environment (physiognomy). This, in turn, affects the availability of critical resources like suitable thermal patches, food, and refuges (Lawton 1983; McCoy and Bell 1991; Meik et al. 2002; Stellatelli et al. 2013). Ultimately, climatic and intrinsic aspects can determine the geographic range of a species (Kearney 2012). Geographic range limits and physiological demands interact on a larger scale. Organisms with high metabolic rates generally exhibit greater thermal tolerance, aligning with the combined hypotheses of thermal plasticity and energy constraints. High sustained energy demands require individuals to forage over larger areas and disperse more widely, which results in lower population densities, larger home ranges, and, consequently, larger geographic range sizes (Pie et al. 2021b). Evaluating how physiological traits limit and interact with geographical ranges can enhance our understanding of the evolution of physiological diversity (Bozinovic and Rosenmann 1988; Spicer and Gaston 1999; McNab 2002; Chown et al. 2004), a complex puzzle (Bozinovic et al. 2011b) named macrophysiology (Gaston et al. 2009).

Reptiles are ectothermic species, making them an ideal model for studying thermal responses at both environmental and physiological levels. Moreover, the relationship between metabolic rate and the environment, particularly with geographic range, has not been extensively studied (Giacometti et al. 2022). Among lizards, the family Liolaemidae represents one of the most ecologically diverse terrestrial vertebrate radiations, encompassing a remarkable breadth of habitats, ranging from hot deserts to cold forests and spanning elevations from sea level to over 5000 m in the Andes (Cerdeña et al. 2021). Their diet varies from being entirely insectivorous (e.g., *Liolaemus pseudoanomalus*; Kozykariski et al. 2011) to mostly herbivorous (*L. poecilochromus*, Valdecantos et al. 2012; *Phymaturus punae*, Córdoba et al. 2015) with some species exhibiting varying degrees of omnivory (Espinoza et al. 2004; Vanhooydonck et al. 2010). Body size in *Liolaemus* species ranges from 45- to 110-mm snout-vent length (SVL; Abdala et al. 2021). Moreover, it was observed that metabolic rates in Liolaemid species may also vary with environmental temperature (Naya et al. 2008). The ecological, physiological, and morphological variability, along with the differences in geographic distribution, make Liolaemid lizards an ideal group for studying geographic patterns in animals. Although they represent one of the most ecologically diverse terrestrial vertebrate radiations, evidence regarding adaptive diversification in phenotypes of *Liolaemus* lizards (the most specious genus among Liolaemids) remains limited or controversial (Edwards et al. 2022). Nevertheless, several studies are focused on uncovering evolutionary relationships between traits. For example, O'Grady et al. (2005) and Pincheira Donoso (2021) found a positive relationship between herbivore proportion diet and gut length, Tulli et al. (2011) observed that claws were positively related to clinging performance, and Cruz et al. (2005) and Moreno Azocar et al. (2013) demonstrated variation in Liolaemid TTR across environmental gradients, with wider ranges and lower critical thermal minimum (CTmin) at cooler temperatures sites, although phylogeny still plays a significant role. Furthermore, Liolaemid species with greater thermal plasticity may achieve broader distribution ranges compared to less plastic species (Bonino et al. 2015). However, several studies have also identified conservatism in many traits. For example, Cruz et al. (2011) found no relationship between standard metabolic rates (SMRs) and other traits (e.g., diet) other than SVL. Liolaemid lizards exhibit different strategies, depending on the phenotypic trait. For example, thermal performance breadth for maximal running speed has been shown to be plastic (Bonino et al. 2015; Moreno Azocar et al. 2020; Stellatelli et al. 2022), whereas morphology appears to be highly conservative (Tulli et al. 2012). Additionally, locomotor performance does not show specialization to the types of substrate where these species occur, aligning with the "jack of all trades and master of none" principle (Tulli et al. 2012). Previous studies on this clade, using a meta-tree and less sophisticated indicators of distribution (e.g., latitudinal range instead of polygons), failed to show relationships between geographic range, thermal biology, and metabolic rates (see Cruz et al. 2005, 2011). The application of new techniques, proper statistical analysis, and the inclusion of more taxa in future studies may help clarify the evolutionary relationships between different traits (Edwards et al. 2022).

Here, we aim to identify the variables that affect metabolic rate in Liolaemid species distributed east of the Andes, and how physiological adaptations and environmental factors interact to shape the distribution patterns within the lizard family Liolaemidae, focusing on the genera *Liolaemus* and *Phymaturus*. In particular, we aimed to (1) investigate the association between standard metabolic rate (SMR) and extrinsic (climate, elevation, vegetation cover) as well as intrinsic (gut length, diet, body size) traits using an ultrametric tree; (2) assess whether metabolic rate, diet (proportion of plant material), thermal plasticity (thermal tolerance range), or climate are associated with the species' distribution ranges (polygons); (3) identify which biological traits have undergone the most rapid evolutionary change and explore whether the potential for these changes can predict current distribution patterns, with consideration of ecological implications. We aim to address several hypotheses: (a) as predicted by the food habits hypothesis (McNab 1986), the SMR of herbivore Liolaemid lizards will be lower than that of omnivores and insectivores; (b) according the metabolic cold adaptation hypothesis (Scholander et al. 1953; Levinton and Monahan 1983), species from colder environments will have the higher metabolic rates, as observed in Tropiduridae species (Giacometti et al. 2022); (c) according to the climatic variability hypothesis (Gaston and Blackburn 2000), species with broader thermal tolerance will be able to achieve greater geographic ranges; and (d) according to the activity hypothesis (Garland 1983), as observed in mammal carnivores (Muñoz-García and Williams 2005), the geographic range of herbivore species will be smaller than that of insectivores Liolaemid lizards.

2 Materials and Methods

2.1 Specimens Collection

We collected 410 specimens from 22 species of *Liolaemus* and 2 of *Phymaturus* in a ~3100 km North–South transect, from Salta to Santa Cruz provinces in Argentina (from 23 °S to 52 °S) and from elevations of 4500 m above sea level to the nearest sea shore (Table 1). We captured the animals by hand or lassoing. During the transportation to the laboratory, lizards were kept in cloth bags. The data we collected correspond to species of four major *Liolaemus* subclades (*chiliensis, boulengeri, lineomaculatus,* and *montanus* clades; Abdala and Quinteros 2014). Although we only sampled 24 species out of the 342 recognized species (Abdala et al. 2021), we covered the four major radiations and nearly 40% of the area of the distribution of the family Liolaemidae (Fig. 1). For the intrinsic data, we collected specimens during different field trips, over several years corresponding to this and previous studies (Cruz et al. 2005, 2009, 2011, 2022; Bonino et al. 2011, 2015; Moreno Azocar et al. 2013, 2020; Stellatelli et al. 2013).

2.2 Laboratory Husbandry

Once specimens were in the lab, and between experimental trials, we released lizards into terraria every three days for four hours. The terraria (1.2 m × 0.6 m × 0.4 m, L × W × H) were divided into five lanes, each 0.12 mm wide. A 150-W infrared bulb provided heat and a parabolic aluminized reflector (PAR) provided illumination. Lizards were fed crickets (*Achaeta domestica*) and mealworms (*Tenebrio molitor*) and given water ad libitum. For omnivorous and

Table 1 List of species studied and number of individuals per species (*n*) per type of measured trait

Species	POL	BM	*n*	SVL	*n*	SMR	*n*	Ctmin	*n*	Ctmax	*n*	TTR	PLP	*n*	RGS	*n*
L. abaucan	742	5.51	(9)	53.2	(17)	0.11	(9)	15.41	(7)	47.3	(7)	31.89	0.48	(4)	1.8	(4)
L. albiceps	451	13.75	(8)	80.74	(24)	0.08	(8)	7.05	(12)	44.33	(12)	37.28	0.64	(12)	4.79	(4)
L. baguali	4515	16.61	(9)	81.69	(14)	0.06	(9)	5.24	(13)	43.52	(13)	38.28	0.48	(10)	2.14	(4)
L. chacoensis	291745	1.30	(8)	45.66	(37)	0.15	(8)	14.93	(12)	44.05	(12)	29.12	0	(3)	2.59	(3)
L. coeruleus	3448	6.75	(4)	63.01	(8)	0.06	(4)	9.22	(20)	43.36	(20)	34.14	0.51	(7)	2.43	(7)
L. cuyanus	277347	14.83	(10)	73.58	(36)	0.12	(10)	11.32	(12)	45.59	(11)	34.27	0.35	(12)	6.32	(6)
L. darwinii	462937	5.673	(10)	54.06	(46)	0.15	(10)	9.38	(40)	44.69	(40)	35.31	0.04	(7)	2.3	(10)
L. elongatus	101803	8.75	(10)	73.87	(13)	0.06	(10)	7.68	(26)	45.1	(20)	37.43	0.2	(10)	2.57	(7)
L. fitzingerii	190059	20.55	(9)	73.36	(14)	0.07	(9)	8.24	(21)	44.45	(18)	36.21	0.82	(6)	4.52	(10)
L. hatcheri	15655	5.14	(6)	57.91	(11)	0.05	(6)	7.02	(7)	43.13	(7)	36.12	0.93	(12)	2.42	(6)
L. irregularis	4495	19.37	(5)	86.67	(24)	0.1	(5)	6.96	(7)	44.98	(7)	38.02	0.64	(9)	5.82	(4)
L. kingii	148431	10.51	(9)	71.71	(12)	0.05	(9)	7.37	(15)	44.22	(15)	36.86	0.91	(11)	10.27	(4)
L. kolengh	5	4.81	(7)	56.05	(16)	0.07	(7)	5.47	(7)	44.05	(7)	38.58	0.84	(28)	2.17	(6)
L. koslowskyi	31026	6.68	(11)	57.37	(36)	0.15	(11)	12.44	(25)	46.28	(25)	33.27	0	(6)	2.56	(9)
L. laurenti	19052	3.7	(7)	55.27	(23)	0.14	(7)	11.5	(16)	47.28	(16)	35.78	0	(4)	2.41	(4)
L. lineomaculatus	348301	3.72	(4)	54.13	(6)	0.06	(4)	7.5	(4)	45.6	(4)	38.1	0.27	(4)	4.21	(3)
L. magellanicus	44102	3.76	(4)	53.33	(8)	0.07	(4)	8.3	(5)	43.59	(5)	35.29	0.8	(4)	0.99	(2)
L. olongasta	37500	4.51	(10)	52.88	(32)	0.18	(10)	12.32	(10)	44.65	(10)	32.33	0.48	(12)	2.46	(3)
L. quilmes	16162	5.12	(10)	58.42	(33)	0.13	(10)	8.26	(10)	44.74	(10)	36.48	0	(8)	2.46	(3)
L. shitan	115	21.16	(12)	89.14	(12)	0.17	(12)	7.48	(10)	45.33	(10)	37.85	0.16	(5)	2.49	(6)
L. xanthoviridis	12911	15.07	(8)	68.8	(14)	0.09	(8)	7.86	(16)	44.48	(16)	36.61	0.85	(7)	3.84	(7)
L. zullyae	672	8.25	(10)	62.65	(14)	0.09	(10)	8.38	(8)	43.95	(8)	35.57	0.71	(6)	1.89	(3)
P. spurcus	1940	19.52	(8)	89.5	(16)	0.07	(8)	6.8	(12)	44.87	(12)	38.07	1	(4)	7.47	(6)
P. tenebrosus	583	23.45	(5)	89.49	(9)	0.1	(5)	9.8	(6)	39.9	(6)	30.1	1	(9)	7.47	(4)

POL polygon (km^2), *BM* Body mass (g), *SVL* snout-vent length (mm), *SMR* standard metabolic rate ($[O_2]$/h g), *CTmin* critical thermal minimum temperature (°C), *Ctmax* critical thermal maximum temperature (°C), *TTR* thermal tolerance range (°C), *PLP* proportion of plant material in the diet, *RGS* intestine surface per unit of SVL

Fig. 1 Map showing the area of distribution of Liolaemidae in South America (light grey) and the area covered by this study (dark grey). (**a**) *Liolaemus irregularis*. (Picture by E. Navarro); (**b**) *L. coeruleus*. (Picture by M. Bonino); (**c**) *L. baguali*. (Picture by M. Bonino); and (**d**) *L. kingii*. (Picture by M. Bonino)

herbivorous species, wild daisies were also offered ad libitum. After feeding, the lizards were returned to their cloth bags to minimize exercise and prevent loss of mass; room temperature was maintained at 22–24 °C. Before each experimental trial (to assess thermal tolerance breadth and standard metabolic rate), lizards were fasted for 48 hours.

2.3 Experimental Trials

Thermal tolerance range, the difference between maximum and minimum critical temperatures (CTmin and CTmax, respectively; TTR = CTmax − CTmin), was recorded for all the species studied here. The critical temperatures are considered as the upper and lower body temperatures, at which the righting response is lost after being set in the supine position (Bonino et al. 2011). To determine the CTmin, lizards ($n \sim 298$) were previously cooled at 20 °C, each individual was placed into a small plastic container (1.2 l) with breathable lids and an ultrathin thermocouple attached to the flank with a hypoallergenic tape. The individual lizard was introduced into a −20 °C freezer (Cruz et al. 2005) with a cooling rate of 1° per minute. After body temperature reached 19 °C, we checked the lizard's response every 10 s (Cruz et al. 2005). Once we observed they lost the righting response, an ultrafine k-type thermocouple was introduced into the cloaca to measure CTmin. To determine the maximum critical temperature (CTmax), lizards ($n \sim 288$) with an initial body temperature of 18 °C were placed into a 15 l bucket with a 0.03-m sand layer on the bottom, a 100 W incandescent light bulb (Phillips, 100 W, Buenos Aires, Argentina) was placed 0.4 m above the sand surface. Body temperatures were monitored every 15 s and every 10 s after observing the onset of panting or escaping response (Cruz et al. 2005), and the righting response was evaluated.

The standard metabolic rate as oxygen (O_2) consumed per gram per hour for specimens from the 24 Liolaemid species ($n \sim 282$) was then quantified. The O_2 consumption was measured individually at each lizard's preferred body temperature (T_{pref}, see Cruz et al. 2014) using a closed respirometry system (see Vleck 1987) and a dynamic injection technique for a constant volume (see Lighton 2008). Individuals were placed into a transparent chamber (148 or 463 ml, depending on the lizard's size, see Cruz et al. 2011). Carbon dioxide (CO_2)- and water (H_2O)-free air was pumped through the chambers using an air pump. The initial air samples were flushed through the O_2 analyzer (see description below). Individuals remained in the chambers for 30–45 min inside a modified incubator with lights off, maintained at the T_{pref} estimated for the species (Cruz et al. 2014). After each trial, the chamber's outlet valve was connected to a 60-ml syringe, the valve was opened, and the gas inside was mixed by pumping the syringe three times. Finally, an air sample was collected with the syringe, and the time was recorded. This procedure was also carried out using a control, and no difference with baseline was observed (for details see Cruz et al. 2011). O_2 concentration was measured with an FC-1B Oxygen Analyzer (Sable Systems, Las Vegas). Typically, air samples produce a similar pattern to those described in Lighton (2008) and Simandle et al. (2001). Subtraction of a differential O_2 concentration of 0.2095 was used to obtain air sample O_2 concentration. The rate of O_2 consumption was calculated according to Vleck (1987, Eq. 9) and Lighton (2008, Eq. 4.18). The total volume of the chamber was estimated by subtracting the volume occupied by each individual (assuming 0.98 g = 1 ml, De Vera and Hayes 1995) and H_2O and CO_2 scrubber volumes from the volume of the corresponding respirometric chamber. Standard metabolic rate (SMR) was later standardized to an ambient temperature of 30 °C using the Q_{10} equation (McNab 2002, p. 47). The ambient temperatures range used to estimate Q_{10} varied between 22 °C and 38 °C. When available, species-specific Q_{10} data were used (Cruz et al. 2011); otherwise, a Q_{10} value of 2.33 was assumed as a valid approximation for Liolaemidae, similar to other squamate reptile (2.4, Andrews and Pough 1985).

2.4 Diet Characterization

Fecal samples ($n \sim 896$) of all studied species were collected from the cloth bags used to transport the lizards to the lab ($n \sim 191$ individuals per species/location). To ensure proper identification, upon arrival at the lab, samples were segregated and stored in separate paper envelopes according to species and capture location (as indicated by the individual cloth bags). Data for males and females were pooled. Additional data were collected from the literature. For the identification of food items in feces, samples were rehydrated in water for 30 min and then analyzed under a microscope (Olympus SZ61, Tokyo, Japan) using a 0.01-mm crosshair scale (wide field 20×/10, Olympus, Tokyo, Japan). Flower buds, seeds, fruits, and arthropod remains were separated and categorized into plant and animal matter. The contribution of each prey type to the diet for each species was estimated in terms of volume. Volumes were calculated based on the presumed volume of entire prey before ingestion. Individual food item volumes were estimated as regular ellipsoids using the formula: volume $= 4/3\ \pi \times [(wt/2) \times 2] \times (L/2)$, where wt is the total item width, and L is the item length (Vitt and Zani 1996; Vitt et al. 1997). To calculate the contribution of different prey types to the total prey volume, we first summed the volumes of all the prey items. The proportion of plant material was then estimated as the ratio of its volume to the total dietary volume.

2.5 Relative Gut Surface

Gastrointestinal tracts (mid-esophagus to anterior to the cloacal aperture) from a subsample of euthanized specimens ($n \sim 108$) were extracted and preserved in 70% ethanol (O'Grady et al. 2005). The gastrointestinal tracts were then dissected into sections separating each organ (e.g., small intestine, large intestine) and the mesenteries were removed (O'Grady et al. 2005). Since the absorptive part of the gut corresponds to the hind (small) and the large gut, only these sections were considered (Kohl et al. 2016). After a longitudinal cut, they were laid flat, and their length and width (to calculate the gut surface area; Srichairat et al. 2018) were measured using a digital caliper (Mitutoyo, Japan, 0.01 mm). SVL was measured to the nearest millimeter for each specimen, which allowed the estimate of the relative gut surface area (RGS = gut surface/SVL).

2.6 Environmental Variables and Polygon Construction

Nineteen bioclimatic variables were obtained from the global meteorological database https://www.worldclim.org at 1 km × 1 km resolution and used in this study. Other variables included elevation (ASTER Global Digital Elevation Model (GDEM) Version 3 (ASTGTM), from https://search.earthdata.nasa.gov/), slope (processed with GIS), and the mean Normalized Difference Vegetation Index (NDVI, from https://www.usgs.gov/centers/eros/science/usgs-eros-archive-land-cover-products-global-land-cover-characterization-glcc) that acts as a relative indicator of vegetation coverage; higher values indicate greater greenness and presumed vigor of vegetation in that area. To reduce the file size and enable analysis with lower computational demand, all layers were clipped to the region of the world that has been accessible to the species via dispersal over relevant periods (study area). Collinearity was controlled between the 22 environmental variables using a pairwise Pearson correlation, and those with $r < 0.75$ were retained, reducing to a set of 12 environmental variables for further analyses. These included Bio2, Mean Diurnal Range [Mean of monthly (max temp − min temp)]; Bio3, Isothermality; Bio4, Temperature Seasonality; Bio5, Maximum Temperature of Warmest Month; Bio6, Minimum Temperature of Coldest Month; Bio9, Mean Temperature of Driest Quarter; Bio12, Annual Precipitation; Bio15, Precipitation Seasonality; Bio19, Precipitation of Coldest Quarter; Elevation; Slope and NDVI.

Minimum convex polygons were used to estimate the distribution area in km^2 of each species. For polygons construction, a dataset of 876 unique localities for 22 *Liolaemus* and 2 *Phymaturus* species from georeferenced localities was compiled during the sampling campaigns, literature, and using the open access database of GBIF (https://www.gbif.org/es). Georeferenced localities were carefully corroborated to identify and exclude any suspicious points due to, for example, being a misidentified species, or imprecisely georeferenced. Minimum convex polygons enclosing all known records of each species were obtained with QGIS 3.16.3 connecting marginal records. This approach provides a more conservative estimate of species distribution compared to methods like Environmental Niche Models (ENMs), such as MaxEnt that, in certain circumstances, tend to overestimate the actual area occupied by the species (García-Roselló et al. 2015; Kramer-Schadt et al. 2013; Pearson et al. 2007).

After running the experiments, individuals were euthanized (following the protocol of the Canadian Council on Animal Care in Science), fixed in formalin, and preserved in ethanol for future housing in the collection of Instituto de Herpetología of the Fundación Miguel Lillo in Tucumán, Argentina. Province authorities who allowed us to collect specimens are S. Montanelli—Chubut, L.B. Ortega—Santa Cruz, M. Failla—Río Negro, F. Lonac—Neuquén, P. Barlanga -Mendoza, L. Fra—Catamarca, A. Ortin—Salta, A. Norino—La Rioja, W. López—Buenos Aires, and M. Jordán—San Juan. This study was supported by CONICET PIPs 2486 and 6287 and ANPCyT through grants PICT 2006-1205, 2015-2471, and 2019-4115 to F.B.C and PICT 2013-3201 to DLMA. Some specimens were housed in the Herpetological Collection at Tucuman, at the Instituto de Herpetología, Fundación Miguel Lillo, Tucumán, Argentina. This research was performed under approvals from the Animal Ethics Committee at the INIBIOMA (CONICET-UNComahue), Río Negro, Argentina (CICUAL RD 009/19).

2.7 Phylogenetic Framework

Since species do not represent independent data points (Felsenstein 1985), phylogeny needs to be taken into account. We used the published phylogeny of Esquerré et al. (2019) as our core tree for the evolutionary relationships of Liolaemidae. We pruned the tree to include only the species for which we had data. For this, *drop.tip* and *extract.clade* of the "ape" package (Paradis and Schliep 2019) were used to obtain a calibrated tree (Fig. 2) which included only the species for which we had data. The function *nnls.tree* in "phangorn" 4.0 (Schliep 2011) was used to test if the tree was ultrametric. The resulting phylogenetic tree is shown in Fig. 2, illustrating the distribution of species on the centroid obtained per each polygon.

2.8 Data Analyses

Because the number of climatic variables remained high (12 variables) even after reduction, a phylogenetically informed principal component analysis (pPCA) was used and the first (PC1) and second (PC2) principal components (Zar 1999) were extracted for further analyses. Initially, some data were adjusted (e.g., Bio4 was divided by 100 to even the data; Bio6 was positivized before log10 transformation to obtain values, which were then converted to negative as previously described). Once the matrix was normalized in terms of magnitude, all climatic variables were transformed to $\log_{10}$ to meet the normality assumption (Sokal and Rohlf 1995). The resulting matrix was standardized using the R "standardize" package (Eager 2017). Finally, the standardized variables were analyzed using *PhyloPCA* in the "phytools" package (Revell 2012) in R, which allowed us to simplify and minimize the number of variables for the subsequent analyses.

The phylogenetic signal of biological traits such as standard metabolic rate (SMR), gut surface by snout-vent length (RGS), thermal tolerance range (TTR), snout-vent length (SVL), body mass (BM), and proportion of plant matter in the diet (PLP) was determined. The phylogenetic signal is considered a measurement of resemblance among species due to the shared evolutionary history (Blomberg et al. 2003). The λ (Pagel 1999), a scaling parameter that weights the influence of phylogeny for explaining the observed data, was calculated. This parameter may vary from zero to one, the first corresponds to independence from phylogeny, and the second an indication of full relatedness among species as the result

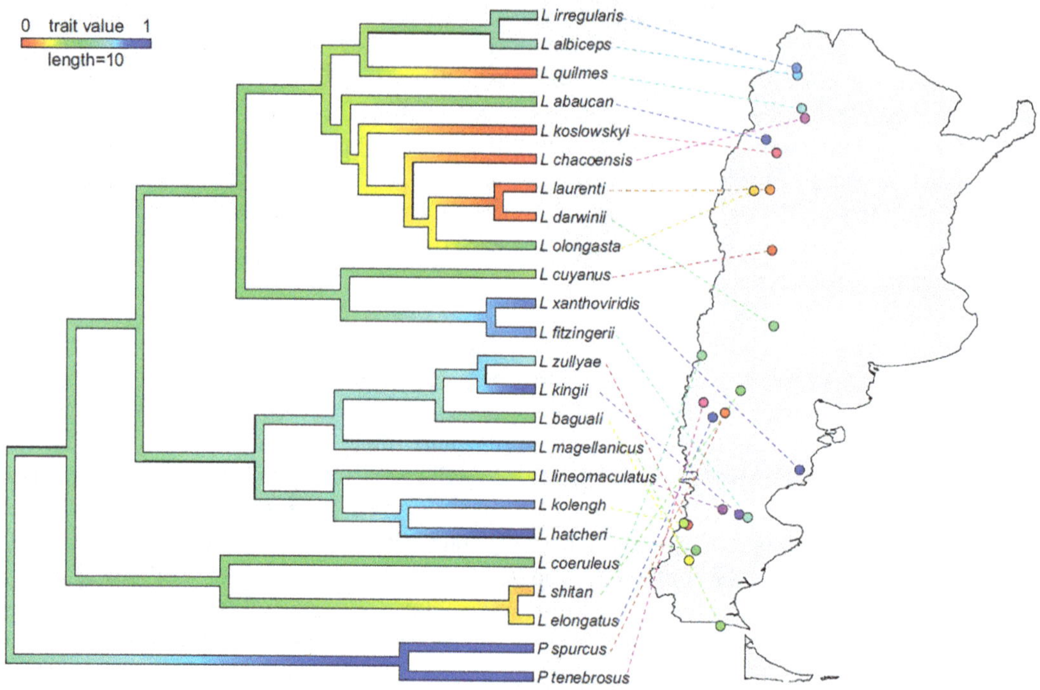

Fig. 2 Traitgram with geographic localization (centroid georeferenced) of the *Liolaemus* species in the study. We used the *contMap* function in the "phytools" package to portray the changes in the proportion of plant material in the diet (PLP) throughout the tree

of phylogenetic structure. A likelihood ratio test was conducted using the *phylosig* function of the R package "phytools" (Revell 2012).

To analyze the predictive variables of the species distribution range, a phylogenetic generalized least squares (PGLS) analysis was performed. Before running the PGLS, the evolutionary model that best fitted the intrinsic data (e.g., PLP, RGS, TTR, SVL) was identified. Three commonly used evolutionary models—Brownian Motion, Ornstein–Uhlenbeck (OU), and Early Burst—were evaluated following the approach of Harmon et al. (2010). The model fitting was performed using the *fitContinuous* function in the "Geiger" package (v2.0.11; Pennell et al. 2014). The Ornstein-Uhlenbeck (OU) model provided the best fit for the three variables (Table 2), and subsequently, all generalized model analyses were performed under the OU model of trait evolution. For the PGLS analysis, the "ape" (Paradis et al. 2004) and "nlme" (Pinheiro et al. 2021) packages were used, employing the *corMartins* and *gls* functions. All analyses were conducted in R v 4.1.3 (R Development Core Team 2024). To examine evolutionary changes in traits across the phylogenetic tree, the phylogenetic ancestral states were reconstructed (Revell 2013). A PLP was plotted on the phylogenetic tree to visualize the evolutionary shifts (Fig. 2) using the *contMap* and *dotTree* functions from the "phytools" package (Revell 2012) which highlighted the most significant changes.

Two models were run, one for SMR as the response variable (complete model: SMR ~ PC1 + PC2 + BM + PLP + GRS), and the other with polygon as the response variable (complete model: Polygon ~ PC1 + PC2 + SMR + SVL + TTR + PLP). PC1 and PC2 (see Table S1) represent principal components 1 and 2, accounting for 74.5% of the total variance. PGLS models were used to simultaneously return the AIC value, likelihood, the phylogenetic signal of model interaction

Table 2 Phylogenetic signal using Pagel's lambda estimate (λ) and P value after permutations

Variable			BrM		OU		EB	
	λ	P	ΔAICc	Wi	ΔAICc	Wi	ΔAICc	Wi
SMR	0.495	0.031	19.74	0.001	0.00	**0.998**	22.37	0.001
SVL	0.917	0.001	0.00	**0.415**	0.49	**0.479**	2.62	0.106
RGS	0.308	0.529	8.32	0.007	0.00	**0.991**	11.05	0.003
PLP	1.026	0.001	0.00	**0.659**	2.27	0.174	2.62	0.167
TTR	0.001	0.999	8.09	0.255	0.00	**0.678**	10.72	0.067
BM	0823	0.025	17.90	0.207	15.35	**0.737**	20.52	0.056

We include Delta Akaike values for small samples for each model (ΔAICc) and the weight of the different models (wAICc) corresponding to the evolutionary models tested (Brownian motion [BrM], Ornstein–Uhlenbeck [OU] and early burst [EB]) of intrinsic biological variables. Boldface indicates the evolutionary model that best fitted the data based on ΔAICc and Wi values. References: *SVL* snout-vent length (mm), *SMR* standard metabolic rate ([O2]/h g), *TTR* thermal tolerance range (°C), *PLP* proportion of plant material in the diet, *RGS* intestine surface per unit of SVL, *BM* body mass (g)

(Pagel's λ), and a table with slope and P values for the intercept and each predictive variable. The most plausible model(s) was selected based on the Akaike information criterion (AIC; Burnham and Anderson 2004), using a $\Delta AICc < 2$ as the threshold to identify the most likely models (Mazerolle 2006). When necessary, data were transformed, for instance, TTR values were $\log_{10}$ transformed, and plant proportion in the diet was transformed using the arcsine of the square root of the proportion. All analyses were performed using R v 4.1.3 (R Development Core Team 2024).

Evolutionary rates of the intrinsic traits (BM, SVL, SMR, TTR, PLP, GRS), expressed as σ^2, were also estimated. The *ratebytree* function in "phytools" (v6.99; Revell 2012) was used for this purpose. The contemporary phylogenetic tree (Fig. 2) was applied to compare the evolutionary rates between traits using the rate parameter, σ^2 (O'Meara et al. 2006; Revell 2008; Revell et al. 2008, 2018). The relationship between pairs of traits and their rate of change (e.g., SVL σ^2 vs SMR σ^2) was then compared.

3 Results

3.1 Phylogenetic Signal

Most intrinsic variables showed a relatively high phylogenetic signal; for example, plant material proportion in the diet, body mass, and snout-vent length (SVL) showed Pagel's λ (λ) values above 0.823 ($P < 0.02$; Table 2), while standard metabolic rate showed a λ value of 0.495 ($P = 0.031$), a moderate and significant estimator (Table 2). However, the thermal tolerance range (TTR) and the relative gut surface area (RGS) showed low or no phylogenetic signal (λ lower than 0.308; $P > 0.529$; Table 2).

3.2 Evolutionary Models

The Ornstein Uhlembeck model was the best-fitted model for four of the five intrinsic (biological) traits [body mass (BM), SVL, GRS, TTR, and standard metabolic rate (SMR); see Table 2], while in the case of plant proportion in the diet, Brownian motion was the best-fitted model. Other models were out of the ΔAIC range by decimal values (Table 2). Thus, Ornstein-Uhlembeck model was considered, and *corMartins* and *gls* functions were used from the libraries in R (R Development Core Team 2024).

3.3 Association Between SMR and Intrinsic and Environmental Variables

The analysis to predict which variables better explain the variation in SMR in the Liolaemid species revealed that the proportion of plants in the diet is negatively associated with SMR (see Table 3). This indicates that a higher intake of plant materials corresponds to a lower SMR. The

Table 3 Best-fitted models (based on their ΔAICc; see Tables S1 and S2) for the phylogenetic generalized least squares (pgls) models run under OU (corMartins) computed with gls in "nlme" package (Pinheiro et al. 2021) for the set of 24 species of Liolaemidae

Models	(Intercept)	Variable	Slope	*P*
SMR~PLP	0.138	PLP	−0.052	0.001*
SMR~PC1 + PLP	0.126	PC1	−0.0115	0.014*
		PLP	−0.0441	0.002*
Pollog~PLP + SMR + SVL + TTR	15.861	PLP	−1.0379	0.189
		SMR	−6.4251	0.449
		SVL	−2.6048	0.447
		TTR	−3.6639	0.718
Pollog~SMR + SVL + TTR	12.679	SMR	0.3424	0.96
		SVL	−4.4456	0.172
		TTR	−0.4097	0.967
Pollog~PC1 + PLP + SMR + SVL + TTR	25.3726	PC1	0.3043	0.208
		PLP	−1.24	0.122
		SMR	−14	0.179
		SVL	−2.839	0.401
		TTR	−9.2677	0.398
Pollog~PLP + SMR + SVL + TTR + PC2	18.314	PC2	−0.2091	0.129
		PLP	−1.1847	0.126
		SMR	−6.831	0.405
		SVL	−1.3833	0.682
		TTR	−6.5137	0.514
Pollog~PLP + SMR + TTR	12.794	PLP	1.2756	0.079*
		SMR	−6.6541	0.428
		TTR	−4.6019	0.644

Abbreviations: *SMR* standard metabolic rate ([O2]/h g), *Pollog* $\log_{10}$ of Polygon area (km2) value, *SVL* snout-vent length (mm), *TTR* thermal tolerance range (°C), *PLP* proportion of plant material in the diet, *PC1* principal component 1, *PC2* principal component 2
Asterisks indicate significant relationships

second most likely model, although beyond the ΔAIC threshold of 2 (see Table S2), included climatic variables and indicated that PC1 is negatively related to SMR; thus, seasonal dry and warm climate is positively associated with SMR (Table 3). However, this model was not considered further to avoid exceeding the ΔAIC threshold.

3.4 Association Between Range Distribution (Polygons), Climate, and Physiology

Our next step was determining whether climate or biological traits predict the polygon surface. This variable better explains the actual geographic range than the northern and southern latitudinal limits of distribution. Five models, out of all possible models, were equally likely (Table S3). Among these, only one model showed that the proportion of plant material in the diet (PLP) was marginally significant in explaining the geographic range polygon (Table 3). Based on this analysis, we can only suggest that an herbivorous diet may limit the geographic range size of these lizards.

3.5 Rate of Evolutionary Change in Physiological and Morphological Traits

Regarding the rate of change (σ^2) in biological traits, RGS exhibited the most rapid evolutionary change among the species studied, followed by

Table 4 Comparison between traits (var) of rate of change (σ^2); $t\sigma$ value and *P*-values

Var1	Var2	$\sigma^2 1$	$\sigma^2 2$	$t\sigma$	*P*
BM	SVL	0.746	0.027	31.53	<0.001*
BM	SMR	0.751	0.599	56.32	<0.001*
BM	PLP	0.747	0.452	3.191	NS
BM	TTR	0.376	0.003	56.21	<0.001*
BM	RGS	0.747	125.91	−53.13	<0.001*
SVL	PLP	0.029	0.501	−45.53	<0.001*
SVL	RGS	0.028	111.65	−130.61	<0.001*
SVL	TTR	0.018	0.003	10.62	<0.005*
SVL	SMR	0.028	0.635	−21.24	<0.001*
SMR	PLP	0.046	0.507	−99.56	<0.001*
SMR	RGS	2.38	187.86	−160.09	<0.001*
SMR	TTR	0.013	0.003	7.84	NS
TTR	PLP	0.003	0.431	−82.71	<0.001*
TTR	RGS	0.003	28.03	−156.43	<0.001*
RGS	PLP	187.88	0.507	53.89	<0.001*

References: *BM* body mas (g), *SVL* snout-vent length (mm), *SMR* standard metabolic rate ([O2]/h g), *TTR* thermal tolerance range (°C), *PLP* proportion of plant material in the diet, *RGS* intestine surface per unit of SVL
Asterisks indicate significant differences between traits

plant proportion in the diet (PLP), BM, and SMR, whereas body size (SVL) and TTR showed slower rates of change (Table 4). The differences between trait pairs were significant in most cases, except for SMR and TTR, and BM and PLP (Table 4).

4 Discussion

This study provides evidence for a diet-related effect on metabolic rate in Liolaemidae lizards. After analyzing 22 *Liolaemus* species and 2 *Phymaturus* species, we observed a negative association between standard metabolic rate and the proportion of plant matter in the diet. These findings contrast with a previous study by Cruz et al. (2011) that reported no effect of diet on metabolic rate in Liolaemidae—a discrepancy likely due to methodological differences. The present study included more species with a broader and more detailed distributional range, used a continuous variable (PLP) to represent diet rather than a categorical classification, and employed an ultrametric tree for data analysis, unlike the metatree analysis (with fixed branch lengths and constructed from two different topologies) used by Cruz et al. (2011).

Our findings not only diverged from Cruz et al. (2011) but also aligned with the conclusions of Muñoz-García and Williams (2005), who emphasized the importance of diet and phylogenetic history in determining metabolic rates. The observed association between metabolic rate and diet supports the food habits hypothesis, which identifies three dietary factors—quality, availability, and predictability—that contribute to variability in metabolic rates (McNab 1986). Specifically, the hypothesis suggests that species or populations relying on food with low energy content and/or poor digestibility tend to evolve lower mass-independent metabolic rates (McNab 2002). Carnivorous species, in contrast, typically exhibit higher metabolic rates due to their consumption of readily digestible, high-energy food sources that are available year-round and contain few chemical deterrents. Herbivores, on the other hand, tend to have lower metabolic rates, as plant material is often less digestible, contains chemical deterrents, fluctuates seasonally in abundance, and includes a higher proportion of non-nutritive fiber (McNab 1986; Wehrle and German 2023).

Previous research has also shown that foraging mode is a good predictor of metabolic rate in Squamata (Dupoué et al. 2017; Stuginski, et al.

2018). In these studies, the energy required to capture prey plays a significant role—species with lower energetic demand for movement may balance out higher expenditure for acquiring more prey. Herbivory, likely the least energetically demanding foraging mode, is also associated with the lowest intake. Thus, it is reasonable that herbivorous species exhibit lower metabolic rates compared to related lizard species with different diets.

We found no significant associations between SMR of the studied Liolaemid species and other intrinsic variables, such as body mass (BM), or relative gut surface (RGS), aside from the influence of diet type discussed previously. Although we observed a positive relationship between RGS and PLP in the diet (slope = 2.44; $P = 0.005$), our phylogenetic generalized least squares models did not reveal a clear pattern. Consistent with our findings, herbivorous Liolaemids have longer guts than their of insectivorous counterparts (Pincheira-Donoso et al. 2021). Additionally, O'Grady et al. (2005) found a significant positive correlation between the percentage of plant matter and small intestine length in species with plant material in their guts.

Some studies have shown a positive relationship between food digestibility and gut length (Secor 2001; Cramp and Franklin 2003, 2005; Sabat et al. 2005; Naya and Bozinovic 2006; Naya et al. 2009). The absence of a clear association between metabolic rate and gut size in our study may be explained by the fact that larger digestive tracts, as seen in some herbivores, can produce similar metabolic output to species with shorter tracts and a more energetic-dense diet (Antinuchi et al. 2007). In this regard, Naya et al. (2008) reported that changes in organ mass, including the gut, were poor predictors of SMR variation in some *Liolaemus* species. It seems that other mechanisms, such as changes in the specific metabolic activity of body tissues (e.g., muscle, heart, kidney, liver) or the gut microbiota, may drive SMR differences and compensate for variation in organ size (Konarzewski and Diamond 1995; Cant et al. 1996; Naya et al. 2008; Ayayee et al. 2018).

Our findings contrast with the associations between SMR and BM reported in Cruz et al. (2011) for Liolaemid lizards. These discrepancies may arise from differences in phylogenetic tree construction methods, the use of more refined techniques in the present study, and our more rigorous analytical approach. Variation in metabolic rate among vertebrates is influenced by factors beyond body size (or BM), including taxonomic affiliation, habitat type, and differences in diet and foraging behavior (Nagy 2005; Hallmann and Griebeler 2018).

Climatic variables, particularly annual precipitation (aridity), thermal variability, and maximum temperatures during hot months (PC1), influence the metabolic rate of the studied Liolaemid species. These findings also differ from those of Cruz et al. (2011), who found no significant effect of climate (based on a climatic index, see Cruz et al. 2005) or elevation range on SMR in Liolaemid lizards. The reduced variability within subclades in Cruz et al. (2011) may explain their lack of observed association, potentially resulting in phylogenetic clustering and lower statistical power to detect adaptive trends, likely due to limited taxon sampling. Additionally, the strong phylogenetic association with geographic/climatic variables reported by Cruz et al. (2011) could have obscured a direct relationship between climate and SMR.

Liolaemid lizards from temperate regions may optimize their physiological performance by aligning their thermal preferences and optimal body temperature ranges with environmental temperature gradients (Stellatelli et al. 2022). In contrast, Liolaemids from environments with fluctuating temperatures are often exposed to suboptimal conditions, while those in stable environments with extreme temperatures face constraints on their physiological capabilities (Laspiur et al. 2021; Valgas et al. 2024). Climate plays a crucial role in physiological responses for example, in lacertid lizards, evaporative water loss and preferred temperature are correlated with bioclimatic parameters, indicating physiological adaptations to climate (García Porta et al. 2019). Many taxa from hot, dry habitats have lower metabolic rates compared to those from milder, mesic environments (e.g., Dunson and Bramham 1981; Mautz 1982; Putnam and Murphy 1982). Since metabolism dictates "the pace of life," reduced metabolic rates in hot and dry habitats are recognized as an energy-efficient strategy to reduce demand and expenditure (Auer et al. 2020). Conversely, increased metabolic rate at colder temperatures may serve as a mechanism of negative compensation for sustain activity (Terblanche et al. 2009; Tsuji 1988).

Numerous reptile species, particularly cold specialists, demonstrate adaptations to cold environments by maintaining high metabolic rates and achieving high physiological performance at relatively low temperatures as result of an efficient thermoregulation (Lourdais et al. 2013; Cecchetto et al. 2020). Although the metabolic cold adaptation hypothesis remains debated (Chown and Gaston 1999), our data do not support it. Liolaemid species from dry habitats, characterized by significant daily and seasonal temperature variations, display higher metabolic rates than species from more humid (less dry), thermally stable habitats. This suggests that the interaction between thermal regime and thermoregulatory behavior drives variation in metabolic rates among Liolaemid species. As Fabrício-Neto et al. (2019) propose, thermal regimes significantly influence metabolic rates, with interspecific differences in thermal biology and life-history strategies shaping the effect.

Contrary to the metabolic cold adaptation hypothesis (Scholander et al. 1953), our results in Liolaemid lizards do not support this idea. In contrast, Giacometti et al. (2022) observed that *Tropidurus* species from colder habitats exhibit higher metabolic rates than those from warmer areas. The role of climate in influencing metabolic rates in ectotherms remains controversial. While some studies, such as those by Scholander et al. (1953) and Addo-Bediako et al. (2002), found limited support for it, Giacometti et al. (2022) suggested that temperature indeed shapes metabolic responses in lizards.

We found that species distribution ranges were likely shaped by diet, as the geographic range polygons showed a marginal and negative relationship with the degree of herbivory in Liolaemid lizards. Contrary to the predictions of the climatic variability hypothesis (Gaston and Blackburn 2000), the distributional range was not explained by SMR, TTR, or climatic variables. Liolaemidae species with herbivorous tendencies typically inhabit arid environments with harsh climates, where food resources are scarce (Valdecantos et al. 2012; Pincheira-Donoso 2021). These environments are often isolated by biogeographic barriers such as mountains along the Andean and sub-Andean regions, historically favoring the emergence of numerous endemic species (Esquerré et al. 2019, 2022).

The relatively low metabolic rate of herbivorous Liolaemid lizards, combined with their limited mobility (as shown in home ranges data from Stellatelli et al. 2016 and Victorica Erostarbe et al. 2022), and the high degree of isolation between suitable habitats, likely explain their restricted geographic distribution compared to insectivorous species. Insectivorous Liolaemids tend to occupy lower-elevation areas with greater habitat continuity, such as steppes and shrublands, in contrast to the isolated valleys and rocky outcrops inhabited by herbivores in the Andean mountains and Patagonia (Abdala et al. 2021).

Herbivory may restrict geographic range expansion, as herbivores require the expansion of plant species into new habitats before they can colonize those areas. Additionally, abiotic conditions (e.g., climate) set species' range limits at higher latitudes, while biotic interactions (e.g., species competition) define range limits at lower latitudes (The north-south hypothesis, MacArthur 1972). Broad comparisons between ecosystems (e.g., temperate vs. tropical systems) and geographic regions have shown variation in how latitude and climate affect species' distribution (Cunningham et al. 2016). However, this pattern may not universally apply, even within a group of ectotherms like the Liolaemid species studied here. Likewise, the great differences between the Northern and Southern Hemispheres in land and ocean masses may modulate climatic effects differently (Cruz et al. 2005).

A comparison of thermal tolerance limits and geographic range boundaries showed that terrestrial ectotherms (reptiles, amphibians, and insects) occupy a larger area than predicted by thermal tolerance alone at the cold boundary but fail to reach their potential range at the warm boundary (Sunday et al. 2012). Thermal tolerance varies considerably across organisms, influencing physiological functions, activity patterns, time and energy budgets, and geographic distribution (Tomlinson 2019; Camacho et al. 2024). For this reason, much attention has been paid in recent years to critical thermal limits (Araújo et al. 2013; Bozinovic et al. 2016; Hoffmann et al. 2013; Sunday et al. 2014, 2019). However, recent evidence suggests that critical thermal limits provide a poor characterization of species' thermal niches in natural environments (Huey and Kearney 2020; Peralta-Maraver and Rezende 2021) because both environmental temperature and thermal tolerance vary across time scales, and their impact on mortality and fecundity are dynamic and cumulative (Rezende et al. 2014, 2020; Alruiz et al. 2022). Understanding the proximate and ultimate mechanisms influencing macroecological patterns will provide insight into how physiology, behavior, and biogeography intersect to shape species' distributions.

The activity hypothesis (Garland 1983; Muñoz-García and Williams 2005) posits that higher metabolic rates in carnivores may be a consequence of the increased energy expenditure required for hunting, capturing, and handling prey (Ricklefs et al. 1996). This could explain why herbivores tend to have smaller ranges and lower metabolic rates (Muñoz-García and Williams 2005). Supporting this, herbivorous Liolaemid species, such as those in the genus *Phymaturus*, have smaller home ranges (0.0063–44.36 m^2; Victorica Erostarbe et al. 2022), while insectivorous *Liolaemus* species have larger home ranges (21–712 m^2), with omnivore species showing intermediate values (41–436 m^2; Stellatelli et al. 2016), consistent with our marginal results.

In conclusion, this study reveals that herbivorous Liolaemid lizards exhibit lower metabolic rates compared to omnivores and insectivores. This adaptation likely represents a trade-off, enhancing resilience in the harsh climatic conditions prevalent throughout the Andean and sub-Andean regions of South America. Herbivorous Liolaemids display a strong association with environments characterized by scarce precipitation, marked thermal fluctuations, and extremely high peak air temperatures (Pincheira-Donoso 2021). Interestingly, herbivorous Liolaemids often display a more restricted geographic distribution compared to their omnivorous and insectivorous counterparts. This observation aligns with the combined effects of ecological traits (passive feeding strategy, limited movement, and low SMR) and historical processes (such as Andean orogeny and glacial cycles) that have fragmented, expanded, and contracted Liolaemidae species distributions (Esquerré et al. 2022). These factors likely facilitated the allopatric evolution of numerous endemic species in isolated areas with distinct climates.

Liolaemid species may exhibit important differences between traits when we compared evolutionary rates of change. For example, RGS showed to be much higher than SVL, SMR, and TTR. However, this trend does not apply straightforwardly to herbivory (PLP), possibly due to the multiple independent origins in different Liolaemid clades (Espinoza et al. 2004; Pincheira-Donoso 2021). The strength of the association between ecophysiological traits may depend on methodological approaches and the specific species included in a study. Variability within subclades of Liolaemidae for certain traits might explain the absence of a clear association. Focusing on species from a few clades with low variability, and likely with high phylogenetic signal, could lead to phylogenetic clustering (Vanhooydonck and Van Damme 1999). As a result, studies with limited sample sizes or lacking of representation from diverse herbivore lineages may have lower statistical power to detect these adaptive trends. For instance, associations between SVL and herbivory (Ocampo et al. 2022, 2024) or between SVL and SMR (Cruz et al. 2011) may be evident in some studies, while absent in others (Pincheira-Donoso 2021; as observed in this study). In this context, our study contributes to the ongoing discussion on how integrating information from both ecophysiology and phylogeny can enhance understanding of macroevolutionary trends.

Acknowledgments We are grateful to Juan Carlos Acosta and his lab members, Cristian Abdala, Gustavo Scrocchi, Romina Semhan, M. Gabriela Perotti, James Schulte, Robert Espinoza, Mario Ruiz Monachesi, Fernando Lobo, and Soledad Valdecantos for their assistance in the field. We also thank the Provincial Fauna offices and Parques Nacionales (Argentine National Parks) for providing the necessary collection permits. This study was supported by multiple Consejo Nacional de Investigaciones Científicas y Técnicas (CONICET) grants (PIP 2846/04; 6287/05 to FBC) and the National Agency for Support to Science and Technology (ANPICyT) through several PICTs grants (PICT FONCyT 098-01867, 2006-01205; 2015-2471 and 2019-04115 to FBC, 2020-2615 DLMA and 2016-0677 to OAS).

References

Abdala CS, Quinteros AS (2014) Los últimos 30 años de estudios de la familia de lagartijas más diversa de Argentina. Actualización taxonómica y sistemática de Liolaemidae. Cuad Herpetol 28:55–82

Abdala CS, Laspiur A, Scrocchi G et al (eds) (2021) Las Lagartijas de la Familia Liolaemidae. Sistemática, Distribución e Historia Natural de Una de las Familias de Vertebrados más Diversas del Cono sur de Sudamérica. Ril Editores, Universidad de Tarapacá, Chile

Addo-Bediako A, Chown SL, Gaston KJ (2000) Thermal tolerance, climatic variability and latitude. Proc Roy Soc Lond B 267:739–745. https://doi.org/10.1098/rspb.2000.1065

Addo-Bediako A, Chown SL, Gaston KJ (2002) Metabolic cold adaptation in insects: a large-scale perspective. Funct Ecol 16:332–338. https://doi.org/10.1046/j.1365-2435.2002.00634.x

Agosta SJ, Bernardo J, Ceballos G et al (2013) A macrophysiological analysis of energetic constraints on geographic range size in mammals. PLoS One 8: e72731. https://doi.org/10.1371/journal.pone.0072731

Alruiz JM, Peralta-Maraver I, Bozinovic F et al (2022) Thermal tolerance in *Drosophila*: repercussions for distribution, community coexistence and responses to climate change. J Anim Ecol 91:655–667. https://doi.org/10.1111/1365-2656.13653

Andrews RA, Pough FH (1985) Metabolism of squamate reptiles: allometric and ecological relationships. Physiol Zool 58:214–231

Angilletta MJ Jr (2001) Variation in metabolic rate between populations of a geographically widespread lizard. Physiol Biochem Zool 74:11–21. https://doi.org/10.1086/319312

Angilletta MJ Jr (2009) Thermal adaptation: a theoretical and empirical synthesis. Oxford University Press, New York

Angilletta MJ, Sears MW (2000) The metabolic cost of reproduction in an oviparous lizard. Funct Ecol 14:39–45. https://doi.org/10.1046/j.1365-2435.2000.00387.x

Antinuchi CD, Zenuto RR, Luna F et al (2007) Energy budget in subterranean rodents: insights from the tuco-tuco *Ctenomys talarum* (Rodentia: Ctenomyidae). In: Kelt DA, Lessa EP, Salazar-Bravo J et al (eds) The quintessential naturalist: honoring the life and legacy of Oliver P. Pearson. University of California Publications in Zoology, Berkeley and Los Angeles, pp 111–139

Araújo MB, Ferri-Yáñez F, Bozinovic F et al (2013) Heat freezes niche evolution. Ecol Lett 16:1206–1219. https://doi.org/10.1111/ele.12155

Arnett AE, Gotelli NJ (1999) Geographic variation in life history traits of the ant lion, *Myrmeleon immaculatus*: evolutionary implications of Bergmann's rule. Evolution 53:180–1188. https://doi.org/10.1111/j.1558-5646.1999.tb04531.x

Ashton KG (2002) Do amphibians follow Bergmann's rule? Can J Zool 80:708–716. https://doi.org/10.1139/z02-049

Auer SK, Solowey JR, Rajesh S et al (2020) Energetic mechanisms for coping with changes in resource availability. Biol Lett 16:20200580. https://doi.org/10.1098/rsbl.2020.0580

Ayayee PA, Ondrejech A, Keeney G et al (2018) The role of gut microbiota in the regulation of standard metabolic rate in female *Periplaneta americana*. PeerJ 6: e4717. https://doi.org/10.7717/peerj.4717

Beck J, Liedtke HC, Widler S et al (2016) Patterns or mechanisms? Bergmann's and Rapoport's rule in moths along an elevational gradient. Community Ecol 17:137–148. https://doi.org/10.1556/168.2016.17.2.2

Bennett AF (1988) Structural and functional determinates of metabolic rate. Am Zool 28:699–708

Bergmann KGC (1847) Über die Verhältnisse der wärmeokönomie der Thiere zu ihrer Grösse. Göttinger Studien 3:595–708

Blackburn TM, Ruggiero A (2001) Latitude, elevation and body mass variation in Andean passerine birds. Glob Ecol Biogeogr 10:245–259. https://doi.org/10.1046/j.1466-822X.2001.00237.x

Blomberg SP, Garland T, Ives AR (2003) Testing for phylogenetic signal in comparative data: behavioral traits are more labile. Evolution 57:717–745. https://doi.org/10.1111/j.0014-3820.2003.tb00285.x

Bonino MF, Moreno Azocar DL, Tulli MJ et al (2011) Running in cold weather: morphology, thermal biology, and performance in the southernmost lizard clade in the world (*Liolaemus lineomaculatus* section: Liolaemini: Iguania). J Exp Zool A 315:495–503. https://doi.org/10.1002/jez.697

Bonino MF, Moreno Azocar DL, Schulte JA II et al (2015) Thermal sensitivity of cold climate lizards and the importance of distributional ranges. Zoology 118: 281–290. https://doi.org/10.1016/j.zool.2015.03.001

Bozinovic F (1993) Fisiología Ecológica de la alimentación y digestión en vertebrados: modelos y teorías. Rev Chil Hist Nat 66:375–382

Bozinovic F, Rosenmann M (1988) Comparative energetics of South American cricetid rodents. Comp Biochem Physiol A 91:195–202. https://doi.org/10.1016/0300-9629(88)91616-7

Bozinovic F, Bastías DA, Boher F et al (2011a) The mean and variance of environmental temperature interact to determine physiological tolerance and fitness. Physiol Biochem Zool 84:543–552. https://doi.org/10.1086/662551

Bozinovic F, Calosi P, Spicer JI (2011b) Physiological correlates of geographic range in animals. Ann Rev Ecol Evol Syst 42:155–179. https://doi.org/10.1146/annurev-ecolsys-102710-145055

Bozinovic F, Medina NR, Alruiz JM et al (2016) Thermal tolerance and survival responses to scenarios of experimental climatic change: changing thermal variability reduces the heat and cold tolerance in a fly. J Comp Physiol B 186:581–587. https://doi.org/10.1007/s00360-016-0980-6

Brown JH, Gillooly JF, Allen AP et al (2004) Toward a metabolic theory of ecology. Ecology 85:771–1789. https://doi.org/10.1890/03-9000

Burnham KP, Anderson D (2004) Multimodel inference: understanding AIC and BIC in model selection. Soc Meth Res 33:261–304. https://doi.org/10.1177/0049124104268644

Camacho A, Rodrigues MT, Jayyusi R et al (2024) Does heat tolerance actually predict animals' geographic thermal limits? Sci Total Environ 917:170165. https://doi.org/10.1016/j.scitotenv.2024.170165

Campbell-Staton SC, Bare A, Losos JB et al (2018) Physiological and regulatory underpinnings of geographic variation in reptilian cold tolerance across a latitudinal cline. Mol Ecol 27:2243–2255. https://doi.org/10.1111/mec.14580

Cant JP, McBride BW, Croom WJJ (1996) The regulation of intestinal metabolism and its impact on whole animal energetics. J Anim Sci 74:2541–2553. https://doi.org/10.2527/1996.74102541x

Careau V, Réale D, Humphries MM, Thomas DW (2010) The pace of life under artificial selection: personality, energy expenditure, and longevity are correlated in domestic dogs. Am Nat 175:753–758. https://doi.org/10.1086/652435

Carter MJ, Cortes PA, Rezende EL (2023) Temperature variability and metabolic adaptation in terrestrial and aquatic ectotherms. J Therm Biol 115:103565. https://doi.org/10.1016/j.jtherbio.2023.103565

Cecchetto NR, Medina SM, Ibargüengoytía NR (2020) Running performance with emphasis on low temperatures in a Patagonian lizard, *Liolaemus lineomaculatus*. Sci Rep 10:1–13. https://doi.org/10.1038/s41598-020-71617-3

Cerdeña J, Farfán J, Quiroz AJ (2021) A high mountain lizard from Peru: the world's highest-altitude reptile. Herpetozoa 34:61–65. https://doi.org/10.3897/herpetozoa.34.61393

Chown SL, Gaston KJ (1999) Exploring links between physiology and ecology at macro-scales: the role of respiratory metabolism in insects. Biol Rev 74:87–120

Chown SL, Gaston KJ, Robinson D (2004) Macrophysiology: large-scale patterns in physiological traits and their ecological implications. Funct Ecol 18:159–167

Clarke A (1993) Seasonal acclimatization and latitudinal compensation in metabolism: do they exist? Funct Ecol 7:139–149

Córdoba MA, Acosta JC, Villavicencio HJ et al (2015) Análisis trófico de *Phymaturus punae* (Iguania: Liolaemidae): variación estacional y sexual en la región más austral de la puna Argentina. Rev Mex Biodiv 86:1004–1013

Cramp RL, Franklin CE (2003) Is re-feeding eficiency compromised by prolonged starvation during aestivation in the green striped burrowing frog, *Cyclorana alboguttata*. J Exp Zool A 300:126–132

Cramp RL, Franklin CE (2005) Arousal and re-feeding rapidly restores digestive tract morphology following aestivation in green-striped burrowing frogs. Comp Biochem Physiol A 142:451–460

Cruz FB, Fitzgerald LA, Espinoza RE et al (2005) The importance of phylogenetic scale in tests of Bergmann's and Rapoport's rules: lessons from a clade of South American lizards. J Evol Biol 18:1559–1574

Cruz FB, Belver L, Acosta JC et al (2009) Thermal biology of *Phymaturus* lizards: evolutionary constraints or lack of environmental variation? Zoology 112:425–432

Cruz FB, Antenucci D, Luna F et al (2011) Energetics in Liolaemini lizards: implications of a small body size and ecological conservatism. J Comp Physiol B 181:373–382

Cruz FB, Moreno Azocar DL, Bonino MF et al (2014) Clima, distribución geográfica y viviparismo en especies de *Liolaemus* (Reptilia; Squamata): Cuando las hipótesis se ponen a prueba. Ecosistemas 23:37–45

Cruz FB, Moreno Azocar DL, Perotti MG et al (2022) The role of climate and maternal manipulation in determining and maintaining reproductive mode in *Liolaemus* lizards. J Zool 317:101–113

Cunningham HR, Rissler LJ, Buckley LB et al (2016) Abiotic and biotic constraints across reptile and amphibian ranges. Ecography 39:1–8

De Vera L, Hayes J (1995) Repeatability of oxygen consumption in *Gallotia galloti*. J Herpetol 29:151–154

Diamond JM (1998) Evolution of biological safety factors: a cost/benefit analysis. In: Weibel ER, Taylor CR, Bolis L (eds) Principles of animal design. Cambridge University Press, Cambridge, pp 21–27

Dietz MW, Piersma T (2007) Red knots give up flight capacity and defended food processing capacity during winter starvation. Funct Ecol 21:899–904

Dillon ME, Wang G, Huey RB (2010) Global metabolic impacts of recent climate warming. Nature 467:704–706

Dunson WA, Bramham CR (1981) Evaporative water loss and oxygen consumption of three small lizards from the Florida Keys: *Sphaerodactylus cinereus, S. notatus,* and *Anolis sagrei*. Physiol Zool 54:253–259

Dupoué A, Brischoux F, Lourdais O (2017) Climate and foraging mode explain interspecific variation in snake metabolic rates. Proc Roy Soc B 284:20172108. https://doi.org/10.1098/rspb.2017.2108

Eager CD (2017) Contrast preservation and constraints on individual phonetic variation. Doctoral thesis. University of Illinois at Urbana-Champaign

Edwards DL, Avila LJ, Martinez L et al (2022) Environmental correlates of phenotypic evolution in ecologically diverse *Liolaemus* lizards. Ecol Evol 12: e9009. https://doi.org/10.1002/ece3.9009

Espinoza RE, Wiens JJ, Tracy CR (2004) Recurrent evolution of herbivory in small, cold-climate lizards: breaking the ecophysiological rules of reptilian herbivory. Proc Nat Acad Sci 101:16819–16824

Esquerré D, Brennan IG, Catullo RA et al (2019) How mountains shape biodiversity: the role of the Andes in biogeography, diversification, and reproductive biology in South America's most species-rich lizard radiation (Squamata: Liolaemidae). Evolution 73:214–230

Esquerré D, Keogh JS, Demangel D et al (2022) Rapid radiation and rampant reticulation: Phylogenomics of South American *Liolaemus* lizards. Syst Biol 71:286–300

Fabrício-Neto A, Gavira RSB, Andrade DV (2019) Thermal regime effects on the resting metabolic rate of rattlesnakes depend on temperature range. J Therm Biol 83:199–205

Felsenstein J (1985) Phylogenies and the comparative method. Am Nat 125:1–15

Finkler MS (2006) Effects of temperature, sex, and gravidity on the metabolism of small-mouthed Salamanders, *Ambystoma texanum*, during the reproductive season. J Herpetol 40:103–106. http://www.jstor.org/stable/4092881

Finkler MS, Cullum KA (2002) Sex-related differences in metabolic rate and energy reserves in spring-breeding small-mouthed Salamanders (*Ambystoma texanum*). Copeia 2002:824–829. http://www.jstor.org/stable/1448163

García Porta J, Irisarri I, Kirchner M et al (2019) Environmental temperatures shape thermal physiology as well as diversification and genome-wide substitution rates in lizards. Nat Commun 10:1–12

García-Roselló E, Guisande C, Manjarrés-Hernández A et al (2015) Can we derive macroecological patterns from primary global biodiversity information facility data? Glob Ecol Biogeogr 24:335–347

Garland T Jr (1983) Scaling the ecological cost of transport to body mass in terrestrial mammals. Am Nat 121: 571–587

Gaston KJ (2003) The structure and dynamics of geographic ranges. Oxford University Press, Oxford

Gaston KJ, Blackburn TM (2000) Pattern and process in macroecology. Blackwell Science, Malden

Gaston KJ, Chown SL, Calosi P et al (2009) Macrophysiology: a conceptual reunification. Am Nat 174:595–612

Giacometti D, Bars-Closel M, Kohlsdorf T et al (2022) Environmental temperature predicts resting metabolic rates in Tropidurinae lizards. J Exp Zool A 337:1039–1052. https://doi.org/10.1002/jez.2656

Gillooly JF, Brown JH, West GB et al (2001) Effects of size and temperature on metabolic rate. Science 293: 2248–2251

Hallmann K, Griebeler EM (2018) An exploration of differences in the scaling of life history traits with body mass within reptiles and between amniotes. Ecol Evol 8:5480–5494

Harmon LJ, Losos JB, Davies J et al (2010) Early bursts of body size and shape evolution are rare in comparative data. Evolution 64:2385–2396

Hoffmann AA, Chown SL, Clusella-Trullas S (2013) Upper thermal limits in terrestrial ectotherms: how constrained are they? Funct Ecol 27:934–949. https://doi.org/10.1111/j.1365-2435.2012.02036.x

Huey RB, Kearney MR (2020) Dynamics of death by heat. Science 369:1163

Karasov WH (1990) Digestion in birds: chemical and physiological determinants and ecological implications. In: Morrison ML, Ralph CJ, Verner J et al (eds) Avian foraging: theory, methodology, and applications studies in avian biology No. 13. Cooper Ornithological Society, Kansas, pp 391–415

Kearney M (2012) Metabolic theory, life history and the distribution of a terrestrial ectotherm. Funct Ecol 26: 167–179

Kohl KD, Brun A, Magallanes M et al (2016) Physiological and microbial adjustments to diet quality permit facultative herbivory in an omnivorous lizard. J Exp Biol 219:1903–1912

Konarzewski M, Diamond J (1995) Evolution of basal metabolic rate and organ masses in laboratory mice. Evolution 49:1239–1248

Kooijman S (2000) Dynamic energy and mass budgets in biological systems. Cambridge University Press, Cambridge

Kozłowski J, Konarzewski M, Czarnoleski M (2020) Coevolution of body size and metabolic rate in vertebrates: a life-history perspective. Biol Rev 95: 1393–1417

Kozykariski ML, Belver LC, Avila LJ (2011) Diet of the desert lizard *Liolaemus pseudoanomalus* (Iguania: Liolaemini) in northern La Rioja Province, Argentina. J Arid Environ 75:1237–1239

Kramer-Schadt S, Niedballa J, Pilgrim JD et al (2013) The importance of correcting for sampling bias in MaxEnt species distribution models. Divers Distrib 19:1366–1379

Laspiur A, Santos JC, Medina SM et al (2021) Vulnerability to climate change of a microendemic lizard species from the central Andes. Sci Rep 11:11653

Lawton JH (1983) Plant architecture and the diversity of phytophagous insects. Annu Rev Entomol 28:23–39

Levinton JS, Monahan RK (1983) The latitudinal compensation hypothesis: growth data and a model of latitudinal growth differentiation based upon energy budgets. Ii. Intraspecific comparisons between subspecies of *Ophryotrocha puerilis* (Polychaeta: Dorvilleidae). Biol Bull 165:699–707. https://doi.org/10.2307/1541472. PMID: 29324019

Lighton JRB (2008) Measuring metabolic rates: a manual for scientists. Oxford University Press

Lourdais O, Guillon M, DeNardo D et al (2013) Cold climate specialization: adaptive covariation between metabolic rate and thermoregulation in pregnant vipers. Physiol Behav 119:149–155

Luna F, Naya H, Naya DE (2017) Understanding evolutionary variation in basal metabolic rate: an analysis in subterranean rodents. Comp Biochem Physiol A 206: 87–94. https://doi.org/10.1016/j.cbpa.2017.02.002

MacArthur RH (1972) Geographical ecology–Harper and Row, New York

Mautz WJ (1982) Correlation of both respiratory and cutaneous water losses of lizards with habitat aridity. J Comp Physiol B 149:25–30

Mazerolle MJ (2006) Improving data analysis in herpetology: using Akaike's Information Criterion (AIC) to assess the strength of biological hypotheses. Amphibia-Reptilia 27:169–180

McCoy ED, Bell SS (1991) Habitat structure: the evolution and diversification of a complex topic. In: McCoy ED, Mushinsky HR (eds) Bell SS. Chapman and Hall, London, pp 3–27

McNab BK (1986) The influence of food habits on the energetics of eutherian mammals. Ecol Monogr 56:1–19

McNab BK (2002) The physiological ecology of vertebrates: a view from energetics. Cornell University Press

Meik JM, Jeo RM, Mendelson JR et al (2002) Effects of bush encroachment on an assemblage of diurnal lizard species in central Namibia. Biol Conserv 106:29–36

Moreno Azocar DL, Vanhooydonck B, Bonino MF et al (2013) Chasing the Patagonian sun: comparative thermal biology of *Liolaemus* lizards. Oecologia 171:773–788

Moreno Azocar DL, Arcos NA, Perotti MG et al (2020) How and when melanic coloration is an advantage for lizards: the case of three closely-related species of *Liolaemus*. Zoology 141:125774

Muñoz MM, Stimola MA, Algar AC et al (2014) Evolutionary stasis and lability in thermal physiology in a group of tropical lizards. Proc Roy Soc B 281: 20132433. https://doi.org/10.1098/rspb.2013.2433

Muñoz-García A, Williams JB (2005) Basal metabolic rate in carnivores is associated with diet after controlling for phylogeny. Physiol Biochem Zool 78:1039–1056. https://doi.org/10.1086/432852

Nagy KA (2005) Field metabolic rate and body size. J Exp Biol 208:1621–1625

Naya DE, Bozinovic F (2006) The role of ecological interactions on the physiological flexibility of lizards. Funct Ecol 20:601–608. https://doi.org/10.1111/j.1365-2435.2006.01137.x

Naya DE, Veloso C, Bozinovic F (2008) Physiological flexibility in the Andean lizard *Liolaemus bellii*: seasonal changes in energy acquisition, storage and expenditure. J Comp Physiol B 178:1007–1015

Naya DE, Veloso C, Bozinovic F (2009) Gut size variation among *Bufo spinulosus* populations along an altitudinal (and dietary) gradient. Ann Zool Fenn 46:16–20

O'Grady SP, Morando M, Avila L et al (2005) Correlating diet and digestive tract specialization: examples from the lizard family Liolaemidae. Zoology 108:201–210

O'Meara BC, Ané C, Sanderson MJ et al (2006) Testing for different rates of continuous trait evolution using likelihood. Evolution 60:922–933

Ocampo M, Pincheira-Donoso D, Sayol F et al (2022) Evolutionary transitions in diet infuence the exceptional diversifcation of a lizard adaptive radiation. BMC Ecol Evol 22:74. https://doi.org/10.1186/s12862-022-02028-3

Ocampo M, Pincheira-Donoso D, Rios RS (2024) Patterns of morphological diversification are influenced by dietary evolution in a highly species-rich lizard radiation. Front Ecol Evol 12:1361799. https://doi.org/10.3389/fevo.2024.1361799

Pagel M (1999) Inferring the historical patterns of biological evolution. Nature 401:877–884

Paradis E, Schliep K (2019) Ape 5.0: an environment for modern phylogenetics and evolutionary analyses in R. Bioinformatics 35:526–528

Paradis E, Claude J, Strimmer K (2004) APE: analyses of phylogenetics and evolution in R language. Bioinformatics 20:289–290

Pearson RG, Raxworthy CJ, Nakamura M et al (2007) Predicting species distributions from small numbers of occurrence records: a test case using cryptic geckos in Madagascar. J Biogeogr 34:102–117

Pennell MW, Eastman JM, Slater GJ et al (2014) Geiger v2. 0: an expanded suite of methods for fitting macroevolutionary models to phylogenetic trees. Bioinformatics 30:2216–2218

Peralta-Maraver I, Rezende EL (2021) Heat tolerance in ectotherms scales predictably with body size. Nat Clim Chang 11:58–63. https://doi.org/10.1038/s41558-020-00938-y

Pettersen AK, Marshall DJ, White CR (2018) Understanding variation in metabolic rate. J Exp Biol 221:166876. https://doi.org/10.1242/jeb.166876

Pie MR, Divieso R, Caron FS (2021a) Do geographic range sizes evolve faster in endotherms? Evol Biol 48:286–292. https://doi.org/10.1007/s11692-021-09537-x

Pie MR, Divieso R, Caron FS et al (2021b) The evolution of latitudinal ranges in reef-associated fishes: heritability, limits and inverse Rapoport's rule. J Biogeogr 48: 2121–2132. https://doi.org/10.1111/jbi.14150

Pincheira-Donoso D (2021) Correlated evolution between herbivory and gastrointestinal tract in a prolific lizard adaptive radiation. Anim Biol 71:233–241. https://doi.org/10.1163/15707563-bja10051

Pinheiro J, Bates D, DebRoy S, et al (2021) nlme: Linear and Nonlinear Mixed Effects Models. R package version 3.1–153. https://CRAN.R-project.org/package=nlme

Plasman M, Bautista A, McCue MD et al (2020) Resting metabolic rates increase with elevation in a mountain-dwelling lizard. Integr Zool 15:363–374. https://doi.org/10.1111/1749-4877.12434. Epub 2020 May 23. PMID: 32306560

Putnam RW, Murphy RW (1982) Low metabolic rate in a nocturnal desert lizard, *Anarbylus switaki* Murphy (Sauria: Gekkonidae). Comp Biochem Physiol 71: 119–123

R Core Team (2024) R: a language and environment for statistical computing. R Foundation for Statistical Computing. https://www.R-project.org/

Rabosky DL (2014) Automatic detection of key innovations, rate shifts, and diversity-dependence on phylogenetic trees. PLoS One 9:e89543. https://doi.org/10.1371/journal.pone.0089543

Revell LJ (2008) On the analysis of evolutionary change along single branches in a phylogeny. Am Nat 172: 140–147

Revell LJ (2012) phytools: an R package for phylogenetic comparative biology (and other things). Methods Ecol Evol 3:217–223

Revell LJ (2013) Ancestral character estimation under the threshold model from quantitative genetics. Evolution 68:743–759

Revell LJ, Harmon LJ, Collar DC (2008) Phylogenetic signal, evolutionary process, and rate. Syst Biol 57: 591–601

Revell LJ, González-Valenzuela LE, Alfonso A et al (2018) Comparing evolutionary rates between trees, clades, and traits. Methods Ecol Evol 9:994–1005

Rezende EL, Castañeda LE, Santos M (2014) Tolerance landscapes in thermal ecology. Funct Ecol 28:799–809. https://doi.org/10.1111/1365-2435.12268

Rezende EL, Bozinovic F, Szilágyi A et al (2020) Predicting temperature mortality and selection in natural *Drosophila* populations. Science 369:1242–1245

Ricklefs RE, Konarzewski M, Daan S (1996) The relationship between basal metabolic rate and daily energy expenditure in birds and mammals. Am Nat 147: 1047–1071

Sabat P, Riveros JM, López-Pinto C (2005) Phenotypic flexibility in the intestinal enzymes of the African clawed frog *Xenopus leavis*. Comp Biochem Physiol A 140:135–139

Schliep MK (2011) Phangorn: phylogenetic analysis in R. Bioinformatics 27:592–593

Scholander PF, Flagg W, Walters V et al (1953) Climatic adaptation in Arctic and tropical poikilotherms. Physiol Zool 26:67–92

Secor SM (2001) Regulation of digestive performance: a proposed adaptive response. Comp Biochem Physiol A 128:565–577

Simandle ET, Espinoza RE, Nussear KE et al (2001) Lizards, lipids, and dietary links to animal function. Physiol Biochem Zool 74:625–640

Šizling AL, Storch D, Keil P (2009) Rapoport's rule, species tolerances, and the latitudinal diversity gradient: geometric considerations. Ecology 90:3575–3586. https://doi.org/10.1890/08-1129.1

Sokal RR, Rohlf FJ (1995) Biometry: the principles and practice of statistics in bio- logical research. W.H. Freeman and Company, New York, NY

Speakman JR, Krol E, Johnson MS (2004) The functional significance of individual variation in basal metabolic rate. Physiol Biochem Zool 77:900–915

Spicer JI, Gaston KJ (1999) Amphipod gigantism dictated by oxygen availability? Ecol Lett 2:397–401

Srichairat N, Taksintum W, Chumnanpuen P (2018) Gross morphological structure of digestive system in water monitor lizard *Varanus salvator* (Squamata: Varanidae). Walailak J Sci Technol (WJST) 15:245–253

Stellatelli OA, Vega LE, Block C et al (2013) Effects on the thermoregulatory efficiency of two native lizards as a consequence of the habitat modification by the introduction of the exotic tree *Acacia longifolia*. J Therm Biol 38:135–142

Stellatelli OA, Block C, Moreno Azocar DL et al (2016) Scale dependency of *Liolaemus* lizards' home range in response to different environmental variables. Curr Zool 62:521–530

Stellatelli OA, Vega LE, Block C et al (2022) Latitudinal pattern of the thermal sensitivity of running speed in the endemic lizard *Liolaemus multimaculatus*. Integr Zool 17:619–637

Stroud JT, Losos JB (2016) Ecological opportunity and adaptive radiation. Ann Rev Ecol, Evol Sys 47:507–532

Stuginski DR, Navas CA, de Barros FC et al (2018) Phylogenetic analysis of standard metabolic rate of snakes: a new proposal for the understanding of interspecific variation in feeding behavior. J Comp Physiol B 188:315–323. https://doi.org/10.1007/s00360-017-1128-z

Sun B, Williams CM, Li T et al (2022) Higher metabolic plasticity in temperate compared to tropical lizards suggests increased resilience to climate change. Ecol Monogr 92:e1512

Sunday JM, Bates AE, Dulvy NK (2012) Thermal tolerance and the global redistribution of animals. Nat Clim Ch 2:686–690. https://doi.org/10.1038/nclimate1539

Sunday JM, Bates AE, Kearney MR et al (2014) Thermal-safety margins and the necessity of thermoregulatory behavior across latitude and elevation. Proceed Nat Acad Sci 111:5610–5615. https://doi.org/10.1073/pnas.1316145111

Sunday J, Bennett JM, Calosi P et al (2019) Thermal tolerance patterns across latitude and elevation. Phil Trans Roy Soc London B 374:20190036. https://doi.org/10.1098/rstb.2019.0036

Terblanche JS, Clusella-Trullas S, Deere JA et al (2009) Directional evolution of the slope of the metabolic rate–temperature relationship is correlated with climate. Physiol Biochem Zool 82:495–503

Tomašových A, Jablonski D, Berke SK et al (2015) Nonlinear thermal gradients shape broad-scale patterns in geographic range size and can reverse Rapoport's rule. Glob Ecol Biogeogr 24:157–167. https://doi.org/10.1111/geb.12242

Tomlinson S (2019) The mathematics of thermal sub-optimality: nonlinear regression characterization of thermal performance of reptile metabolic rates. J Therm Biol 81:49–58

Tsuji JS (1988) Seasonal profiles of standard metabolic rate of lizards (*Sceloporus occidentalis*) in relation to latitude. Physiol Zool 61:230–240

Tulli MJ, Abdala V, Cruz FB (2011) Relationships among morphology, clinging performance and habitat use in Liolaemini lizards. J Evol Biol 24:843–855

Tulli MJ, Abdala V, Cruz FB (2012) Effects of different substrates on sprint performance of lizards. J Exp Biol 215:774–784

Valdecantos MS, Arias F, Espinoza RE (2012) Herbivory in *Liolaemus poecilochromus*, a small, cold-climate lizard from the Andes of Argentina. Copeia 2012: 203–210

Valgas AAN, Cubas GK, de Oliveira DR et al (2024) Ecophysiological responses of *Liolaemus arambarensis* juveniles to experimental temperature variations. Comp Biochem Physiol A 290:111577

Vanhooydonck B, Van Damme R (1999) Evolutionary relationships between body shape and habitat use in lacertid lizards. Evol Ecol Res 1:785–805

Vanhooydonck B, Cruz FB, Abdala CS et al (2010) Sex-specific evolution of bite performance in *Liolaemus* lizards (Iguania: Iguanidae): the battle of the sexes. Biol J Linn Soc 101:461–475

Victorica Erostarbe AE, Fava GA, Acosta JC (2022) Restricted use of space in an endemic lizard of the Andes: addressing the effects of intrinsic and environmental factors. Behav Ecol Sociobiol 76:15

Vitt LJ, Zani PA (1996) Ecology of the South American lizard *Norops chrysolepis* (Polychrotidae). Copeia 1996:56–68

Vitt LJ, Zani PA, Caldwell JP et al (1997) Ecology of whiptail lizards (*Cnemidophorus*) in the Amazon Region of Brazil. Copeia 1997:745–757

Vleck D (1987) Measurement of O_2 consumption, CO_2 production, and water vapor production in a closed system. J Appl Physiol 62:2103–2106

Wehrle BA, German D (2023) Reptilian digestive efficiency: past, present, and future. Comp Biochem Physiol A 277:111369

White CR, Kearney MR (2013) Determinants of interspecific variation in basal metabolic rate. J Comp Physiol B 183:1–26

Wiens JJ, Ackerly DD, Allen AP et al (2010) Niche conservatism as an emerging principle in ecology and conservation biology. Ecol Lett 13:1310–1324

Wiersma P, Nowak B, Williams JB (2012) Small organ size contributes to the slow pace of life in tropical birds. J Exp Biol 215:1662–1669

Zanotto F, Gouveia S, Simpson S et al (1997) Nutritional homeostasis in locusts: is there a mechanism for increased energy expenditure during carbohydrate overfeeding? J Exp Biol 200:2437–2448. https://doi.org/10.1242/jeb.200.18.2437. PMID: 9320360

Zar JH (1999) Biostatistical analysis. Prentice-Hall, New Jersey

Part III

Ecological and Physiological Adaptations

Physiological Ecology of Andean Amphibians Associated with Freezing Temperatures

Juan Manuel Carvajalino-Fernández and Carlos A. Navas

1 Amphibians of the High Andes

Although this section does not intend to fully describe the diversity of Andean amphibians, it does intend to highlight that the diversity of amphibians found above 2000 m asl is not negligible relative to global amphibian diversity. Following the taxonomical system proposed by Frost (2024a), several species of the families Bufonidae, Centrolenidae, Craugastoridae, Dendrobatidae, Hylidae, Microhylidae, Leptodactylidae, Plethodontidae, Strabomantidae, and Telmatobiidae, among others, have evolved physiological and ecological adaptations to the harsh condition of the high Andean Mountain. High Andean amphibians occur in two main regions, the high Andes tropical region, distributed in northwest of South America from eastern Venezuela, passing through Colombia, Ecuador, Peru, and to Bolivia, and the high Andes subtropical region (>25° S) of Argentina and Chile.

Armesto and Señaris (2017) summarized the diversity of frogs in the northern of Andes, including localities in Colombia, Ecuador, and Venezuela. The authors argue that the Strabomantidae family is dominant from 2,000 to 4,000 m asl, where it is mainly represented by the genus *Pristimantis* (Fig. 1). Genera of Bufonidae such as *Atelopus* and *Osornophryne* also live above 3,000 m asl. In addition, species of Telmatobiidae live above 2,000 m asl, while Microhylidae and Leptodactylidae disappear above 3,000 m asl. Among Dentrobatids, *Hyloxalus subpunctatus* is the only dendrobatid that occurs as high as 4,020 m asl, while the species with the highest record in the northern Andes is the bufonid *Osornophryne bufoniformis*, which inhabits the Central Cordillera of Colombia and the Eastern Cordillera of Ecuador from 2,800 to 4,700 m asl (Acosta 2000).

Despite the efforts and rigorous work of Armesto and Señaris (2017), information gaps remain, particularly if we consider the diversity of Venezuela. Saboya and Urbina-Cardona (2023) studied the distributions of amphibians in the high Colombian Andes, reporting 142 species (95 endemic) from the paramo region (>3,000 m asl). No reviews are available of the high mountain amphibian's species in Ecuador, Peru, or Bolivia, so the following species richness numbers must be underestimates, considering that total amphibian richness of these countries are 679, 667, and 252 species, respectively (Frost 2024). In Ecuador, the initiative "Bioweb" of the Pontifical Catholic University identifies

J. M. Carvajalino-Fernández (✉)
Laboratory of Adaptations to Extreme Environments and Global Change Biology, Departamento de Biología, Facultad de Ciencias, Universidad Nacional de Colombia – Sede Bogotá, Bogotá, Colombia
e-mail: jmcarvajalinof@unal.edu.co

C. A. Navas
Department of Physiology, Biosciences Institute, University of São Paulo, São Paulo, SP, Brazil

J. N. Urbina-Cardona et al. (eds.), *Andean Herpetofauna*, Sustainable Development Goals Series,
https://doi.org/10.1007/978-3-032-00074-3_11

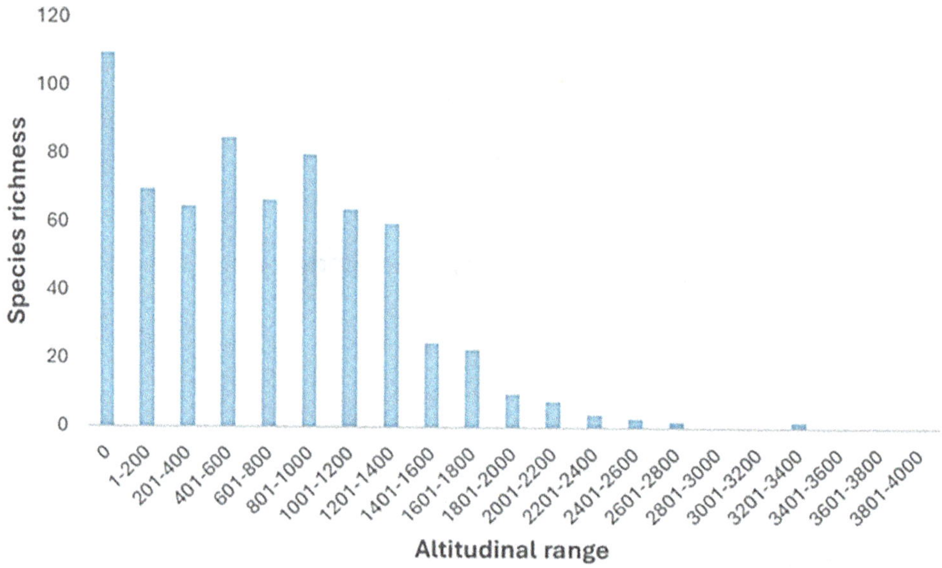

Fig. 1 Altitudinal range of amphibians of the northern Andes. (Adapted from: Armesto and Señaris 2017)

83 species of amphibians with a distribution above 2,000 m asl (Ron et al. 2024): Bufonidae with the genera *Osornophyne* and *Atelopus*, Centrolenidae, Dendrobatidae with genus *Hyloxalus*, Stabomantidae with genus *Pristimantis*, and Hemiphractidae with the genus *Gastrotheca*; which is also present in the paramos of southern Colombia, Peru, Bolivia, and Argentina. Glass frogs are also well represented in Ecuador, with 16 species above 2,000 masl, and with representatives of the "*buckleyi* group" occurring as high as 3,500 m asl (Guayasamin et al. 2020). Finally, Ecuador shares with Peru, Bolivia, Chile, and Argentina the Andean high mountain lake family Telmatobiidae, which contains 64 species. The Ecuadorian endemic paramo frog *Telmatobius niger* is currently considered extinct (Barbour and Noble 1920; Merino-Viteri et al. 2010).

Ecuador, Peru, and Bolivia share similarities in their high mountain species richness. Just referring to threatened species, Peru registers more than 46 species reaching or surpassing 2,000 m (Catenazzi and von May 2014). In Bolivia and Peru, small frogs in the genera *Bryophryne*, *Microkayla*, *Noblella*, *Oreobates*, and *Qosqophryne* are noted for living from 2,500 to 4,000 m asl (Burrowes et al. 2020; De la Riva et al. 2000; Padial et al. 2012; Catenazzi et al. 2020). Finally, Peru and Bolivia share the most diversity of the unique and interesting group of aquatic groups *Telmatobius*, with 25 species in Peru and 15 species in Bolivia (Catenazzi and von May 2014). In this group, we highlight the species living at the highest elevation worldwide, *T. marmoratus* at 5,400 m in southern Peru (Seimon et al. 2006), where it occurs along with *Pleurodema marmoratum* (Leptodactylidae).

In the Andean subtropical and temperate regions, Argentina and Chile share similar components of their amphibian faunas, recording 177 species and 59 species, respectively. Although less diverse than their tropical counterpart, the subtropical and temperate Andes include fascinating cases of adaptation to high altitude and seasonality. The Andean subtropical and temperate species experience low-temperature and freezing periods like the Nearctic temperate species, but boast higher species richness, having, just in Chile, more than 16 species living above 2,000 m asl (Díaz-Páez and Ortiz 2003). Among several species groups, we want to highlight the species of the genera *Pleurodema*, *Alsodes*, and *Rhinella*. These groups are interesting for including species such as *Pleurodema thaul* and *P. bufoninum*, which have population at very low latitudes and exposed to the harsh winters of the southern tip of the continent. On the other hand, the genus *Alsodes* is a mountain endemic group, with more than 19 species, some with overwintering tadpoles (Logares and Ubeda 2006; Blotto et al. 2012; Frost 2024b). Finally, *Rhinella spinulosa* has a wide latitudinal distribution in the Andes. This species figures among the few that have been studied in detail regarding hydric and thermal balance (Sinsch 1984; Lambrinos and Kleier 2003; Sanabria et al. 2015; Alveal et al. 2016; Alveal et al. 2022).

2 Habitats at Different Scales and Their Importance in the Thermal Balance of High Andean Amphibians

Two different thermal regimes are present in the high Andes with respect to temperatures changes, first the regime of high Andean tropical regions where the temperature range can span 50°C, from -10°C at dawn to 40°C at midday (Carvajalino-Fernández et al. 2011). The second regime is found in the Andean subtropical and temperate regions, where both altitude and seasonality influence the frequency and the magnitude of low temperature (below 4°C) and freezing periods (Coronato and Bisigato 1998). Low and freezing temperatures are common in both regimes, but their frequency and magnitude differ between regions (Reider et al. 2021).

Setting aside such differences, a common consideration in both cases is the thermal landscape and the availability of thermal refuges (Gonzáles-del-Pliego et al. 2020). The diversity of potential thermal refuges and microhabitats is vital for the community structure of small animals (as are most Andean amphibians) (Dobrowski 2011; Navas et al. 2013). The characterization of microhabitats for tropical Andean amphibians has been studied since the 1990s. Navas (1996b) showed that species of nocturnal paramo frogs segregate by microhabitat type (Fig. 2). Carvajalino-Fernández et al. (2011) followed up on Navas's work by describing the microhabitats used during times of activity and lower temperatures in *P. nervicus* (Fig. 2). In terms of exposure to freezing, the conclusion was that, independently of microhabitat used, during some periods of the year *P. nervicus* face high risk of freezing (Carvajalino-Fernández et al. 2011).

3 Amphibian Behaviors Associated with Freezing Temperatures Risk

In the high tropical Andes, daily temperatures exhibit extreme variability (Navas 2006) and may include subzero temperatures (Carvajalino-Fernández et al. 2011). Reider et al. (2021) highlighted the importance of the freeze–thaw cycles in the high tropical Andes, identifying three adaptative strategies for low temperatures: thermal tolerance breadth, use of buffered microhabitats, and behavioral thermoregulation. During the night, when freezing risk is high, some frog species remain active, such as *Pristimantis* frogs, which may vocalize at temperatures below 3°C (Navas 1996a). Low temperatures generally reduce locomotion and metabolic rate in frogs, but in paramo species *Atelopus lozanoi*, *Hyloxalus subpunctatus*, *Pristimantis bogotensis*, and *Dendrosophus labialis* locomotion (swimming or walking) exhibits reduced thermal dependence below 10°C (Navas 1996b, c).

An interesting and anecdotal report of frogs exposed to freezing risk comes from the Sierra Nevada de Santa Marta, the highest mountainous massif of Colombia, located in the north of the country 21 km south of the Caribbean Sea and isolated from the Andes (Cardona and Ojeda 2011). In 2001, the ornithologist and explorer Ralf Strewe found adults of the toad *Atelopus carrikeri* waking on hail, during the day (Fig. 3). The report is fascinating because it suggests *A. carrikeri* as model species to study freezing resistance. Freezing tolerance is uncommon among toads in the Bufonidae (Voituron et al. 2009; Pratihar and Kundu 2011; Carvajalino-Fernández et al. 2013; Rodríguez et al. 2019; de Amaral et al. 2024).

The diversity of amphibians that maintain activity at low temperatures may be high in

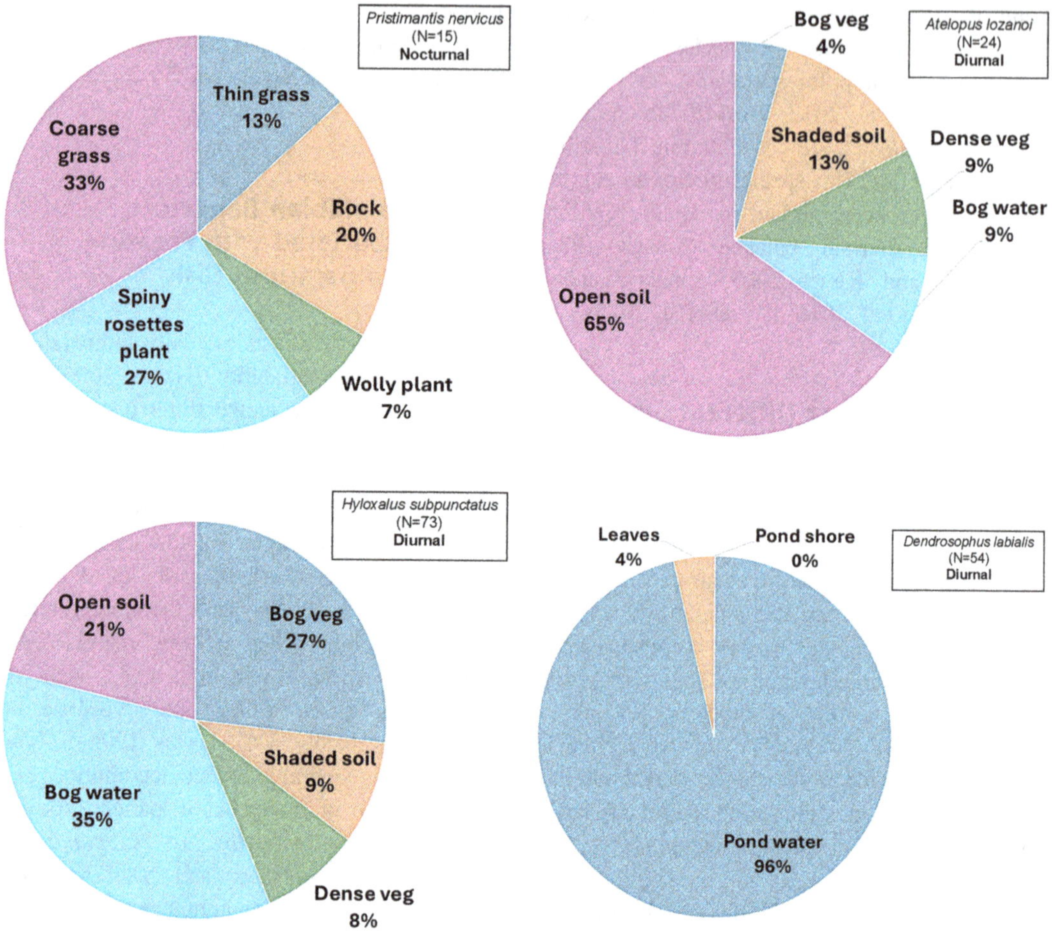

Fig. 2 Microhabitats used by the paramo frog during activity. (Adapted from Navas 1996b; Carvajalino-Fernández et al. 2011)

Andean high tropical regions, for example in the family Strabomantidae. Many species likely avoid activity at low temperatures or under freezing conditions, entering a state of lethargy or dormancy. Above 3000 m asl, the toad *Osornophryne percrassa* decreases its activity and shelters in bed mosses and other wet subtracts in the paramo at low temperatures. During lethargy periods, the animals adopt a posture to conserve water and maintain temperature with legs and ventral flanks very rigid (Fig. 4). To emerge from their lethargic state, these animals required a 20-minute recovery period following handling and exposure to direct light at 24 °C (Personal communication, Professor Teddy Angarita of the National University of Colombia).

In the subtropical Andean countries (Argentina and Chile), the use of hibernacula and the diminution of activity are the most frequent strategies associated with low temperatures during autumn or winter (Cei 1980; Valenzuela-Sánchez et al. 2015). In some cases, rainfall or snow might be more important than temperature to determine activity patterns. For example, the microendemic leptodactylid *Atelognathus nitoi* is restricted to the area surrounding the pond Laguna Verde in the Nahuel Huapi National Reserve, northwestern Patagonia, a semiaquatic ecosystem than allows the species to maintain activity during almost all seasons; with rainfall being the principal factor determining their behavior (Ubeda et al. 1999).

Fig. 3 Individual of *Atelopus carrikeri* waking on hail at 4300 m asl in the "Paramo de la Cuenca Alta de Rio Frio in the Sierra Nevada de Santa Marta—Colombia." (Photo credit for Ralf Strewe)

Fig. 4 Postures of *Osornophryne percrassa*, during lethargy periods; notice the rigidity and absence of movement in legs and ventral. (Photo credit for Teddy Angarita)

4 Physiological Mechanisms of Amphibians Associated with Survival at Low and Freezing Temperatures: What Do We Know About the Andean Species?

Two physiological strategies are compatible with freezing in adult amphibians, freeze avoidance, and freeze tolerance (Costanzo and Lee 2013), which are not mutually exclusive (Storey 2006; Rozsypal 2018). Freeze avoidance is associated with avoidance of cell freezing, via biochemical protection, with ice forming in body fluids. This strategy can be seen in Antarctic and artic fishes (Devries and Cheng 2000; Duman 2014), molluscs (Ansart and Vernon 2003), insects (Bale 1987; Teets et al. 2023), amphibians (Costanzo et al. 1999; de Amaral et al. 2024), and reptiles (Lowe et al. 1971; Wharton 2012). Freeze tolerance and freeze avoidance rely on the production of antifreeze proteins and accumulation of low weigh biomolecules which decrease the cryoscopy point of the fluid (Costanzo and Lee 2013). The most common antifreeze molecules are carbohydrates, such as glucose or glycerol, which act as a cryoprotectants at high concentrations to significantly depress the freezing point and reduce the amount of ice formed during freezing (Storey 1997; Storey and Storey 2017). Not much is known about freeze avoidance in Andean amphibians, but we highlight reports in juvenile and adults of toad *Rhinella spinulosa*, which showed supercooling temperatures in the order of −1.5 °C (Rodríguez et al. 2019; Sanabria et al. 2015).

Freeze tolerance results from ecological and genetically based physiological mechanisms that promote controlled ice grow of body fluids while preserving fluidity inside the cells (Storey and Storey 2013, 2017). Historically the description of amphibian freeze tolerance originated in the 1980s by Janet Margaret Storey and Kennet Storey (University of Carlenton), and Jon Constanzo (Miami University). Likewise, most studies are concentrated only in a handful species in Northern America and the northern Palearctic regions, including *Epidalea calamita*, *Rana parvipalmata*, *Pseudacris crucifer*, *Pseudacris triseriata*, *R. arvalis*, and *Lithobates sylvaticus* (Storey and Storey 2013; Edwards et al. 2000; Higgins and Swanson 2013; do Amaral et al. 2018; Shekhovtsov et al. 2022; de Amaral et al. 2024).

The physiological state associated with freeze tolerance ensures cryoprotection, maintenance of osmotic and redox balance, and preservation of cell viability (including their genetic controls). The mechanisms used are not limited to cell biochemistry, though, and relevant additional elements include choice of hibernacles, gut symbiosis with bacteria, membrane phospholipid modulation, factors that interact with the production, accumulation, and distribution of cryoprotectants in specific organs and tissues, and the production of nucleation proteins (INP) (Hazel 1989; Storey and Storey 2013, 2017; Costanzo and Lee 2005; Costanzo and Lee 2008, 2013).

According to their physicochemical properties, cryoprotectants are split in colligative or non-colligative. Colligative cryoprotectants include simple solutes such as glucose and glycerol, derived primarily from glycogen stores or hydrolysis of triglyceride (Davenport 1992; Dieni et al. 2012). Non-colligative anti-freeze protection relies on hysteresis, which is concentration independent and offers freezing temperatures lower than melting counterparts (Storey and Storey 2013). Non-colligative hysteresis occurs when INP are produced, mainly during winter (Wolanczyk et al. 1990), to control ice formation in the extracellular fluids (Storey and Storey 2013). In addition, other molecular mechanism such as post-translational regulation of antioxidant enzymes may maintain the redox balance during freezing events (Giraud-Billoud et al. 2019, 2024).

Although not Andean species, the two well-studied frogs *L. sylvaticus* and *H. versicolor* illustrate anti-freezing mechanism in anurans. In both species, glycerol and glucose protect against the formation of ice and stabilize membranes by reducing the fraction of extracellular body ice and cellular water loss (Costanzo et al. 1993, 2013). Urea becomes an additional colligative cryoprotectant in continuous production due to protein degradation, particularly where the role of glucose is either nonexistent or limited (de Amaral et al. 2024). Other biochemical and molecular mechanisms that maintain cell and organismal integrity during freezing events are associated with redox balance (Dawson et al. 2015; Dawson and Storey 2016; Giraud-Billoud et al. 2019), epigenetic regulation, cell signaling, and transcription factor regulation (Kiss et al. 2011; Storey and Storey 2017).

What northern species exposed to winter share with Andean counterparts in their resistance or tolerance to cold remains to be discovered. Glucose accumulation has been reported in paramo species *Pristimantis nevicus* and *P. bogotensis*, which can elevate blood glucose concentration >120%, reaching levels of > 6.1 mmol/L, and can resist body ice grown for up to 63% of their total body water (Carvajalino-Fernández et al. 2021). In both species, these physiological mechanisms occur every night during activity (Fig. 5), in contrast with Nearctic species such as *L. sylvatica* in which protection is maximized during winter coinciding with strong metabolic depression.

Some subtropical Andean species also accumulate biomolecules during the autumn or winter, possibly via a cold acclimation mechanism similar to that in Nearctic counterparts. Rodríguez et al. (2019) report glucose concentration rising during experimental freezing in the toad *R. spinulosa* (increase of 139%, maximum value 8,18+2,52 mmol/mL).

The modulation of lipid concentrations in ectotherm is another strategy for maintaining the homeoviscous state of membranes at low temperatures. Particularly, cholesterol influences the physical and functional properties of the membrane (Hazel 1989). For example, during the Patagonian autumn, the tonchek spiny-chest frog *Alsodes gargola* displays higher basal cholesterol concentrations than the gray four-eyed frog *Pleurodema bufoninum*, likely related to the constant low temperatures and sporadic freezing temperatures that characterize alpine Andean-Patagonian forests during autumn (Fig. 6; Carvajalino-Fernandez et al., in review).

The frog *P. bufoninum* presents another mechanism, presumably associated with freezing survival during the Patagonian autumn. We refer

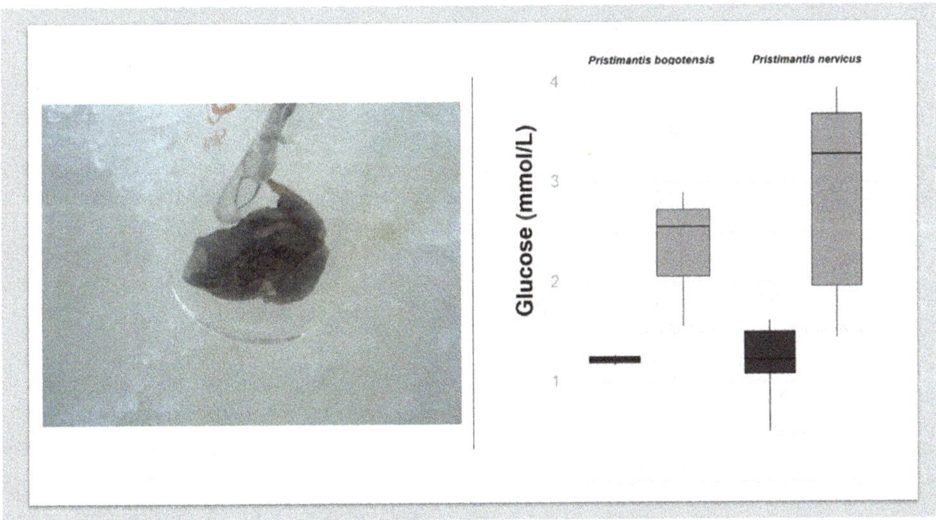

Fig. 5 Differences in glucose concentration during experimental freezing. Individual of *Pristimantis nervicus* in a calorimetric glass cell during a freezing period exposition on ice with salt (**a**). Glucose concentration of *P. bogotensis* ($N = 4$) and *P. nervicus* ($N = 7$) at noon (12:00 pm, light grey boxplot) and dawn (4:00 am, black boxplot). (Adapted from Carvajalino-Fernández et al. 2021)

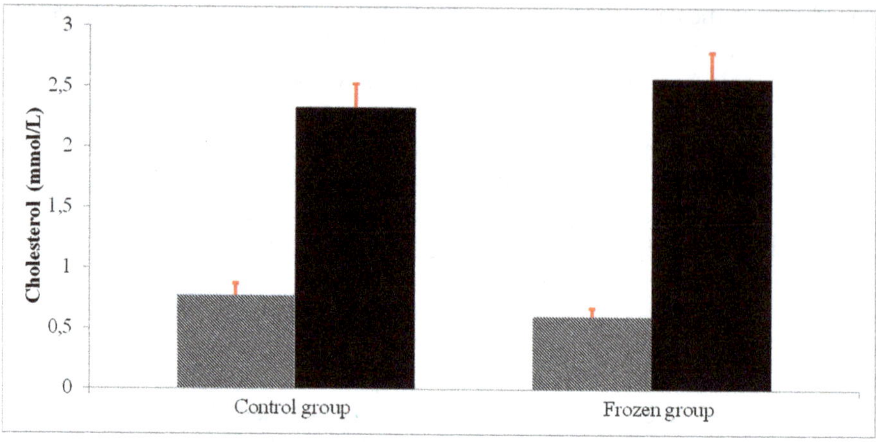

Fig. 6 Effect of freezing on cholesterol concentrations in *Alsodes gargola* (black bar) and *Pleurodema bufoninum* (degraded bar). Mean + S.E.M (* denote $p < 0.05$). (Figures based on data from Carvajalino-Fernández 2017)

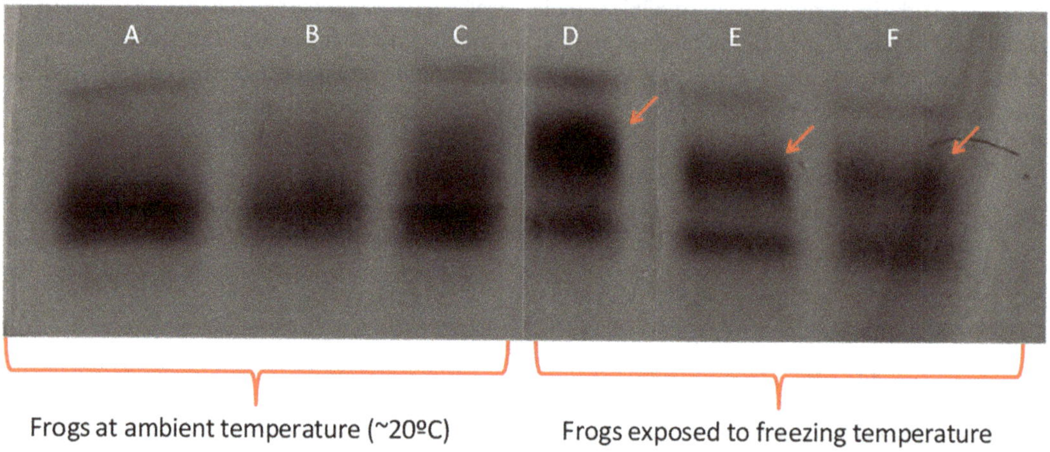

Fig. 7 Protein electrophoresis made on supernatant of homogenized body tissues of *Pleurodema bufoninum*. Arrows indicate protein production in frozen group. (Figures based on data from Carvajalino-Fernández 2017)

here to the production of proteins compatible with ice nucleation (Fig. 7; Carvajalino-Fernandez et al., under review). This preliminary result is promising and relatable with Nearctic frogs' hibernation (Le Tri et al. 2019). The distribution range of *P. bufoninum* is similar in latitudinal extent to that of *Lithobates sylvaticus* in the Northern Hemisphere, exhibiting a notably broad range (Ferraro and Casagranda 2009). Therefore, it is plausible that the freezing resistance and associated mechanisms in *P. bufoninum* vary along its distribution range.

5 Conclusion

The ability of small ectothermic and terrestrial vertebrates to colonize habitats above 2,000 m asl is often linked to the evolution of resistance to low and freezing temperatures. Amphibian species in the Andean cordilleras are a prime example, with many lineages having independently evolved in high-altitude environments, leading to the remarkable biodiversity seen in these mountains. Regardless of whether freezing risks is determined by circadian or circannual rhythms, high-elevation and tropical Andean ecosystems involves daily cycles of constant protective baselines, and even reactive or circadian defenses (Carvajalino-Fernández et al. 2021). For instance, in the Andean high tropical regions, exposure to freezing conditions may last only minutes or hours, which can lead to some species becoming less tolerant to cold spells and thus restricted to lower elevations or less exposed habitats (Navas et al. 2013; Carvajalino-Fernández et al. 2021).

Given the current scenarios of global climate change, with peak low and freezing temperatures expected to become more frequent throughout the year in some Andean regions (IPCC 2023), it is critical to study the freeze tolerance of amphibians. These studies are necessary to create scenarios for community reconfiguration in habitats facing new thermal extremes. The interest in understanding the adaptations and ecological and physiological mechanisms associated with the colonization of the Andes has been growing since the works of Monasterio (1980) and Sinsch (1984), which inspired further work by Navas in the 1990s (Navas 1996a, b, c, 1997). Thus, it is essential to adapt methodologies used for exploring cold adaptation in high-latitude amphibians to study physiological mechanisms in their Andean counterparts. Information on the activity of cryoprotectants, lipid modulation within cell membranes, and strategies for maintaining osmotic and redox balance, along with gene activation and posttranscriptional modifications, must be contextualized within the ecophysiological realities described for the amphibians along the latitudinal range of the Andes, particularly in high tropical and subtropical regions, where frogs might face freezing risk every day.

References

Acosta G (2000) Ranas, salamandras y caecilias (Tetrapoda: Amphibia) de Colombia. Biota Colombiana 1(3):289–319

Alveal N, Díaz-Páez H, Ortiz JC (2016) Thermal tolerance in the andean toad *Rhinella spinulosa* (Anura: Bufonidae) at three sites located along a latitudinal gradient in Chile. J Therm Biol 60:237–247

Alveal N, Rivera R, Pinochet J et al (2022) Potential current and future distribution of the andean toad *Rhinella spinulosa* Wiegmann, 1834 (Anura: Bufonidae): is the species vulnerable to climate change? South Am J Herpetol 25(1):1–11

Ansart A & Vernon P (2003) Cold hardiness in molluscs. Acta Oecologica 24:95–102

Armesto LO, Señaris JC (2017) Anuros del norte de los Andes: patrones de riqueza de especies y estado de conservación. Papéis Avulsos de Zoologia 57(39): 491–526

Bale JS (1987) Insect cold hardiness: freezing and supercooling-an ecophysiological perspective. J Insect Physiol 33(12):899–908

Barbour T, Noble GK (1920) Some amphibians from northwestern Peru, with a revision of the genera *Phyllobates* and *Telmatobius*. Bull Museum Compar Zool 1920:395–427

Blotto B, Nuñez JJ, Basso NG, Úbeda CA, Wheeler WC, Faivovich J (2012) Phylogenetic relationships of a Patagonian frog radiation, the *Alsodes* + *Eupsophus* clade (Anura: Alsodidae), with comments on the supposed paraphyly of *Eupsophus*. Cladistics 29(2): 113–131

Burrowes PA, Navas CA, Jiménez-Robles O et al (2020) Climatic heterogeneity in the bolivian Andes: are frogs trapped? South Am J Herpetol 18(1):1–12

Cardona A, Ojeda G (2011) Special volume: geological evolution of the Sierra Nevada de Santa Marta and adjacent basins, colombian caribbean region. J S Am Earth Sci 29(4):761–763

Carvajalino-Fernández JM (2017) Tolerância ao congelamento em algumas espécies de anuros de América do Sul: mecanismos e estratégias. Tese (Doutorado em Fisiologia Geral) – Instituto de Biociências, Universidade de São Paulo, São Paulo. https://doi.org/10.11606/T.41.2017

Carvajalino-Fernández JM, Bonilla-Gomez MA, Navas CA (2011) Freezing risk in tropical high-elevation anurans: an assessment based on the andean frog *Pristimantis nervicus* (strobomantidae). South Am J Herpetol 6:73–78

Carvajalino-Fernández JM, Jeckel AM, Indicatti RP (2013) *Melanophryniscus moreirae* (Amphibia, Anura, Bufonidae): dormancy and hibernacula use during cold season. Herpetol Brasileira 2:61–62

Carvajalino-Fernández JM, Bonilla-Gómez M, Giraldo-Gutiérrez L (2021) Freeze tolerance in neotropical frogs: an intrageneric comparison using *Pristimantis* species of high elevation and medium elevation. J Trop Ecol 37:118–125

Costanzo JP and Lee RE Jr (2005) Cryoprotection by urea in a terrestrially hibernating frog. J. Exp. Biol. 208,4079–4089

Costanzo JP, and RE Lee (2013) Avoidance and tolerance of freezing in ectothermic vertebrates. The Journal of Experimental Biology216, 1961–7

Costanzo JP, RE. Lee, and PH. Lortz. 1993 Glucose concentration regulates freeze tolerance in the wood frog Rana sylvatica. The Journal of experimental biology181, 245–55

Catenazzi A, von May R (2014) Conservation status of amphibians in Peru. Herpetol Monogr 28(1):1–23

Catenazzi A, Mamani L, Lehr E, von May R (2020) A new genus of terrestrial-breeding frogs (Holoadeninae, Strabomantidae, Terrarana) from southern Peru. Diversity 12(5):184

Cei JM (1980) Amphibians of Argentina. Monitore zoologico italiano N S Monografia 2. xii + 609 pp

Coronato F, Bisigato A (1998) A temperature pattern classification in Patagonia. Int J Climatol 18(7): 765–773

Costanzo JP, Lee RE (2008) Urea loading enhances freezing survival and postfreeze recovery in a terrestrially hibernating frog. J Exp Biol 211:2969–2975

Costanzo JP (2019) Overwintering adaptations and extreme freeze tolerance in a subarcticpopulation of the wood frog, Rana sylvatica. J Comp Physiol B 189:1–15.https://doi.org/10.1007/s00360018-1189-7

Costanzo JP, Lee RE (2013) Avoidance and tolerance of freezing in ectothermic vertebrates. J Exp Biol 216(11):1961–1967

Davenport J (1992) Animal life at low temperature. Chapman & Hall

Dawson NJ, Storey KB (2016) A hydrogen peroxide safety valve: the reversible phosphorylation of catalase from the freeze-tolerant north American wood frog, Rana sylvatica. Biochim Biophys Acta Gen Subj 1860:476–485

Dawson NJ, Katzenback BA, Storey KB (2015) Free-radical first responders: the characterization of CuZnSOD and MnSOD regulation during freezing of the freeze- tolerant north American wood frog, Rana sylvatica. Biochim Biophys Acta Gen Subj 1850:97–106

de Amaral M, Carvajalino-Fernández JM, Nicieza AG et al (2024) Urea and glucose modulation during freezing exposure in three temperate frogs reveals specific targets in relation to climate. J Therm Biol 121:103854

De la Riva I, Köhler J, Lötters S et al (2000) Ten years of research on Bolivian amphibians: updated checklist, distribution, taxonomic problems, literature and iconography. Revista Española de Herpetología 14:19–164

Devries AL, Cheng CHC (2000) Freezing avoidance strategies differ in Antarctic and Arctic fishes. Am Zool 40(6):997–997

Díaz-Páez H, Ortiz JC (2003) Assessment of the conservation status of amphibians in Chile. Rev Chil Hist Nat 76(3):509–525

Dieni Ca, Bouffard MC, Storey KB (2012) Glycogen synthase kinase-3: cryoprotection and glycogen metabolism in the freeze-tolerant wood frog. J Exp Biol. 215: 543–551

Duman JG (2014) An early classic study of freeze avoidance in marine fish. J Exp Biol 217(6):820–823

Do Amaral, MCF, J Frisbie DL, Goldstein, and CM Krane (2018) The cryoprotectant system of Cope's gray treefrog, Dryophytes chrysoscelis: responses to cold acclimation, freezing, 405 and thawing. J. Comp. Physiol. B.188, 611–621

Dobrowski SZ (2011) A climatic basis for microrefugia: The influence of terrain on climate.Global Change Biology 17:1022–1035. https://doi.org/10.1111/j.1365-2486.2010.02263.x

Edwards JR, Koster KL, Swanson DL (2000) Time course for cryoprotectant synthesis in the freeze-tolerant chorus frog, Pseudacris triseriata. Comp Biochem Physiol A Mol Integr Physiol. Mar;125(3):367-75. doi: 10.1016/s1095-6433(00)00168-9. PMID: 10794966

Ferraro DP, Casagranda MD (2009) Geographic distribution of the genus Pleurodema in Argentina (Anura: Leiuperidae). Zootaxa 2024:33–55

Frost DR (2024a) Amphibian species of the world: an online reference. Version 6.2 (date of access). Electronic database accessible at https://amphibiansoftheworld.amnh.org/index.php. American Museum of Natural History, New York. https://doi.org/10.5531/db.vz.0001

Frost DR (2024b) «Alsodes». Amphibian species of the world: an online reference. Version 6.2. (en inglés). Nueva York, EEUU: Museo Americano de Historia Natural

Giraud-Billoud M, Rivera-Ingraham GA, Moreira DC et al (2019) Twenty years of the 'preparation for oxidative stress' (POS) theory: Ecophysiological advantages and molecular strategies. Comparative biochemistry and physiology. Part A Mol Integr Physiol 234:36–49

Giraud-Billoud M, Moreira DC, Minari M et al (2024) REVIEW: evidence supporting the 'preparation for oxidative stress' (POS) strategy in animals in their natural environment. Comp Biochem Physiol A Mol Integr Physiol 293:111626

González-Del-Pliego P, Scheffers BR, Freckleton RP, Basham EW, AraújoMB, Acosta-Galvis AR, Medina Uribe CA, Haugaasen T, Edwards DP (2020) Thermal tolerance and the importance of microhabitats for Andean frogs in the context of land use and climate change. J Anim Ecol (11):2451–2460. https://doi.org/10.1111/1365-2656.13309

Guayasamin, Cisneros-Heredia DF, McDiarmid RW, Pena P, Hutter CR (2020) "Glassfrogs of Ecuador: Diversity, Evolution, and Conservation." Diversity (Basel), 12(6), 222

Hazel JR (1989) Cold adaptation in ectotherms: regulation of membrane function and cellular metabolism. In: Wang LCH (ed) Animal adaptation to cold. Springer, Berlin Heidelberg, Berlin, Heidelberg, pp 1–50

Higgins SA, Swanson DL (I 2013) Urea is not a universal cryoprotectant among hibernating anurans: evidence from the freeze-tolerant boreal chorus frog (Pseudacris maculata). Comp Biochem Physiol A Mol Integr Physiol. Feb;164(2):344-50. doi: 10.1016/j.cbpa.2012.11.002. Epub 2012 Nov 9. PMID:23142424

IPCC (2023) Climate change 2023: synthesis report. In: Core Writing Team, Lee H, Romero J (eds) Contribution of working groups I, II and III to the sixth assessment report of the intergovernmental panel on climate change. IPCC, Geneva, Switzerland, pp 1–34

Kiss AJ, TJ Muir, RE Lee, and JP Costanzo (2011) Seasonal variation in the hepatoproteome of the dehydrationand freeze-tolerant wood frog, Rana sylvatica. International Journal of Molecular Sciences12, 8406–8414

Lambrinos JG, Kleier C (2003) Thermoregulation of juvenile Andean toads (*Bufo spinulosus*) at 4300 m. J Therm Biol 28(1):15–19

Le Tri D, Childers CL, Adam MK et al (2019) Characterization of ice recrystallization inhibition activity in the novel freeze-responsive protein Fr10 from freeze-tolerant wood frogs, *Rana sylvatica*. J Therm Biol 84: 426–430

Logares R, Ubeda C (2006) First insights into the overwintering biology of *Alsodes gargola* frogs and tadpoles inhabiting harsh Andean-Patagonian alpine environments. Amphibia-Reptilia 27(2):263–267

Lowe CH, Lardner PJ, Halpern EA (1971) Supercooling in reptiles and other vertebrates. Comp Biochem Physiol A Physiol 39(1):125–135

Merino-Viteri A, Coloma LA, Ron S (2010) *Telmatobius niger*. In: IUCN 2012. IUCN Red List of Threatened Species. Version 2012.2. http://www.iucnredlist.org. Downloaded on 9 July 2024

Monasterio M (1980) Las formaciones vegetales de los páramos Venezolanos. En: M. Monasterio (ed.). Estudios ecológicos en los páramos. Universidad de los Andes, Mérida: 47–92

Navas CA (1996a) The effect of temperature on the vocal activity of tropical anurans: a comparison of high and low elevation species. J Herpetol 30(4):488–497

Navas CA (1996b) Implications of microhabitat selection and patterns of activity on the thermal ecology of high elevation Neotropical anurans. Oecologia 108:617–626

Navas CA (1996c) Thermal dependency of field locomotor and vocal performance of high-elevation anurans in the tropical Andes. J Herpetol 30(4):478–487

Navas CA (1997) Thermal extremes at high elevations in the Andes: physiological ecology of frogs. J Therm Biol 22(6):467–477

Navas CA (2006) Patterns of distribution of anurans in high Andean tropical elevations: insights from integrating biogeography and evolutionary physiology. Integr Comp Biol 46(1):82–91

Navas CA, Carvajalino-Fernández JM, Saboyá-Acosta LP et al (2013) The body temperature of active amphibians along a tropical elevation gradient: patterns of mean and variance and inference from environmental data. Funct Ecol 27:1145–1154

Padial JM, Chaparro JC, Castroviejo-Fisher S, Guayasamin JM, Lehr E, Delgado AJ, Vaira M, Teixeira M, Aguay R, de la Riva I (2012) A revision of species diversity in the Neotropical genus *Oreobates* (Anura: Strabomantidae), with the description of three new species from the Amazonian slopes of the Andes. Am Mus Novit 3752:1–55

Pratihar S, Kundu JK (2011) Hibernation in *Duttaphrynus melanostictus*, a life in the cold lane. Herpetol Bull 117:10–14

Reider KE, Larson DJ, Barnes BM et al (2021) Thermal adaptations to extreme freeze–thaw cycles in the high tropical Andes. Biotropica 53(1):296–306

Rodríguez CY, Bustos DA, Sanabria E (2019) Adaptation of the Andean toad *Rhinella spinulosa* (Anura: Bufonidae) at low temperatures: the role of glucose as cryoprotectant. Physiol Biochem Zool 92(5):473–480

Ron SR, Merino-Viteri A, Ortiz DA (2024) Anfibios del Ecuador. Version 2024.0. Museo de Zoología, Pontificia Universidad Católica del Ecuador. https://bioweb.bio/faunaweb/amphibiaweb, fecha de acceso 3 de julio, 2024

Rozsypal J, Košťál (2018) Supercooling and freezing as eco-physiological alternatives rather than mutually exclusive strategies: a case study in *Pyrrhocoris apterus*. J Insect Physiol 111: 53–62

Saboyá LP, Urbina-Cardona JN (2023) Current state of knowledge of Páramo amphibians in Colombia: Spatio temporal trends and information gaps to be strengthened for effective conservation. Trop Conserv Sci 16: 19400829231169984

Sanabria EA, Rodríguez CY, Vergara C et al (2015) Thermal ecology of the post- metamorphic Andean toad (*Rhinella spinulosa*) at elevation in the Monte desert, Argentina. J Therm Biol 52:52–57

Seimon TA, Seimon A, Daszak P (2006) Upward range extension of Andean anurans and chytridiomycosis to extreme elevations in response to tropical deglaciation. Glob Chang Biol 12:1–12

Shekhovtsov SV, Bulakhova NA., Tsentalovich YP, Zelentsova EA., Meshcheryakova EN, Poluboyarova TV, Berman DI (2022) Metabolomic Analysis Reveals That the Moor Frog Rana arvalis Uses Both Glucose and Glycerol as Cryoprotectants. Animals , 12, 1286. https://doi.org/10.3390/ani12101286

Sinsch U (1984) Behavioral thermoregulation of the Andean toad (*Bufo spinulosus*) at high altitudes. Oecologia 80:32–38

Storey KB (1997) Organic solutes in freezing tolerance. Compar Biochem Physiol A Physiol 117:319–326

Storey KB (2006) Reptile freeze tolerance: metabolism and gene expression. Cryobiology 52:1–16

Storey KB, Storey JM (2013) Molecular biology of freezing tolerance. Compr Physiol 3(3):1283–1308

Storey KB, Storey JM (2017) Molecular physiology of freeze tolerance in vertebrates. Physiol Rev 97(2): 623–665

Teets NM, Marshall KE, Reynolds JA (2023) Molecular mechanisms of winter survival. Annu Rev Entomol 68:1

Ubeda C, Zagarese H, Diaz M (1999) First steps towards the conservation of the microendemic Patagonian frog *Atelognathus nitoi*. Oryx 33(1):59–66

Valenzuela-Sánchez A, Cunningham AA, Soto-Azat C (2015) Geographic body size variation in ectotherms: effects of seasonality on an anuran from the southern temperate forest. Front Zool 12:37–43

Voituron Y, Barré H, Ramløv H et al (2009) Freeze tolerance evolution among anurans: frequency and timing of appearance. Cryobiology 58(3):241–247

Wolanczyk JP, Storey KB, Baust JG (1990) Ice nucleating activity in the blood of the freezetolerant frog, Rana sylvatica. Cryobiology. Jun;27(3):328–35. https://doi.org/10.1016/0011-2240(90)90032-y. PMID:237941

Wharton DA (2012) Supercooling and freezing tolerant animals. In: Wilson P (ed) Supercooling. InTech. ISBN: 978-953-51-0113-0

Thermal Biology of Squamate Reptiles in the High Mountains of the Northern Andes

Martha Lucía Calderón-Espinosa,
Miguel Ángel Méndez-Galeano, and Alexandra Montoya-Cruz

1 Introduction

1.1 The Mountains of the Northern Andes

Mountain regions (>1000 m a.s.l.; Holdridge 1967; Blyth et al. 2002) cover approximately a quarter of the Earth's continental surface, hosting a wide variety of climatic conditions and biomes (Blyth et al. 2002). Each mountain system exhibits unique characteristics, with some harboring greater biodiversity than others (Myers et al. 2000; Blyth et al. 2002; Chan et al. 2024). One of the main mountainous regions of the planet is the Andean system, being also the longest (Gassner 1973). Notably, the tropical Andes, stretching from 10°N in western Venezuela to 17–18°S in northwestern Bolivia (Sarmiento 1986), rank as one of the most biodiverse regions on Earth and a critical biodiversity hotspot (Myers et al. 2000).

Within the tropical Andes lies a subregion known as the Northern Andes, stretching from 10°N in western Venezuela to 6°S at the Huancabamba Depression in Peru, representing the equatorial Andean region by definition (Gassner 1973; Ríos-Touma et al. 2014). This subregion is the most biodiverse within the entire Andean system, hosting unique ecosystems such as the paramos (Blyth et al. 2002). Its exceptional biodiversity results from the region's distinct geological and climatological history, spanning from the elevation of the different mountainous areas that make up the Northern Andes (Gassner 1973; Gregory-Wodzicki 2000), to the climatic changes resulting from the glacial and interglacial periods, which triggered pulses of diversification known as "The Flickering Connectivity System" (Flantua and Hooghiemstra 2018; Flantua et al. 2019).

1.2 What Happens in the Tropical Environments of the Northern Andes?

Due to meteorological, orogenic, and astronomical phenomena, regions in the middle or temperate latitudes exhibit pronounced seasonality. In contrast, tropical zones (hereafter referring to the strip between the Tropics of Cancer and Capricorn: 23°26′10.0″ N–23°26′10.0″ S) generally maintain greater climatic stability and receive higher solar radiation, with annual fluctuations primarily associated with changes in rainfall. This results in greater annual temperature variation in temperate zones compared to tropical ones (Fig. 1; Sarmiento 1986). However, while tropical zones are more stable in terms of annual temperature variation, they exhibit significant daily

M. L. Calderón-Espinosa · M. Á. Méndez-Galeano (✉) · A. Montoya-Cruz
Grupo de Morfología y Ecología Evolutiva, Instituto de Ciencias Naturales, Universidad Nacional de Colombia, Bogotá, Colombia
e-mail: miamendezga@unal.edu.co

J. N. Urbina-Cardona et al. (eds.), *Andean Herpetofauna*, Sustainable Development Goals Series,
https://doi.org/10.1007/978-3-032-00074-3_12

fluctuations (Ghalambor et al. 2006). In general, tropical high-altitude regions experience greater daily temperature variation than temperate high-altitude zones (Sarmiento 1986; Ghalambor et al. 2006). Remarkably, the daily temperature variation in tropical high-altitude areas can be comparable to the mean annual temperature variation in temperate regions—a daily equivalent of temperate seasonality (Fig. 1; Sarmiento 1986; Ghalambor et al. 2006).

The interaction between elevation and latitude, and their effects on temperature, exhibits complex patterns (Janzen 1967; Ghalambor et al. 2006; Muñoz and Bodensteiner 2019). In general, temperate zones experience greater seasonal fluctuations in environmental temperatures compared to tropical zones, regardless of elevation. Similarly, as elevation increases, environmental temperatures decrease regardless of latitude (Fig. 1; Sarmiento 1986; Ghalambor et al. 2006). Nevertheless, as latitude increases, both elevation and environmental temperatures decline, resulting in a consistent pattern of low temperatures with high seasonal fluctuation at extreme latitudes (Fig. 1; Holdridge 1967). Additionally, with two-thirds of the Earth's continental mass located in the Northern hemisphere, about half of the world's mountainous zones are concentrated in the temperate region of this hemisphere (Blyth et al. 2002), primarily represented by the Himalayan range and the Tibetan Plateau. This is significant, as it implies that most of the world's highest elevation zones experience pronounced annual fluctuations associated with temperate seasonality (Ghalambor et al. 2006). Finally, biodiversity generally declines with increasing elevation, primarily due to the reduction in energy availability for organisms (Navas 2002).

In this regard, the tropical environments of the Northern Andes exhibit the following: (1) low annual fluctuations in environmental temperatures, (2) high daily fluctuations in environmental temperatures, and (3) a decrease in environmental temperature with increasing elevation, similar to other mountainous regions worldwide. These factors underscore the unique combination of thermal characteristics in the Northern Andes, which have distinct effects and consequences for the reptiles that inhabit the region, a topic that will be explored in the following sections.

1.3 The Squamate Reptiles of the Northern Andes

The reptile diversity in high-elevation zones of the Northern Andes does not surpass that of amphibians (Navas 2002). As outlined in the previous section, this region is characterized by a decrease in environmental temperature and atmospheric pressure with increasing elevation, along with high solar radiation and significant daily fluctuations in environmental temperature, which can resemble the seasonal temperature variations of temperate zones. These climatic peculiarities of the Northern Andean mountains give rise to unique, yet-to-be-described patterns of thermal traits in tropical species, within the broader latitudinal and elevational context of reptile thermal biology worldwide.

In the mountains of the Northern Andes, lizard and snake species with diverse life history traits are distributed, though the thermal biology of only a few species is well understood (Navas 2002). This region harbors species of heliothermic arboreal lizards of the genus *Anolis*, terrestrial counterparts of the genus *Stenocercus*, semifossorial (fossorial and terrestrial) heliothermic and diurnal thigmothermic lizards such as *Anadia bogotensis* (Peters 1862) and crepuscular-nocturnal thigmothermic species like the thickhead ground snake *Atractus crassicaudatus* (Duméril et al. 1854), among others (Navas 2002). The natural history traits of these species offer valuable insights into the behavioral and physiological strategies organisms employ to overcome the challenges of the mountain ecosystems in the Northern Andes.

In this chapter, we will address the following topics: (1) the characterization of thermoregulation strategies in *A. crassicaudatus* and several species of the *Anolis heterodermus* group; (2) various thermal traits in species of the genus *Stenocercus*, as well as in *Anolis tolimensis*

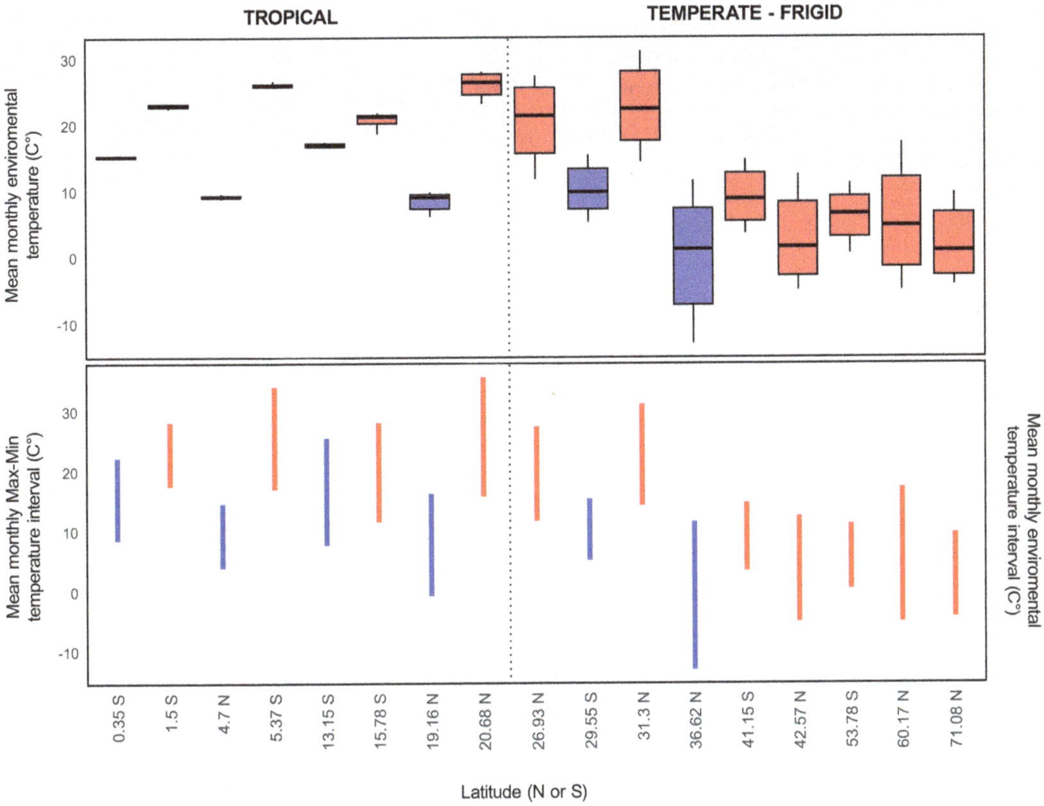

Fig. 1 (Top) Distribution of average monthly environmental temperatures and (bottom) ranges of average monthly maximum and minimum environmental temperatures (tropical) or average monthly environmental temperatures (temperate and boreal zones) from 17 different locations (most related to squamate reptile thermal biology studies) on a latitudinal gradient (0–72° N-S). The distinction is made between lowlands (red) and highlands (blue), dividing tropical latitudes from temperate and frigid with a dotted line. Note in (A) the expansion of the range of environmental temperatures experienced during the year at higher latitudes due to seasonality, as well as the difference in temperatures in high areas. From a certain latitude (40°), both high and low areas present low temperatures in wide ranges. It can also be observed that the highest latitudes in the tropics begin to present a certain degree of seasonality. (Data source: WorldClim (worldclim.org). References of the locations: 0.35 S: Guerra-Correa et al. (2020), 1.5 S: Altamirano-Benavides et al. (2019), 4.7 N: Méndez-Galeano et al. (2020), 5.37 S: Pontes da Silva et al. (2018), 13.15 S: No study, 15.78 S: Caetano et al. (2020), 19.16 N: Fierro-Estrada et al. (2022), 20.68 N: Martínez-Méndez et al. (2015), 26.93 N: Lara-Reséndiz et al. (2015), 29.55 S: Laspiur et al. (2021), 31.3 N: Lara-Reséndiz et al. (2019), 36.62 N: Jiang et al. (2023), 41.15 S: Kubisch et al. (2016), 42.57 N: Ortega et al. (2017), 53.78 S: Ibargüengoytía et al. (2010), 60.17 N: Herczeg et al. (2006, 2008), 71.08 N: No study)

Werner, 1916; (3) locomotor performance and thermal tolerances in *A. crassicaudatus* and the *A. heterodermus* group; and, finally, (4) the thermal dynamics of *A. bogotensis* nests. Table 1 provides the terminology used throughout this chapter, which is now widely recognized in the study of reptile thermal biology and other ectothermic organisms.

2 Thermoregulation in the Mountains of the Northern Andes

Anolis richteri (referred to as *Anolis heterodermus* before Moreno-Arias et al. 2023) was the first species from the Northern Andes mountains for which the thermoregulation strategy was characterized (Méndez-Galeano and Calderón-Espinosa 2017). This diurnal, arboreal

Table 1 Variables and metrics mentioned throughout this chapter

Variable or metric	Calculation/measure	Definition
Field body temperature (T_b)[a]	Using a fast-read thermometer or thermal imaging cameras	Internal or surface temperature of an organism, reflecting the balance between heat gain and loss
Preferred temperature (T_{pref})[b]	Recording body temperature in thermal gradients. Could be described in terms of an average value or by a given interquartile interval	The body temperature that an ectothermic animal actively selects or maintains under conditions free from ecological or physiological constraints
Operative temperature (T_e)[c]	Obtained from environmental temperature records using biophysical models (null models) previously calibrated with a linear regression	Temperature which an organism would experience in microsites when it was in thermal equilibrium with the environment
Critical thermal minimum (CT_{min}) and maximum (CT_{max})[a]	Usually measuring the upper and lower body temperature at which an organism loses its righting response	Upper and lower body temperature values at which an organism lose their body functions
Thermal tolerance breadth (TTB)[d]	$TTB = CT_{max} - CT_{min}$	Temperature interval within which individuals can perform
Thermal performance curves (TPC)[a]	Constructed by modeling from a physiological trait, which is measured across a range of controlled temperatures, and usually constrained among the thermal tolerance breadth (TTB)	A function that describes the physiological sensitivity to variations in body temperature or the impact of these variation on physiological traits
Optimal temperature (T_{opt})[a]	It can be extracted from a thermal performance curve (TPC)	The temperature at which organisms maximize performance in a measured physiological trait
Habitat thermal quality index (d_e)[b]	$d_e = \lvert T_{pref} - T_e \rvert$ For more details on the calculation, see Hertz et al. (1993)	Estimates how far an organism T_b ′s deviates from a preferred temperature (T_{pref}) interval
Thermoregulatory accuracy index (d_b)[b]	$d_b = \lvert T_{pref} - T_b \rvert$ For more details on the calculation, see Hertz et al. (1993)	Estimates how far an organism T_e ′s deviates from a preferred temperature (T_{pref}) interval
Thermoregulatory effectiveness index (E)[b]	$E = 1 - (d_b/d_e)$ 0 to 1 equal from thermoconformity to thermoregulation	Ectotherm effectiveness in maintaining its T_b's on a T_{pref} interval given a certain habitat T_e's availability
Thermal safety margin (TSM)[e]	$TSM = T_{opt}$ − environmental temperature Small values mean higher susceptibility to slight temperature changes	Vulnerability of an organism to perform optimally with changes in environmental temperature
Warming tolerance (WT)[e]	$WT = CT_{max}$ − environmental temperature Small values mean higher vulnerability or threat to slight temperature changes	"...average amount of environmental warming an ectotherm can tolerate before performance drops to fatal levels" (Deutsch et al. 2008, p. 6669)
Thermal-niche buffer index (TNB)[f]	$TNB = T_{pref} - T_e$ Higher values equal to high buffering capacity	Potential of a species to resist or tolerate environmental disturbances

References:
[a]Taylor et al. (2020)
[b]Hertz et al. (1993)
[c]Dzialowski (2005)
[d]Sunday et al. (2011)
[e]Deutsch et al. (2008)
[f]Ibargüengoytía et al. (2021)

lizard achieves moderate thermoregulation in a low thermal quality environment through heliothermy, primarily between 11:00 a.m. and 1:00 p.m. (Méndez-Galeano and Calderón-Espinosa 2017), at an elevation of 2650 meters in the eastern mountain range of Colombia.

Additionally, the thermoregulation of this species is more pronounced during the rainy season, as body temperatures (T_b) and preference (T_{pref}) remain unchanged compared to the dry season, despite a decrease in operative temperatures (T_e) in the rainy season (Méndez-Galeano and Calderón-Espinosa 2017). Therefore, it seems to be a behavioral compensation mechanism for the effects of rainfall on the species' thermal niche, similar to the behavioral adjustments observed in other *Anolis* species to cope with the impact of elevation on the thermal quality of their environment (Hertz and Huey 1981; Hertz 1981; Hertz et al. 1993; Méndez-Galeano and Calderón-Espinosa 2017).

Méndez-Galeano et al. (2020) subsequently characterize the thermoregulation of *A. heterodermus* and *Anolis tequendama* (referred to as *Anolis heterodermus* before Moreno-Arias et al. 2023) at elevations of 3400 m and 2200 m a.s.l., respectively. Both of these *Anolis* species, along with *A. richteri*, belong to the *A. heterodermus* group (Moreno-Arias et al. 2023) and are diurnal, arboreal lizards native to the eastern mountain range of Colombia. Despite being evaluated at different elevations, the three species share the same preferred body temperature (T_{pref}), all inhabiting regions above 2000 m a.s.l. (Méndez-Galeano and Calderón-Espinosa 2017; Méndez-Galeano et al. 2020). While the body temperature (T_b) of all three species is closer to T_{pref} than to the operative temperatures (T_e), the habitats of *A. heterodermus* and *A. tequendama* are colder than that of *A. richteri*, with T_e being lower and more distant from T_{pref}. This results in less precise thermoregulation for *A. heterodermus* and *A. tequendama*, which adopt a more thermoconformist strategy (Fig. 2a–c; Méndez-Galeano et al. 2020).

Contrary to expectations, the habitat of *A. tequendama* exhibits a lower environmental temperature (T_e) compared to the habitat of *A. richteri*, despite *A. tequendama* being found at a lower elevation (Fig. 2a, b). This discrepancy arises because *A. tequendama* inhabits the Andean cloud forest ecosystem, which is characterized by greater thermal stability, less habitat fragmentation, and lower solar radiation than the Andean shrub ecosystem that defines the habitat of *A. richteri* (Méndez-Galeano et al. 2020). Nevertheless, the thermal fluctuation with respect to elevation follows the expected pattern: as elevation increases, there is a greater range of T_e, indicating greater thermal variability (Fig. 2a–c; Méndez-Galeano et al. 2020).

The subpáramo and páramo are the ecosystems with the highest altitudes and lowest temperatures along the elevational gradient inhabited by species of the *Anolis heterodermus* group. Among these, *A. heterodermus* is one of the species in the group that inhabits these environments. Populations of this species in these regions exhibit the highest thermoregulatory imprecision and the most pronounced thermoconformism within the group (Méndez-Galeano et al. 2020). However, these ecosystems also experience higher solar radiation and significant thermal fluctuations. On sunny days, improved thermal conditions enable this species to engage in the most active thermoregulation observed in the group (Méndez-Galeano et al. 2020). This makes *A. heterodermus* a particularly labile and thermo-opportunistic species, adapting its thermoregulatory strategies dynamically on a day-to-day basis.

The studied populations of *A. richteri* and *A. heterodermus* are located in a specific region of the eastern mountain range of the Colombian Andes known as the Cundiboyacense plateau (Moreno-Arias et al. 2023). In this region, they share their habitat with the thickhead ground snake *Atractus crassicaudatus*. This small snake species (total length: 400–440 mm) exhibits fossorial, crepuscular, and nocturnal habits and is commonly found under shelters such as rocks, bricks, wood, or logs, or actively moving through grass (Paternina-Cruz and Calderón-Espinosa 2022). Research by Paternina-Cruz and Calderón-Espinosa (2022) indicates that during the dry season, *A. crassicaudatus* employs an active thermoregulation strategy, whereas during the wet season, it shifts to thermoconformity. This behavioral change is primarily driven by significant seasonal variations in preferred temperatures (T_{pref}) and fluctuations in thermal availability throughout the day (Fig. 2d, e).

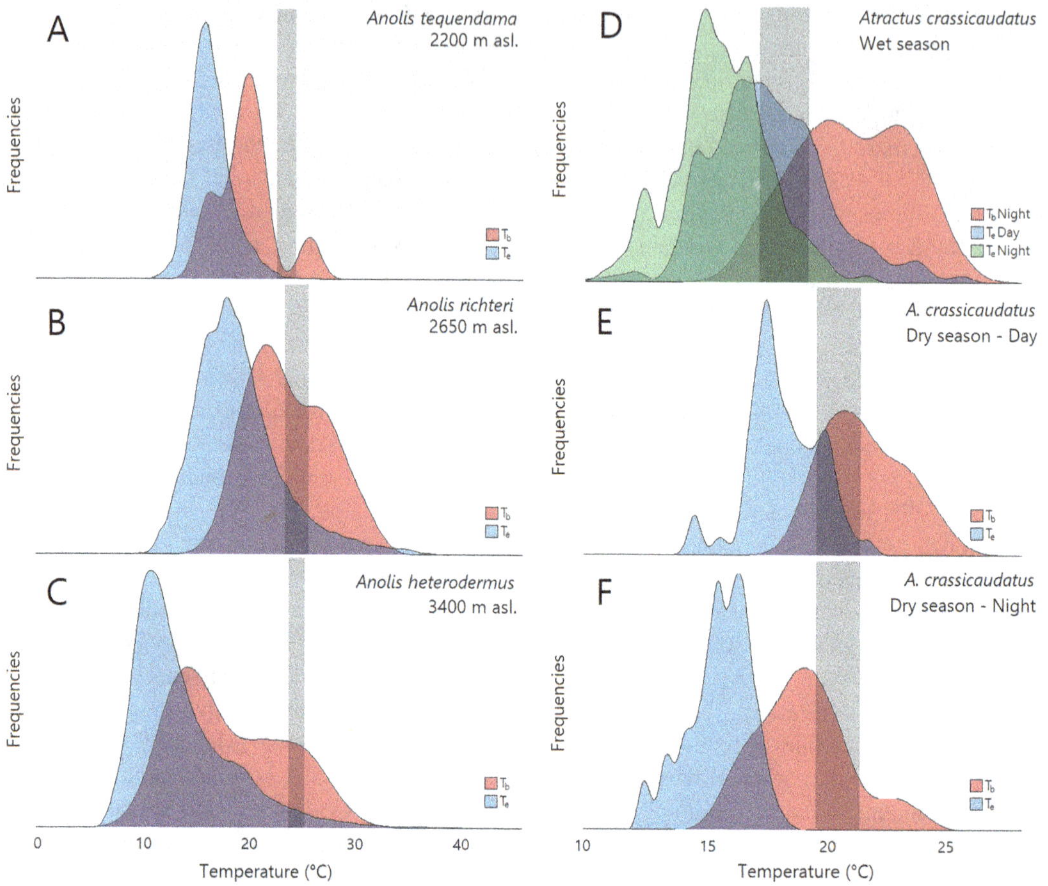

Fig. 2 (**a–c**) Distribution of body (T_b) (red) and operative (T_e) (blue) temperatures for (**a**) *Anolis tequendama*, (**b**) *Anolis richteri*, and (**c**) *A. heterodermus*. Original data were obtained from Méndez-Galeano et al. (2020). (**d–f**) Distribution of bodys (T_b) and operative (T_e) temperatures for the thickhead ground snake *Atractus crassicaudatus*. Original data were obtained from Paternina-Cruz and Calderón-Espinosa (2022). The gray band in each diagram corresponds to the interquartile range (25–75%) of T_{pref}. Note that during the rainy season for *A. crassicaudatus* (**d**), the body temperatures during the crepuscular and nocturnal activity period of the species (in red) mostly coincide with the distribution of operative temperatures during the day (in green) than during the night (in blue), because the refuges where the snake lives provide ideal temperatures during the day and twilight for its crepuscular and nocturnal activity. On the other hand, during the dry season, the body temperatures (in red) and operative temperatures (in blue) of the day (**e**) decrease consistently at night (**f**), with the crepuscular and nocturnal body temperatures in (**f**) aligning with the daytime operative temperatures in (**e**). Finally, the seasonal difference in the range of T_{pref} is evident

This seasonal variation in T_{pref} reveals a mechanism of acclimatization to changing environmental conditions in a short time (from one climatic season to another). This plasticity may be influenced by the availability of refuges, which could be limited during the wet season due to the flooding typical of the highlands. Snakes interact directly with the thermal environment while taking shelter, primarily under rocks or concrete slabs, as observed in the university campus population (Castillo-Rivera and Calderón-Espinosa pers. comm.). The thermal conditions of these microhabitats vary throughout the day and across seasons. For example, during the dry season, microsites exposed to direct sunlight or partial shade provide temperatures that are more stable and closer to the preferred range (T_{pref}) particularly at twilight, when snake activity typically begins. In contrast, during the wet season, these same microenvironments tend to show greater temperature variability, with notably lower temperatures around twilight (Paternina-Cruz and Calderón-Espinosa 2022). Given two key conditions of the wet season—the potential limitation of refuges due to soil flooding and the fluctuating thermal supply with lower temperatures at the onset of activity—these factors could drive a decrease in T_{pref}. However, the influence of reproductive condition and sex on the thermal traits of this species remains unexplored.

The observed plasticity in T_{pref} in the thickhead ground snake contrasts with findings from studies suggesting that thermal traits, such as T_{pref} and CT_{max}, are not influenced by the thermal environment during early embryonic development in reptiles, as indicated by the meta-analysis of Zhang et al. (2023). This is particularly evident in experimental studies conducted on oviparous species. However, these authors suggest evaluating the potential effects of environmental variation on thermal traits at different ontogenetic stages. In *A. crassicaudatus*, the observed seasonal plasticity in T_{pref} in adults indicates that, at least during this ontogenetic phase, individuals adjust their preferences in response to environmental variation. Similarly, studies on lizards have documented seasonal plasticity in preferred temperature (Díaz et al. 2006, Karameta et al. 2023, Domínguez-Guerrero et al. 2019), or variation in this trait in response to thermal conditions in different environments (Pafilis et al. 2019), often interpreted as plasticity. However, Huey et al. (2003) argued that T_{pref} is a conserved trait, due to the Bogert effect (see Sect. 4.1). Indeed, as mentioned earlier, seasonal plasticity in T_{pref} is not characteristic of other species inhabiting highland areas, such as *Anolis* species from the *A. heterodermus* group. In other words, the mechanisms employed by the squamate reptile fauna in the region to adapt to seasonal environmental changes differ. While in the thickhead ground snake there is a seasonal behavioral and physiological adjustment in T_{pref}, in *Anolis* species the adjustment is completely behavioral.

In addition to the seasonal adjustment in T_{pref}, the fluctuation throughout the day of T_b and the environmental thermal offer, recorded in the dry season, allowed to characterize a subtle variation in the thermoregulation strategy between day (photophase) and night (scotophase): moderately active thermoregulation during the day (Fig 2e) and highly active thermoregulation during the night (Fig 2f). Individuals are less accurate in thermoregulation during the day, when thermal quality is better compared to the night (with T_e closer to T_{pref} during the day). However, accuracy in thermoregulation increases at night (with T_b closer to T_{pref}).

Active thermoregulation in nocturnal species has been observed in tropical geckos during their active phase (Norberg and Schwarzkopf 2019) and in geckos from arid regions during both activity and inactivity (Lara-Reséndiz et al. 2013). Additionally, moderate thermoregulation at night has been documented in the nocturnal elapid snake *Hoplocephalus bungaroides* (Webb and Shine 1998). In nocturnal organisms, thermoregulation during the active phase is primarily achieved through conduction from environmental surfaces that retain heat accumulated through daytime radiation. Therefore, the type of refuge plays a significant role in diurnal thermoregulation (Kearney and Predavec 2000; Kearney 2002). The thickhead ground snake may use the refuge's temperature during the afternoon to offset the cooler ambient temperatures (Castillo-Rivera and Calderón-Espinosa pers. comm.; Paternina-Cruz and Calderón-Espinosa 2022). *Atractus crassicaudatus* appears to have a strong dependence on shelters, such as rocks, where it takes refuge during the day. Remaining under such shelters may help protect it from predators such as the Great Thrush (*Turdus fuscater*), which has been observed actively hunting and preying on snakes (Díaz and García-Mieles, 2025). The role of shelters in thermoregulation for nocturnal species, both during periods of activity and inactivity, is thoroughly explored in the referenced studies. Consequently, understanding the thermal dynamics of *A. crassicaudatus'* shelters and their relationship to the species' thermal biology is crucial for explaining the observed variation in its thermoregulation strategy.

It remains unclear whether thermal traits and thermoregulatory strategies in *A. crassicaudatus* vary with geographical distribution, particularly with elevation. However, given the documented high plasticity in T_{pref}, variations in T_{b}, and environmental thermal availability, it is likely that elevation influences the thermal biology of this species, as observed in other squamate reptiles from the Northern Andes. Data on T_{b} from two locations within the Cundiboyacense highlands at similar elevations reveal significant differences, particularly in maximum T_{b} values. This suggests that variations in vegetation cover or other ecological factors may influence the species' thermal biology. Nonetheless, the current information is insufficient to establish definitive geographical patterns for the thickhead ground snake.

We conclude that *Atractus crassicaudatus* exhibits plasticity in thermoregulation across multiple temporal scales, not limited to seasonal changes. Throughout the day, these snakes adjust to temperature variations in microsites, likely influenced by ecological constraints.

Table 2 presents values for various traits and/or thermal biology indices for species of the *Anolis heterodermus* group, the thickhead ground snake *Atractus crassicaudatus*, and two species of the genus *Stenocercus*. A key observation is that the body temperatures of the *Stenocercus* species are significantly higher than those of the *Anolis* lizards or the *Atractus crassicaudatus* snake. For instance, *Stenocercus trachycephalus* has a T_{b} range between 23.5 and 28.9 °C, and *Stenocercus lache* exhibits a body temperature of 29 °C, both at elevations comparable to those of *Stenocercus guentheri* and *Stenocercus festae* (2700–3100 m a.s.l.; Barragán-Forero 2016). This suggests that natural history influences the thermal biology of reptiles inhabiting the Northern Andes mountains. Specifically, terrestrial and heliothermic species tend to have higher T_{b}, while the heliothermic arboreal and semifossorial thigmothermic species exhibit lower T_{b}.

3 Thermal Ecophysiology Traits in the Squamate Reptiles of the Northern Andes Mountains

The critical thermal minimum and maximum (CT_{min} and CT_{max}) in squamate reptiles of the Northern Andes mountains have been recorded only for *Stenocercus festae* and *S. guentheri* (Andrango et al. 2016; Guerra-Correa et al. 2020), species of the *Anolis heterodermus* group (Méndez-Galeano et al. 2020), *Anolis tolimensis* (Pinzón-Barrera et al. 2024), and the thickhead ground snake *Atractus crassicaudatus* (Paternina-Cruz 2016).

In *S. guentheri* and *A. tolimensis*, critical thermal minimum (CT_{min}) and maximum (CT_{max}) have been recorded for adults and juveniles, as well as for adult males and females. Notable differences have been observed between sexes in both species (though only in CT_{min} for *A. tolimensis*) and between age groups in *S. guentheri* (Andrango et al. 2016; Pinzón-Barrera et al. 2024). According to Pinzón-Barrera et al. (2024), the geographical variation in CT_{min} for *A. tolimensis* appears to result from a complex interaction between sex and size. Similarly, Andrango et al. (2016) suggest that ontogenetic differences in CT_{min} and CT_{max} may lead to distinct microhabitat use between juveniles and adults, as well as varying thermal sensitivities to global warming, with juveniles being more vulnerable. These findings indicate that thermal traits are influenced by various factors, including the environmental context encountered by individuals, which can vary not only between populations of the same species but also among individuals within the same population.

Telemeco and Gangloff (2020) stress the importance of describing thermoregulatory behavior and performance curves across all ontogenetic stages while considering environmental variability among populations. Additionally, they highlight the need to quantify the selection pressures and heritability of thermal traits to more accurately predict the effects of thermal variation on population dynamics. Building on this, it is

Table 2 Body (T_b), operative (T_e), and preferred (T_{pref}) temperatures, as well as the thermoregulation indices of Hertz et al. (1993) obtained from the studies of thermoregulation in squamate reptiles of the northern Andes mountains

Species	Locality	T_b (°C) (interval) (N)	T_{pref} (°C) (25–75% interquartile interval) (N)	T_e (°C) (interval)	d_b (°C)	d_e (°C)	E	References
Anolis tequendama	Chicaque (2200 m)	19.8 (15.9–25.8) (10)	23.5 (22.7–24.5) (17)	16.2 (11–33)	3.3	6.5	0.49	Méndez-Galeano and Calderón-Espinosa (2017), Méndez-Galeano et al. (2020)
Anolis ritcheri	Tabio (2650 m)	23.8 (16.6–31.9) (98)	24.6 (23.5–25.7) (33)	19.1 (8.7–38.4)	2.2	5.1	0.57	
Anolis heterodermus	Matarredonda (3400 m)	18.1 (11–28.9) (52)	24.5 (23.8–25.3) (16)	13.7 (5.9–44.3)	6.2	10.4	0.4	
Atractus crassicaudatus	Bogotá (2630 m)	19.98 (15.9–24.9) (108)	19.43 (18.3–20.2) (30)	15.8 (3.1–41.4)	1.65	3.08	0.68	Paternina-Cruz and Calderón-Espinosa (2022)
Stenocercus guentheri	Jerusalem RP (2278 m)	31.5 (24.3–34.7) (17)	31.8 (27–35.1) (27)	20.9	–	–	–	Guerra-Correa et al. (2020)
	Calacalí (2950 m)	32.3 (25.1–35.6) (26)	22.6 (19.8–24.2) (26)	16.9	–	–	–	
Stenocercus festae	La Paz (3100 m)	28.2 (16–34.8) (20)	22.7 (14.5–31.9) (20)	13.2	–	–	–	

The range of values is shown for each variable, as well as the sample size. The study locations are also shown also crucial to characterize thermal traits separately for males and females, as seen in *A. tolimensis*. This differentiation can help elucidate how thermal traits respond to variables such as elevation and associated climatic and environmental factors.

Additionally, as noted by Telemeco and Gangloff (2020), the logistical challenges of recording thermal traits in the embryonic stage likely contribute to the scarcity of studies on this topic. This underscores the importance of characterizing the thermal environments where embryos develop (see Sect. 5).

With respect to elevation, the *A. tolimensis* population at higher altitude (2200 m a.s.l.) exhibits lower CT_{min} values compared to populations at lower elevations. Furthermore, Guerra-Correa et al. (2020) find differences in CT_{min}, CT_{max} and T_{pref} between two populations of *S. guentheri* at different elevations (2278 and 2950 m a.s.l.) and a population of *S. festae* at 3100 m. These differences appear to be influenced by both elevation and phylogenetic constraints. In this last study, T_{pref} had a considerably higher value in the lower population (2278 m a.s.l.), CT_{max} differed between species, being higher in *S. festae*, and CT_{min} was lower in the two highest elevations (2950 and 3100 m a.s.l.).

A markedly different scenario is observed in species of the *Anolis heterodermus* group, where no interspecific or elevational differences are found in CT_{min}, CT_{max}, T_{pref} or in locomotor performance as a function of the T_b (Méndez-Galeano and Calderón-Espinosa 2017; Méndez-Galeano et al. 2020). This contrasts particularly with the elevational variation in CT_{min} in other species of the genus *Anolis* (Muñoz et al. 2014; Pinzón-Barrera et al. 2024). Méndez-Galeano et al. (2020) suggest that this may result from thermal niche conservatism, where the climatic conditions during glacial periods in the Northern Andes led the ancestors of these species to adapt to low temperatures (lower CT_{min}). During interglacial periods, this trait would be preserved due to the lack of selective pressure to increase CT_{min}. However, this hypothesis requires evaluation through phylogenetic studies (Muñoz et al. 2014) and consideration of the intrapopulation variability of these traits, as well as the phenotypic plasticity (acclimatization) they may exhibit (reviewed in Bodensteiner et al. 2020).

It remains unclear why thermal preference and locomotor performance as a function of body temperature do not vary between species of the *Anolis heterodermus* group. This could be attributed to thermal niche conservatism, as mentioned above, or to the strictly arboreal habits and slow movements of these species. Méndez-Galeano and Calderón-Espinosa (2017) also note that *A. richteri*, despite showing no differences in T_{pref} between sexes, ages, or climatic seasons, exhibits considerable intrapopulation variation in this trait. This could indicate a form of phenotypic plasticity, which, while not yet fully understood, may significantly influence the thermoregulation of this species. This same phenomenon could also be occurring in *A. heterodermus* and *A. tequendama*.

The thermal performance curves (TPC) for *Anolis* species in the *heterodermus* group were estimated based on locomotion (sprint speed) and displayed the typical left-skewed shape of such curves. Performance dropped more sharply as temperatures increased, although it still declined when temperatures deviated from the optimal performance temperature (T_{opt}), which ranged between 24.9 and 25.2 °C for the three species (Fig. 3).

On the other hand, Paternina-Cruz and Cristancho (com. pers.) characterized the TPC of the species using the ventilation rate, which is derived from O_2 consumption. The T_{opt} value extracted from this curve was 12 °C, much lower than that observed in the species of the *Anolis heterodermus* group. There is no available data on possible elevational variation in CT_{min}, CT_{max}, or T_{pref} for *A. crassicaudatus*. However, given the seasonal plasticity in thermal preferences and the species' high dependence on the thermal dynamics of its refuges, it is highly likely that both thermal preference and critical temperatures vary in relation to altitudinal distribution.

Table 3 presents a range of thermal ecophysiological traits for squamate reptiles from the

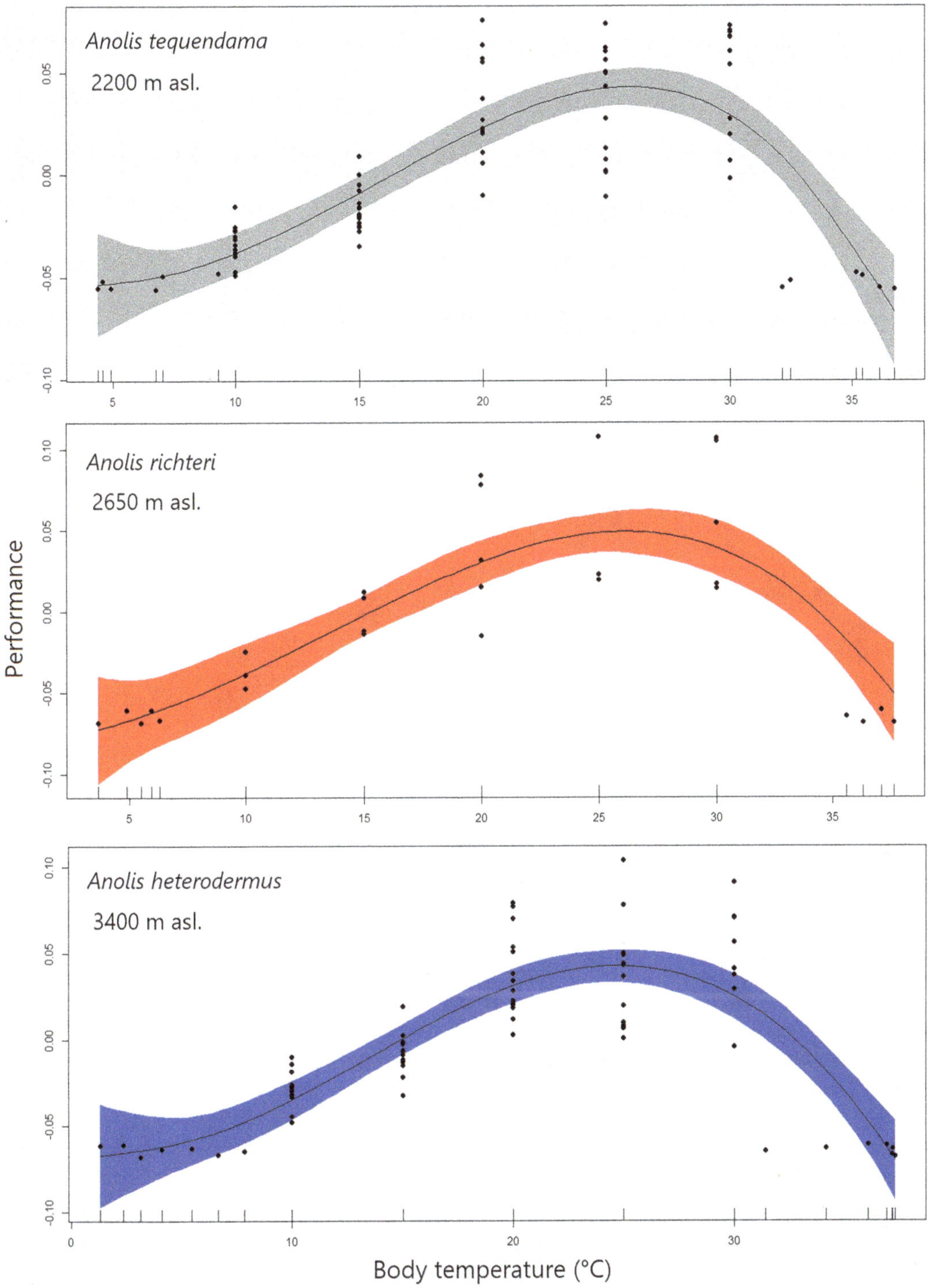

Fig. 3 Thermal performance curves (TPCs) based on locomotion for the species of the *Anolis heterodermus* group. Note the values of the critical thermal minimum (CT_{min}) and maximum (CT_{max}) at the ends of the curves. T_{opt} *A. tequendama* = 25.2 °C; T_{opt} *A. richteri* = 25.2 °C, T_{opt} *A. heterodermus* = 24.9 °C. (Original data were obtained from Méndez-Galeano et al. (2020))

Table 3 Critical thermal minimum (CT_{min}) and maximum (CT_{max}), thermal tolerance range (TTB), thermal safety margin (TSM), warming tolerance (WT), and thermal niche buffer (TNB) obtained from thermoregulation studies in squamate reptiles from the northern Andes mountains

Species	Locality	CT_{min} (°C) (*N*)	CT_{max} (°C) (*N*)	TTB (°C)	TSM (°C)	WT (°C)	TNB (°C)	References
Anolis tequendama	Chicaque (2200 m)	6.2 (6)	34.7 (6)	28.5	**9**	**18.5**	**7.3**	Méndez-Galeano et al. (2020)
Anolis ritcheri	Tabio-Gachancipá (2650 m)	5.3 (6)	36.7 (6)	31.4	**6.1**	**17.6**	**5.5**	
Anolis heterodermus	Matarredonda (3400 m)	4.4 (7)	35.8 (7)	31.4	**11.2**	**22.1**	**10.8**	
Anolis tolimensis	Chicaque (2200 m)	5.5 (9)	34.2 (9)	28.7	–	**12.2**	–	Pinzón-Barrera et al. (2024)
	Anzoátegui (2100 m)	8.8 (10)	33 (10)	24.2	–	**16**	–	
	Subia-Silvania (1900 m)	8.3 (8)	34.2 (16)	25.9	–	**12.2**	–	
	Llanitos (1600 m)	8.3 (7)	33.9 (6)	25.6	–	**11.9**	–	
Atractus crassicaudatus	Sibaté/UN Campus (2550 m)	3.9	–	–	**−3.8**	–	**3.6**	Paternina-Cruz and Calderón-Espinosa (2022)
Stenocercus guentheri	Jerusalem RP (2278 m)	7.9 (25)	43.8 (25)	35.9	26.2	19.2	**20.9**	Guerra-Correa et al. (2020)
	Calacalí (2950 m)	6.5 (26)	44.4 (26)	37.9	18.2	27.3	**5.7**	
Stenocercus festae	La Paz (3100 m)	5.6 (21)	46.8 (21)	41.2	21.5	39.9	**9.5**	

The sample size for CT_{min} and CT_{max} is shown. Values not directly extracted from the studies, but calculated from the data within them, are shown in bold. For the species of the genus *Stenocercus*, the WT and TSM values were calculated by Guerra-Correa et al. (2020) from only the operative temperature (T_e) average of models located under shade during the hottest hours of the day and from the T_e minimum hourly average of models located under shade, respectively, while for the species of the *Anolis heterodermus* group and for *Atractus crassicaudatus*, they were calculated using the T_e total average in the period of activity of the species; for all cases, the T_e was used as a proxy for the environmental temperature (see Guerra-Correa et al. (2020)). For the species of the *Anolis heterodermus* group and for *A. crassicaudatus*: $TSM = T_{opt} - T_e$; for the species of the genus *Stenocercus*: $TSM = T_{pref} - T_e$ (see details of the use of T_{pref} as a proxy for T_{opt} in Guerra-Correa et al. (2020)); for the species of the *Anolis heterodermus* group and the species of the genus *Stenocercus*: $WT = CT_{max} - T_e$; for *Anolis tolimensis*: $WT = CT_{max}$ − environmental temperature

Northern Andes mountains, including new metrics for certain species. It provides values for the thermal safety margin (TSM) and warming tolerance (WT) in *Stenocercus* species and the *Anolis heterodermus* group, the TSM for *Atractus crassicaudatus*, and the thermal niche buffer (TNB) index for all species except *A. tolimensis*. Nearly all the evaluated lizards exhibit positive and high values for these parameters, indicating that they can tolerate temperature increases without reaching lethal body temperatures (WT), experience enhanced physiological performance with rising temperatures (as they inhabit colder environments relative to their thermal physiology, TSM), and possess a high capacity to withstand environmental thermal fluctuations (TNB).

In contrast, *A. crassicaudatus* and *A. richteri* have a moderate capacity to withstand thermal fluctuations in their environment (TNB = 3.63–5.5 °C). Importantly, *A. crassicaudatus* shows a low but negative TSM value (−3.8 °C), indicating that, with its environmental temperature (T_e) being higher and closer to its T_{opt}, this species may experience a decline in physiological performance as environmental temperatures increase. This finding is particularly interesting, as it highlights a clear difference in these parameters between the evaluated snake and the lizard species.

4 Patterns of Thermal Biology in Squamate Reptiles from the Northern Andes Mountains and Other Regions of the World

How does the thermal biology of squamate reptiles from the Northern Andes align with patterns observed in other species worldwide? What similarities and differences can be found in their thermal biology when compared to other types of herpetofauna? To address these questions, we must first examine how these organisms have adapted to environmental temperature conditions globally. As mentioned in Sect. 1, temperature decreases with elevation and toward extreme latitudes, while in temperate zones, it shows marked seasonal variation. In contrast, in tropical regions, temperature variation occurs primarily during the day (Fig. 1). Several hypotheses have been proposed to explain how these temperature patterns influence life on Earth, particularly the thermal biology of organisms like reptiles. We will begin this section by addressing two key hypotheses supported by different lines of evidence: the Bogert effect and the Janzen hypothesis.

4.1 The Bogert Effect and the Janzen Hypotheses

The first hypothesis, known as "the Bogert effect," proposes that certain thermal physiological traits in ectotherms remain unchanged in evolutionarily related species, even when they inhabit vastly different environments or appear not to align with the environmental conditions of their habitats (Huey et al. 2003). In essence, this hypothesis suggests that the ability of ectothermic organisms to regulate their body temperature behaviorally acts as a buffer, reducing the need for physiological adaptation (Huey et al. 2003).

On the other hand, "the Janzen hypothesis" (Janzen 1967) posits that in tropical regions, the lack of pronounced thermal seasonality means that temperature regimes remain relatively stable throughout the year. Consequently, different elevations within the tropics exhibit distinct temperature regimes, promoting patterns of isolation, zonation, and diversification (Janzen 1967; Ghalambor et al. 2006), making this hypothesis one of the main explanations for the higher biodiversity observed in tropical regions compared to temperate zones (Ghalambor et al. 2006).

In this context, the Bogert effect suggests that closely related reptile species inhabiting different environmental conditions would not exhibit substantial differences in their thermal traits. Conversely, Janzen's hypothesis predicts that reptiles from tropical regions would display thermal traits specific to particular elevations, with narrow thermal tolerance breadths (TTB) and little to no overlap between species from different elevations, as well as low or negligible warming tolerances (WT). In contrast, reptiles from temperate zones would exhibit wider TTBs, partial or total overlap between species with different altitudinal ranges, and higher WTs. As we will explore, the validity of these predictions in squamate reptiles depends on the specific thermal traits considered and the unique characteristics of each region, including the mountains of the Northern Andes. These will serve as a starting point for comparison, beginning with thermoregulation patterns.

4.2 Patterns in Thermoregulation Strategies

The behavioral strategies observed in species of the *Anolis heterodermus* group and the thickhead ground snake (*Atractus crassicaudatus*) range from thermoconformity to active thermoregulation. In both lizards and the snake, individuals employ behavioral adjustments that are plastic across different temporal scales and respond rapidly to environmental variations. However, this plasticity is exclusively behavioral in the *Anolis heterodermus* species group, whereas it involves both behavioral and physiological components in the thickhead ground snake.

For both cases, the thermal environment predominantly consists of low temperatures, offering limited opportunities for thermoregulation, particularly for diurnal and heliothermic lizard species. In *Anolis* species, individuals actively seek temperatures close to their preferred body temperature (T_{pref}), which, on average, are higher than the ambient temperatures available. In other words, lizards strive to warm themselves despite limited opportunities to do so. However, when environmental conditions significantly reduce the availability of temperatures near their T_{pref}, individuals shift to thermoconformity as an alternative strategy.

Climatic variation in the mountains of the northern Andes occurs on multiple temporal scales: throughout the day, between different days, and across distinct climatic seasons. Consequently, the observed behavioral plasticity aligns with at least two temporal scales: between cold, cloudy days and not so cold, clear days, as well as between dry and rainy seasons.

In the lizards of the *Anolis heterodermus* group, the variation in thermoregulatory strategies is more pronounced between days, as seen in *A. heterodermus*, than between seasons, as observed in *A. richteri*. This pattern corresponds to the fact that, in the tropical mountains of the northern Andes, daily thermal fluctuations are greater than monthly fluctuations (Fig. 1).

Behavioral plasticity in thermoregulation strategies has also been observed in other lizard species, including the tropical nocturnal gecko *Gehyra dubia* (Nordberg and Schwarzkopf 2019). In this species, individuals actively thermoregulate during the night and early morning hours when the available operative temperatures differ significantly from their thermal preferences, by seeking out warm microclimates. However, when daytime temperatures align more closely with their thermal preferences, the geckos shift their strategy to thermoconformity.

At night, in the high-elevation environments of the tropical Andes, operative temperatures drop to levels well below the preferred temperatures of both diurnal and crepuscular-nocturnal species (Méndez-Galeano et al. 2020; Paternina-Cruz and Calderón-Espinosa 2022). However, the fossorial and thigmothermic nature of the thickhead ground snake (*Atractus crassicaudatus*), combined with its lower preferred temperature (T_{pref}) compared to the diurnal lizards of the region, enables it to engage in active thermoregulation even during the night.

In this species, the thermal dynamics of its refuges play a critical role, allowing the snake to maintain its body temperature close to its preferred range, even as external temperatures drop significantly (Fig. 2e, f). Moreover, the variation in T_{pref} characteristic of *A. crassicaudatus* supports a plastic thermoregulation strategy that adjusts to changes in thermal availability across different climatic periods. In contrast, the plasticity of its thermoregulatory behavior over the course of a single day arises from variations in both the environmental thermal offer and the snake's body temperature.

Compared to high-elevation environments in temperate zones, thermoregulation strategies in some species show certain similarities. For example, diurnal species such as *Iberolacerta aurelioi* (Ortega et al. 2017), certain *Liolaemus* species from the Puna (Valdecantos et al. 2013), and the gecko *Quedenfeldtia trachyblepharus* (Bouazza et al. 2016) inhabit cold environments with low thermal quality, yet they exhibit high effectiveness in thermoregulation. For *Quedenfeldtia trachyblepharus*, thermoregulation effectiveness is highest at the beginning of the activity season, when thermal quality is lower. This contrasts with species from the *A. heterodermus* group, whose individuals, as previously mentioned, tend to thermoconformism when the thermal availability significantly decreases.

At low-elevation sites, thermoregulation also becomes a challenge. In extremely cold, high-latitude environments with average temperatures below the preferred temperature (T_{pref}), as observed in *Liolaemus sarmientoi* and *Liolaemus magellanicus* (Ibargüengoytia et al. 2010), as well as *Darevskia praticola* (Ćorović and Crnobrnja-

Isailović 2018), individuals exhibit limited thermoregulatory capabilities and tend to adopt a thermoconformist strategy. Specifically, in this case, the thermoregulation strategy is like that observed in species of the *A. heterodermus* group. In contrast, low-elevation environments in temperate zones, such as deserts and arid island areas, present different thermoregulatory challenges. In these environments, individuals actively regulate their body temperature to prevent overheating from high daytime temperatures. This is seen in species like *Callisaurus draconoides* (Pérez-Delgadillo et al. 2021), *Phrynosoma cornutum* and *Phrynosoma modestum* (Lara-Reséndiz et al. 2015), *Liolaemus scapularis* (Salva et al. 2021), and *Podarcis erhardii* (Pafilis et al. 2019). Additionally, individuals in these environments may employ other behavioral strategies, such as increasing panting frequency (Salva et al. 2021; Villavicencio et al. 2007; Martori et al. 2002).

Thermoregulation strategies in high-elevation tropical species have primarily been studied in those distributed toward the more extreme latitudes of the tropical belt. As shown in Fig. 1, these regions experience a degree of thermal seasonality, distinguishing them from areas closer to the equator, such as the northern Andes. In tropical mountain species such as *Barisia imbricata* (Fierro-Estrada et al. 2019), altitudinal variation in thermoregulation strategies has been observed at sites with very low thermal quality (2200 and 3700 m). When thermal conditions slightly improve, these species tend to engage in more active thermoregulation. Similarly, *Sceloporus grammicus* exhibits active thermoregulation at elevations between 2600 and 4150 m a.s.l. However, as thermal quality improves at higher elevations, its thermoregulatory efficiency slightly decreases without completely losing its regulatory behavior (Díaz de la Vega-Pérez et al. 2019).

Finally, in low-elevation tropical environments, species exhibit a range of strategies from thermoconformity to active thermoregulation. This variation appears to be influenced by changes in operative temperatures; however, other ecological factors, such as interspecific competition, predation, and species' natural history, also play a key role in shaping their behavioral responses to thermal conditions. For example, Piantoni et al. (2019) found that in certain *Tropidurus* species from the Cerrado, Brazil, operative temperature quality varies between forested sites and urban areas. Similarly, the thermoregulatory efficiency of *Tropidurus* populations decreases as their distribution approaches the equator, where they inhabit both urban and forested environments. In these tropidurines, variation in thermoregulatory behavior occurs alongside physiological adjustments, such as shifts in the lower and upper limits of T_{pref}. The combination of factors such as the thermal heterogeneity of microsites and high environmental temperatures lead to different strategies and in some cases to changes in the activity schedule, in order to avoid overheating. Species like the Komodo dragon exhibit varying responses to the thermal environment depending on body size and thermal availability (Harlow et al. 2010). In forest environments, where thermal conditions are more favorable, individuals tend to be thermoconformers—both small and large individuals—while medium-sized individuals exhibit moderate thermoregulation. In all cases, medium-sized dragons are more efficient at thermoregulating, confirming the influence of specific traits on behavioral adjustment to the thermal environment.

As seen in other regions and elevations, thermoregulation strategies in the tropical lowlands fluctuate throughout the year, in response to climatic seasons. This is exemplified by *Takydromus kuehnei* (Tao et al. 2021), where individuals shift from thermoconformity to highly efficient thermoregulation in response to changes in thermal availability.

In species with crepuscular and/or nocturnal activity, such as the thickhead ground snake, it is difficult to identify clear patterns in thermoregulation. While many snake species exhibit nocturnal behavior (Huey 1982), there are few studies that characterize their thermoregulation strategies (Webb and Shine 1998; Anderson et al. 2005). In the nocturnal elapid snake *Haplocephalus bungariodes* in Australia, Webb and Shine (1998) estimate that individuals transition from thermoconformity to moderate thermoregulation in the afternoon, with thermoregulatory efficiency increasing at night during the snake's activity period. Throughout most of the day, the snakes avoid temperatures near their preferred range, but as evening approaches, they seek to raise their body temperature in preparation for their active phase. Similarly, in *A. crassicaudatus*, a shift in thermoregulation strategy occurs throughout the day, with thermoregulatory efficiency increasing toward the night. In both of these species, the refuge plays a key role in shaping thermal dynamics during both day and night.

As observed in other nocturnal squamate reptiles, such as the gecko *Christinus marmoratus* (Kearney 2002), seasonal variation in environmental temperature directly influences refuge selection and availability. This has also been suggested for the thickhead ground snake, where behavioral and/or physiological adjustments to seasonal changes are closely linked to the thermal conditions experienced while the snake is hidden beneath rocks or logs.

As demonstrated, the behavioral and physiological adjustments that define a species' thermoregulation strategy are the result of the interaction between various factors, including the species' environment and specific natural history traits. While some environments, such as forests, have less variable operative temperatures, other ecological factors and species-specific characteristics interact with the thermal environment, potentially limiting thermoregulation, as seen in *Tropidurus*. This observation leads to the conclusion that variation in behavioral and physiological responses to the thermal environment—across different elevations and in both temperate and tropical regions—is likely the norm, rather than the exception, in squamate reptiles.

4.3 Thermal Tolerances and Temperature Variations Across Latitude and Elevation

Research on the variation of thermal ecophysiological traits in ectotherms with increasing elevation or latitude has primarily focused on warming tolerance (WT) as well as the critical thermal minimum (CT_{min}) and maximum (CT_{max}), which, in turn, has been interpreted in terms of the thermal tolerance breadth (TTB). Particularly, the studies conducted by Sunday et al. (2011, 2014, 2019) show that CT_{max} does not tend to vary with latitude or elevation in ectotherms, while CT_{min} decreases with increasing latitude and elevation in terrestrial ectotherms, and WT decreases with decreasing elevation and latitude in reptiles.

From these trends, we can make additional conjectures. For example, the fact that CT_{max} does not change significantly with elevation or latitude, but WT does, suggests that it is the environmental temperatures (T_{max}) or maximum operative temperatures ($T_{e,max}$) that are driving the observed pattern in WT. This is due to the fact that environmental temperatures are higher at lower elevations and latitudes.

On the other hand, although CT_{max} generally does not show latitudinal or altitudinal variation, CT_{min} does, which implies that the thermal tolerance range (TTB) also increases with elevation or latitude. It is worth clarifying that the calculation of WT in Sunday et al. (2014) was based on maximum operative temperatures, resulting in negative values. This indicates that, to this day, reptiles in general do not exhibit significant tolerance to warming and face a high thermal risk in the tropics.

A decrease in WT values as latitude decreases aligns with other studies highlighting the vulnerability of tropical reptile species, which appears to be associated with narrower TTBs, but especially with higher environmental temperatures in the tropical region (Deutsch et al. 2008; Tewksbury et al. 2008). Similarly, the increase in TTB with increasing latitude is consistent with the expectations of Janzen's hypothesis. However, the work of Sunday et al. (2014, 2019) shows that, since changes in TTB arise from variations in CT_{min} rather than CT_{max}, this implies that the TTBs of species at different elevations in the tropical region overlap significantly, contradicting the prediction of Janzen's hypothesis. This is clearly seen in the work of Muñoz et al. (2014), where body temperature (T_b) and

CT_{max} do not vary intra- or interspecifically in species of *Anolis* from the *cybotes* group as elevation increases, but of CT_{min} values do decrease. In this regard, Fig. 4 compiles some thermal tolerance data from neotropical lizards, including data from the work of Muñoz et al. (2014), as well as species from the *Anolis* and *Stenocercus* genera in the mountains of northern Andes, among others, where this pattern can be observed.

In this regard, Muñoz and Bodensteiner (2019), based again on species of *Anolis* from the *cybotes* group, elegantly explain that, due to a clear elevational stratification associated with low environmental temperatures (e.g., minimum operative temperatures of the *Anolis cybotes* group), this has had an effect on the evolution of CT_{min}, which has clearly generated zonation in the altitudinal distribution of tropical reptile species, thereby reaffirming Janzen's hypothesis. However, they also clarify that this stratification in temperature mainly occurs at night, while diurnal environmental temperature ranges overlap across different elevations. As mentioned earlier, it is important to note that daily variations in environmental temperatures in the high tropical zone can match the magnitude of seasonal variation in temperate regions (Ghalambor et al. 2006) (Fig. 1). In this context, Muñoz and Bodensteiner (2019) add that adaptation to specific climatic conditions may occur only in certain traits like CT_{min}, while other traits may be less labile or more conserved, as seen with T_b or CT_{max}. This is because, with the broad thermal availability generated by environmental temperature variation, regardless of elevation, reptiles will make behavioral adjustments to thermoregulate, thus reducing selective pressure on traits like T_b or CT_{max}, as postulated by the Bogert effect. In contrast, at night, when behavioral adjustments cannot be made during inactivity hours, traits like CT_{min} in diurnal species will show adaptations to cold environments, such as those at high elevations or extreme latitudes.

Figure 5 shows the TTBs of several species of squamate reptiles at different elevations and latitudes. The previously mentioned patterns are evident, such as lower CT_{min} in species from high elevations or extreme latitudes, following the variation pattern of environmental temperature across the planet (Fig. 1). A similar magnitude can also be observed between the intervals of species from high tropical elevations and species

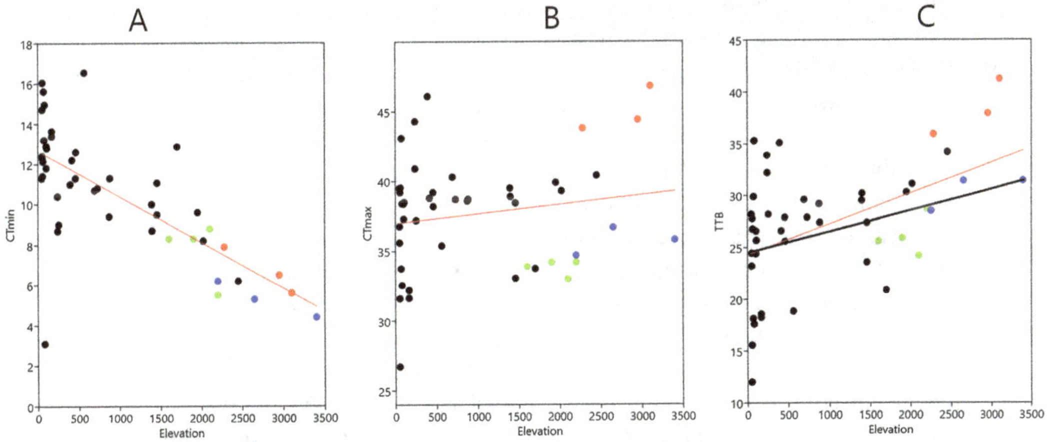

Fig. 4 Variation of the CT_{min} (**a**), CT_{max} (**b**), and TTB (**c**) with respect to elevation for squamate reptiles in the neotropics. Blue points: group *Anolis heterodermus* (Méndez-Galeano et al. 2020); green points: *Anolis tolimensis* (Pinzón-Barrera et al. 2024); red points: *Stenocercus guentheri* and *S. festae* (Guerra-Correa et al. 2020); black points: other works (Muñoz et al. 2014; Van Berkum et al. 1986; Arenas-Moreno et al. 2018, 2021; Muñoz-Nolasco et al. 2023). Regression trends are shown. The relationship shown in B is not significant. In C, the red trend line corresponds to the relationship taking into account all the data, while the black line corresponds to the relationship without taking into account the data from *S. guentheri* and *S. festae*

from high elevations in temperate regions, as well as species from extreme latitudes. This aligns with the equivalence between the daily temperature variation in high tropical zones and the seasonal temperature variation in temperate zones (Fig. 1). However, it is important to note that this may also be due to reptiles in high temperate zones or extreme latitudes becoming inactive due to brumation during winter (Vitt and Caldwell 2014). On the other hand, we observe that the variation in CT_{max} does not show a clear relationship with elevation and latitude (Fig. 5), likely due to behavioral adjustments in line with the Bogert effect. Information on the species shown in Fig. 5 can be found in Table 4.

In Fig. 5, it is noteworthy that the natural history of organisms could also influence these patterns. As observed, most of the lizards whose thermal biology traits have been characterized are terrestrial and diurnal. However, *Anolis heterodermus* (Fig. 5 and Table 4C), while diurnal, are arboreal. On the other hand, *Xantusia sherbrookei* (Fig. 5 and Table 4H) is a nocturnal desert-dwelling species. Despite inhabiting high mountains, *A. heterodermus* exhibits lower CT_{max} values compared to other diurnal, terrestrial lizards from high mountain regions (e.g., *Stenocercus*, *Liolaemus*, *Phrynocephalus*). Similarly, although *X. sherbrookei* is a desert-dwelling terrestrial species, it appears to have a lower CT_{max} and a narrower TTB compared to, for example, *Callisaurus draconoides* (Fig. 5 and Table 4J), a diurnal desert-dwelling terrestrial species. This is relevant, as it reflects how distinct highland tropical arboreal lizards are, inhabiting vegetative covers such as forests or shrubs, which provide cooler temperature conditions (Huey and Webster 1976; Huey et al. 1989, 2009). Similarly, nocturnal species engage in their activities at lower temperatures compared to diurnal species. In both cases, it is possible that cold environments could be removing a potential selective pressure to increase tolerance to high temperatures. Additionally, it is interesting to consider this pattern from the perspective of diurnal terrestrial species from highland areas, as high CT_{max} values imply a wider TTB. This can be seen in the case of *Stenocercus guentheri* and *S. festae* (Fig. 4C), where, although the general trend of an increase in TTB as CT_{min} decreases with elevation remains, these species accentuate this pattern with high CT_{max} values.

The squamate reptiles of the northern Andes mountains, such as species from the *Anolis* and *Stenocercus* genera, as well as reptiles from tropical highlands in general (*Barisia imbricata*, *Liolaemus aparicioi*), exhibit wide thermal tolerance ranges (TTB) and low CT_{min} values. These are comparable to those seen in highland species from temperate zones (*Liolaemus*, *Phrynocephalus*) or species from extreme latitudes (*Zootoca vivipara*, *Liolaemus lineomaculatus*).

Diurnal terrestrial species from highlands, regardless of latitude, typically exhibit high CT_{max} values. It is also interesting to note that the variation in TTB and CT_{max} associated with elevation occurs even at mid-latitudes (Fig. 5), aligning with the global temperature pattern (Fig. 1). This may be due to the fact that many regions within this belt of the planet experience the driest and hottest conditions, such as the desert areas of Africa, North America, and Australia (Prăvălie 2016; Huang et al. 2016).

It is also important to remember that traits such as T_{pref}, CT_{min}, or CT_{max} may exhibit phenotypic plasticity in response to seasonal variations (Gvoždík and Castilla 2001; Domínguez-Guerrero et al. 2019). Finally, it is crucial to emphasize the need for a deeper understanding of these patterns, considering factors such as phenotypic plasticity, and the inclusion of new evidence from high-elevation arboreal species, fossorial or semifossorial species, and nocturnal species. Additionally, it is essential to go beyond thermal tolerances and expand the knowledge on metrics like T_{pref}, TPC, TNB, and vulnerability indices such as TSM and WT.

Despite the many similarities between the squamate reptiles of the northern Andes mountains and other species from high-altitude or extreme-latitude regions, an important difference is often overlooked: oviparity. Unlike high-elevation species from genera like *Liolaemus*, *Sceloporus*, and *Phrynocephalus*, or lizards from extreme latitudes such as *Zootoca vivipara* and

Table 4 Some representative studies on thermal tolerance in squamate reptiles on the planet, taking into account the variation in altitude and latitude, indicating for each one the species, the country, the location, the latitude, the elevation, the critical thermal minimum (CT_{min}) and maximum (CT_{max}), the range of thermal tolerance (TTB), and the reference

Letter	Species	Country	Locality	Latitude (N or S)	Elevation (meters)	CT_{min} (°C)	CT_{max} (°C)	TTB (° C)	References
A	*Stenocercus guentheri*	Ecuador	Calacalí	0	2950	6.5	44.4	37.9	Guerra-Correa et al. (2020)
B	*Mabuya nigropunctata*	Brazil	Caxiuanã National Forest, Pará	1°44′ S	35	13.2	43.1	29.9	Diele-Viegas et al. (2018)
C	*Anolis heterodermus*	Colombia	Ecological Park Matarredonda, Choachí, Cundinamarca	4°33′ S	3350	4.4	35.8	31.4	Méndez-Galeano et al. (2020)
D	*Kentropyx calcarata*	Brazil	Canguçu Research Center, Pium, Tocantins	9°56′ S	150	12.8	43.18	30.38	Pontes da Silva et al. (2018)
E	*Liolaemus aparicioi*	Bolivia	Taypichullo, Mecapaca, La Paz Valley, La Paz	16°38′ S	3000	6.41	44.73	38.32	Miranda-Calle et al. (2021)
F	*Sceloporus macdougalli*	Mexico	Santa Cruz Bamba, Oaxaca	16°01′ N	60	12.13	39.54	27.41	Muñoz-Nolasco et al. (2023)
G	*Barisia imbricata*	Mexico	Paso de Cortés, Mexico state	19°05′ N	3700	1.89	37	35.11	Fierro-Estrada et al. (2022)
H	*Xantusia sherbrookei*	Mexico	Oasis of La Purísima, South Baja California	26°11′ N	89	12.87	37.48	24.61	Arenas-Moreno et al. (2021)
I	*Liolaemus ruibali*	Argentina	Don Carmelo Reserve, Ullum, San Juan	30°56′ S	3166	6.62	44.65	38.03	Gómes Alés et al. (2018)
J	*Callisaurus draconoides*	Mexico	San Felipe Desert, Baja California	31°11′ N	72	13.26	43.3	30.04	Pérez-Delgadillo et al. (2021)
K	*Phrynocephalus vlangalii*	China	Maduo, Qinghai	34°55′ N	4250	1.1	45.8	44.7	Wu et al. (2018)
L	*Psammodromus algirus*	Spain	Navacerrada, Madrid	40°43′ N	1200	6.8	43.1	36.3	Herrando-Pérez et al. (2019)
M	*Zootoca vivipara*	Republic of Belarus	Radun, Vóranava, Grodno	49°53′ N	600	1.9	43.9	42	Gvoždík and Castilla (2001)
N	*Liolaemus lineomaculatus*	Argentina	El Calafate, Lago Argentino, Santa Cruz	50°15′ S	450	2.97	42.4	39.43	Cecchetto et al. (2020)

The letter of each study corresponds to those presented in Fig. 5

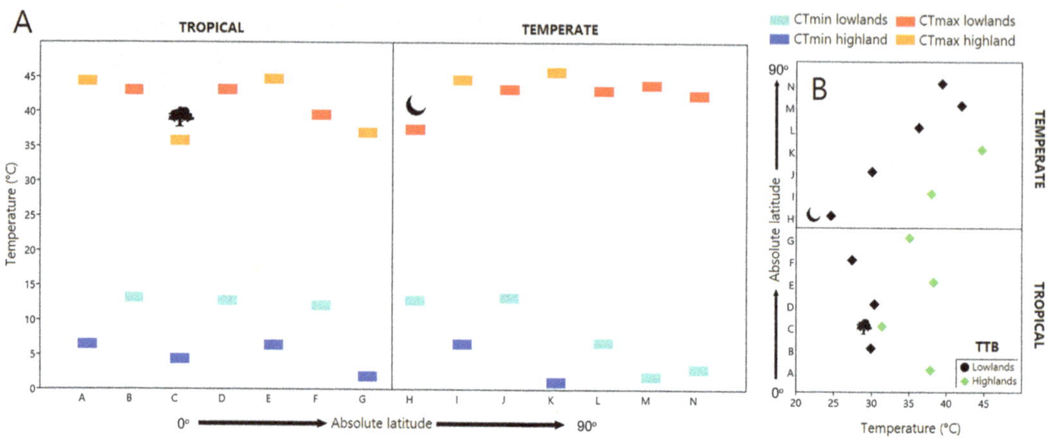

Fig. 5 Values of CT_{min} and CT_{max} (**a**), as well as TTB (**b**) of some species of lizards in high and lowlands, as well as at different latitudes. The information for each letter (A–N) is shown in Table 4. The distinction is made between the tropical and temperate zone. The moon and tree icons correspond to the nocturnal and arboreal species, respectively

Liolaemus magellanicus, species from tropical highlands like those in the genera *Stenocercus* and *Anolis* are exclusively oviparous (Torres-Carvajal 2007; Vitt and Caldwell 2014; Pruett et al. 2022). Viviparity is known to be a reproductive strategy in response to harsh environmental conditions, particularly low temperatures (Wang et al. 2014; Ma et al. 2018; Zimin et al. 2022). This raises the question: Why do not species in tropical highlands, such as those in the northern Andes, employ this strategy? One key to answering this lies in understanding the thermal and water dynamics that embryos experience in their nests, particularly in species that inhabit high elevations. The following section will highlight the main findings regarding the thermal ecology of communal nests in the high-Andean species *Anadia bogotensis*.

5 Thermal Ecology of Communal Nests in *Anadia bogotensis*

To understand how an ectothermic organism copes with an environment that is thermally challenging, it is crucial to characterize its response to temperature throughout its ontogeny, from embryo to adulthood. In the context of global warming, elevated temperatures can have varying effects on different stages of an organism's life cycle (Levy et al. 2015). Specifically, warming may enhance the fitness of individuals at one stage of development, while simultaneously reducing it at another, particularly for species from temperate regions (Levy et al. 2015).

In oviparous species, where part of the embryonic development occurs in nests, the thermal and hydric environment of the nest influences various factors, from sex determination in species with temperature-dependent sex determination (TSD) to the rate of development, hatching success, phenotype, performance, behavior, and physiological characteristics (reviewed in Booth 2006; Noble et al. 2018; Gatto and Reina 2022), effects that can persist into adulthood (Noble et al. 2018). Characterizing the temperature and humidity of nests forms the basis for experiments that analyze the effect of different values of these variables on key aspects of an organism's life history, which, in turn, influence the persistence of populations and species in nature. This knowledge is crucial for predicting the impact of changes in incubation temperature on life history traits of populations, as demonstrated in studies like that of Dayananda et al. (2016) on the nocturnal gecko *Oedura lesueurii*.

Oviparous species employ several strategies to evade or minimize the negative effects of temperature and desiccation during the nesting phase. These strategies include altering the timing of the nesting season, selecting specific oviposition sites (as reviewed in Refsnider et al. 2015), and constructing nests at varying depths. Such behaviors have been observed in species of *Sceloporus* (Telemeco et al. 2017), which are found in seasonal environments, as well as in monitor lizards (Doody et al. 2020). However, Telemeco et al. (2017) demonstrated that the evolution of the ability to modify both the site and depth of the nest (phenotypic plasticity in nest site choice) does not always align with the expectations of global warming scenarios. Furthermore, the evolution of oviposition behaviors that enhance embryo survival in nests requires that the population has previously faced conditions that threaten embryo survival, such as temperatures nearing the minimum and maximum critical thresholds for embryonic development (Telemeco et al. 2017). These evolutionary responses are thus shaped by the environmental context in which organisms have evolved. In environments with high daily temperature fluctuations, such as the mountains of the northern Andes, the factors influencing the selection of nesting sites and the characteristics of the nests would likely include not only maximum and minimum temperature values but also daily thermal variation. Additionally, the interaction between these temperature variables and other physical characteristics of the nest may play a significant role. Predicting the impact of extreme conditions, such as those projected in global warming scenarios, on embryo survival in nests is a complex challenge. In addition to thermal conditions, hydric factors and the potential interactions between these variables also influence females' decisions regarding when and where to lay their eggs.

In 2014, the thermal ecology, hydric conditions, and some physical variables of communal nests of the gymnophthalmid *Anadia bogotensis* were studied (Calderón-Espinosa et al. 2018). This neotropical lizard exhibits traits such as reduced clutch size (one or two eggs per reproductive event), oviposition in communal nests, and asynchrony in oviposition events, which are integral to its life history. These traits deserve special attention when studying the effect of temperature and humidity on embryonic development, both because they pose a methodological challenge in experimental studies and because the environmental conditions of a nest simultaneously impact multiple individuals within the population. Females of this species lay their eggs beneath rocks, fallen logs, frailejón remnants, or vegetation (Medina-Rangel 2013). Similarly, *Riama striata*, another gymnophthalmid lizard with communal nesting behavior that inhabits open areas of the Eastern Cordillera of the Colombian Andes, deposits its eggs in nests covered by rocks. Although *R. striata* and *A. bogotensis* have overlapping elevational distributions, *R. striata* extends to lower elevations while sharing similar life history traits with *A. bogotensis*.

Monitoring the temperature and humidity of 17 nests over nearly a year revealed significant variation in both factors across nests, months, and throughout the day. In the soil of some nests, humidity ranged from 4.5% to 48%, while in the air layer beneath the rock of one nest, it fluctuated between 1% and 100%. The presence of dehydrated eggs in some nests suggests that the embryo's position relative to the ground and the air column beneath the rock may impact its survival. This factor likely interacts with nest temperature, collectively influencing egg hatching success.

Nest temperatures follow the same fluctuation pattern as air temperature, ranging from 9.6 to 17.6 °C during the coldest month and 9.8 to 21.9 °C during the warmest month. While extreme temperature events were recorded—dropping to a minimum of 1.5 °C and peaking at 40.8 °C—these occurred only for a single day. However, if such extremes increase in frequency, duration, or intensity, they could significantly impact population persistence, as suggested by Murali et al. (2023) for vertebrates under climate change scenarios. Additionally, even a single extreme maximum temperature event of high magnitude could compromise embryo survival (Angilletta et al. 2013).

The average minimum and maximum temperatures of the nests, measured beneath the rock, correlated with the minimum and maximum air temperatures and followed the same daily fluctuation pattern. However, in some nests, temperatures fell below or rose above ambient levels, a pattern that could not be explained by nest structure alone.

The thermal variation, measured using the coefficient of variation (CV), indicated that fluctuations are much greater during the daytime heating phase of the nest compared to the cooling phase at night. Additionally, the CV during the heating phase is similar across nests and may or may not differ from the CV of the ambient temperature. In contrast, the CV during the cooling phase differs between nests with viable eggs and those without, as well as between each nest type and the CV of the environmental temperature. The maximum ambient temperature affects the average fluctuation of the nest temperature, while the volume of the rock covering the nest acts as a buffer, reducing the average thermal fluctuation in nests with larger rocks. This demonstrates that the rock plays a crucial role in slightly attenuating the daily temperature variation to which the embryos are exposed. By characterizing a larger number of nests or experimentally evaluating nests with different types and sizes of covers (such as rocks), we may uncover insights into how thermal fluctuations influence nest selection. Interestingly, the number of viable eggs appears to be linked to temperature fluctuations during the cooling phase of the nest, suggesting that, rather than high temperatures, low temperatures may play a more significant role in guiding the choice of oviposition site, should females exhibit any nest selection behavior in this species. The variation in nest depth observed in *Sceloporus tristichus* is an adaptive response to avoid low temperatures that can negatively affect offspring fitness (Telemeco et al. 2017). Therefore, investigating the effects of different temperatures and fluctuations during egg incubation in *Anadia bogotensis* will help us understand the key factors that have shaped the evolution of nesting behavior in this species.

The embryos in *A. bogotensis* nests are exposed to minimum and maximum temperatures typical of highland environments from the moment of oviposition until the completion of embryonic development, which can last 6–7 months (Clavijo and Fajardo 1981; Medina-Rangel 2013), occurring without any anatomical alterations (Jerez et al. 2020). The wide thermal fluctuations experienced during development in oviparous species have also been observed in other lizard species with communal nests, such as the nocturnal gecko *Oedura lesueurii* and the skink *Bassiana duperreyi* (Dayananda et al. 2016; Dissanayake et al. 2021; Pike et al. 2010). While other species in the study area, such as *Anolis heterodermus* and *Stenocercus trachycephalus*, do not appear to lay eggs under rocks or use communal nests, much remains unknown about their natural history. This gap in knowledge limits the ability to conduct comparative studies on thermal ecology among sympatric species in the neotropical highlands of the northern Andes.

6 Conclusions

Despite variations in temperature regimes across latitude, altitude, and different temporal scales (daily vs. annual), the thermal biology of squamate reptiles in the northern Andes mountains generally aligns with that of species inhabiting extreme latitudes and high elevations in temperate regions. Shared traits include a marked plasticity in thermoregulation strategies, broad thermal tolerance ranges associated with an enhanced ability to withstand low temperatures, and a greater tolerance to warming. However, differences are observed in aspects of their biology, such as reproductive modes (oviparity vs. viviparity) and the ability to undergo brumation, which reflect the unique thermophysiological adaptations of each group. We propose that natural history traits, such as activity patterns and habitat type (arboreal, terrestrial, fossorial), play a key role in shaping thermal tolerance to high temperatures and thermoregulation strategies.

The observed similarities underscore the influence of low temperatures and environmental thermal fluctuations as key determining factors, both spatially and temporally, in the biology of squamate reptiles across the globe.

References

Altamirano-Benavides MA, Domínguez-Guerrero SF, Muñoz-Nolasco FJ et al (2019) Thermal ecology and risk of extinction due to climate change of *Gonatodes concinnatus* (Squamata: Sphaerodactylidae), an endemic lizard of the western Amazon. Rev Mex Biodiv 90:e902824

Anderson NL, Hetherington TE, Coupe B et al (2005) Thermoregulation in a nocturnal, tropical, arboreal snake. J Herpetol 39(1):82–90

Andrango MB, Sette C, Torres-Carvajal O (2016) Short-term predicted extinction of Andean populations of the lizard *Stenocercus guentheri* (Iguanidae: Tropidurinae). J Therm Biol 62:30–36

Angilletta MJ Jr, Zelic MH, Adrian GJ et al (2013) Heat tolerance during embryonic development has not diverged among populations of a widespread species (*Sceloporus undulatus*). Conserv Physiol 1(1):cot018

Arenas-Moreno DM, Santos-Bibiano R, Muñoz-Nolasco FJ et al (2018) Thermal ecology and activity patterns of six species of tropical night lizards (Squamata: Xantusiidae: *Lepidophyma*) from Mexico. J Therm Biol 75:97–105

Arenas-Moreno DM, Lara-Resendiz RA, Domínguez-Guerrero SF et al (2021) Thermoregulatory strategies of three reclusive lizards (genus *Xantusia*) from the Baja California peninsula, Mexico, under current and future microenvironmental temperatures. J Exp Zool A Ecol Integr Physiol 335(5):499–511

Barragán-Forero A (2016) Thermal ecology of *Stenocercus trachycephalus* and *S. lache* (Squamata: Tropiduridae) and assessment of the risk of extinction associated with global warming in the Andes of Colombia. Dissertation, National University of Colombia

Blyth S, Groombridge B, Lysenko I, Miles L, Newton A (2002) Mountain Watch -environmental change and sustainable development in mountains. UNEP-WCMC, Cambridge, UK

Bodensteiner BL, Agudelo-Cantero GA, Arietta AA et al (2020) Thermal adaptation revisited: how conserved are thermal traits of reptiles and amphibians? J Exp Zool Part A: Ecol Integr Physiol 335(1):173–194

Booth DT (2006) Influence of incubation temperature on hatchling phenotype in reptiles. Physiol Biochem Zool 79(2):274–281

Bouazza A, Slimani T, El Mouden H et al (2016) Thermal constraints and the influence of reproduction on thermoregulation in a high-altitude gecko (*Quedenfeldtia trachyblepharus*). J Zool 300(1):36–44

Caetano GH, Santos JC, Godinho LB et al (2020) Time of activity is a better predictor of the distribution of a tropical lizard than pure environmental temperatures. Oikos 129(7):953–963

Calderón-Espinosa ML, Jerez A, Medina-Rangel GF (2018) Living in the extremes: the thermal ecology of communal nests of the Highland Andean Lizard *Anadia bogotensis* (Squamata: Gymnophthalmidae). Curr Herpetol 37(2): 158–171

Cecchetto NR, Medina SM, Ibargüengoytía NR (2020) Running performance with emphasis on low temperatures in a Patagonian lizard, *Liolaemus lineomaculatus*. Sci Rep 10(1):14732

Chan WP, Lenoir J, Mai GS et al (2024) Climate velocities and species tracking in global mountain regions. Nature 629(8010):114–120

Clavijo JJ, Fajardo AP (1981) Contribution to the knowledge of the biology of *Anadia bogotensis* (Peters) (Sauria, Teiidae). Undergraduate thesis, National University of Colombia, Bogotá,132 pp

Ćorović J, Crnobrnja-Isailović J (2018) Aspects of thermal ecology of the meadow lizard (*Darevskia praticola*). Amphib-Reptil 39(2):229–238

Dayananda B, Gray S, Pike D, Webb JK (2016) Communal nesting under climate change: fitness consequences of higher incubation temperatures for a nocturnal lizard. Glob Change Biol 22(7):2405–2414

Deutsch CA, Tewksbury JJ, Huey RB et al (2008) Impacts of climate warming on terrestrial ectotherms across latitude. Proc Natl Acad Sci 105(18):6668–6672

Díaz G, García-Mieles S (2025) Depredación de *Atractus crassicaudatus* (Squamata: Colubridae) por *Turdus fuscater* (Passeriformes: Turdidae) en la sabana de Bogotá, Colombia. Rev latinoam herpetol 8(1): e1120–21.

Díaz de la Vega-Pérez AH, Barrios-Montiel R, Jiménez-Arcos VH et al (2019) High-mountain altitudinal gradient influences thermal ecology of the Mesquite Lizard (*Sceloporus grammicus*). Can J Zool 97(8): 659–668

Díaz JA, Iraeta P, Monasterio C (2006) Seasonality provokes a shift of thermal preferences in a temperate lizard, but altitude does not. J Therm Biol 31(3): 237–242

Diele-Viegas LM, Vitt LJ, Sinervo B et al (2018) Thermal physiology of amazonian lizards (reptilia: Squamata). PLoS One 13(3):e0192834

Dissanayake DS, Holleley CE, Georges A (2021) Effects of natural nest temperatures on sex reversal and sex ratios in an Australian alpine skink. Sci Rep 11(1): 20093

Domínguez-Guerrero SF, Muñoz MM, de Jesús Pasten-Téllez D et al (2019) Interactions between thermoregulatory behavior and physiological acclimatization in a wild lizard population. J Therm Biol 79:135–143

Doody JS, McGlashan J, Fryer H et al (2020) Plasticity in nest site choice behavior in response to hydric conditions in a reptile. Sci Rep 10(1):16048

Duméril AMC, Bibron G, Duméril AHA (1854) Erpétologie générale ou Histoire Naturelle complète des Reptiles. Vol. 7 (partie 1). Paris, xvi + 780 S

Dzialowski EM (2005) Use of operative temperature and standard operative temperature models in thermal biology. J Therm Biol 30(4):317–334

Fierro-Estrada N, González YGG, Miles DB et al (2019) Thermoregulation of the lizard *Barisia imbricata* at altitudinal extremes. Amphib-Reptil 40(3):349–360

Fierro-Estrada NF, de la Cruz FRM, Miles DB et al (2022) An Ecophysiological Assessment of Extinction Risk from Climate Change of the lizard *Barisia imbricata* (Squamata: Anguidae). Preprint: https://doi.org/10.21203/rs.3.rs-1399449/v1

Flantua SG, Hooghiemstra H (2018) Historical connectivity and mountain biodiversity. In: Hoorn C, Perrigo A, Antonelli A (eds) Mountains, climate and biodiversity, 1 st edn. Wiley–Blackwell, Oxford, pp 171–185

Flantua SG, O'Dea A, Onstein RE et al (2019) The flickering connectivity system of the north Andean páramos. J Biogeogr 46(8):1808–1825

Gansser A (1973) Facts and theories on the Andes: twenty-sixth William Smith lecture. J Geol Soc 129(2):93–131

Gatto CR, Reina RD (2022) A review of the effects of incubation conditions on hatchling phenotypes in non-squamate reptiles. J Comp Physiol B 192(2): 207–233

Ghalambor CK, Huey RB, Martin PR et al (2006) Are mountain passes higher in the tropics? Janzen's hypothesis revisited. Integr Comp Biol 46(1):5–17

Gómez Alé R, Acosta JC, Astudillo V et al (2018) Effect of temperature on the locomotor performance of species in a lizard assemblage in the Puna region of Argentina. J Comp Physiol B 188:977–990

Gregory-Wodzicki KM (2000) Uplift history of the Central and Northern Andes: a review. Geol Soc Am Bull 112(7):1091–1105

Guerra-Correa ES, Merino-Viteri A, Andrango MB et al (2020) Thermal biology of two tropical lizards from the Ecuadorian Andes and their vulnerability to climate change. PLoS One 15(1):e0228043

Gvoždík L, Castilla AM (2001) A comparative study of preferred body temperatures and critical thermal tolerance limits among populations of *Zootoca vivipara* (Squamata: Lacertidae) along an altitudinal gradient. J Herpetol 35:486–492

Harlow HJ, Purwandana D, Jessop TS et al (2010) Body temperature and thermoregulation of Komodo dragons in the field. J Therm Biol 35(7):338–347

Herczeg G, Gonda A, Saarikivi J et al (2006) Experimental support for the cost–benefit model of lizard thermoregulation. Behav Ecol Sociobiol 60:405–414

Herczeg G, Herrero A, Saarikivi J et al (2008) Experimental support for the cost–benefit model of lizard thermoregulation: the effects of predation risk and food supply. Oecologia 155:1–10

Herrando-Pérez S, Ferri-Yáñez F, Monasterio C et al (2019) Intraspecific variation in lizard heat tolerance alters estimates of climate impact. J Anim Ecol 88(2): 247–257

Hertz PE (1981) Adaptation to altitude in two West Indian anoles (Reptilia). J Zool 195(1):25–37

Hertz PE, Huey RB (1981) Compensation for altitudinal changes in the thermal environment by some *Anolis* lizards on Hispaniola. Ecology 62(3):515–521

Hertz PE, Huey RB, Stevenson RD (1993) Evaluating temperature regulation by field-active ectotherms: the fallacy of the inappropriate question. Am Nat 142(5): 796–818

Holdridge LR (1967) Life zone ecology. Tropical Science Center, San Jose, Costa Rica

Huang J, Yu H, Guan X et al (2016) Accelerated dryland expansion under climate change. Nat Clim Change 6(2):166–171

Huey RB (1982) Temperature, physiology, and the ecology of reptiles. In: Gans C, Pough FH (eds) Biology of the reptilia. Academic Press, London, pp 25–91

Huey RB, Webster TP (1976) Thermal biology of *Anolis* lizards in a complex fauna: the *cristatellus* group on Puerto Rico. Ecology 57(5):985–994

Huey RB, Peterson CR, Arnold SJ (1989) Hot rocks and not-so-hot rocks: retreat-site selection by garter snakes and its thermal consequences. Ecology 70(4):931–944

Huey RB, Hertz PE, Sinervo B (2003) Behavioral drive versus behavioral inertia in evolution: a null model approach. Am Nat 161(3):357–366

Huey RB, Deutsch CA, Tewksbury JJ, Vitt LJ, Hertz PE, Álvarez Pérez HJ, Garland Jr T (2009) Why tropical forest lizards are vulnerable to climate warming. Proc R Soc B Biol Sci 276(1664): 1939-1948

Ibargüengoytía NR, Medina SM, Fernández JB et al (2010) Thermal biology of the southernmost lizards in the world: *Liolaemus sarmientoi* and *Liolaemus magellanicus* from Patagonia, Argentina. J Therm Biol 35(1):21–27

Ibargüengoytía NR, Medina M, Laspiur A et al (2021) Looking at the past to infer into the future: thermal traits track environmental change in Liolaemidae. Evolution 75(10):2348–2370

Janzen DH (1967) Why mountain passes are higher in the tropics. Am Nat 101(919):233–249

Jerez A, Bonilla-Garzón A, Castellanos-Montilla DS (2020) Developmental stages in *Anadia bogotensis* : contributions to the understanding of the body plan evolution in Gymnophthalmoidea (Squamata). Cuad Herpetol 34(2):1–12

Jiang ZW, Ma L, Mi CR et al (2023) Distinct responses and range shifts of lizard populations across an elevational gradient under climate change. Glob Change Biol 29(10):2669–2680

Karameta E, Gavriilidi I, Sfenthourakis S, Pafilis P (2023) Seasonal variation in the thermoregulation pattern of an insular agamid lizard. Animals 13:3195.https://doi.org/10.3390/ani13203195

Kearney M (2002) Hot rocks and much-too-hot rocks: seasonal patterns of retreat-site selection by a nocturnal ectotherm. J Therm Biol 27:205–218. https://doi.org/10.1016/S0306-4565(01)00085-7

Kearney M, Predavec M (2000) Do nocturnal ectotherms thermoregulate? A study of the temperate gecko *Christinus marmoratus*. Ecology 81(11):2984–2996

Kubisch EL, Fernández JB, Ibargüengoytía NR (2016) Vulnerability to climate warming of *Liolaemus pictus* (Squamata, Liolaemidae), a lizard from the cold temperate climate in Patagonia, Argentina. J Comp Physiol B 186:243–253

Lara-Reséndiz RA, Arenas-Moreno DM, Méndez-de La Cruz FR (2013) Diurnal and nocturnal thermoregulation of the lizard *Phyllodactylus bordai* (Gekkota: Phyllodactylidae) in a semi-arid region of central Mexico. Rev Chil Hist Nat 86(2):127–135

Lara-Reséndiz RA, Gadsden H, Rosen PC et al (2015) Thermoregulation of two sympatric species of horned lizards in the Chihuahuan Desert and their local extinction risk. J Therm Biol 48:1–10

Lara-Reséndiz RA, Galina-Tessaro P, Pérez-Delgadillo AG et al (2019) Effects of climate change on a widely distributed thermophilic lizard species (*Dipsosaurus dorsalis*): an ecophysiological approach. Rev Mex Biodiv 90:e902888

Laspiur A, Santos JC, Medina SM et al (2021) Vulnerability to climate change of a microendemic lizard species from the central Andes. Sci Rep 11(1):11653

Levy O, Buckley LB, Keitt TH et al (2015) Resolving the life cycle alters expected impacts of climate change. Proc Roy Soc B Biol Sci 282(1813):20150837

Ma L, Buckley LB, Huey RB et al (2018) A global test of the cold-climate hypothesis for the evolution of viviparity of squamate reptiles. Glob Change Biol 27(6): 679–689

Martínez-Méndez N, Mejía O, de la Cruz FRM (2015) The past, present and future of a lizard: the phylogeography and extinction risk of *Sceloporus serrifer* (Squamata: Phrynosomatidae) under a global warming scenario. Zool Anz 254:86–98

Martori R, Aun L, Orlandini S (2002) Temporal thermal relationships in a population of *Liolaemus koslowskyi*. Cuad Herpetol 16(1):33–45

Medina-Rangel GF (2013) *Anadia bogotensis* (Bogota Anadia lizard) nesting. Herpetol Rev 44:312–313

Méndez-Galeano MA, Calderón-Espinosa ML (2017) Thermoregulation in the Andean lizard *Anolis heterodermus* (Squamata: Dactyloidae) at high elevation in the Eastern Cordillera of Colombia. Iheringia Sér Zool 107:e2017018

Méndez-Galeano MA, Paternina-Cruz RF, Calderón-Espinosa ML (2020) The highest kingdom of *Anolis* : thermal biology of the Andean lizard *Anolis heterodermus* (Squamata: Dactyloidae) over an elevational gradient in the Eastern Cordillera of Colombia. J Therm Biol 89:102498

Miranda-Calle AB, Pacheco LF, Aparicio J et al (2021) Thermoregulation of *Liolaemus aparicioi* (Iguania: Liolaemidae) along a 1000 m elevational gradient in La Paz Valley, La Paz. Bolivia J Therm Biol 99: 102940

Moreno-Arias RA, Méndez-Galeano MA, Beltrán I et al (2023) Revealing anole diversity in the highlands of the Northern Andes: new and resurrected species of the *Anolis heterodermus* species group. Vertebr Zool 73: 161–188

Muñoz MM, Stimola MA, Algar AC et al (2014) Evolutionary stasis and lability in thermal physiology in a group of tropical lizards. Proc Roy Soc B Biol Sci 281: 20132433. https://doi.org/10.1098/rspb.2013.243310.1098/rspb.2013.2433

Muñoz MM, Bodensteiner B (2019) Janzen's hypothesis meets the Bogert effect: connecting climate variation, thermoregulatory behavior, and rates of physiological evolution. Integr Org Biol 1(1): oby002.

Muñoz-Nolasco FJ, Arenas-Moreno DM, Gandarilla-Aizpuro FJ et al (2023) Physiological ecology and vulnerability to climate change of a microendemic, habitat-specialist lizard in a tropical dry forest of Mexico. Clim Change Ecol 5:100066

Murali G, Iwamura T, Meiri S et al (2023) Future temperature extremes threaten land vertebrates. Nature 615(7952):461–467

Myers N, Mittermeier RA, Mittermeier CG et al (2000) Biodiversity hotspots for conservation priorities. Nature 403(6772):853–858

Navas CA (2002) Herpetological diversity along Andean elevational gradients: links with physiological ecology and evolutionary physiology. Comp Biochem Physiol A Mol Integr Physiol 133(3):469–485

Noble DW, Stenhouse V, Schwanz LE (2018) Developmental temperatures and phenotypic plasticity in reptiles: a systematic review and meta-analysis. Biol Rev 93(1):72–97

Nordberg EJ, Schwarzkopf L (2019) Heat seekers: a tropical nocturnal lizard uses behavioral thermoregulation to exploit rare microclimates at night. J Therm Biol 82: 107–114

Ortega Z, Mencía A, Pérez-Mellado V (2017) Wind constraints on the thermoregulation of high mountain lizards. Int J Biometeorol 61:565–573

Pafilis P, Herrel A, Kapsalas G et al (2019) Habitat shapes the thermoregulation of Mediterranean lizards introduced to replicate experimental islets. J Therm Biol 84:368–374

Paternina-Cruz RF (2016) Thermoregulation strategy and risk of extinction of *Atractus crassicaudatus* (Squamata: Dipsadidae) associated with global warming. Dissertation, National University of Colombia

Paternina-Cruz RF, Calderón-Espinosa ML (2022) Withstanding cold nights at high elevation: thermal strategy of *Atractus crassicaudatus* , a nocturnal tropical snake endemic to the eastern Andes of Colombia. Rev mex Biodiv 93:e933705

Peters W (1862) Über Cercosaura und die mit dieser Gattung verwandten Eidechsen aus Südamerika. Abh. Berlin Akad. 1862:165–225

Pérez-Delgadillo AG, Lara-Reséndiz RA, Valdez-Villavicencio JH et al (2021) Thermal ecology of a thermophilic lizard *Callisaurus draconoides* through a latitudinal gradient. J Arid Environ 195:104634

Piantoni C, Curcio FF, Ibargüengoytía NR et al (2019) Implications of climate change on the habitat shifts of tropical lizards. Austral Ecol 44(7):1174–1186

Pike DA, Webb JK, Shine R (2010) Nesting in a thermally challenging environment: nest-site selection in a rock-dwelling gecko, *Oedura lesueurii* (Reptilia: Gekkonidae). Biol J Linn Soc 99(2):250–259

Pinzón-Barrera C, Suárez-Ayala N, Carrillo-Chávez LM et al (2024) Unveiling critical thermal limits of *Anolis tolimensis* (Squamata, Anolidae) across an elevational landscape. Curr Herpetol 43(1):115–134

Pontes-da-Silva E, Magnusson WE, Sinervo B et al (2018) Extinction risks forced by climatic change and intra-specific variation in the thermal physiology of a tropical lizard. J Therm Biol 73:50–60

Prăvălie R (2016) Drylands extent and environmental issues. A global approach. Earth-Sci Rev 161:259–278

Pruett JE, Hall JM, Tiatragul S et al (2022) Nesting in *Anolis* lizards: an understudied topic in a well-studied clade. Front Ecol Evol 10:821115

Refsnider JM, Palacios MG, Reding DM et al (2015) Effects of a novel climate on stress response and immune function in painted turtles (*Chrysemys picta*). J Exp Zool Part A Ecolo Genet Physiol 323(3): 160–168

Ríos-Touma B, Acosta R, Prat N (2014) The Andean Biotic Index (ABI): revised tolerance to pollution values for macroinvertebrate families and index performance evaluation. Rev Biol Trop 62:249–273

Salva AG, Robles CI, Tulli MJ (2021) Thermal biology of *Liolaemus scapularis* (Iguania: Liolaemidae) from argentinian northwest. J Therm Biol 98:102924

Sarmiento G (1986) Ecological features of climate in high tropical mountains. In: High altitude tropical biogeography. Oxford University Press, Oxford, pp 11–45

Sunday JM, Bates AE, Dulvy NK (2011) Global analysis of thermal tolerance and latitude in ectotherms. Proc R Soc B 278(1713):1823–1830

Sunday JM, Bates AE, Kearney MR et al (2014) Thermal-safety margins and the necessity of thermoregulatory behavior across latitude and elevation. Proc Natl Acad Sci USA 111:5610–5615

Sunday JM, Bennett JM, Calosi P et al (2019) Thermal tolerance patterns across latitude and elevation. Philos Trans R Soc B 374(1778):20190036

Tao S, Cheng K, Li X et al (2021) The thermal biology of *Takydromus kuehnei* indicates tropical lizards from high elevation have not been severely threatened by climate change. Front Ecol Evol 9:767102. https://doi.org/10.3389/fevo.2021.767102

Taylor EN, Diele-Viegas LM, Gangloff EJ et al (2020) The thermal ecology and physiology of reptiles and amphibians: a user's guide. J Exp Zool A Ecol Integr Physiol 335(1):13–44

Telemeco RS, Gangloff EJ (2020) Introduction to the special issue–beyond CTmax and CTmin: advances in studying the thermal limits of reptiles and amphibians. J Exp Zool A Ecol Integr Physiol 335(1):5–12

Telemeco RS, Fletcher B, Levy O et al (2017) Lizards fail to plastically adjust nesting behavior or thermal tolerance as needed to buffer populations from climate warming. Glob Change Biol 23(3):1075–1084

Tewksbury JJ, Huey RB, Deutsch CA (2008) Putting the heat on tropical animals. Science 320(5881): 1296–1297

Torres-Carvajal O (2007) A taxonomic revision of South American *Stenocercus* (Squamata: Iguania) lizards. Herpetol Monogr 21(1):76–178

Valdecantos S, Martínez V, Lobo F et al (2013) Thermal biology of *Liolaemus* lizards from the high Andes: being efficient despite adversity. J Therm Biol 38(3): 126–134

Van Berkum FH (1986) Evolutionary patterns of the thermal sensitivity of sprint speed in *Anolis* lizards. Evolution 40(3):594–604

Villavicencio H, Acosta J, Marinero J et al (2007) Thermal ecology of a population of the lizard, *Liolaemus pseudoanomalus* in western Argentina. Amphib-Reptil 28(1):163–165

Vitt LJ, Caldwell JP (2014) Herpetology: an introductory biology of amphibians and reptiles, 4th edn. Academic Press, San Diego

Wang Z, Lu HL, Ma L et al (2014) Viviparity in high-altitude *Phrynocephalus* lizards is adaptive because embryos cannot fully develop without maternal thermoregulation. Oecologia 174:639–649

Werner F (1916) Bemerkungen über einige niedere Wirbeltiere der Anden von Kolumbien mit Beschreibungen neuer Arten. Zool. Anz. 47:301-311

Webb JK, Shine R (1998) Thermoregulation by a nocturnal elapid snake (*Hoplocephalus bungaroides*) in southeastern Australia. Physiol Zool 71(6):680–692

Wu Q, Dang W, Hu YC et al (2018) Altitude influences thermal ecology and thermal sensitivity of locomotor performance in a toad-headed lizard. J Therm Biol 71: 136–141

Zhangv RY, Wild KH, Pottier P et al (2023) Developmental environments do not affect thermal physiological traits in reptiles: an experimental test and meta-analysis. Biol Lett 19(5):20230019

Zimin A, Zimin SV, Shine R et al (2022) A global analysis of viviparity in squamates highlights its prevalence in cold climates. Glob Ecol Biogeogr 31(12):2437–2452

Thermal Adaptations in Tropical Andean Amphibians: Exploring Spatial Variability and Global Change Impacts

Pol Pintanel, Andrés Merino-Viteri, and Miguel Tejedo

1 Introduction

Temperature is a critical environmental factor that governs the physiological and biochemical processes of organisms, impacting a range of factors including performance, reproductive success, growth rates, and survival (Angilletta 2009). Environmental temperatures, therefore, play a pivotal role in shaping the distribution, abundance, and phenology of species (e.g., Brown et al. 2016). In tropical montane regions, species are particularly sensitive to climate change due to the presence of stable climatic zonations throughout the year. These zones act as potential thermal barriers across different elevations, which hinder vertical migration and limit the potential for species to shift their ranges in response to changing temperatures. Additionally, high-elevation ecosystems experience pronounced circadian thermal variability, with dramatic day-to-night fluctuations that impose unique challenges for ectothermic species. These combined factors foster the development of highly specialized thermal niches, promoting increased geographical isolation and reducing the capacity for distributional shifts and range expansions (Janzen 1967; Colwell et al. 2008). Understanding these adaptations is vital, especially in the context of global climate change, which poses significant threats to biodiversity.

P. Pintanel
Departament de Ecología Evolutiva, Estación Biológica de Doñana, EBD-CSIC, Sevilla, Spain

A. Merino-Viteri
Laboratorio de Ecofisiología, Facultad de Ciencias Exactas, Naturales y Ambientales, Pontificia Universidad Católica del Ecuador, Quito, Ecuador

M. Tejedo (✉)
Departamento de Ecología Evolutiva, Estación Biológica de Doñana, EBD-CSIC, Sevilla, Spain
e-mail: tejedo@ebd.csic.es

Amphibians, being ectothermic vertebrates with limited mobility, are particularly exposed to temperature changes. This makes them excellent models for studying the evolution of thermal physiology along elevational gradients in the tropical Andes. Andean amphibians occupy a broad elevational range of over 5000 m in equatorial regions, encompassing diverse terrestrial and aquatic environments, and encounter varying thermal pressures across these habitats (Reider et al. 2021; Ron et al. 2024). Studying their thermal ecology is essential for predicting their future survival and informing effective conservation strategies.

The field of ecophysiology of tropical Andean amphibians is relatively new, with significant advancements occurring since the late twentieth century. The early studies in the 1970s and 1980s investigated thermoregulation in high-elevation amphibians, providing crucial baseline data on how these species cope with fluctuating temperatures and cold environments (Valdivieso and Tamsitt 1974; Pearson and Bradford 1976; Sinsch 1989), and highlighting the unique thermal adaptations in this group compared to reptiles

J. N. Urbina-Cardona et al. (eds.), *Andean Herpetofauna*, Sustainable Development Goals Series,
https://doi.org/10.1007/978-3-032-00074-3_13

(Pearson and Bradford 1976). Similarly, later, Lüddecke (1997) demonstrated how *Dendropsophus molitor* has adapted to the unique thermal conditions of its high-elevation habitat through behavioral thermoregulation, specific habitat selection, and seasonal activity patterns.

It was not until the 1990s that comparative studies saw a significant surge, spearheaded by Carlos A. Navas (University of São Paulo). Navas conducted a series of influential studies that further explored the thermal ecology of high-elevation Neotropical anurans. He examined the thermal dependency on the vocal (Navas 1996a) and locomotor performance (Navas 1996b) in high-elevation species. These studies provided insights into the physiological constraints imposed by low temperatures on the locomotor and vocal capabilities of these amphibians, emphasizing the importance of thermal adaptation in their survival and reproductive success. He also investigated how microhabitat selection and activity patterns influenced thermal regulation, showing that these behaviors were crucial for coping with thermal variability (Navas 1996c). His works on metabolic physiology, locomotor performance, and thermal niche breadth revealed the capacity of these high-elevation anurans to perform optimally across a broad range of temperatures in comparison to their lowland counterparts, identifying the adaptive mechanisms that enable survival in harsh thermal environments (Navas 1996d, 1997). The broader thermal niche observed in high-elevation anurans is likely an adaptation to the greater thermal variability at these elevations, where temperatures can fluctuate dramatically between day and night (Navas et al. 2013). Temperatures are especially pronounced in high-elevation páramos, where nocturnal temperatures often approach freezing, and daytime temperatures can rise sharply due to solar radiation. This daily oscillation contrasts with the more stable thermal conditions observed in lowland tropical forests, where the buffering effect of dense canopy cover limits both diurnal and nocturnal extremes. This pronounced variability in circadian temperatures in high-elevation environments creates selective pressures for physiological and behavioral adaptations, fostering broader thermal tolerances in species inhabiting these regions. This aligns with Janzen's hypothesis (Janzen 1967), which suggests that thermal gradients in tropical mountains create physiological barriers that limit species dispersal, which may promote diversification (Ghalambor et al. 2006).

Subsequent research on the thermal adaptations of tropical Andean amphibians has significantly expanded our understanding of these species' responses to varying thermal environments. Following previous research, numerous studies have examined the thermal tolerance and physiological limits of these amphibians in greater detail, incorporating both laboratory and field observations. Researchers have explored the importance of thermal niche breadth and its evolutionary implications in the context of diverse amphibian populations across elevational gradients (Navas 2002, 2006; Navas et al. 2008, 2013; von May et al. 2017, 2019; Pintanel et al. 2019, 2021, 2022, 2024). Additionally, research on thermal plasticity in Andean amphibians has emphasized their capacity to adjust to their thermal tolerance to temporal scales (Bernal et al. 2017; Burrowes et al. 2020; Turriago et al. 2022, 2023), and the role of microhabitats in thermal regulation under changing climatic conditions (e.g., Pintanel et al. 2019, 2021, 2022; González-del-Pliego et al. 2020; Turriago et al. 2022). Many studies have focused on the thermal ecology of specific groups, such as the high-elevation direct-developing frogs (Carvajalino-Fernández et al. 2011, 2021; Carvajalino-Fernández and Navas 2025, Chapter "Physiological Ecology of Andean Amphibians Associated with Freezing Temperatures" of this volume; von May et al. 2017; Pintanel et al. 2019; González-del-Pliego et al. 2020, 2023), amphibian larvae at interspecific (Pintanel et al. 2020, 2022, 2024) and intraspecific levels (Páez-Vacas and Funk 2024) and even anuran embryos (Bernal and Lynch 2013; Turriago et al. 2015), and, finally, exploring biochemical adaptations to freezing risk in direct developing frogs of high páramos (Carvajalino-Fernández et al. 2011, 2021; Carvajalino-Fernández and Navas 2025,

Chapter "Physiological Ecology of Andean Amphibians Associated with Freezing Temperatures" of this volume).

Overall, this body of work underscores the complexity of thermal adaptation mechanisms and highlights the importance of conserving the diverse microhabitats that support these unique amphibian populations. The ongoing research continues to contribute valuable insights into how tropical Andean amphibians may cope with future climate scenarios, emphasizing the need for comprehensive conservation strategies to protect these ecologically significant species.

2 The Climates in Tropical Andes: Axes of Temperature Variation and Physiological Barriers

2.1 Axes of Temperature Variation

The tropical Andes rising dramatically from sea level to 6768 m a.s.l. at Huascarán ranges, Perú, within relatively short geographical distances, creates an extraordinary array of climates along their steep mountain gradients. The region contains most of Earth biomes, from hot, perhumid lowland tropical forest to tundra-like páramos and shelter an extraordinary biodiversity of amphibians (Herzog et al. 2011; Ron et al. 2024). These climatic gradients, coupled with a multitude of microclimates, possibly have led to the evolution of a wide variety of thermal adaptations making the region an invaluable natural laboratory for studying amphibian thermal physiology.

The ecosystems of the tropical Andes exhibit low thermal variation with nearly absent seasonal and interannual variability. Consequently, there are only two primary sources of temperature variation in the region (Fig. 1, inbox). The first axis of temperature variation is the elevational adiabatic lapse rate, air temperature decreases at an average rate of about 0.6 °C/100 m elevation, with slight variation, depending on orientation and humidity (Sarmiento 1986). This promotes regional temperature variation. The second axis of temperature variation is the microenvironmental variation, which fuels additional temperature variation at the local scale, such as forest versus open habitats due to recent anthropogenic landscape conversion in terrestrial environments (e.g., Nowakowski et al. 2018a, b; Pintanel et al. 2019) and streams versus ponds in aquatic habitats (e.g., Pintanel et al. 2021, 2022, 2024). This wide range of thermal conditions makes the region an ideal setting for analyzing the diversity of thermal tolerances, their spatial patterns of variation, and the main drivers of observed variability.

Forests play a critical role as thermal buffers for biodiversity, particularly in tropical montane regions where canopy cover alleviates thermal extremes, mitigating the effects of climate change by moderating diurnal temperature fluctuations (Pintanel et al. 2019). The dense canopy acts as an insulator, shielding understory habitats from rapid temperature increases during the day and retaining heat at night (De Frenne et al. 2021), and maintains higher humidity levels (De Frenne et al. 2019). Such thermally buffered microhabitats are particularly vital in regions experiencing rapid land-use changes, as they offer refuges that sustain species unable to tolerate the extreme thermal conditions of anthropogenic open habitats, which can amplify thermal extremes and alter local thermal dynamics (Nowakowski et al. 2018a, b). In fact, the capacity of forests to mediate thermal variability has been shown to surpass the pace of macroclimatic warming, effectively acting as refugia for biodiversity under changing climates (De Lombaerde et al. 2022; De Frenne et al. 2019, 2021).

Furthermore, forest structure contributes to fine-scale thermal heterogeneity, with primary forests offering the most effective buffering due to their complex canopy layers (De Lombaerde et al. 2022). Vertical stratification adds another layer of microclimatic variation, creating gradients from the cooler, more stable conditions near the forest floor to the hotter, more variable environments in the canopy. This vertical layering drives divergence in thermal tolerances among species, with arboreal species adapting to higher temperatures and greater variability, while ground-dwelling species show narrower tolerance

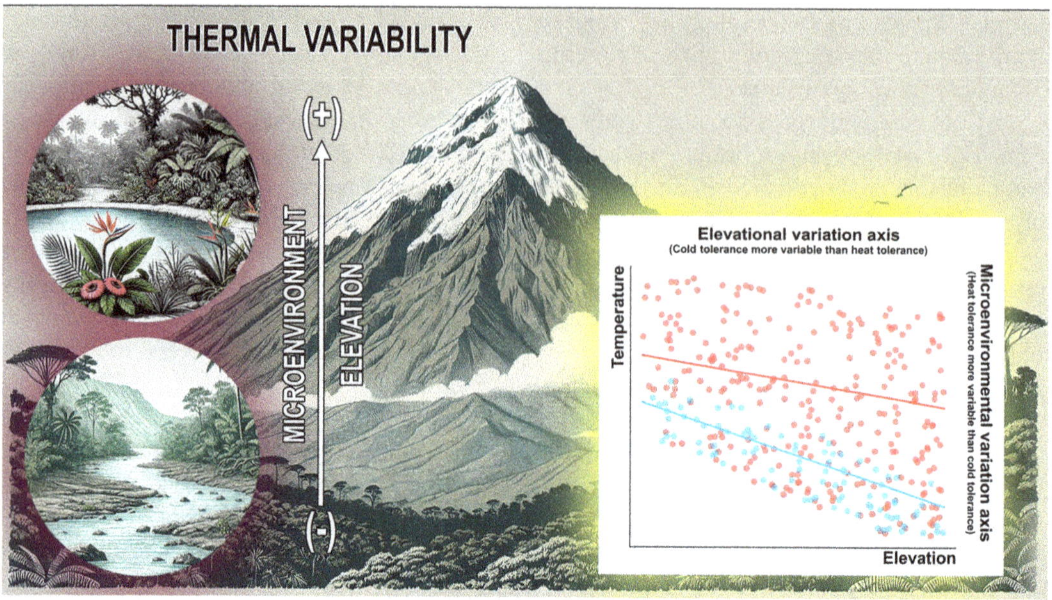

Fig. 1 Hypothetical variation of extreme temperatures; t_{max} (red) and t_{min} (blue), along an aseasonal tropical elevational gradient. The ecosystems of the tropical Andes exhibit low thermal variation with nearly absent seasonal and inter-annual variability. We postulate two major axes of temperature variation (inbox, based on Pintanel et al. 2019). The first axis is the elevational adiabatic lapse rate, by which air temperature decreases at an average rate of about 0.6 °C/100 m elevation. The second axis is microenvironmental variation, which fuels additional temperature variation at the local scale, such as forest versus open habitats in terrestrial environments (e.g., Pintanel et al. 2019) and streams versus ponds in aquatic habitats (e.g., Pintanel et al. 2021, 2022, 2024). Background images obtained with Open AI, DALL-E, July 6, 2024

ranges suited to stable microclimates (Leahy et al. 2022; Nakamura et al. 2022). Secondary forests, although less complex, still provide critical thermal refugia as they recover structure and microhabitat diversity over time. Conservation strategies that prioritize forest preservation and promote secondary forest regrowth are essential to maintain these thermally buffered microhabitats (González-del-Pliego et al. 2016, 2020).

2.2 Physiological Barriers

In the early 1800s, naturalist and first biogeographer Alexander von Humboldt, through his observations in the Andes of Ecuador, concluded that species were distributed across distinct elevational zones, proposing climate as the primary driver of this pattern (Fig. 1; von Humboldt 1849; von Humboldt and Bonpland 2009). Following Humboldt's definition of the elevational zonation hypothesis, Daniel H. Janzen (1967) observed that there was lower temperature overlap across elevation bands in tropical mountains compared to temperate ones. He predicted that climate stratification in aseasonal tropical mountains creates physiological barriers that limit organismal dispersal. These barriers are particularly significant for ectotherms (such as amphibians and reptiles) due to their high dependence on temperature for physiological performance. Thus, ectothermic amphibians face limited dispersal across tropical elevational gradients because moving to different elevations would expose them to temperatures outside their native thermal environments.

A physiological barrier can be understood through the thermal performance curve (TPC), which describes how an organism's fitness-

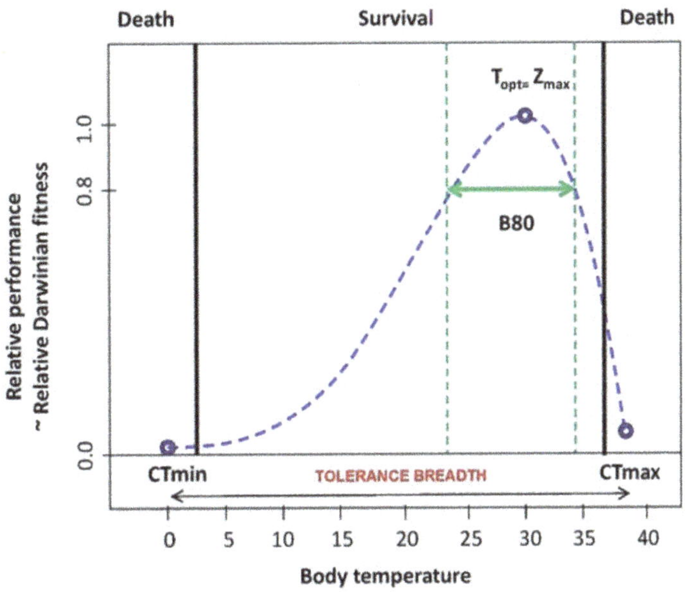

Fig. 2 Hypothetical thermal performance curve (TPC). An operational framework to define a physiological thermal barrier that constrains organismal dispersal when environmental temperatures do not allow survival or drastically reduce fitness. See main text for the explanation of the parameters. The Body Temperature X-axis presents hypothetical, yet realistic values, in degrees Celsius

related traits (e.g., growth, development, survival) vary with temperature (Huey and Stevenson 1979). The TPC (Fig. 2) defines both the temperature ranges at which an organism's fitness is zero (beyond their critical thermal limits, CT_{min} and CT_{max}) and then cannot survive, and the optimal temperature conditions (T_{opt}) where the organism achieves maximal fitness (Z_{max}). Suboptimal temperature ranges (B_i) are those where the organism can survive but at reduced performance levels. These ranges can be quantified as percentages of maximal performance, such as B_{80}, which is the temperature range where organisms express 80% of their maximal performance. Therefore, TPC parameters provide a useful framework to define Janzen's physiological barrier.

Janzen's hypothesis originally predicted that tropical areas, with low-temperature variability, would favor thermal specialists, while temperate regions, with more variable climates, would favor thermal generalists. This argument can also be applied to explain physiological diversity along local environmental gradients (Sheldon et al. 2018). In the tropical Andes, lowland and mountain forests, with their thermal buffering capacity, are less thermally variable than open environments, such as natural grasslands or anthropogenic habitats like pastures, potreros, and crops established within the last few centuries. Consequently, Janzen's climatic variability hypothesis predicts that forest ectotherms should exhibit narrower tolerance breadths than species in open habitats across the elevational gradient. This means that physiological barriers to dispersal will be stronger for thermally specialized forest species, which will also have narrower elevational ranges compared to open-habitat species (Ghalambor et al. 2006; Sheldon et al. 2018, see Pintanel et al. 2019 for a test).

Restoring landscape connectivity emerges as a vital strategy to mitigate the compounded effects of habitat modification and climate change. The integration of thermal biology into landscape ecology highlights the critical role of microclimatic refugia in maintaining biodiversity across fragmented landscapes. For instance, microclimate mapping reveals that secondary forests, when strategically connected to primary forests, can enhance thermal landscape continuity (González-del-Pliego et al. 2016; Nowakowski et al. 2018a, b), enabling species to track favorable thermal conditions without large-scale migrations (Samira Geres et al. 2024). Moreover, conservation efforts focused on ecological corridors and secondary forest regrowth not only support biodiversity but also create opportunities for thermally sensitive species to adapt locally to gradual warming (González-del-Pliego et al. 2016, 2020). These strategies are particularly pertinent in the tropical Andes, where topographic and thermal barriers amplify the impacts of climate change.

This chapter explores the remarkable diversity in thermal tolerances and sensitivity of fitness-related traits among amphibian communities in the Ecuadorian tropical Andes. To understand how physiological diversity varies across spatial (elevation) and local (habitat) scales, and whether this variation correlates with local temperatures experienced by organisms, we begin by examining two major macrophysiological hypotheses (Gaston et al. 2009), such as Brett's hypothesis, which predicts that elevational climatic gradients promote less geographic variation in upper thermal limits compared to lower thermal limits (Brett 1956), and Janzen's hypothesis, which posits that narrow local habitat variation promoted thermal specialists, with narrower thermal breadth, whereas thermal generalist, with broader thermal breadth, will dominate habitats with broader temperature variation (Janzen's hypothesis, thought 3 in Janzen 1967, and assumption 2 in Sheldon et al. 2018), (Fig. 1, inbox). Thermal physiological traits may serve as barriers to organismal dispersion (i.e., Janzen's hypothesis) and promote reproductive isolation, potentially accounting for the extraordinary diversity of frogs in this region (Ghalambor et al. 2006). Next, we calculate vulnerability indexes to ongoing warming using tolerance and optimal trait data alongside current and projected environmental temperatures. For this purpose, we extracted thermal tolerance data from studies conducted on amphibians in the Tropical Andes. To ensure comprehensive coverage, we cross-checked our dataset against one of the most extensive databases on heat tolerance in amphibians (Pottier et al. 2022). Only studies with substantial data in terms of species richness (>10 species) and elevational distribution (>1000 m) were included. We excluded redundant datasets and non-peer-reviewed studies to maintain consistency and prevent potential conflicts with the publication of those works (i.e., Bernal Castro 2019). Additionally, we explore how biotic interactions could shape species' thermal physiology, highlighting their importance for understanding local thermal adaptation and synergistic effects on their vulnerability. For this purpose, we studied the mismatch in physiological tolerance at the community level between predators (dragonfly naiads) and prey (tadpoles) across elevational and local habitat scales.

In summary, this chapter delves into the spatial variation in physiological tolerances of Ecuadorian Andean amphibians, highlighting the intricate interplay of elevational gradients, microclimatic conditions, and biotic interactions. These factors collectively shape the thermal niches and distribution patterns of these species. As global temperatures continue to rise, understanding these adaptive mechanisms is crucial for predicting the impacts of climate change on biodiversity and developing effective conservation strategies.

3 Spatial Variation in Physiological Tolerances in Tropical Andes Amphibians: Spatial Patterns, Drivers, and Evolutionary Potential

The tropical Andes of Ecuador exhibit steep climatic gradients, housing a diverse array of ecosystems from lowland mountain forests to high-elevation páramos, streams, wetlands, and anthropogenically disturbed environments. This environmental richness has fostered the evolution of an exceptionally diverse amphibian community, making it a global hotspot with over 677 species, including 325 endemics, in just 250,000 km^2 (Ron et al. 2024).

Studying how elevation influences the variation in cold (i.e., critical thermal minimum or CT_{min}) and heat (i.e., critical thermal maximum or CT_{max}) tolerance provides insights into the directional patterns of organismal dispersal across mountain gradients. According to macrophysiological principles (Chown and Gaston 2016), it is expected that climatic gradients promote less geographic variation in heat tolerance compared to cold tolerance—a concept known as the "heat-invariant hypothesis" or "Brett's hypothesis" (Brett 1956; Chown and Gaston 2016). This pattern, where heat tolerance remains relatively constant while cold tolerance varies more significantly, has been observed in other ectotherms across both latitudinal and elevational gradients (Muñoz et al. 2014; Sunday et al. 2014), reinforcing the idea that species' thermal limits are more flexible at colder extremes than at warmer ones.

We examined the thermal tolerance for 75 tadpole species (Pintanel et al. 2022) and 21 adult species of adult *Pristimantis* (Pintanel et al. 2019), ranging from a combined 23 to 4236 m a.s.l., inhabiting a wide range of local microenvironments. In both aquatic larval and terrestrial frogs, the findings supported Brett's heat-invariant hypothesis, indicating that elevation induced more variation in cold tolerance compared to heat tolerance, similar to the findings of von May et al. (2017) in terrestrial Peruvian *Pristimantis* (see Fig. 3). While CT_{max} remained relatively stable across elevations, CT_{min} values varied strongly suggesting a greater evolutionary lability in cold tolerance compared to heat tolerance. Conversely, species exhibited higher heat tolerances at similar elevations in open ponds and terrestrial environments compared to streams and forest habitats in aquatic and terrestrial environments, respectively (Pintanel et al. 2019, 2022), thus suggesting an invert Brett's heat invariant rule across finer spatial scales (e.g. across microenvironments, Pintanel et al. 2025). Therefore, species in more thermally variable habitats exhibited broader thermal breadths than those in more stable stream environments, thus confirming Janzen's climatic variability hypothesis, showing that species in thermally variable open habitats had higher CT_{max} and broader thermal breadths than those in more thermally stable habitats. These results highlight the influence of microhabitat thermal variability on physiological tolerance limits and the evolutionary potential of these amphibians to adapt to changing climates.

The contrasting pattern in thermal niche variation between cold (labile) and heat (conservative) tolerances with elevation reflects the spatial pattern of peak maximum (t_{max}) and minimum (t_{min}) environmental temperatures across habitats. Unlike maximum temperatures, minimum temperatures did not vary significantly across habitats, making it impossible for frogs to mitigate cold extremes at higher elevations through thermoregulation. Consequently, frogs cannot shield themselves against extreme minimum temperatures and must instead adapt physiologically. This necessity for physiological adaptation likely explains why CT_{min} is highly variable across elevations and evolves more rapidly than CT_{max} (Muñoz et al. 2014; Muñoz and Bodensteiner 2019). This "static" nature of heat tolerance poses a significant challenge for tropical amphibians as global temperatures rise, particularly for lowland species already living near their upper thermal limits. These species may need to shift their ranges upslope or seek cooler microhabitats to survive (Samira Geres et al. 2024), underscoring the importance of thermal refugia in conservation strategies.

We further explored the relationship between thermal physiology, elevational distribution, and thermal stress in dendrobatid frogs (Pintanel et al. 2024). We used thermal performance curves (TPCs) and environmental temperature models to predict current and future exposure to thermal variation. Our findings showed significant correlations between thermal physiology metrics—such as optimum temperature and cold resistance—and elevational distributions. However, the thermal physiology models predicted lower distributions than those observed for high-elevation species. This discrepancy highlights the complex relationship between environmental temperatures and the physiological tolerances of tropical Andean amphibians. The findings suggest that while lowland species thrive in thermally optimal environments, highland species inhabit suboptimal thermal conditions, which may limit their distribution. However, the study also posits that highland species could benefit from the warming climate, but potential negative biotic interactions may further constrain their distribution (see below, Sect. 4). This complex dynamic underscores the necessity of understanding thermal tolerances and their evolutionary potential in

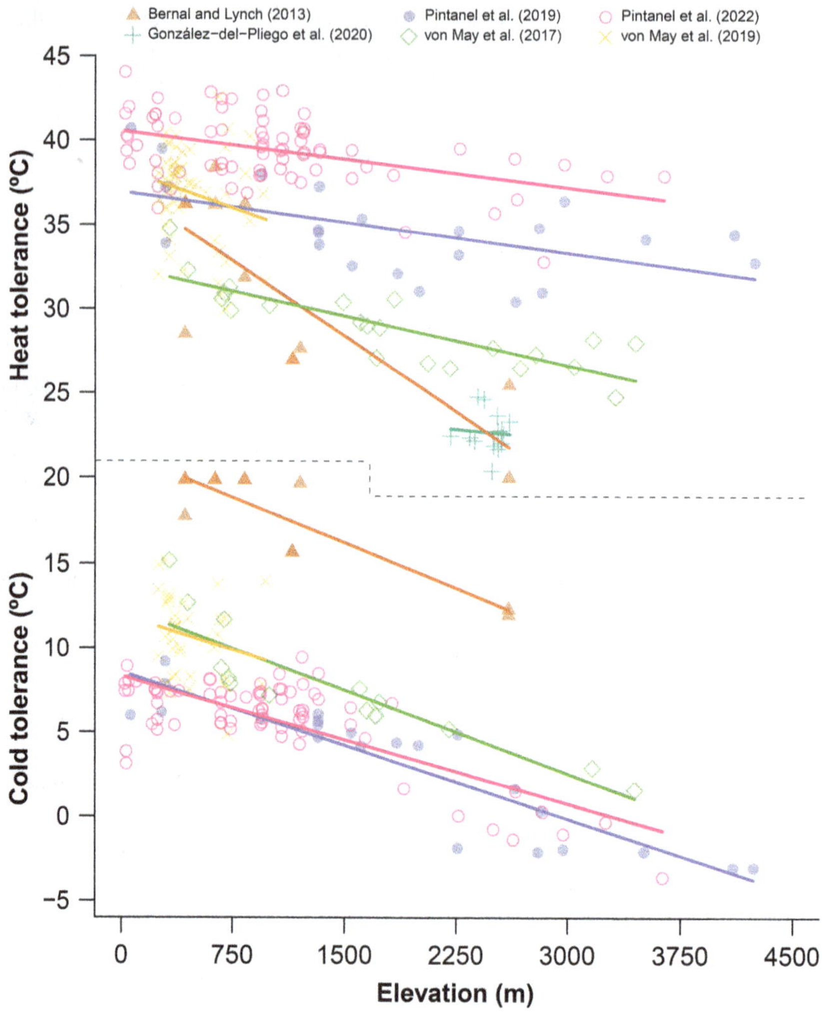

Fig. 3 Heat and cold thermal tolerance in various species of tropical Andean amphibians at different ontogenetic stages, from own results and other sources extracted from the literature: Colombian embryo species (Bernal and Lynch 2013), Ecuadorian larval species (Pintanel et al. 2022), Ecuadorian *Pristimantis* adults (Pintanel et al. 2019), Colombian *Pristimantis* adults (González-del-Pliego et al. 2020), Peruvian *Pristimantis* adults (von May et al. 2017), and Peruvian adult species (von May et al. 2019). Regression lines are derived from raw data. The methodology for obtaining thermal tolerances differs between articles. Most thermal tolerance estimates were based on CT_{max} and CT_{min} obtained employing dynamic ramping methods (Lutterschmidt and Hutchison 1997) with the exception of Bernal and Lynch (2013) tolerances based on TPCs of embryo survival for 12 frogs' species. In this case, thermal tolerances were equated to temperatures for a survival higher than 70% (see their Fig. 1). For the latter source, data were obtained using the online program WebPlotDigitizer. González-del-Pliego et al. (2020) does not provide cold tolerance data

predicting species responses to climate change, suggesting that factors beyond thermal physiology, such as habitat buffering or biotic interactions, may play crucial roles in species' elevational ranges.

The evolutionary potential of tropical Andean amphibians to adapt to global warming is further complicated by their microhabitat preferences. Species inhabiting open habitats are more exposed to extreme temperature fluctuations, leading to higher heat tolerances and broader thermal breadths. In contrast, species in forested areas or streams, which experience more stable thermal environments, have narrower thermal tolerances. This microhabitat-specific thermal adaptation suggests that habitat fragmentation and changes in land use could significantly impact the ability of these species to cope with climate change. Edge effects, which alter microhabitat temperatures, further compound these challenges by exacerbating thermal stress on amphibian populations (Posse-Sarmiento and Banks-Leite 2024).

It is important to emphasize the importance of habitat-driven thermal variation, which could provide climatic refuges for species most vulnerable to thermal shocks. This spatial thermal heterogeneity along elevational gradients is crucial in shaping the thermal tolerance diversity observed among these amphibians. The potential for genetic isolation due to limited dispersal ability in forest species could lead to higher speciation rates in tropical mountainous populations, particularly among terrestrial breeding frogs. These findings underscore the critical role of habitat and microclimatic conditions in shaping the thermal niches and evolutionary trajectories of Andean amphibians.

4 Patterns of Spatial Physiological Tolerances in Interacting Species

Climate change may significantly impact ecological communities. Biotic interactions can either mitigate or exacerbate the effects of climate change, depending on the relative thermal sensitivities of the species involved. We examined the thermal tolerance in predatory dragonfly naiads and amphibian larvae, providing insights into how thermal mismatches between predators and prey vary across spatial gradients (Pintanel et al. 2021). Predatory dragonfly naiads exhibit much higher heat tolerances than their tadpole prey, a trend that remains stable across different habitats and elevations. This consistent thermal advantage suggests that dragonfly larvae can maintain predatory pressure even as temperatures rise, potentially increasing their impact on prey populations under climate warming scenarios. However, the effects of climate change on biotic interactions are not always straightforward. For instance, increased temperatures can enhance predator performance, leading to higher predation rates. However, if prey species also benefit from warmer conditions through increased growth rates, the net effect on predator–prey dynamics might be neutral or even beneficial to prey (Amundrud and Srivastava 2020). Therefore, the interplay between direct physiological effects and indirect biotic interactions is complex and requires further and more comprehensive investigation.

5 Can Tropical Andean Amphibians Take the Heat?

In the preceding sections, we have discussed how environmental conditions in the Andes influence the variation in physiological thermal limits that may constrain organismal distribution. Here, we turn our focus from the spatial to the temporal aspect, examining how Andean amphibians would respond to global change, particularly the challenges posed by rising temperatures over time. Understanding both the spatial and temporal dimensions of thermal adaptation is crucial for predicting how these species will fare in a warming world.

Tropical amphibians face a significant risk to rising temperatures, given their exposure near to their physiological optimum (Deutsch et al. 2008; Huey et al. 2009) and thermal limits (Duarte et al. 2012; Pintanel et al. 2019, 2022; von May et al. 2019). As discussed previously, the relatively static nature of CT_{max} poses a critical challenge, especially for lowland species already living near their upper thermal limits (Duarte et al. 2012; Catenazzi et al. 2014; Gutiérrez-Pesquera et al. 2016; Pintanel et al. 2019, 2022; von May et al. 2017, 2019).

Amphibians in the Tropical Andes are particularly susceptible to climate warming especially at lowlands since environmental maximum temperatures (t_{max}) are closer to their upper thermal limits showing low values of warming tolerances ($CT_{max} - t_{max}$) (e.g., Duarte et al. 2012). This may indicate that, as global temperatures rise, amphibians at lower elevations may experience significant thermal stress at all ontogenetic stages, embryo, larvae, and adults (Bernal and Lynch 2013; Pintanel et al. 2019, 2021, 2022; von May et al. 2017) (see Fig. 4). Our results on physiological distribution models in dendrobatids, which often predict lower than observed distributions for high-elevation species, suggest that factors beyond thermal physiology, such as habitat buffering or biotic interactions, should play crucial roles in determining species' elevational ranges (Pintanel et al. 2024). The findings indicate that as temperatures continue to rise, Andean amphibians may be forced to shift their elevational ranges, adapt their behaviors, or seek cooler microhabitats to survive, underscoring the critical importance of thermal refugia in conservation strategies.

Furthermore, studies on predator–prey dynamics highlight another layer of vulnerability. Predatory dragonfly larvae generally have much higher heat tolerances than their tadpole amphibian prey. This thermal mismatch can exacerbate the impact of warming, as predators can maintain their physiological performance at higher temperatures, potentially increasing predation pressure on amphibian populations (Pintanel et al. 2021). This dynamic underscores the multifaceted challenges that tropical Andean amphibians face in a warming climate, emphasizing the need for comprehensive conservation strategies that consider both direct thermal stress and indirect ecological interactions.

In summary, the ability of tropical Andean amphibians to "take the heat" is limited by several factors: their inherent thermal physiological constraints, the necessity for physiological adaptations, their limited mobility, the rapid pace of temperature increases that outstrip their capacity for evolutionary adaptation, complex ecological interactions, and anthropogenic pressures. Additionally, the synergistic interactions among these factors further complicate their survival prospects. As global temperatures rise, understanding these adaptive mechanisms and vulnerabilities is crucial for predicting the impacts of climate change on these species and developing effective conservation strategies to ensure their survival.

6 Future Perspectives in Conservation Physiology in the Tropical Andes

To ensure the long-term conservation of tropical Andean amphibians, a multifaceted approach is required, combining immediate conservation actions with comprehensive future research. One critical measure is the continuous monitoring of environmental temperatures and habitat conditions in terrestrial and aquatic habitats. Establishing long-term monitoring programs will help track changes and assess the real-time impacts of climate change on amphibian populations. Therefore, this could allow stakeholders to identify areas where temperature extremes are naturally mitigated and provide critical safe havens for vulnerable species. Detailed microclimate studies can inform targeted conservation efforts by highlighting specific areas in need of stricter environmental regulations necessary to protect natural habitats and control human activities, or in need of restoration or rehabilitation. As a last resort, cool environments (high páramos, upper cloud forests) can result in ecological traps due to the presence of chytrid fungus (e.g., Seimon et al. 2007). A possibility by creating artificial thermal refugia for targeted amphibian populations might be a critical strategy. For instance, artificial saunas, or hotspot shelters, stimulate frog resistance to emerging diseases, such as chytridiomycosis (Waddle et al. 2024).

Building on current knowledge, future research in thermal physiology will play a critical role in identifying the thermal thresholds and responses of tropical Andean amphibians to ongoing climatic change. Improving the geographic coverage of physiological data is essential, as current data are often biased toward specific regions and taxa, limiting our ability to make robust inferences about global patterns (Chown and Gaston 2016). This coverage should also target the effects of microclimate variability on physiological traits, as current models often rely on macroclimate data that may not accurately reflect the environmental conditions experienced by organisms (Gutiérrez-Pesquera et al. 2016; Bovo et al. 2018). This includes understanding how daytime and nighttime temperature changes in highland versus lowland species affect organisms' fitness and survival (see Navas 1996d; Rueda-Solano et al. 2016). Physiological experimental studies attempting to mimic natural conditions or daily thermal fluctuations are increasingly being conducted in the Andes (e.g., Turriago et al. 2015, 2022, 2023; Bernal et al. 2017). Although challenging, these approaches are essential for scaling physiological data from laboratory studies to field conditions, thereby enhancing our ability to predict the ecological consequences of physiological variation under realistic environmental scenarios.

Analogous to micro- and macro-environmental temperatures, future physiological studies should focus into on thermal adaptation at the population or individual scales rather than at the species'. These studies should consider the spatial and temporal scales relevant to the organisms' life history to accurately predict species' responses to environmental changes. For instance, a study with Iberian lizards suggests that Brett's hypothesis inverts at intraspecific scales, with heat tolerance being more variable than cold tolerance (Herrando-Pérez et al. 2020). Recently, Pintanel et al. (2025) has postulated an analogous invert Brett's hypothesis at the inter-specific level, across microenvironments for Andean amphibian species and other taxa, see Section 3 above). Conversely, Páez-Vacas and Funk (2024) found that while CT_{max} was static, CT_{min} decreased with elevation in the Ecuadorian poison frog species, *Epipedobates anthonyi*, across populations in two 1500 elevational gradients (spatial scales). Additionally, understanding the role of stage-specific thermal tolerances (temporal scales) provides insights into the vulnerability to global warming of complex life-cycle species, such as amphibians, allowing us to identify the most vulnerable ontogenetic stages (e.g., Turriago et al. 2015; Enriquez-Urzelai et al. 2019) and thereby allocate our conservation efforts more effectively (see Fig. 4).

Ecological interactions play a pivotal role in shaping biodiversity and ecosystem dynamics. Therefore, it is not surprising that thermal physiology mediates ecological interactions at multiple levels, including within and between species and across community assemblages. Understanding the impact of physiological variation on competitive dynamics, host–pathogen and predator–prey relationships, and mutualistic interactions along environmental gradients is crucial. Additionally, it is essential to examine how environmental stressors like climate change modify these interactions by altering species' physiological performance. Research should also focus on the cascading effects of physiological changes on ecosystem functions and services, especially considering biodiversity loss and habitat modification. By integrating physiological data with ecological network models, we can achieve a deeper understanding of how individual traits affect broader ecological processes. For instance, such studies may help explain why highland tropical species with broad (e.g., generalists) thermal tolerance breadths (Navas 1996d, 1997; Pintanel et al. 2022) do not migrate to lower elevations, despite having similar sensitivities to their lowland counterparts (Pintanel et al. 2024).

Furthermore, exploring the interaction between multiple abiotic environmental factors and their impact on fitness over the Andes will provide insights into the complex nature of physiological responses to climate change. Future studies should examine how different combinations of environmental variables influence physiological plasticity, adaptation, and overall organismal fitness—or performance as a proxy for fitness—in both natural and experimental settings. Amphibians during terrestrial life are governed by hydric conditions due to their permeable skin, so reduced moisture and water availability become major constraints of frog activity

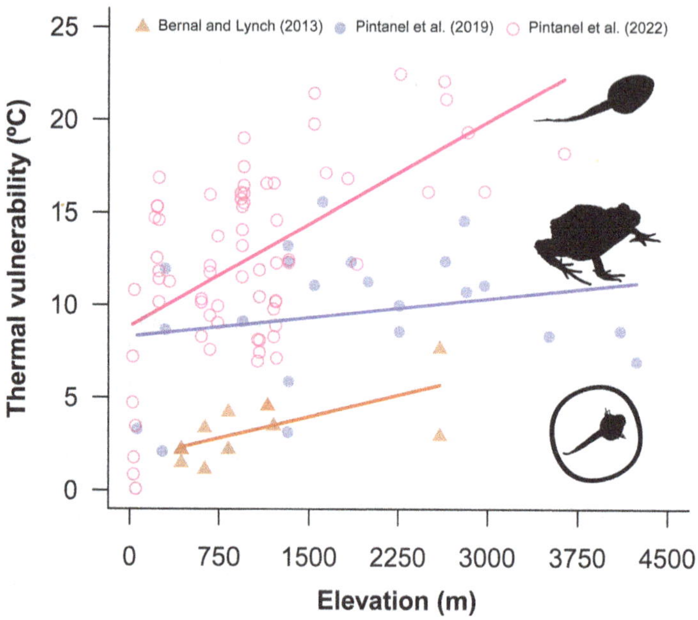

Fig. 4 Thermal vulnerability defined by the index warming tolerance (e.g., $CT_{max} - t_{max}$) among Andean amphibians at different ontogenetic stages: embryos (Bernal and Lynch 2013), larvae (Pintanel et al. 2022), and *Pristimantis* adults (Pintanel et al. 2019). Regression lines derived from raw data. In the case of Bernal and Lynch (2013), warming tolerance was equated to the thermal tolerances for a survival higher than 70% minus the mean maximum temperatures of the environment. Data were obtained using the online program WebPlotDigitizer

and fitness. Further studies examining water economy adaptations in tropical Andean amphibians are badly needed, since its fundamental role for understanding species distribution and their ability to disperse (Bovo et al. 2018). For instance, Cruz-Piedrahita et al. (2018), examined evaporative water loss among 17 Colombian lowland anuran species and populations through a hydric gradient, and found that arboreal species inhabiting more xeric habitats exhibited lower evaporative water loss, whereas higher levels of water uptake were noted for terrestrial species. Galindo and Bernal (2018) found that thermal selection in a terrestrial lungless salamander was dependent on the hydric environment. Additionally, it is important to address the behavior and limited mobility of Andean amphibians (but see Seimon et al. 2007; Sinsch 2014), which restrict their ability to thermoregulate or migrate in response to changing temperatures. Research on movement ecology, ancestral reconstructions of amphibian distribution throughout the Andes, and the development of predictive physiological models may offer initial insights into their potential responses to global change. Anthropogenic pressures, such as habitat conversion and pollution, interact synergically with climate change, further exacerbating the vulnerability of amphibian populations and restricting their mobility (e.g., Nowakowski et al. 2018a, b). This information should improve strategic decision-making in conservation efforts, including the creation of corridors that facilitate movement between habitats, enabling species to find suitable microclimates.

Genetic studies are instrumental in understanding the adaptive capacity of different species concerning thermal physiology. Incorporating molecular techniques to measure thermal thresholds at which cellular or protein damage occurs can provide a more precise understanding of thermal tolerance. This approach should be integrated into macroecological and physiological research to correlate molecular, cellular, and whole-organism levels of biological organization (Pörtner et al. 2006; Somero 2011). Differentiating evolutionary adaptation and phenotypic plasticity under changing climates still remains a challenging yet crucial aspect of understanding species' responses (Merilä and Hendry 2014). In this context, amphibians may constitute an ideal group of terrestrial vertebrates for studying thermal adaptation through natural selection experiments in the field or laboratory. This is because they are ectotherms, highly sensitive to temperature changes, and some species, such as Andean dendrobatid frogs (e.g., *Epipedobates* spp. or *Hyloxalus* spp.), have relatively short generation times, facilitating studies of evolutionary processes over multiple generations within a feasible timeframe, reproduce easily in artificial conditions and lay many eggs, with some species reproducing weekly.

Ex situ conservation strategies, including captive breeding programs, offer a last line of defense against extinction for critically endangered species. Furthermore, these centers provide a valuable opportunity to conduct research on thermal physiology within controlled environments and avoiding any adverse effects on wild populations. The physiological data generated in these centers are instrumental in assessing species' vulnerability to climate change and supporting the restoration of populations in natural habitats. For instance, a study on *Agalychnis spurrelli* demonstrated that both captive-bred and wild-caught tadpoles exhibited similar thermal tolerances, indicating that captive-bred populations can be effectively used to investigate thermal physiological adaptations and predict species' resilience to global warming (Pintanel et al. 2020).

References

Amundrud SL, Srivastava DS (2020) Thermal tolerances and species interactions determine the elevational distributions of insects. Glob Ecol Biogeogr 29:1315–1327. https://doi.org/10.1111/geb.13106

Angilletta MJ (2009) Thermal adaptation: a theoretical and empirical synthesis. Oxford University Press, Oxford

Bernal Castro EA (2019) Influence of environment on thermal ecology of direct-developing frogs (Anura: Craugastoridae: *Pristimantis*) in the eastern Andes of Colombia. (Master), Uniandes, Bogotá. Retrieved from hdl.handle.net/1992/44204

Bernal MH, Lynch JD (2013) Thermal tolerance in anuran embryos with different reproductive modes: relationship to altitude. Sci World J 2013:183212. https://doi.org/10.1155/2013/183212

Bernal MH, Turriago JL, Villa-Navarro FA (2017) Impact of daily variable temperatures in life-history traits of tropical anurans. Rev Biol Trop 65:55–63. https://doi.org/10.15517/rbt.v65i1.2049

Bovo R, Navas C, Tejedo M, Valença S, Gouveia S (2018) Ecophysiology of amphibians: information for best mechanistic models. Diversity 10:118. https://doi.org/10.3390/d10040118

Brett JR (1956) Some principles in the thermal requirements of fishes. Q Rev Biol 31:75–87. https://doi.org/10.1086/401257

Brown CJ, O'Connor MI, Poloczanska ES, Schoeman DS, Buckley LB, Burrows MT, Duarte CM, Halpern BS, Pandolfi JM, Parmesan C, Richardson AJ (2016) Ecological and methodological drivers of species' distribution and phenology responses to climate change. Glob Change Biol 22:1548–1560. https://doi.org/10.1111/gcb.13184

Burrowes PA, Navas CA, Jiménez-Robles O, Delgado P, De la Riva I (2020) Climatic heterogeneity in the Bolivian Andes: are frogs trapped? South Am J Herpetol 18:1–12. https://doi.org/10.2994/SAJH-D-18-00047.1

Carvajalino-Fernández JM, Navas CA (2025) Physiological ecology of Andean amphibians associated with freezing temperatures. In: Urbina-Cardona JN, Navas CA, Carla Piantoni, Alessandro Catenazzi, y Nicolás Urbina-Cardona

Carvajalino-Fernández JM, Bonilla Gomez MA, Navas CA (2011) Freezing risk in tropical high-elevation anurans: an assessment based on the Andean frog *Pristimantis nervicus* (Strobomantidae). South Am J Herpetol 6:73–78. https://doi.org/10.2994/057.006.0205

Carvajalino-Fernández JM, Bonilla Gomez MA, Giraldo-Gutierréz L, Navas CA (2021) Freeze tolerance in neotropical frogs: an intrageneric comparison using *Pristimantis* species of high elevation and medium elevation. J Trop Ecol 37:118–125. https://doi.org/10.1017/S026646742100016X

Catenazzi A, Lehr E, Vredenburg VT (2014) Thermal physiology, disease, and amphibian declines on the eastern slopes of the Andes. Conserv Biol 28:509–517. https://doi.org/10.1016/j.biocon.2013.09.010

Chown SL, Gaston KJ (2016) Macrophysiology—progress and prospects. Funct Ecol 30:330–344. https://doi.org/10.1111/1365-2435.12510

Colwell RK, Brehm G, Cardelús CL, Gilman AC, Longino JT (2008) Global warming, elevational range shifts, and lowland biotic attrition in the wet tropics. Science

322:258–261. https://doi.org/10.1126/science.1162547

Cruz-Piedrahita C, Navas CA, Crawford AJ (2018) Life on the edge: a comparative study of ecophysiological adaptations of frogs to tropical semiarid environments. Physiol Biochem Zool 91:740–756. https://doi.org/10.1086/695705

De Frenne P, Zellweger F, Rodríguez-Sánchez F, Scheffers BR, Hylander K, Luoto M et al (2019) Global buffering of temperatures under forest canopies. Nat Ecol Evol 3:744–749. https://doi.org/10.1038/s41559-019-0842-1

De Frenne P, Lenoir J, Luoto M, Scheffers BR, Zellweger F, Aalto J et al (2021) Forest microclimates and climate change: importance, drivers and future research agenda. Glob Chang Biol 27:2279–2297. https://doi.org/10.1111/gcb.15569

De Lombaerde E, Vangansbeke P, Lenoir J, Van Meerbeek K, Lembrechts J, Rodríguez-Sánchez F et al (2022) Maintaining forest cover to enhance temperature buffering under future climate change. Sci Total Environ 810:151338. https://doi.org/10.1016/j.scitotenv.2021.151338

Deutsch CA, Tewksbury JJ, Huey RB, Sheldon KS, Ghalambor CK, Haak DC, Martin PR (2008) Impacts of climate warming on terrestrial ectotherms across latitude. Proc Natl Acad Sci USA 105:6668–6672. https://doi.org/10.1073/pnas.0709472105

Duarte H, Tejedo M, Katzenberger M, Marangoni F, Baldo D, Beltrán JF, Martí DA, Richter-Boix A, Gonzalez-Voyer A (2012) Can amphibians take the heat? Vulnerability to climate warming in subtropical and temperate larval amphibian communities. Glob Chang Biol 18:412–421. https://doi.org/10.1111/j.1365-2486.2011.02518.x

Enriquez-Urzelai U, Sacco M, Palacio AS, Pintanel P, Tejedo M, Nicieza AG (2019) Ontogenetic reduction in thermal tolerance is not alleviated by earlier developmental acclimation in *Rana temporaria*. Oecologia 189:385–394. https://doi.org/10.1007/s00442-019-04342-y

Galindo CA, Cruz EX, Bernal MH (2018) Evaluation of the combined temperature and relative humidity preferences of the Colombian terrestrial salamander *Bolitoglossa ramosi* (Amphibia:Plethodontidae) Can. J Zool 96:1230–123512. https://doi.org/10.1139/cjz-2017-0330

Gaston KJ, Chown SL, Calosi P, Bernardo J, Bilton DT, Clarke A et al (2009) Macrophysiology: a conceptual reunification. Am Nat 174:595–612. https://doi.org/10.1086/605982

Ghalambor CK, Huey RB, Martin PR, Tewksbury JJ, Wang G (2006) Are mountain passes higher in the tropics? Janzen's hypothesis revisited. Integr Comp Biol 46:5–17. https://doi.org/10.1093/icb/icj003

González-del-Pliego P, Scheffers BR, Basham EW, Woodcock P, Wheeler C, Gilroy JJ et al (2016) Thermally buffered microhabitats recovery in tropical secondary forests following land abandonment. Biol Conserv 201:385–395. https://doi.org/10.1016/j.biocon.2016.07.038

González-del-Pliego P, Scheffers BR, Freckleton RP, Basham EW, Araújo MB, Acosta-Galvis AR, Medina Uribe CA, Haugaasen T, Edwards DP (2020) Thermal tolerance and the importance of microhabitats for Andean frogs in the context of land use and climate change. J Anim Ecol 89:2451–2460. https://doi.org/10.1111/1365-2656.13309

González-del-Pliego P, Freckleton RP, Scheffers BR, Basham EW; ,Acosta-Galvis AR, Medina Uribe CA, Haugaasen T, Edwards DP (2023) Phylogeny and morphology determine vulnerability to global warming in *Pristimantis* frogs. Land 12: 130. doi:https://doi.org/10.3390/land12010130

Gutiérrez-Pesquera LM, Tejedo M, Olalla-Tárraga MÁ, Duarte H, Nicieza A, Solé M (2016) Testing the climate variability hypothesis in thermal tolerance limits of tropical and temperate tadpoles. J Biogeogr 43: 1166–1178. https://doi.org/10.1111/jbi.12700

Herrando-Pérez S, Monasterio C, Beukema W, Gomes V, Ferri-Yáñez F, Vieites DR et al (2020) Heat tolerance is more variable than cold tolerance across species of Iberian lizards after controlling for intraspecific variation. Funct Ecol 34:631–645. https://doi.org/10.1111/1365-2435.13507

Herzog SK, Martínez R, Jørgensen PM, Tiessen H (2011) Climate change and biodiversity in the tropical Andes. Inter-American Institute for Global Change Research (IAI) and Scientific Committee on Problems of the Environment (SCOPE), Montevideo

Huey RB, Stevenson RD (1979) Integrating thermal physiology and ecology of ectotherms: discussion of approaches. Am Zool 19:357–366. https://doi.org/10.1093/icb/19.1.357

Huey RB, Deutsch CA, Tewksbury JJ, Vitt LJ, Hertz PE, Álvarez Pérez HJ, Garland T (2009) Why tropical forest lizards are vulnerable to climate warming. Proc R Soc B Biol Sci 276:1939–1948. https://doi.org/10.1098/rspb.2008.1957

Janzen DH (1967) Why mountain passes are higher in the tropics. Am Nat 101:233–247. https://doi.org/10.1086/282487

Leahy L, Scheffers BR, Williams SE, Andersen AN (2022) Arboreality drives heat tolerance while elevation drives cold tolerance in tropical rainforest ants. Ecology 103:e03549. https://doi.org/10.1002/ecy.3549

Lüddecke H (1997) Colonization of the eastern Andes of Colombia by anurans: evidence from natural history data of *Hyla labialis*. Salamandra 33:11–132

Lutterschmidt WI, Hutchison VH (1997) The critical thermal maximum: data to support the onset of spasms as the definitive end point. Can J Zool 75:1553–1560. https://doi.org/10.1139/z97-782

Merilä J, Hendry AP (2014) Climate change, adaptation, and phenotypic plasticity: the problem and the evidence. Evol Appl 7:1–14. https://doi.org/10.1111/eva.12137

Muñoz MM, Bodensteiner BL (2019) Janzen's hypothesis meets the Bogert effect: connecting climate variation, thermoregulatory behavior and rates of physiological evolution. Integr Org Biol 1:1–12. https://doi.org/10.1093/iob/oby002

Muñoz MM, Stimola MA, Algar AC, Conover A, Rodriguez AJ, Landestoy MA, Bakken GS, Losos JB (2014) Evolutionary stasis and lability in thermal physiology in a group of tropical lizards. Proc R Soc Lond B Biol Sci 281:20132433. https://doi.org/10.1098/rspb.2013.2433

Nakamura A, Ashton LA, Scheffers BR, Kitching RL (2022) Editorial: understanding patterns and mechanisms of forest canopy diversity and ecosystem functions in a changing world. Front For Glob Change 5:944981. https://doi.org/10.3389/ffgc.2022.944981

Navas CA (1996a) The effect of temperature on the vocal activity of tropical anurans: a comparison of high and low-elevation species. J Herpetol 30:488. https://doi.org/10.2307/1565691

Navas CA (1996b) Thermal dependency of field locomotor and vocal performance of high-elevation anurans in the tropical Andes. J Herpetol 30:478. https://doi.org/10.2307/1565690

Navas CA (1996c) Implications of microhabitat selection and patterns of activity on the thermal ecology of high elevation neotropical anurans. Oecologia 108:617–626. https://doi.org/10.1007/bf00329034

Navas CA (1996d) Metabolic physiology, locomotor performance, and thermal niche breadth in neotropical anurans. Physiol Zool 69:1481–1501. https://doi.org/10.1086/physzool.69.6.30164271

Navas CA (1997) Thermal extremes at high elevations in the Andes: physiological ecology of frogs. J Therm Biol 22:467–477. https://doi.org/10.1016/S0306-4565(97)00065-X

Navas CA (2002) Herpetological diversity along Andean elevational gradients: links with physiological ecology and evolutionary physiology. Comp Biochem Physiol A 133:469–485. https://doi.org/10.1016/S0306-4565(01)00094-8

Navas CA (2006) Patterns of distribution of anurans in high Andean tropical elevations: insights from integrating biogeography and evolutionary physiology. Integr Comp Biol 46:82–91. https://doi.org/10.1093/icb/icj001

Navas CA, Gomes FR, Carvalho JE (2008) Thermal relationships and exercise physiology in anuran amphibians: integration and evolutionary implications. Comp Biochem Physiol A 151:344–362. https://doi.org/10.1016/j.cbpa.2007.07.003

Navas CA, Carvajalino-Fernández JM, Saboyá-Acosta LP, Rueda-Solano LA, Carvajalino-Fernández MA (2013) The body temperature of active amphibians along a tropical elevation gradient: patterns of mean and variance and inference from environmental data. Funct Ecol 27:1145–1154. https://doi.org/10.1111/1365-2435.12106

Nowakowski AJ, Frishkoff LO, Agha M, Todd BD, Scheffers BR (2018a) Changing thermal landscapes: merging climate science and landscape ecology through thermal biology. Curr Landsc Ecol Rep 3:57–72. https://doi.org/10.1007/s40823-018-0034-8

Nowakowski AJ, Watling JI, Thompson ME, Brusch GA, Catenazzi A, Whitfield SM et al (2018b) Thermal biology mediates responses of amphibians and reptiles to habitat modification. Ecol Lett 21:345–355. https://doi.org/10.1111/ele.12901

Páez-Vacas MI, Funk WC (2024) Thermal limits along tropical elevational gradients: poison frog tadpoles show plasticity but maintain divergence across elevation. J Therm Biol 120:103815. https://doi.org/10.1016/j.jtherbio.2024.103815

Pearson OP, Bradford DF (1976) Thermoregulation of lizards and toads at high altitudes in Peru. Copeia 1976:155–170. https://doi.org/10.2307/1443786

Pintanel P, Tejedo M, Ron SR, Llorente GA, Merino-Viteri A (2019) Elevational and microclimatic drivers of thermal tolerance in Andean *Pristimantis* frogs. J Biogeogr 46:1664–1675. https://doi.org/10.1111/jbi.13596

Pintanel P, Tejedo M, Merino-Viteri A, Almeida-Reinoso-F, Gutiérrez-Pesquera LM (2020) Critical thermal limits do not vary between wild-caught and captive-bred tadpoles of *Agalychnis spurrelli* (Anura: Hylidae). Diversity 12:43. https://doi.org/10.3390/d12020043

Pintanel P, Tejedo M, Salinas-Ivanenko S, Jervis P, Merino-Viteri A (2021) Predators like it hot: thermal mismatch in a predator-prey system across an elevational tropical gradient. J Anim Ecol 90:1985–1995. https://doi.org/10.1111/1365-2656.13516

Pintanel P, Tejedo M, Merino-Viteri A, Almeida-Reinoso-F, Salinas-Ivanenko S, López-Rosero AC, Llorente GA, Gutiérrez-Pesquera LM (2022) Elevational and local climate variability predicts thermal breadth of mountain tropical tadpoles. Ecography 2022:e05906. https://doi.org/10.1111/ecog.05906

Pintanel P, Tejedo M, Camacho A, Enriquez-Urtzelai U, Llorente GA, Merino-Viteri A (2024) Physiological thermal niches, elevational ranges and thermal stress in dendrobatid frogs: an integrated approach. J Biogeogr 51:1880–1893. https://doi.org/10.1111/jbi.14860

Pintanel P, Tejedo M, Enriquez-Urzelai U, Domínguez-Guerrero SF, & Muñoz MM (2025) High thermal variation in maximum temperatures invert Brett's heat-invariant rule at fine spatial scales Abstract Ecology 106(6). https://doi.org/10.1002/ecy.v106.6 10.1002/ecy.70124

Pörtner HO, Bennett AF, Bozinovic F, Clarke A, Lardies MA, Lucassen M et al (2006) Trade-offs in thermal adaptation: the need for a molecular to ecological integration. Physiol Biochem Zool 79:295–313. https://doi.org/10.1086/499986

Posse-Sarmiento V, Banks-Leite C (2024) The effects of edge influence on the microhabitat, diversity and life-history traits of amphibians in western Ecuador. J Trop Ecol 40:e7. https://doi.org/10.1017/S026646742400004X

Pottier P, Lin H-Y, Oh RRY, Pollo P, Rivera-Villanueva AN, Valdebenito JO et al (2022) A comprehensive database of amphibian heat tolerance. Sci Data 9:600. https://doi.org/10.1038/s41597-022-01704-9

Reider KE, Larson DJ, Barnes BM, Donnelly MA (2021) Thermal adaptations to extreme freeze–thaw cycles in the high tropical Andes. Biotropica 53:296–306. https://doi.org/10.1111/btp.12875

Ron SR, Merino-Viteri A, Ortiz DA (2024) Anfibios del Ecuador. Version 2024.0. Museo de Zoología, Pontificia Universidad Católica del Ecuador. https://bioweb.bio/faunaweb/amphibiaweb, fecha de acceso 3 de julio, 2024

Rueda-Solano LA, Navas CA, Carvajalino-Fernández JM, Amézquita A (2016) Thermal ecology of montane *Atelopus* (Anura: Bufonidae): a study of intrageneric diversity. J Therm Biol 58:91–98. https://doi.org/10.1016/j.jtherbio.2016.04.007

Samira Geres L, Blüthgen N, Müller J, Seifert L, Seibold S, Bässler C (2024) To the top or into the dark? Relationships between elevational and canopy cover distribution shifts in mountain forests. Ecography e07449. https://doi.org/10.1111/ecog.07449

Sarmiento G (1986) Ecological features of climate in high tropical mountains. In: Vuilleumier F, Monasterio M (eds) High altitude tropical biogeography. Oxford University Press, New York, pp 11–46

Seimon TA, Seimon A, Daszak P, Halloy SRP, Schloegel LM, Aguilar CA et al (2007) Upward range extension of Andean anurans and chytridiomycosis to extreme elevations in response to tropical deglaciation. Glob Chang Biol 13:288–299. https://doi.org/10.1111/j.1365-2486.2006.01278.x

Sheldon KS, Huey RB, Kaspari M, Sanders NJ (2018) Fifty years of mountain passes: a perspective on Dan Janzen's classic article. Am Nat 191:553–565. https://doi.org/10.1086/697046

Sinsch U (1989) Behavioural thermoregulation of the Andean toad (*Bufo spinulosus*) at high altitudes. Oecologia 80:32–38. https://doi.org/10.1007/bf00789928

Sinsch U (2014) Movement ecology of amphibians: from individual migratory behaviour to spatially structured populations in heterogeneous landscapes. Can J Zool 92:491–502. https://doi.org/10.1139/cjz-2013-0028

Somero GN (2011) Comparative physiology: a "crystal ball" for predicting consequences of global change. Am J Physiol Regul Integr Comp Physiol 301:R1–R14. https://doi.org/10.1152/ajpregu.00719.2010

Sunday JM, Bates AE, Kearney MR, Colwell RK, Dulvy NK, Longino JT, Huey RB (2014) Thermal-safety margins and the necessity of thermoregulatory behavior across latitude and elevation. Proc Natl Acad Sci 111:5610–5615. https://doi.org/10.1073/pnas.1316145111

Turriago JL, Parra CA, Bernal MH (2015) Upper thermal tolerance in anuran embryos and tadpoles at constant and variable peak temperatures. Can J Zool 93:267–272. https://doi.org/10.1139/cjz-2014-0254

Turriago JL, Tejedo M, Hoyos JM, Bernal MH (2022) The effect of thermal microenvironment in upper thermal tolerance plasticity in tropical tadpoles. Implications for vulnerability to climate warming. J Exp Zool A 337:746–759. https://doi.org/10.1002/jez.2632

Turriago JL, Tejedo M, Hoyos JM, Camacho A, Bernal MH (2023) The time course of acclimation of critical thermal maxima is modulated by the magnitude of temperature change and thermal daily fluctuations. J Therm Biol 114:103545. https://doi.org/10.1016/j.jtherbio.2023.103545

Valdivieso D, Tamsitt JR (1974) Thermal relations of the neotropical frog, *Hyla labialis* (Anura: Hylidae). Royal Ontario Museum Life Sciences. Occasional Papers 26. https://doi.org/10.5962/bhl.title.60691

von Humboldt A (1849) Aspects of nature in different lands and different climates; with scientific elucidations. trans. Elizabeth JL. Longman, Brown, Green and John Murray, Sabine. https://doi.org/10.5962/bhl.title.151955

von Humboldt A, Bonpland A (2009) Essay on the geography of plants. In: Jackson ST (ed) Translated by Sylvie Romanowski. Chicago University Press, Chicago/London

von May R, Catenazzi A, Corl A, Santa-Cruz R, Carnaval AC, Moritz C (2017) Divergence of thermal physiological traits in terrestrial breeding frogs along a tropical elevational gradient. Ecol Evol 7:3257–3267. https://doi.org/10.1002/ece3.2929

von May R, Catenazzi S-CR, Gutierrez AS, Moritz C, Rabosky DL (2019) Thermal physiological traits in tropical lowland amphibians: vulnerability to climate warming and cooling. PLoS One 14:e0219759. https://doi.org/10.1371/journal.pone.0219759

Waddle AW, Clulow S, Aquilina A, Sauer EL, Kaiser SW, Miller C et al (2024) Hotspot shelters stimulate frog resistance to chytridiomycosis. Nature 631:344–349. https://doi.org/10.1038/s41586-024-07582-y

Part IV

Conservation Challenges and Emerging Threats

Emerging Infectious Diseases of Amphibians in the Andean Region

Sandra V. Flechas, Andrea Paz, and Jenny Urbina

1 Introduction

The second global amphibian assessment confirms that the worldwide conservation crisis this group of vertebrates faces has not abated. According to the report, 41% of amphibian species are within threatened categories (Critically Endangered, Endangered, Vulnerable), more than any other group of vertebrates (Luedtke et al. 2023). Records of extinct species continue to emerge, from 23 species in 1980 to 37 species in 2022. Although different criteria must be met for a species to be considered extinct, several amphibian species have not been observed for decades and there is no evidence of surviving populations; therefore they have been categorized as Critically Endangered and labeled as possibly extinct (PE) (Luedtke et al. 2023). The number of species in this category (PE) has increased by an order of magnitude over the last few years, from 24 species in 1980 to 185 in 2022.

S. V. Flechas (✉)
Bichos.team, Bogotá, Colombia
e-mail: vickyflechas@gmail.com

A. Paz
Département de Sciences Biologiques, Université de Montréal, Montréal, QC, Canada
e-mail: andrea.paz.velez@umontreal.ca

J. Urbina
Department of Fisheries, Wildlife, and Conservation Sciences, Oregon State University, Corvallis, OR, USA
e-mail: urbinagj@oregonstate.edu

The Tropical Andes, spanning over a million and a half square kilometers and encompassing seven countries, is one of the most diverse regions in the world for amphibians (Myers et al. 2000). The number of species reported in the Andes exceeds 1,000 (AmphibiaWeb 2024) with around 800 endemic species, more diversity than most other biogeographic regions (Hutter et al. 2017). The Andes hotspot faces multiple threats to its biodiversity. Different drivers such as deforestation caused by the increasing need for agricultural areas, energy, and water (Armenteras et al. 2017), and increasing mining of diverse minerals of gold, silver, and copper (Wallace and Hall-Wallace 2003) lead to the degradation of soil and water sources which can also affect amphibians (Peluso et al. 2023).

Another critical threat to amphibian diversity, either independently or in synergy with other factors and threats, is emerging infectious diseases (EIDs). Despite its significant impact, only a few diseases have been thoroughly studied in the Andes, resulting in a substantial gap in our understanding. Recent studies report disease as the predominant driver of status deterioration not only in the Andes from 1980 to 2022 but in other places around the world (Luedtke et al. 2023). During the past two decades, our understanding of EIDs, their causative pathogens, and their impact and effects on hosts has improved substantially. This progress is attributable to the development of new analytical and diagnostic techniques, which enable accurate diagnosis and

J. N. Urbina-Cardona et al. (eds.), *Andean Herpetofauna*, Sustainable Development Goals Series,
https://doi.org/10.1007/978-3-032-00074-3_14

facilitate effective interventions. In addition, the significance of environmental factors in exacerbating infections or increasing the prevalence of pathogens has become increasingly apparent. Research has expanded to investigate the effects of climate change, alongside continued exploration of host–pathogen interactions. The Tropical Andes are one of the areas where the greatest number of endangered species are concentrated and the EIDs are one of the main threats. Understanding the impact of diseases on amphibians is crucial for conservation actions that focus on identifying and mitigating disease impacts and informing policy decisions to protect amphibians.

2 Amphibian Diseases: An Overview

Amphibians can be affected by various pathogens and parasites, including viruses, fungi, bacteria, protozoa, helminths, and arthropods (Densmore and Green 2007). Over the past few years, our understanding of diseases affecting amphibians has grown significantly. This surge of knowledge is mainly due to the efforts of pathologists and veterinarians who have gathered much information. These data enable us to identify the clinical signs and pinpoint the causative agent responsible for the observed pathology. Moreover, the advancement and development of new molecular tools have markedly enhanced our capacity to quickly and accurately identify infectious agents. This has facilitated accurate disease diagnosis, particularly in cases when the cause is not so obvious or when the clinical signs closely resemble one another. Despite these advancements, a need remains for rapid tests, for many pathogens. Such tests would facilitate and expedite decision-making processes regarding effective treatments, thereby reducing the risk of mortality.

Infectious diseases affecting amphibians have emerged as a significant concern in conservation biology (Daszak et al. 2000). Many populations and species have vanished worldwide due to EIDs; however, gauging the approximate number of affected species or the impact of each disease remains a challenge. Moreover, many mortality events of wild amphibians have been attributed to a specific etiological agent, based mainly on observing clinical signs and the isolation of bacteria that has led to erroneous conclusions. As a result, mass mortality events were attributed to an agent, while the true cause remained unrecognized and unaddressed. The similarity of clinical signs complicates accurate diagnosis. This underscores the statement by Pessier (2017) that "*the need for veterinary pathologists to direct and or provide support for studies into the amphibian disease has never been greate*r."

In the following sections, we concisely present a review of the diseases and infectious agents reported in amphibians. First, we will encompass general aspects of each disease, including clinical signs, diagnostic tests, and potential treatments. When available, we will provide a broader context of these findings, explicitly focusing on Andean species, and identifying existing knowledge gaps and areas that require further research.

2.1 Helminths

In South American amphibians, the most commonly reported *helminths* are nematodes and gastrointestinal roundworms belonging to the families Cosmocercidae, Kathlaniidae, Molineidae, Physalopteridae, and lungworms of Rhabdiasidae (Campião et al. 2015). After nematodes, trematodes are the next most common parasites. Although they infect a broad range of families, they are most frequently reported in aquatic species of the genera *Lithobates* and *Pseudis* (Campião et al. 2015). Trematode metacercariae have been associated with severe pathology in amphibians. Among these parasites, *Ribeiroia ondatrae* is widely distributed, and it is responsible for limb abnormalities in amphibians. Affected individuals may exhibit a range of deformities, from extra limbs to missing or partially missing hind limbs or cutaneous fusions (Johnson et al. 2001; Blaustein and Johnson 2003). Malformations caused by *Ribeiroia* are mostly detrimental to survival. Severe malformations impair mobility, hinder food intake, and increase susceptibility to predators and parasites (Blaustein and Johnson 2003; Johnson and McKenzie 2009). Despite our improved understanding of the type of parasites affecting amphibians, reports remain scarce, considering the great diversity of amphibian species in the Andes (Gómez et al. 2020).

2.2 Bacterial Infections

Multiple bacterial pathogens can infect the skin of amphibians, leading to diseases such as ulcerative dermatitis. These diseases can weaken amphibians, making them more susceptible to additional infections. One of the most reported diseases is the red leg syndrome (RLS), also known as red leg disease or bacterial dermatosepticemia. This disease is characterized by vasodilatation and hyperemia, especially in the extremities and the flanks (Densmore and Green 2007; Pasteris et al. 2006). Legs present subcutaneous edema and areas of hemorrhage within the skeletal muscle. In addition, the liver, spleen, and kidney are considerably enlarged (Mauel et al. 2002). Other clinical signs include anorexia, swelling, edema, coelomic effusions, epidermal erosions, ulcers, sloughing, or necrosis (Densmore and Green 2007). Red leg syndrome has been reported in a wide range of anurans and caudate species both in captivity and in the wild. The etiological agents of RLS are mainly members of the Enterobacteriaceae, including *Proteus vulgaris*, *Pr. mirabilis*, *Pseudomonas aeruginosa*, *Aeromonas hydrophila*, *Streptococcus iniae*, *Chryseobacterium meningosepticum*, *Chr. indulgences*, *Citrobacter freundii*, *Edwardsiella tarda*, and some strains of *Staphylococcus epidermis* (Mauel et al. 2002). Diagnosis is primarily made through the isolation and culture of bacteria for subsequent identification (Densmore and Green 2007). Treatment involves supportive therapy, stressor removal, administration of systemic antibiotics or antifungals (preferably determined by culture and sensitivity tests), and topical preparations on ulcerated skin (Drake et al. 2010). Antibiotic therapy has been demonstrated to decrease mortalities during outbreaks (Mauel et al. 2002).

Mycobacteriosis is a disease predominantly reported in adult individuals, caused by bacteria from the genus *Mycobacterium* (Martinho and Heatley 2012). At least nine mycobacterial species have been identified and described in amphibians including *M. marinum*, *M. chelonei*, *M. fortuitum*, *M. gordonae*, *M. liflandii*, *M. ranae*, *M. xenopi*, *M. abscessus*, *M. avium*, *M. thamnospheos*, and *M. szulgai* (Fremont-Rahl et al. 2011). These bacteria, ubiquitous in aquatic environments or as soil saprophytes, are commonly found in nature and are likely to be reported in facilities. Most reports of amphibian mycobacteriosis come from captive animals, and the prevalence of infections in wild individuals is unknown. Mycobacteriosis can manifest as a mild infection with few or no clinical signs, or as a severe infection with a high mortality rate. It is characterized by skin lesions often called "fish tank granuloma" (Ferreira et al. 2006). Clinical signs may vary according to mycobacterial species, host species, and environmental conditions. Indicators of illness can include weight loss, lethargy, anorexia, bloating, and dermatitis. Mycobacteriosis can be challenging to diagnose due to its chronic and often nonspecific clinical signs. Diagnostic methods include fluid analysis, biopsy, histopathology, bacterial cultures, and molecular diagnostics through PCR.

Chlamydia, a significant intracellular bacterial pathogen, affects a wide range of vertebrates, including humans, wild and domesticated mammals as well as various species of amphibians, reptiles, and fishes (Borel et al. 2018; Corsaro and Venditti 2004; Reed 2000). The species most identified are *Chlamydophila psittaci* and *C. pneumoniae* (Eisenberg et al. 2020; Newcomer et al. 1982; Pessier 2007). Notably, *C. pneumoniae* has been found in humans, horses, koalas, and amphibians, where it has been reported as a potential cause of lethal systemic diseases (Berger et al. 1999; Bodetti et al. 2002; Corsaro and Venditti 2004). Chlamydiosis ranges from asymptomatic infection to severe and fatal diseases. Clinical signs of the disease include petechiation and sloughing of skin, abdominal swelling, accumulation of excess fluid in lymphatic sacs, lethargy, and cutaneous depigmentation. However, the characteristics of induced inflammatory changes depend on the bacterial species, the host, and the tissue infected (Borel et al. 2018). Due to the great similarity of the clinical signs of chlamydiosis with red leg syndrome and ranaviral infectious, accurate diagnosis of the disease is difficult. Diagnostic procedures include cell culture, histopathologic examination, immunofluorescence, immunohistochemistry, electron microscopy, and PCR analysis (Bodetti et al. 2002; Pessier 2007; Reed 2000; Wilcke et al. 1983).

Bacterial disease associated with *Flavobacteria* has been reported in wild and captive amphibians, including anurans and caudates (Densmore and Green 2007). In addition, these bacteria have been reported as opportunistic zoonotic pathogens affecting birds, fish, turtles, frogs, cats, dogs, and humans (Green et al. 1999; Olson et al. 1992). Clinical signs include swollen abdomen, widespread edema, respiratory distress, inability to dive, lethargy, congestion of web vessels, and petechial hemorrhage. The clinical signs of infection by Flavobacteria closely resemble the clinical signs of the RLS (Densmore and Green 2007). Pathogenic species reported in amphibians include *Flavobacterium oderans*, *F. indologenes*, and *F. meningosepticum* (Green et al. 1999; Olson et al. 1992). Bacterial culture and molecular analysis through polymerase chain reaction (Green et al. 1999; Mauel et al. 2002; Olson et al. 1992) are used to diagnose Flavobacteriosis.

2.3 Viruses

At least six groups of *viruses* have been reported in amphibians, including members of caliciviruses, herpesviruses, and iridoviruses. Among the iridoviruses, members of the genus *Ranavirus* (*Rv*) are perhaps the most studied group known to have an impact on amphibians (Hemingway et al. 2009). Ranaviruses are large (150–170 nm), icosahedral, double-stranded DNA viruses that infect amphibians, reptiles, and fish (Hyatt et al. 2000; Johnson et al. 2008; Stöhr et al. 2013). These viruses are known to infect a wide range of amphibian species worldwide, causing systemic disease that affects many internal organs and causes marked necrosis. Ranaviruses are well known for causing severe disease outbreaks, particularly in aquatic environments (Chinchar 2002). *Rv* is one of the five genera of the family Iridoviridae known to have caused mass mortality events in amphibian populations worldwide (Daszak et al. 1999). The clinical signs include systemic hemorrhages, erythema on the lips, regurgitation, bleeding, muscle abnormalities, and ulceration (Gray et al. 2009). They have been associated with die-offs in larval and adult amphibians in the Americas, Europe, and Asia, and wild and cultured fish populations worldwide (Cunningham et al. 1996; Duffus et al. 2015; Fox et al. 2006; Une et al. 2009).

Ranaviruses are primarily transmitted through direct contact between infected and susceptible individuals, as well as through environmental contamination (Johnson and Wellehan 2005). Diagnosis is typically performed using molecular assays such as polymerase chain reaction (PCR) analysis to detect viral DNA (Miller et al. 2025). They can persist in water and soil for extended periods, making aquatic environments a significant reservoir for infection. Several species of ranaviruses have been identified affecting amphibian species including, *Frog virus 3-FV3* (Duffus et al. 2008), *Bohle iridovirus* (Cullen et al. 1995), *Ambystoma tigrinum virus* (Bollinger et al. 1999; Jancovich et al. 2001; Picco et al. 2007), *Rana esculenta virus* (Stöhr et al. 2013), *Rana grylio virus* (Zhang et al. 2001), *Tiger frog virus* (He et al. 2002), *Rana catesbeiana virus Z* (Majji et al. 2006), *common midwife toad virus* (Mavian et al. 2012), and *Chinese giant salamander virus* (Geng et al. 2011). In South America, the presence of *Rv* has been documented in Argentina (Fox et al. 2006), Brazil (Mazzoni et al. 2009; Ruggeri et al. 2019), Colombia (Flechas et al. 2023), Chile (Peñafiel-Ricaurte et al. 2023; Soto-Azat et al. 2016), Ecuador (Urgiles et al. 2021), Peru (Warne et al. 2016), Uruguay (Duffus et al. 2015), and Venezuela (Zupanovic et al. 1998).

2.4 Mycotic Diseases

Chytridiomycosis is caused by two fungal species, *Batrachochytrium dendrobatidis* (*Bd*) discovered in 1998 (Berger et al. 1998) and *Batrachochytrium salamandrivorans* (*Bsal*) discovered in 2013 (Martel et al. 2013). *Bd* is a highly pathogenic, virulent, and readily transmissible and panzootic fungus with low specificity among amphibians (Berger et al. 1998; Longcore et al. 1999; Gower et al. 2013) and widely distributed across all continents (Olson et al. 2021). *Bd* is implicated in epizootic mass mortality on a global scale and is linked to species declines, local population extinctions, and species extinctions (Daszak et al. 2003; Scheele et al. 2019). Five distinct lineages have been recognized including BdGPL, BdCAPE, BdASIA-1, BdASIA-2/BdBRAZIL, and Bd-Asia 3 (O'Hanlon et al. 2018; Byrne et al. 2019; Azat 2021). *Bd* is an aquatic organism with a two-stage life cycle, including a phase with an aquatic, motile, and uniflagellate zoospore that invades keratinized tissues, and a sessile phase in which the zoospore becomes encysted and the zoosporangium develops, which upon maturing releases the motile zoospores back into the environment. Single-celled *Bd* attacks the superficial keratinized areas of the skin of juvenile and adult amphibians, and the mouthparts of tadpoles (Longcore et al. 1999). In susceptible individuals, *Bd* infects the epidermis of postmetamorphic individuals, causing damage of the epidermal cells, and disrupting the osmoregulatory function, leading to electrolyte depletion, and osmotic imbalance, leading to death (Voyles et al. 2009). In other species, the fungus inhibits the immune response (Fites et al. 2013). However, some species tolerate infection and can act as pathogen reservoirs (Reeder et al. 2012; Schloegel et al. 2010). *Bd* emerged worldwide during the twentieth century and is considered responsible for the decline of at least 501 amphibian species (6.5% of described amphibian species); of these, 90 (18%) species are presumed extinct in the wild (Scheele et al. 2019).

A second species, *B. salamandrivorans* (*Bsal*), appears to be native to Asia, where it is endemic, and as a result of globalization and lack of biosecurity, protocols have recently been introduced to Europe, resulting in rapid declines in populations of European fire salamanders (Martel et al. 2014; Stegen et al. 2017). Although it is currently restricted to Europe and Asia, it is considered a major threat to salamander species in biodiversity hotspots (García-Rodríguez et al. 2022). *Bsal* invades keratinized amphibian skin, leading to superficial erosions and numerous ulcerations across the body of susceptible host species; then, overgrowth of secondary bacteria may result in lethal septicemia (Bletz et al. 2018). Clinical signs of chytridiomycosis caused by *Bsal* also include lethargy, focal lesions, skin sloughing, hemorrhage, and anorexia (Martel et al. 2013). It is essential to recognize that chytridiomycosis associated with *Bsal* is not limited to urodele host species (Towe et al. 2021), and the pathogen has also been detected in anurans (Gray et al. 2023; Nguyen et al. 2017). Clinical signs associated with *Bd* and *Bsal* include excessive skin shedding and reddened or discolored skin. In *Bd* infections, this results in erythema and disruptions in respiration and osmoregulation, while in *Bsal* infections, it leads to deep epidermal ulcerations that are subsequently colonized by other microorganisms (Sabino-Pinto et al. 2018).

2.5 Protozoan Parasites

Another disease associated with massive mortality events, mainly in premetamorphic stages, is known as Severe Perkinsea Infection (SPI). This disease is caused by a protozoan belonging to the phylum Perkinsozoa, which forms a distinct group referred to as the Pathogenic Perkinsea Clade of Frogs (Isidoro-Ayza et al. 2019). The risk of susceptible species to SPI has become evident only recently. Given the pathological and epidemiological similarities between this disease and other diseases caused by bacteria, viruses, and fungi, it is very likely that there has been a misdiagnosis and an underestimation of the cases of *Perkinsea* infections in other regions of the world. The infected tadpoles showed no gross morphological symptoms of disease (Chambouvet et al. 2015). Internally, infected tadpoles present accumulation of fluids in the coelomic cavity, enlargement of the liver, spleen, and kidneys, and lesions in the pancreas. In addition, changes in coloration or multifocal hemorrhages in the organs are associated with SPI in tadpoles (Isidoro-Ayza et al. 2019). For the Andes, we found only one recently published study reported on the presence of *Perkinsea* in Ecuador (Urgiles et al. 2021). This finding undoubtedly sets off alarm bells and increases the risk for susceptible species as this parasite has been associated with mass mortality events in other regions.

The purpose of this chapter is to present the results of the research carried out on pathogens and emerging infectious diseases in the Andes. With the information gathered, we will be able to identify information gaps and therefore possibilities for future research that will allow us to advance in the knowledge of the geographic and taxonomic distribution of infectious agents and pave the way for action that will enable us to generate concrete strategies to reduce the risk of extinction of species due to emerging infectious diseases.

3 Amphibian Diseases: The Case of the Andes

We obtained data from 66 papers on viral, fungal, bacterial, and parasitic diseases affecting amphibian species in Andean countries. The studies we reviewed were published between 2005 and 2023, although some publications include results of *Bd* diagnostic tests performed on specimens deposited in museum collections or analyze the presence of *Ranavirus* from tissue samples preserved in specialized collections. The samples analyzed date back to the year 1863, the most recent are from 2019.

At first, we consulted Soto-Azat's (2022) comprehensive article, which provides an extensive review of *Batrachochtryrium dendrobatidis* occurrence in South America. Since this study covered the entire continent, we examined the original references, specifically focusing on records related to species found in the Andes—the central theme of this chapter. Additionally, we looked for the original research papers to extract detailed information for each record. This included diagnostic techniques employed, species studied, number of individuals sampled, count of positive cases, infection intensity, and infection prevalence (calculated if not explicitly reported). We also noted the locality, country, and IUCN status associated with each record.

Our study includes data on individuals from 335 species, along with 168 records identified only at the genus level. These species are distributed across 23 taxonomic families, including three records from two caecilian families and ten records from the salamander family Plethodontidae. The compiled database contains 1501 unique entries, defined as combinations of species and studies (species × study). Most records correspond to species in the family Bufonidae, followed by those in Telmatobiidae. To gather comprehensive information on amphibian diseases in the Andes, we conducted an extensive search focusing on emerging infectious diseases and pathogens. This effort yielded records from seven countries: Argentina, Bolivia, Chile, Colombia, Ecuador, Peru, and Venezuela.

To map the distribution of sampling and disease across the Andes, we used published coordinates whenever available. When coordinates were not provided, we contacted individual authors and/or georeferenced localities from available descriptions. Challenges arose particularly with reports of helminth infections due to limited locality information. Therefore, we manually georeferenced several reports using published maps or identifying their nearest urban center. Not all reviewed articles defined the Andes in the same way and not all had precise locality descriptions preventing us from georeferencing them. Given these challenges, we could not filter further than species studied in each selected paper and thus a few localities are outside the Andes delimitation. Our review identified significant gaps in disease sampling across the Andean countries, for example, northern and central Peru exhibit substantial gaps across all diseases. Given the information deficiency in certain regions, there is a huge opportunity to guide future monitoring efforts that can address existing geographic and taxonomic gaps comprehensively.

From the 1501 records analyzed, 1213 (80.8%) were tested for the presence of the fungal pathogen *Bd,* while 208 (13.9%) were screened for *Rv*. Reports of helminth parasite infections accounted for 45 (2.99%) records, bacterial infections for 8 (0.53%) records, and protist infections specifically *Perkinsea* for 22 (1.46%) records. Additionally, there were 5 (0.33%) records indicating coinfection by *Bd* and *Rv*. Infections by other agents than *Bd* and *Rv* have received considerably less attention in Andean amphibian species. Given the predominant focus on *Bd* and *Rv* in sampling efforts, we created separate distribution maps for these two pathogens (Figs. 1 and 2).

Studies on bacteria, other viruses, helminths, and protists have been reported for amphibians in other latitudes; however, for Andean species, the studies are notably sparse, scarce, or nonexistent and geographical sampling has been limited. For instance, bacterial infections were reported at a single site, where *Aeromonas caviae* and Enterobacteria from the genera *Enterobacter*, *Serratia*, and *Hafnia* were identified (Enciso et al. 2008). Similarly, sampling for *Perkinsea* has been limited, with reports indicating positive findings in sampled sites but not across all species (Urgiles et al. 2021). These isolated reports suggest these disease agents might be present in other localities across the Andes.

Out of the 335 sampled species reported in the evaluated studies, 162 are categorized as threatened according to the IUCN. Specifically, 31 species are Critically Endangered (CR), 57 are Endangered (EN), 47 are Vulnerable (VU), 27 are listed as Near Threatened category (NT), 163 species are categorized as Least Concern (LC), and 10 species are listed as Data Deficient (DD). For 168 entries, it was not possible to determine their category because they were reported only to genus level or as sp., aff., or cf., representing 11.2% of the compiled records in our review. Among these entries, 35 individuals tested positive for *Bd*, two for *Rv*, and three for *Perkinsea.* Accurate species-level identification is crucial when designing effective species management plans.

Helminth infections in South American amphibians are reported as adult nematodes parasitizing the gastrointestinal tract, with digenetic trematodes being the most reported (Campião et al. 2014). Studies documenting helminths have been reported in Ecuador, Colombia, and Peru, although most studies lack precise geographical information (Campião et al. 2014; Chero et al. 2023; Goldberg and Bursey 2003; Gómez et al. 2020; McAllister et al. 2010a, b). In Peru, comprehensive reviews on the presence of helminths have primarily focused on Amazonian species (McAllister et al. 2010a, b). Recently, Chero et al. (2023) published an update expanding the knowledge on helminth diversity in Peruvian amphibians. In their review, 83 distinct helminth taxa were associated with 78 species of anurans, marking a 176.7% increase compared to previous studies. Despite this progress, the evaluation covers only a small fraction of Peru's amphibian species, indicating opportunities for further exploration. There remains a substantial gap in understanding the diversity of helminth parasites and their impact on amphibian populations across the Andes, where studies are notably scarce.

Batrachochytrium dendrobatidis is the most extensively sampled pathogen in the Andes, with 321 species tested for its presence. Sampling efforts for *Bd* have been geographically concentrated in the Colombian and Bolivian Andes (Acevedo et al. 2016; Burrowes and De La Riva 2017; De La Riva and Burrowes 2014; Flechas et al. 2017; Rueda-Solano et al. 2016; Urbina and Galeano 2011; Vasquez-Ochoa et al. 2012; Viáfara-Vega 2020) (Fig. 1). Positive *Bd* samples (red points) are widespread throughout the region, highlighting the extensive sampling, distribution, and prevalence. In addition, 63 individuals that could only be identified at the genus level were included in the examination. *Bd* diagnosis has been carried out using different

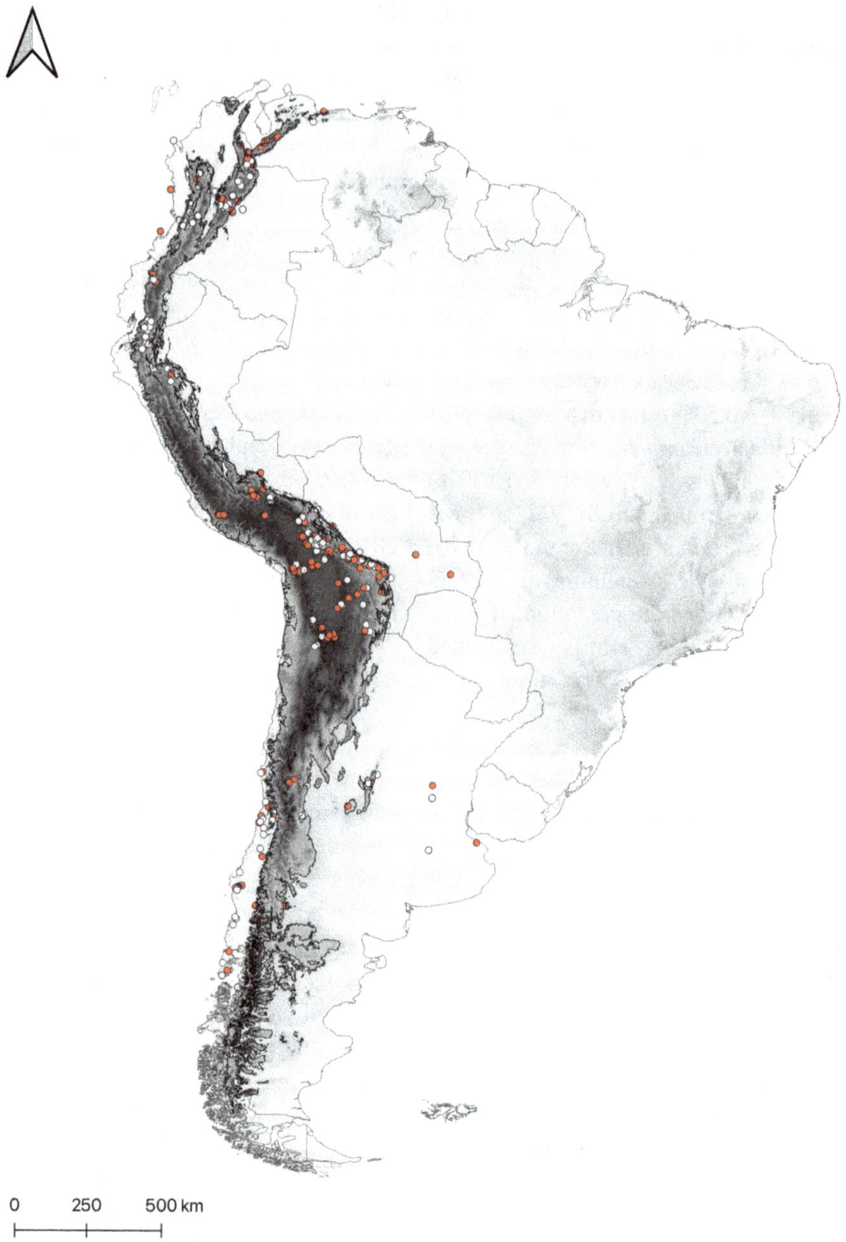

Fig. 1 Map of *Batrachochytrium dendrobatidis* sampling with coordinates or locality description in the reviewed studies [some localities are approximates, some could not be georeferenced]. Red points represent localities where at least one sample was positive, white points are localities where no sample was positive. In the map of South America, shades of gray represent elevation with lower elevation in light gray and higher elevation in darker shades of gray and lines represent country boundaries. Elevation information was obtained from a 30 m resolution digital elevation model from the shuttle radar topography mission (SRTM, Farr et al. 2007). The delimitation shown corresponds to the Tropical Andes hotspot as defined by Conservation International (Mittermeier et al. 2004) a classification largely based on global plant endemism patterns

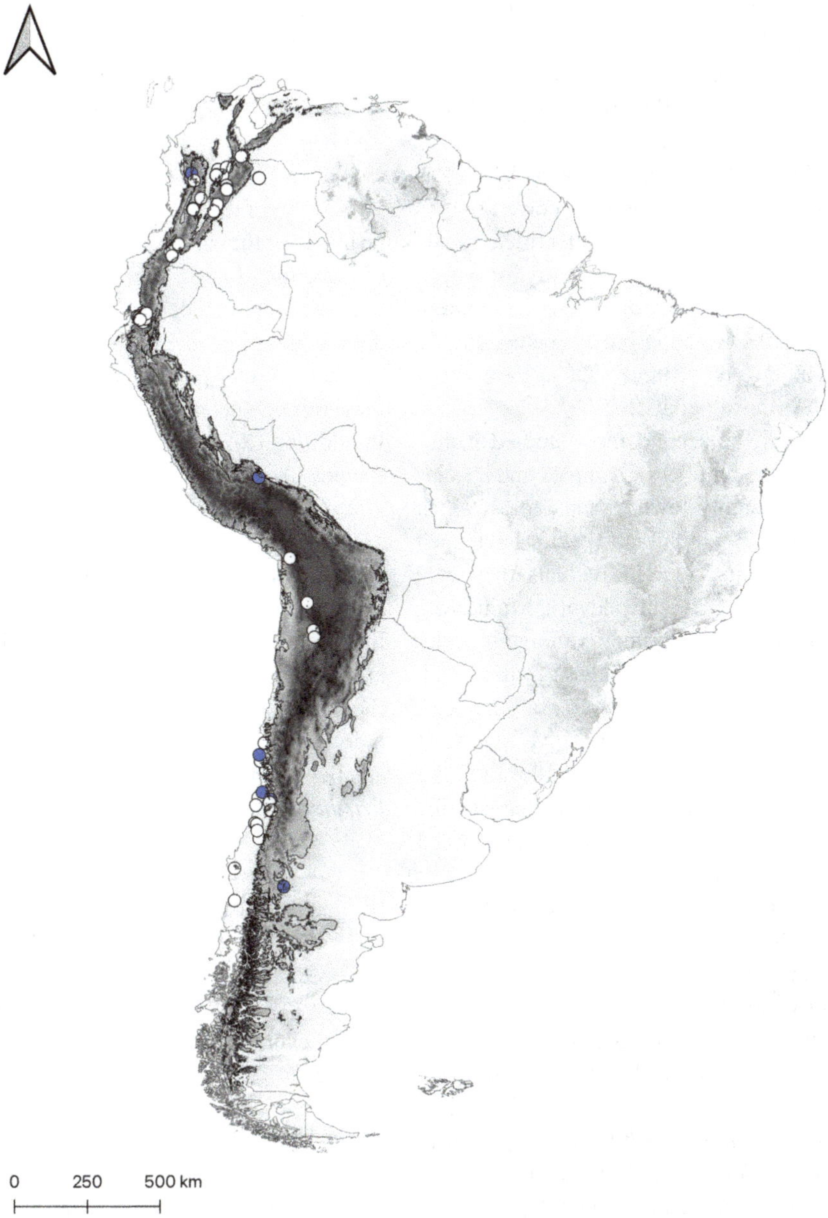

Fig. 2 Map of *Ranavirus* sampling with coordinates or locality description in the reviewed studies [some localities are approximates, some could not be georeferenced]. Blue points represent localities where at least one sample was positive, white points are localities where no sample was positive. In the map of South America, shades of gray represent elevation with lower elevation in light gray and higher elevation in darker shades of gray. Elevation information was obtained from a 30 m resolution digital elevation model from the shuttle radar topography mission (SRTM, Farr et al. 2007). The delimitation shown corresponds to the Tropical Andes hotspot as defined by Conservation International (Mittermeier et al. 2004) a classification largely based on global plant endemism patterns region

techniques, including molecular methods such as conventional PCR, quantitative PCR, and histological analysis. Out of the 321 tested species, 209 tested positive for *Bd*. According to IUCN assessments, of the positive species, 20 species are categorized as CR, 32 as EN, 30 as VU, 11 as NT, 84 as LC, 5 as DD, and 27 correspond to individuals who could not be identified to the species level. Only a single study surveyed *Batrachochytrium salamandrivorans* in southern Peru, without detecting the pathogen in 131 salamanders across three sites at the foothill of the Andes (Parrott et al. 2017).

Ranavirus is the second most studied pathogen, although the number of samples and species tested is considerably lower compared to *Bd*. In South America, 79 amphibian species have been evaluated (including four individuals that were identified only to the genus level). Monitoring efforts for *Rv* have been primarily geographically focused on Colombia and Chile (Flechas et al. 2023; Peñafiel-Ricaurte et al. 2023; Soto-Azat et al. 2013) (Fig. 2), with few positive reports of individuals infected across the Andes. Of the species sampled, 17 have tested positive for the virus. Among these, six species are categorized under threat levels according to the IUCN (CR: 2, EN: 1, VU: 3), while 10 are considered LC. Two individuals could not be categorized due to being only reported at the genus level (*Pristimantis* and *Osornophryne*).

In South America, Chile is at the forefront of *Rv* monitoring. Nearly 1700 samples from five species including *Calyptocephalella gayi*, *Pleurodema thaul*, *Rhinella spinulosa*, *Telmatobufo bullocki*, and *Xenopus laevis* have been evaluated, revealing an infection prevalence of 1.0% ($N = 1694$, CI = 0.6–1.6%). All positive samples belong to the introduced African clawed frog (*X. laevis*, Pipidae) (Peñafiel-Ricaurte et al. 2023; Soto-Azat et al. 2016). This species was first introduced in Chile in 1973 (Jaksic 1998), near the Santiago International Airport, and subsequently spread to various water bodies around Santiago and beyond (Lobos and Jaksic 2005). This species represents a significant threat to native amphibians as it acts as a reservoir host, potentially facilitating the virus' spread to new areas (Peñafiel-Ricaurte et al. 2023)

In Colombia, at least 58 species belonging to the families Aromobatidae, Bufonidae, Centrolenidae, Craugastoridae, Dendrobatidae, Hemiphractidae, Hylidae, Leptodactylidae, Ranidae, and Strabomantidae have been evaluated for the presence of *Rv* (Flechas et al. 2023). Out of the 272 samples analyzed, 12 tested positive for the virus, indicating a low prevalence of infection at 4.4% (CI = 2.3–7.5%). The positive samples include six native species, three of them belonging to the genus *Pristimantis*, two species of the genus *Leptodactylus,* and one unidentified species of the genus *Osornophryne*. In addition, *Rv* has been detected in the invasive species *Aquarana catesbeiana* (Flechas et al. 2023). In Ecuador, *Rv* has been studied in species of the genus *Pristimantis,* with only three positive samples or 0.7% ($N = 426$, CI = 0.1–2.0%) (Ramírez Piedra and Villalta Sanmartín 2020). In Argentina, the prevalence of the viral infection is considerably high in *Atelognathus patagonicus* (Fox et al. 2006). In Perú, *Rv* was detected in individuals of *Telmatobius marmoratus*, *Rhinella manu*, *Boana gladiator*, and five species of *Pristimantis* (Warne et al. 2016).

4 Ecological Consequences

The evaluation of the virulence and demographic impacts of EIDs in Andean amphibians highlights critical challenges due to limited data availability. Pathogens such as *Batrachochytrium dendrobatidis* (*Bd*), *Ranavirus* (*Rv*), helminths, protists, and bacteria have variable effects on hosts, ranging from impaired individual fitness (e.g., reduced reproductive capacity) to severe population-level outcomes, including declines and extinctions (Fisher et al. 2012; Duffus et al. 2015). Co-infections, such as those involving *Bd* and *Rv*, may amplify disease severity, potentially due to immune suppression mechanisms induced by *Bd*, which increase susceptibility to secondary infections and exacerbate overall impacts (Fisher et al. 2012). These interactions further complicate host–pathogen dynamics, often reducing genetic diversity, altering population structures, and disrupting ecological relationships (Luedtke et al. 2023).

The implications extend beyond individual species, significantly affecting Andean amphibian assemblages. Lips et al. (2006) in Panama highlight how chytridiomycosis-driven extirpations can result in shifts in assemblages, favoring opportunistic species and reducing overall biodiversity. These changes can alter ecological functions such as nutrient cycling, predation dynamics, and competition within amphibian communities. In the Andes, where high-altitude ecosystems harbor endemic species with limited ranges, pathogen-induced declines exacerbate the risk of losing entire assemblages. For instance, *Bd* infections in the Andes have been associated with the collapse of montane stream amphibian populations, leading to the disappearance of species that play keystone roles in aquatic and terrestrial food webs (Luedtke et al. 2023). The loss of such functional diversity may reduce ecosystem resilience and hinder recovery following disturbances such as habitat degradation or climate shifts.

Pathogen spillover further complicates these assemblages, as infections spread among sympatric species, increasing overall disease prevalence and stress on communities already fragmented by deforestation and pollution. Synergistic interactions between EIDs and environmental pressures, such as warming temperatures, increase the risk of ecosystem-level collapses, underscoring the urgency of incorporating disease ecology into conservation strategies to maintain both species diversity and ecological integrity (Pounds et al. 2006).

5 Perspective for the Andes

5.1 Overall Disease Bias

The Andes represent one of the richest biogeographic regions for amphibians (Hutter et al. 2017). However, despite numerous studies confirming the presence of pathogens in the area, there remains a critical gap in our understanding of how these pathogens affect amphibian populations and their conservation status. Moving beyond mere detection, researchers in the region must transition from diagnosis to proactive measures. Given that the Andes host the planet's greatest amphibian diversity, there exists a profound responsibility to safeguard this richness. Thus, a more comprehensive understanding of pathogens' impact on amphibian fauna is urgently needed, along with the development of rapid and efficient diagnostic tools. Additionally, research efforts must prioritize species that act as asymptomatic carriers, serving as models to elucidate mechanisms of disease resistance that could be explored and applied to vulnerable species. The Andean region not only harbors diverse species with varied life histories, behaviors, and physiologies but also holds promise as a potential reservoir for solutions to mitigate the impacts of emerging infectious diseases currently devastating amphibian populations.

The bias toward researching diseases like chytridiomycosis and ranavirus may be attributed to the fact that *Bd* has been convincingly linked to mass die-offs, population declines, and species extirpations worldwide, while *Rv* is responsible for seasonal mass die-offs in temperate regions. The prevalence of certain diseases may also be influenced by their distribution and the number of researchers studying them. Diseases with broader distributions and those that affect a larger number of species may attract more attention and resources for research. On the other hand, diseases caused by organisms such as helminths, bacteria, and protists may receive less attention due to limited information and the lack of well-developed monitoring and diagnostic methods. This could lead to an incomplete understanding of the impact of these diseases on amphibian populations in the Andes. Currently, the most diagnosed pathogens correspond to *Batrachochytrium dendrobatidis* and *Ranavirus*, and for these two pathogens, we have a better overview of host species, geographic distribution, and in some cases, we have data on infection prevalence and infection intensity.

Diseases caused by bacteria have been, in general, less studied, and specifically for the Andes, information on infected species is very limited; most of the reports come from facilities keeping animals in captivity. Clinical signs for several bacterial diseases are vague or are not unique or specific to a particular disease making the diagnosis more difficult. Therefore, applying an adequate treatment that reduces mortality in the affected group of individuals is almost impossible. It is also important to note that when individuals are infected and sick, they undergo immunosuppression which makes them more susceptible to subsequent infections by other infectious agents, making proper diagnosis even more difficult. Diagnosis and correct identification of the pathogen often requires the use of different methods that may include cytopathological analysis, histopathology, bacterial cultures, and biochemical or polymerase chain reaction (PCR) tests. This highlights the need to improve our methodologies to make a rapid and accurate diagnosis to find a correct and timely treatment.

5.2 Opportunities from the Field and Biological Collections

Including pathogen sampling in research projects is essential for gaining a comprehensive understanding of the status of amphibian populations across multiple localities. Many of these projects are conducted in challenging environments with limited access or in politically unstable regions, often involving significant economic and personal costs. In this context, proactive sampling using noninvasive techniques is particularly valuable. Employing noninvasive methods that require minimal materials (e.g., gloves, swabs, storage tubes, and markers) ensures cost-effective sampling. Even if immediate analysis is not planned, such samples create opportunities for future research when resources become available. We encourage researchers in other areas of herpetology to collaborate and help bridge the existing information gaps in the disease ecology of Andean amphibians. Working in the Andes is not easy; their broad extension and complex topography include multiple habitat types that are not always easy to access. Additionally, funding to support research in the area can be limited. Leveraging museum specimens can then improve our knowledge on disease ecology in this region (Nachman et al. 2023). Each specimen can contain information about a particular pathogen in a place and time. As specimens represent an invaluable sort of information, during the last years several digitization initiatives have been put in place to improve the visibility and use of the information provided for specimens.

Digitization increases visibility and facilitates connections among researchers, museums, and different entities that can use the associated data of specimens. Museum specimens can inform epidemiology, helping to infer when a particular infectious agent arrived at a specific locality, its geographic distribution, and how the disease/pathogen is spreading (Vredenburg et al. 2019). Changes in prevalence and information about disease dynamics can also be obtained from biological collections (Burrowes and De La Riva 2017). Even information about co-infection has been obtained from museum collections (Karwacki et al. 2021). During the last decades, novel technologies and advances in molecular biology using biological collections have advanced fields such as taxonomy, systematics, and environmental and ecological disciplines.

Biological collections have biases and limitations, from type of specimen (common species vs rare species), size, and time of collection, among others but their potential to provide information is considerable. Planning and having clear collection strategies can be better used in a field like disease ecology. Prioritizing areas where collection gaps are notorious and include information associated with the specimen can replace opportunistic collecting that generally lacks details of the specimens. The collection of tissues that can be analyzed for the presence of pathogens can be improved using standard methods that prevent degradation and facilitate the use of downstream information coming from the specimen. Additionally, given that many molecular diagnostic assays are noninvasive, improving the processes and methodologies for storing DNA extracts is key to enhancing their long-term research value.

5.3 Standardized and Open-Access Data

Our revision of all available data for the presence of amphibian diseases in the Andean countries showed an overall lack of data availability. Even published reports often do not contain locality information or provide rather vague and incomplete descriptions and very seldom include precise coordinates of sampled areas. Furthermore, reports are not standardized with respect to sampling effort, diagnostics criteria, and sampling methods. Furthermore, even for diseases like chytridiomycosis and ranavirus where standard methods exist, inconsistencies in reporting pathogen load can complicate comparisons between species and hinder efforts to assess the overall impact of these diseases on amphibians in the region. Standardizing reporting units and methods could help address this issue and improve the comparability of research findings. These data shortcomings severely handicap our understanding of disease distribution and impact potential conservation, mitigation, and management plans. We encourage the community to think of better ways to share standardized data on disease occurrence by using open-access database models already existing such as the Global Biodiversity Information Facility (GBIF) used for species presences. It could be potentially useful to add a disease item to the GBIF reports (or more than one) so they could complement accounts of species presences and facilitate data sharing for disease experts. To report *Bd* cases and facilitate their accessibility, we suggest researchers upload their data on the Amphibian Disease Portal developed by AmphibiaWeb and the US Forest Service.

Research efforts must be made between well-studied and lesser-known diseases in amphibians. We need to continue research on well-known pathogens such as *Bd* and *Rv*, which can help us understand the dynamics of these diseases and potentially inform and refine management strategies. Lesser-known diseases must also be considered particularly for those with previous evidence or when affecting threatened species. Ensuring coverage of multiple diseases and joint efforts among researchers will provide expertise and facilitate an efficient use of resources. Also, researching multiple diseases aligns with approaches like One Health, which recognizes the importance of studying connections among environment, animal, and human health (Mackenzie and Jeggo 2019). Understanding multiple diseases will advance our basic scientific knowledge and strengthen conservation plans and efforts for mitigating disease impact on populations of threatened amphibians.

6 Conclusion

This chapter attempts to combine existing data from studies published in peer-reviewed sources. However, information in non-peer-reviewed sources was not considered and needs to be reviewed and compiled (e.g., reports, theses). Furthermore, considering other emerging diseases that can affect amphibians is of the utmost importance. As we suggest in this chapter, we need to improve data on disease ecology by (1) synthesizing and integrating already available data, (2) leveraging natural history collections for historical data, (3) integrating disease sampling into standard herpetology fieldwork (regardless of intent), (4) targeting diseases beyond *Batrachochytrium dendrobatidis* and *Ranavirus* in a standardized way, and (5) standardizing reporting parameters (geographical data, disease parameters) into publicly available databases. All these strategies should be established throughout the Andes to improve our knowledge of diseases affecting amphibians. Comprehensive assessments will strengthen our knowledge and inform management decisions, facilitate potential implementation of conservation strategies to protect vulnerable species from a broader spectrum of threats beyond *Bd* and *Rv*.

Acknowledgments We would like to thank Alessandro Catenazzi, Claudio Soto-Azat, and Ignacio de la Riva for sharing information on their surveys conducted in Peru, Chile, and Bolivia that enabled us to complete the database. We also thank Paulo Murillo for the provision of GIS files.

References

Acevedo AA, Martínez MP, Armesto LO, Solano-Florez-L, Silva-Perez K, Lizcano DJ (2016) Detection of *Batrachochytrium dendrobatidis* in Amphibians from Northeastern Colombia

AmphibiaWeb (2024) Information on amphibian biology and conservation. http://amphibiaweb.org/

Armenteras D, Espelta JP, Rodríguez N, Retana J (2017) Deforestation DYNAMICS AND DRIVERS IN DIFFERENT FOREST TYPES in Latin America: three decades of studies (1980–2010). Glob Environ Chang 46(September):139–147. https://doi.org/10.1016/j.gloenvcha.2017.09.002

Azat C (2021) Not just a pathogen: The importance of recognizing genetic variability to mitigate a wildlife pandemic. Mol Ecol Resour 21:1410–1412. https://doi.org/10.1111/1755-0998.13348

Berger L, Speare R, Daszak P, Green DE, Cunningham AA, Goggin CL, Slocombe R et al (1998) Chytridiomycosis causes amphibian mortality associated with population declines in the rain forests of Australia and Central America. Proc Natl Acad Sci 95(15):9031–9036. https://doi.org/10.1073/pnas.95.15.9031

Berger L, Volp K, Mathews S, Speare R, Timms P (1999) *Chlamydia pneumoniae* in a free-ranging giant barred frog (*Mixophyes Iteratus*) from Australia. J Clin Microbiol 37(7):2378–2380. https://doi.org/10.1128/JCM.37.7.2378-2380.1999

Blaustein AR, Johnson PTJ (2003) The complexity of deformed amphibians. Front Ecol Environ 1(2):87–94

Bletz MC, Kelly M, Sabino-Pinto J, Bales E, Van Praet S, Bert W, Boyen F et al (2018) Disruption of skin microbiota contributes to salamander disease. Proc R Soc B Biol Sci 285(1885):20180758. https://doi.org/10.1098/rspb.2018.0758

Bodetti TJ, Jacobson E, Wan C, Hafner L, Pospischil A, Rose K, Timms P (2002) Molecular evidence to support the expansion of the hostrange of *Chlamydophila pneumoniae* to include reptiles as well as humans, horses, koalas and amphibians. Syst Appl Microbiol 25(1):146–152. https://doi.org/10.1078/0723-2020-00086

Bollinger TK, Mao J, Schock D, Brigham RM, Chinchar VG (1999) Pathology, isolation, and preliminary molecular characterization of a novel iridovirus from tiger salamanders in Saskatchewan. J Wildl Dis 35(3):413–429. https://doi.org/10.7589/0090-3558-35.3.413

Borel N, Polkinghorne A, Pospischil A (2018) A review on chlamydial diseases in animals: still a challenge for pathologists? Vet Pathol 55(3):374–390. https://doi.org/10.1177/0300985817751218

Burrowes PA, De La Riva I (2017) Unraveling the historical prevalence of the invasive Chytrid fungus in the Bolivian Andes: implications in recent amphibian declines. Biol Invasions 19(6):1781–1794. https://doi.org/10.1007/s10530-017-1390-8

Byrne AQ et al. (2019) Cryptic diversity of a widespread global pathogen reveals expanded threats to amphibian conservation. Proc Natl Acad Sci 116(41):20382–7. https://doi.org/10.1073/pnas.1908289116

Campião KM, Morais DH, Dias OT, Aguiar A, Toledo G, Tavares LER, Da Silva RJ (2014) Checklist of helminth parasites of amphibians from South America. Zootaxa 3843(1). https://doi.org/10.11646/zootaxa.3843.1.1

Campião KM, Ribas ACA, Morais DH, Da Silva RJ, Tavares LER (2015) How many parasites species a frog might have? Determinants of parasite diversity in South American Anurans. PLoS One 10(10): e0140577. https://doi.org/10.1371/journal.pone.0140577

Chambouvet A, Gower DJ, Jirků M, Yabsley MJ, Davis AK, Leonard G et al (2015) Cryptic infection of a broad taxonomic and geographic diversity of tadpoles by Perkinsea protists. Proc Natl Acad Sci 112(34): E4743–E4751. https://pnas.org/doi/full/10.1073/pnas.1500163112

Chero JD, Cruces CL, Cacique ER, Ponce JA, Iannacone J, Alvariño L, Sanchez L, Sáez G, Lopez J, Da Silva RJ (2023) A comprehensive update on helminth parasite biodiversity and richness in Peruvian amphibians. Diversity 15(12):1169. https://doi.org/10.3390/d15121169

Chinchar VG (2002) Ranaviruses (Family Iridoviridae): emerging cold-blooded killers. Arch Virol 147(3): 447–470. https://doi.org/10.1007/s007050200000

Corsaro D, Venditti D (2004) Emerging Chlamydial infections. Crit Rev Microbiol 30(2):75–106. https://doi.org/10.1080/10408410490435106

Cullen BR, Owens L, Whittington RJ (1995) Experimental infection of Australian Anurans (*Limnodynastes terraereginae* and *Litoria latopalmata*) with Bohle Iridovirus. Dis Aquat Org 23:83–92. https://doi.org/10.3354/dao023083

Cunningham AA, Langton TES, Bennett PM, Lewin JF, Drury SEN, Gough RE, MacGregor SK (1996) Pathological and microbiological findings from incidents of unusual mortality of the common frog (*Rana temporaria*). Philos Trans R Soc Lond B Biol Sci 351(1347):1539–1557. https://doi.org/10.1098/rstb.1996.0140

Daszak P, Cunningham AA, Hyatt AD (2000) Emerging infectious diseases of wildlife – threats to biodiversity and human health. *Science* 287(5452):443–449. https://doi.org/10.1126/science.287.5452.443

Daszak P, et al (1999) Emerging Infectious Diseases and Amphibian Population Declines. Emerg Infect Dis 5(6):735–48. https://www.ncbi.nlm.nih.gov/pmc/articles/PMC2640803/

Daszak P, Cunningham AA, Hyatt AD (2003) Infectious Disease and Amphibian Population Declines. *Divers Distrib* 9(2):141–150. https://doi.org/10.1046/j.1472-4642.2003.00016.x

De La Riva I, Burrowes PA (2014) A new species of *Psychrophrynella* (Anura: Craugastoridae) from the Cordillera Real, Department La Paz, Bolivia. Zootaxa 3887(4):459. https://doi.org/10.11646/zootaxa.3887.4.4

Densmore CL, Green DE (2007) Diseases of amphibians. ILAR J 48(3):235–254. https://doi.org/10.1093/ilar.48.3.235

Drake GJ, Koeppel K, Barrows M (2010) Disinfectant (F10SC) nebulisation in the treatment of 'Red Leg Syndrome' in amphibians. Vet Rec 166(19):593–594. https://doi.org/10.1136/vr.b4839

Duffus ALJ, Pauli BD, Wozney K, Brunetti CR, Berrill M (2008) Frog Virus 3-like infections in aquatic amphibian communities. J Wildl Dis 44(1):109–120. https://doi.org/10.7589/0090-3558-44.1.109

Duffus ALJ, Waltzek TB, Stöhr AC, Allender MC, Gotesman M, Whittington RJ, Hick P, Hines MK, Marschang RE (2015) Distribution and host range of Ranaviruses. In: Ranaviruses. Springer, pp 9–57

Eisenberg T, Fawzy A, Kaim U, Nesseler A, Riße K, Völker I, Hechinger S et al (2020) Chronic wasting associated with *Chlamydia pneumoniae* in three ex situ breeding facilities for tropical frogs. Antonie Van Leeuwenhoek 113(12):2139–2154. https://doi.org/10.1007/s10482-020-01483-6

Enciso MA, Mendoza AP, Chávez G (2008) Rapid survey on amphibian skin diseases in a mountain forest at the Northern Andes of Peru. FrogLog 87:4–6

Farr TG, Rosen PA, Caro E, Crippen R, Duren R, Hensley S et al (2007) The shuttle radar topography mission. Rev Geophys 45(2):2005RG000183. https://doi.org/10.1029/2005RG000183

Ferreira R, De Souza-Fonseca L, Afonso AM, Gomes Da Silva M, Saad MH, Lilenbaum W (2006) A report of Mycobacteriosis caused by *Mycobacterium marinum* in Bullfrogs (*Rana Catesbeiana*). Vet J 171(1): 177–180. https://doi.org/10.1016/j.tvjl.2004.08.018

Fisher MC, Henk DA, Briggs CJ, Brownstein JS, Madoff LC, McCraw SL, Gurr SJ (2012) Emerging fungal threats to animal, plant, and ecosystem health. Nature 484(7393):186–194

Fites JS, Ramsey JP, Holden WM, Collier SP, Sutherland DM, Reinert LK, Gayek AS et al (2013) The invasive Chytrid fungus of amphibians paralyzes lymphocyte responses. Science 342(6156):366–369. https://doi.org/10.1126/science.1243316

Flechas SV, Paz A, Crawford AJ, Sarmiento C, Acevedo AA, Arboleda A, Bolívar-García W et al (2017) Current and predicted distribution of the pathogenic fungus *Batrachochytrium dendrobatidis* in Colombia, a Hotspot of Amphibian Biodiversity. Biotropica 49(5): 685–694. https://doi.org/10.1111/btp.12457

Flechas SV, Urbina J, Crawford AJ, Gutiérrez K, Corrales K, Castellanos LA, González MA, Cuervo AM, Catenazzi A (2023) First evidence of Ranavirus in native and invasive amphibians in Colombia. Dis Aquat Org 153:51–58. https://doi.org/10.3354/dao03717

Fox SF, Greer AL, Torres-Cervantes R, Collins JP (2006) First case of Ranavirus-associated morbidity and mortality in natural populations of the South American Frog *Atelognathus patagonicus*. Dis Aquat Org 72(1): 87–92. https://doi.org/10.3354/dao072087

Fremont-Rahl JJ, Ek C, Williamson HR, Small PLC, Fox JG, Muthupalani S (2011) *Mycobacterium liflandii* outbreak in a research colony of *Xenopus (Silurana) tropicalis* frogs. Vet Pathol 48(4):856–867. https://doi.org/10.1177/0300985810388520

García-Rodríguez A, Basanta DM, García-Castillo M, Zumbado-Ulate H, Neam K, Rovito S, Searle CL, Parra-Olea G (2022) Anticipating the potential impacts of *Batrachochytrium salamandrivorans* on neotropical salamander diversity. Biotropica 54:157–169

Geng Y, Wang KY, Zhou ZY, Li CW, Wang J, He M, Yin ZQ, Lai WM (2011) First report of a Ranavirus associated with morbidity and mortality in farmed Chinese giant salamanders (*Andrias davidianus*). J Comp Pathol 145(1):95–102. https://doi.org/10.1016/j.jcpa.2010.11.012

Goldberg SR, Bursey CR (2003) Helminths of two Anuran species, Atelopus Spurrelli (Bufonidae) and *Dendrobates histrionicus* (Dendrobatidae), from Colombia, South America. Parasitol Int 52(3): 251–253. https://doi.org/10.1016/S1383-5769(03)00013-8

Gower DJ et al. (2013) Batrachochytrium dendrobatidis Infection and Lethal Chytridiomycosis in Caecilian Amphibians (Gymnophiona). EcoHealth 10(2): 173–83. http://link.springer.com/10.1007/s10393-013-08319

Gómez G, Sánchez L, Ñacari LA, Espínola-Novelo JF (2020) Nematode parasites from six species of marsupial *Gastrotheca* (Anura: Hemiphractidae) frogs from the Peruvian Andean Highlands. Pac Sci 74(1):65. https://doi.org/10.2984/74.1.5

Gray MJ, Miller DL, Hoverman JT (2009) Ecology and pathology of amphibian Ranaviruses. Dis Aquat Org 87(December):243–266. https://doi.org/10.3354/dao02138

Gray MJ, Carter ED, Piovia-Scott J, Cusaac JPW, Peterson AC, Whetstone RD, Hertz A et al (2023) Broad host susceptibility of North American amphibian species to *Batrachochytrium salamandrivorans* suggests high invasion potential and biodiversity risk. Nat Commun 14(1):3270. https://doi.org/10.1038/s41467-023-38979-4

Green SL, Bouley DM, Tolwani RJ, Waggie KS, Lifland BD, Otto GM, Ferrell JE (1999) Identification and management of an outbreak of *Flavobacterium meningosepticum* infection in a colony of South African clawed frogs (*Xenopus Laevis*). 214(12):1833–1838

He JG, Lü L, Deng M, He HH, Weng SP, Wang XH, Zhou SY, Long QX, Wang XZ, Chan SM (2002) Sequence analysis of the complete genome of an Iridovirus isolated from the tiger frog. Virology 292(2): 185–197. https://doi.org/10.1006/viro.2001.1245

Hemingway V, Brunner J, Speare R, Berger L (2009) Viral and bacterial diseases of amphibians. In: Viral and bacterial diseases in amphibians, vol 8. Surrey Beatty & Sons, Australia, pp 2963–2985

Hutter CR, Lambert SM, Wiens JJ (2017) Rapid diversification and time explain amphibian richness at different scales in the Tropical Andes, Earth's most biodiverse hotspot. Am Nat 190(6):828–843. https://doi.org/10.1086/694319

Hyatt AD, Gould AR, Zupanovic Z, Cunningham AA, Hengstberger S, Whittington RJ, Kattenbelt J, Coupar BEH (2000) Comparative studies of piscine and amphibian Iridoviruses. Arch Virol 145(2):301–331. https://doi.org/10.1007/s007050050025

Isidoro-Ayza M, Grear DA, Chambouvet A (2019) Pathology and case definition of severe Perkinsea infection of frogs. Vet Pathol 56(1):133–142. https://doi.org/10.1177/0300985818798132

Jaksic FM (1998) Vertebrate invaders and their ecological impacts in Chile. Biodivers Conserv 7(11):1427–1445. https://doi.org/10.1023/A:1008825802448

Jancovich JK, Davidson EW, Seiler A, Jacobs BL, Collins JP (2001) Transmission of the *Ambystoma Tigrinum* virus to alternative hosts. Dis Aquat Org 46:159–163. https://doi.org/10.3354/dao046159

Johnson PTJ, McKenzie VJ (2009) Effects of environmental change on helminth infections in amphibians: exploring the emergence of Ribeiroia and Echinostoma infections in North America. In: The biology of Echinostomes. Springer, New York

Johnson AJ, Wellehan JFX (2005) Amphibian virology. Vet Clin N Am Exot Anim Pract 8(1):53–65. https://doi.org/10.1016/j.cvex.2004.09.001

Johnson PTJ, Lunde KB, Haight RW, Bowerman J, Blaustein AR (2001) *Ribeiroia ondatrae* (Trematoda: Digenea) infection induces severe limb malformations in western toads (*Bufo boreas*). Can J Zool 79(3): 370–379

Johnson AJ, Pessier AP, Wellehan JFX, Childress A, Norton TM, Stedman NL, Bloom DC et al (2008) Ranavirus infection of free-ranging and captive box turtles and tortoises in the United States. J Wildl Dis 44(4):851–863. https://doi.org/10.7589/0090-3558-44.4.851

Karwacki EE, Martin KR, Savage AE (2021) One hundred years of infection with three global pathogens in frog populations of Florida, USA. Biol Conserv 257 (May):109088. https://doi.org/10.1016/j.biocon.2021.109088

Lips KR, Brem F, Brenes R, Reeve JD, Alford RA, Voyles J, Carey C, Livo L, Pessier AP, Collins JP (2006) Emerging infectious disease and the loss of biodiversity in a Neotropical amphibian community. Proc Natl Acad Sci USA 103(9):3165–3170

Lobos G, Jaksic FM (2005) The ongoing invasion of African Clawed Frogs (*Xenopus laevis*) in Chile: causes of concern. Biodivers Conserv 14(2):429–439. https://doi.org/10.1007/s10531-004-6403-0

Longcore JE, Pessier AP, Nichols DK (1999) *Batrachochytrium dendrobatidis* Gen. et Sp. Nov., a Chytrid pathogenic to amphibians. Mycologia 91(2): 219–227. https://doi.org/10.1080/00275514.1999.12061011

Luedtke JA, Chanson J, Neam K, Hobin L, Maciel AO, Catenazzi A, Borzée A, Hamidy A, Aowphol A et al (2023) Ongoing declines for the world's amphibians in the face of emerging threats. 622:308–314. https://doi.org/10.1038/s41586-023-06578-4

Mackenzie JS, Jeggo M (2019) The one health approach – why is it so important? Trop Med Infect Dis 4(2):88. https://doi.org/10.3390/tropicalmed4020088

Majji S, LaPatra S, Long SM, Sample R, Bryan L, Sinning A, Chinchar VG (2006) *Rana Catesbeiana* Virus Z (RCV-Z): a novel pathogenic Ranavirus. Dis Aquat Org 73(November):1–11. https://doi.org/10.3354/dao073001

Martel A, Spitzen-van Der Sluijs A, Blooi M, Wim Bert R, Ducatelle MC, Fisher AW et al (2013) *Batrachochytrium salamandrivorans* Sp. Nov. causes lethal Chytridiomycosis in amphibians. Proc Natl Acad Sci 110(38):15325–15329. https://doi.org/10.1073/pnas.1307356110

Martel A, Blooi M, Adriaensen C, Van Rooij P, Beukema W, Fisher MC, Farrer RA et al (2014) Recent introduction of a Chytrid fungus endangers Western Palearctic Salamanders. Science 346(6209):630–631. https://doi.org/10.1126/science.1258268

Martinho F, Heatley JJ (2012) Amphibian mycobacteriosis. Vet Clin N Am Exot Anim Pract 15(1):113–119. https://doi.org/10.1016/j.cvex.2011.09.001

Mauel MJ, Miller DL, Frazier KS, Hines ME (2002) Bacterial pathogens isolated from cultured bullfrogs (*Rana Castesbeiana*). J Vet Diagn Invest 14(5): 431–433. https://doi.org/10.1177/104063870201400515

Mavian C, López-Bueno A, Balseiro A, Casais R, Alcamí A, Alejo A (2012) The genome sequence of the emerging common midwife toad virus identifies an evolutionary intermediate within Ranaviruses. J Virol 86(7):3617–3625. https://doi.org/10.1128/JVI.07108-11

Mazzoni R, de Mesquita AJ, Fleury LFF, de Brito WME, Nunes IA, Robert J, Morales H et al (2009) Mass mortality associated with a frog Virus 3-like Ranavirus infection in farmed tadpoles *Rana catesbeiana* from Brazil. Dis Aquat Org 86(3):181–191. https://doi.org/10.3354/dao02096

McAllister CT, Bursey CR, Freed PS (2010a) Helminth parasites of selected amphibians and reptiles from the Republic of Ecuador. Comp Parasitol 77(1):52–66. https://doi.org/10.1654/4402.1

McAllister CT, Bursey CR, Freed PS (2010b) Helminth parasites of amphibians and reptiles from the Ucayali Region, Peru. J Parasitol 96(2):444–447

Miller DL, Pessier AP, Hick P, Whittington RJ, Forzán MJ (2025) Pathology and diagnostics. In: Gray MJ, Chinchar VG (eds) Ranaviruses. Springer, Cham. https://doi.org/10.1007/978-3-031-64973-8_8

Mittermeier RA, Robles-Gil P, Hoffman M, Pilgrim JD, Brooks TB, Mittermeier CG, Lamoreux JL (2004) Biodiversity hotspots revisited. http://www.biodiversityhotspots.org/xp/Hotspots/resources/maps.xml. Accessed Jan 2018

Myers N, Mittermeier RA, Mittermeier CG, Da Fonseca GAB, Kent J (2000) Biodiversity hotspots for conservation priorities. Nature 403(6772):853–858. https://doi.org/10.1038/35002501

Nachman MW, Beckman EJ, Bowie RCK, Cicero C, Conroy CJ, Dudley R, Hayes TB et al (2023) Specimen collection is essential for modern science. PLoS Biol 21(11):e3002318. https://doi.org/10.1371/journal.pbio.3002318

Newcomer CE, Simmons JL, Wilcke BW Jr, Nace GW (1982) Spontaneous and experimental infections of *Xenopus laevis* with *Chlamydia psittaci*. Lab Anim Sci 32(6):680–686

Nguyen TT, Nguyen TV, Ziegler T, Pasmans F, Martel A (2017) Trade in wild Anurans vectors the Urodelan Pathogen *Batrachochytrium salamandrivorans* into Europe. Amphibia-Reptilia 38(4):554–556. https://doi.org/10.1163/15685381-00003125

O'Hanlon S et al. (2018) Recent Asian origin of chytrid fungi causing global amphibian declines. – Science 360:621–627

Olson ME, Gard S, Brown M, Hampton R, Morck DW (1992) *Flavobacterium indologenes* infection in leopard frogs. JAVMA 201(11):1766–1770. https://avmajournals.avma.org/view/journals/javma/201/11/javma.1992.201.11.1766.xml

Olson DH, Ronnenberg KL, Glidden CK, Christiansen KR, Blaustein AR (2021) Global patterns of the fungal pathogen *Batrachochytrium dendrobatidis* support conservation urgency. Front Vet Sci 8(July):685877. https://doi.org/10.3389/fvets.2021.685877

Parrott J, Shepack A, Burkart D, LaBumbard B, Scimè P, Baruch E, Catenazzi A (2017) Survey of pathogenic chytrid fungi in salamanders from three mountain ranges in Europe and the Americas. EcoHealth 14: 296–302. https://doi.org/10.1007/s10393-016-1188-7

Pasteris SE, Bühler MI, Nader-Macías ME (2006) Microbiological and histological studies of farmed-bullfrog (*Rana catesbeiana*) tissues displaying Red-Leg Syndrome. Aquaculture 251(1):11–18. https://doi.org/10.1016/j.aquaculture.2005.05.007

Peluso J, Chehda AM, Olivelli MS, Ivanic FM, Coll CSP, Gonzalez F, Valenzuela L et al (2023) Metals, pesticides, and emerging contaminants on water bodies from agricultural areas and the effects on a native amphibian. Environ Res 226(June):115692. https://doi.org/10.1016/j.envres.2023.115692

Peñafiel-Ricaurte A, Price SJ, Leung WTM, Alvarado-Rybak M, Andrés Espinoza-Zambrano C, Valdivia AAC, Azat C (2023) Is *Xenopus laevis* introduction linked with *Ranavirus* incursion, persistence and spread in Chile? PeerJ 11(February):e14497. https://doi.org/10.7717/peerj.14497

Pessier AP (2007) Cytologic diagnosis of disease in amphibians. Vet Clin N Am Exot Anim Pract 10(1): 187–206. https://doi.org/10.1016/j.cvex.2006.10.006

Picco AM, Brunner JL, Collins JP (2007) Susceptibility of the endangered California Tiger Salamander, *Ambystoma californiense*, to Ranavirus infection. J Wildl Dis 43(2):286–290. https://doi.org/10.7589/0090-3558-43.2.286

Pounds JA, Bustamante MR, Coloma LA, Consuegra JA, Fogden MPL, Foster PN, La Marca E, Masters KL, Merino-Viteri A, Puschendorf R, Ron SR, Sánchez-Azofeifa GA, Still CJ, Young BE (2006) Widespread amphibian extinctions from epidemic disease driven by global warming. Nature 439(7073):161–167

Ramírez Piedra ER, Villalta Sanmartín CI (2020) Prevalencia e Intensidad de Tres Enfermedades Infecciosas En Ranas Del Género *Pristimantis* (Anura: Strabomantidae), En Varias Localidades de Las Provincias Azuay y Cañar al Sur de Los Andes Ecuatorianos. Universidad Azuay, Cuenca

Reed K (2000) *Chlamydia pneumoniae* infection in a breeding colony of African Clawed Frogs (*Xenopus tropicalis*). Emerg Infect Dis 6(2):196–199. https://doi.org/10.3201/eid0602.000216

Reeder NMM, Pessier AP, Vredenburg VT (2012) A reservoir species for the emerging amphibian pathogen *Batrachochytrium dendrobatidis* thrives in a landscape decimated by disease. Edited by Anastasia P. Litvintseva. PLoS One 7(3):e33567. https://doi.org/10.1371/journal.pone.0033567

Rueda-Solano LA, Flechas SV, Galvis-Aparicio M, Rocha-Usuga AA, Rincón-Barón EJ, Cuadrado-Peña B, Franke-Ante R (2016) Epidemiological surveillance and amphibian assemblage status at the Estación experimental de San Lorenzo, Sierra Nevada de Santa Marta, Colombia. Amphib Reptile Conserv 10(1)

Ruggeri J, Ribeiro LP, Pontes MR, Toffolo C, Candido M, Carriero MM, Zanella N, Sousa RLM, Toledo LF (2019) Discovery of wild amphibians infected with Ranavirus in Brazil. J Wildl Dis 55(4):897–902. https://doi.org/10.7589/2018-09-224

Sabino-Pinto J, Veith M, Vences M, Steinfartz S (2018) Asymptomatic infection of the fungal pathogen *Batrachochytrium salamandrivorans* in captivity. Sci Rep 8(1):11767

Scheele BC, Pasman F, Skerratt LF, Berger L, An Martel W, Beukema AA et al (2019) Amphibian fungal Panzootic causes catastrophic and ongoing loss of biodiversity. Science 363(6434):1459–1463. https://doi.org/10.1126/science.aav0379

Schloegel LM, Ferreira CM, James TY, Hipolito M, Longcore JE, Hyatt AD, Yabsley M et al (2010) The North American Bullfrog as a reservoir for the spread of *Batrachochytrium dendrobatidis* in Brazil. Anim Conserv 13(s1):53–61. https://doi.org/10.1111/j.1469-1795.2009.00307.x

Soto-Azat C, Valenzuela-Sánchez A, Clarke BT, Busse K, Ortiz JC, Barrientos C, Cunningham AA (2013) Is Chytridiomycosis driving Darwin's frogs to extinction? PLoS One 8(11):e79862. https://doi.org/10.1371/journal.pone.0079862

Soto-Azat C, Peñafiel-Ricaurte A, Price SJ, Sallaberry-Pincheira N, García MP, Alvarado-Rybak M, Cunningham AA (2016) *Xenopus laevis* and emerging amphibian pathogens in Chile. EcoHealth 13(4): 775–783. https://doi.org/10.1007/s10393-016-1186-9

Soto-Azat C, Alvarado-Rybak M, Solano-Iguaran JJ, Velasco A, Valenzuela-Sánchez A, Flechas SV et al (2022) Synthesis of *Batrachochytrium dendrobatidis* infection in South America: amphibian species under risk and areas to focus research and disease mitigation. Ecography 2022(7):e05977

Stegen G, Pasmans F, Schmidt BR, Rouffaer LO, Van Praet S, Schaub M, Canessa S et al (2017) Drivers of salamander extirpation mediated by *Batrachochytrium salamandrivorans*. Nature 544(7650):353–356. https://doi.org/10.1038/nature22059

Stöhr AC, Hoffmann A, Papp T, Robert N, Pruvost NBM, Reyer HU, Marschang RE (2013) Long-term study of an infection with Ranaviruses in a group of edible frogs (*Pelophylax Kl. esculentus*) and partial characterization of two viruses based on four genomic regions. Vet J 197(2):238–244. https://doi.org/10.1016/j.tvjl.2013.02.014

Towe AE, Gray MJ, Carter ED, Wilber MQ, Ossiboff RJ, Ash K, Bohanon M, Bajo BA, Miller DL (2021) *Batrachochytrium salamandrivorans* can devour more than salamanders. J Wildl Dis 57(4). https://doi.org/10.7589/JWD-D-20-00214

Une Y, Sakuma A, Matsueda H, Nakai K, Murakami M (2009) Ranavirus outbreak in North American Bullfrogs (*Rana catesbeiana*), Japan, 2008. Emerg Infect Dis 15(7):1146. https://doi.org/10.3201/eid1507.081636

Urbina JC, Galeano SP (2011) *Batrachochytrium dendrobatidis* detected in amphibians of the Central Andean Cordillera of Colombia. Herpetol Rev 42(4): 558–560

Urgiles VL, Ramírez ER, Villalta CI, Siddons DC, Savage AE (2021) Three pathogens impact terrestrial frogs from a high-elevation tropical hotspot. EcoHealth 18(4):451–464. https://doi.org/10.1007/s10393-021-01570-8

Vasquez-Ochoa A, Bahamón-Carmona P, Prada-Salcedo LD, Franco-Correa M (2012) Detection and quantification of *Batrachochytrium dendrobatidis* in amphibians of the Central and Eastern Andean, Orinoco and Amazon Regions of Colombia. Herpetotropicos 8(1–2):13–21

Viáfara-Vega RA (2020) Detección de *Batrachochytrium dendrobatidis* (Chytridiomycota) en localidades del Valle del Cauca por técnica de PCR. Revista Latinoamericana de Herpetología 3(1):81. https://doi.org/10.22201/fc.25942158e.2020.1.93

Voyles J, Young S, Berger L, Campbell C, Voyles WF, Dinudom A, Cook D et al (2009) Pathogenesis of Chytridiomycosis, a cause of catastrophic amphibian declines. Science 326(5952):582–585. https://doi.org/10.1126/science.1176765

Vredenburg VT, McNally SVG, Sulaeman H, Butler HM, Yap T, Koo MS, Schmeller DS et al (2019) Pathogen invasion history elucidates contemporary host pathogen dynamics. Edited by J. Kerby. PLoS One 14(9): e0219981. https://doi.org/10.1371/journal.pone.0219981

Wallace TC, Hall-Wallace MK (2003) Minerals of the Andes: Emeralds, Gold, and Silver from the Sky. *Rocks Miner* 78(1):12–38. https://doi.org/10.1080/00357529.2003.9926689

Warne RW, LaBumbard B, LaGrange S, Vredenburg VT, Catenazzi A (2016) Co-infection by Chytrid fungus and Ranaviruses in wild and harvested frogs in the Tropical Andes. PLoS One 11(1):e0145864. https://doi.org/10.1371/journal.pone.0145864

Wilcke BW, Newcomer CE, Anver MR, Simmons JL, Nace GW (1983) Isolation of *Chlamydia psittaci* from naturally infected African Clawed Frogs (*Xenopus laevis*). Infect Immun 41(2):789–794. https://doi.org/10.1128/iai.41.2.789-794.1983

Zhang QY, Xiao F, Li ZQ, Gui JF, Mao J, Chinchar VG (2001) Characterization of an Iridovirus from the cultured pig frog *Rana grylio* with lethal syndrome. Dis Aquat Org 48:27–36. https://doi.org/10.3354/dao048027

Zupanovic Z, Musso C, Lopez G, Louriero C-L, Hyatt AD, Hengstberger S, Robinson AJ (1998) Isolation and characterization of Iridoviruses from the giant toad *Bufo marinus* in Venezuela. Dis Aquat Org 33: 1–9. https://doi.org/10.3354/dao033001

Diversity and Extinction Risk of Colombian Andean Amphibians Across Life Regions

Pamela González-del-Pliego and Vinicius Augusto Galvão Bastazini

1 Introduction

Owing to its extreme topological diversity, the Andean range is characterised by rapid habitat transitions, all co-occurring within relatively small areas (Etter et al. 2006; Sánchez-González and Navarro-Sigüenza 2009). This topographic complexity leads to very high levels of biodiversity, endemicity and rates of species turnover, and hence a level of beta diversity that is potentially unparalleled elsewhere in the world (Orme et al. 2005; Lamoreux et al. 2006; Tobar-Suárez et al. 2022). The Colombian Andes present very high biodiversity and endemism levels, particularly of amphibians (Acosta-Galvis 2023; Re:wild et al. 2023). Indeed, the Cordillera Central of Colombia harbours the highest anuran species richness within the Northern Andes (Ecuador, Colombia and Venezuela) (Armesto and Señaris 2017). Furthermore, 73% of the anuran richness from the Northern Andes is concentrated in Colombia (Armesto and Señaris 2017), the second richest country in amphibian species worldwide (892 species) (Acosta-Galvis 2023), but also ranking first for the number of threatened amphibian species (301 species) (Re:wild et al. 2023).

The Colombian Andes host a diverse range of ecosystems, spanning from tropical rainforests to alpine environments. They are divided into four distinct life zones based on elevation: the Tropical region (100–1000 m), the sub-Andean region (1000–2350 m), the Andean region (2350–3500 m), and the Paramo region, which lies above 3500 m. The Colombian Andes, in particular, are one of the ecosystems that has most been impacted by habitat change since the sixteenth century (Etter et al. 2008), with the rate of change accelerating in recent times. In the eastern Cordillera of the Colombian Andes, for example, 51% of the original vegetation has been converted to a different land use since 1989 (Armenteras et al. 2003). The Colombian Andes are heavily populated, with most of the human population in Colombia (~70%) being concentrated in this region. The region is also under continuous and intense anthropogenic pressure and highly threatened by agricultural development, with agricultural activities (including illegal activities (Fjeldså et al. 2005)), cattle grazing and infrastructure development (i.e. urbanisation, dam construction, and roads) being among the main causes of habitat loss and fragmentation (Etter and Wyngaarden 2000; Myers et al. 2000).

Supplementary Information: The online version contains supplementary material available at https://doi.org/10.1007/978-3-032-00074-3_15.

P. González-del-Pliego (✉)
Rui Nabeiro Biodiversity Chair, MED – Mediterranean Institute for Agriculture, Environment and Development & CHANGE–Global Change and Sustainability Institute, Institute for Advanced Studies and Research, University of Évora, Évora, Portugal
e-mail: pamelagp@uevora.pt

V. A. G. Bastazini
Independent Researcher, Nivelles, Belgium

J. N. Urbina-Cardona et al. (eds.), *Andean Herpetofauna*, Sustainable Development Goals Series,
https://doi.org/10.1007/978-3-032-00074-3_15

The high levels of intense productive activities and population concentration diminish Andean biodiversity (Rodríguez et al. 2004), including that of amphibians. This is because the transformation of primary forest into agricultural areas is associated with a decrease in amphibian abundance (Roach et al. 2020), species richness, a change in community composition (Gilroy et al. 2015; Basham et al. 2016) and alterations in their environment, which affect their ability to cope with thermal extremes (González-del-Pliego et al. 2020, 2023). Biological richness, a high level of endemism and a high level of anthropogenic threat make the Colombian Andes one of the world's major conservation priorities.

By exploring changes in biodiversity and their extinction risk across life regions, we can obtain new insights into biodiversity patterns, changes in community composition and threat status and the potential responses of biodiversity to human impacts. However, since biodiversity responses to global changes are scale-dependent (Chase et al. 2018), accounting for these dependencies becomes a crucial step to fully understand the broad consequences of human disturbances on biodiversity (Chase et al. 2018). Decomposing diversity into within-community (i.e. alpha diversity) and between-community (i.e. beta diversity) components thus becomes an essential step in seeking to properly address the effects of global changes on biodiversity (Hekkala and Roberge 2018; Hillebrand et al. 2018; Bastazini et al. 2021). Moreover, by examining beta diversity, we can explore distinct processes such as species replacement (i.e. species turnover) and changes in species richness through species loss or gain (i.e. nestedness) across environmental gradients (Baselga et al. 2012). This approach provides a deeper understanding of how human impacts affect biodiversity (Hekkala and Roberge 2018; Hillebrand et al. 2018; Bastazini et al. 2021) and offers invaluable guidelines for selecting appropriate spatial scales for conservation planning (Legendre et al. 2005; Bergamin et al. 2017). Here we aim to answer: (1) What are the differences in species richness and beta diversity and the number of threatened species across the four life regions? and (2) How do human impacts affect each life region and threatened species? We address these questions with direct reference to the Colombian Andes predicting that (1) species richness will peak at intermediate elevations as shown for several Andean species and (2) human density and agricultural activities will increase the probability of threat of amphibians.

2 Methods

We obtained the elevational distribution of amphibians from the Colombian Andes from Acosta-Galvis (2023) and their range size from González-del-Pliego et al. (2019). From the latter database, the geographical range of 469 species was obtained from the IUCN and AmphibiaWeb (2024). For 10 species, spatial polygons were created from coordinates available from the literature and from AmphibiaWeb (see González-del-Pliego et al. 2019 for further details). For one species, range size was obtained from (Moura et al. 2024) (see Supplementary Material Table S15.1).

For each species, we obtained the following:

1. Human density: Average within-range human population density in 2017, quantified as the number of people per square kilometre. Data derived from (Goldewijk et al. 2017; Moura et al. 2024).
2. Proportion of cropland area: Proportion of species range map covered by cropland area, identical to FAO's category 'Arable land and permanent crops' in 2017. Data derived from HYDE v. 3.2 (Goldewijk et al. 2017; Moura et al. 2024).
3. Proportion of pasture area: Proportion of species range map covered by cropland, defined as grazing land with an aridity index >0.5, assumed to be more intensively managed (converted in climate models) in 2017. Data derived from HYDE v. 3.2 (Goldewijk et al. 2017; Moura et al. 2024).

Our resulting database contained 480 amphibian species. Each species was assigned a threat status according to the IUCN red list (IUCN 2023). We also categorised species into non-threatened (LC and NT) and threatened species (VU, EN and CR), with two species categorised as not evaluated and 34 species as data-deficient and thus these two groups were excluded in analysis where the threat status was required (see Supplementary Material Table S15.1 for the full list of species).

We categorised each species into the different life regions based on their elevational distribution: Tropical region (100–1000 m); sub-Andean region (1000–2350 m); Andean region (2350–3500 m) and Paramo region (with elevations higher than 3500 m). Species could belong to more than one life region (see Supplementary Material Table S15.1). While we used present-day elevation ranges of amphibians, elevation per se does not determine the spatial distribution of amphibians; rather they respond to temperature, moisture or other environmental variables that can be subject to seasonal variations. We therefore assessed the specific ecology of any species bordering between life regions (Acosta-Galvis 2023; AmphibiaWeb 2024). For example, the elevational distribution recorded for *Elachistocleis pearsei* is from 0 to 1015 m. Although at this elevation this species would be considered as belonging to the Tropical and sub-Andean region, this species is found mainly in lowlands (Bernal et al. 2017), and was thus categorised exclusively in the Tropical region.

To understand the variation of the species composition of assemblages, we calculated beta diversity using the Jaccard index. We also partitioned beta diversity into its turnover (differences in composition caused by species replacements) and nestedness (differences in species composition caused by species gain/loss) components (Baselga et al. 2012). This allows us to quantify to what extent differences in beta diversity are related to differences in turnover and to differences in species richness. We also obtained the contribution to turnover and the contribution to richness for each of the life regions. Beta diversity calculations were obtained using the 'adespatial' R package (Dray et al. 2023).

We performed individual ANOVAs and Tukey post hoc tests using range size, human density, proportion of cropland area and proportion of pasture area as our response variables to assess whether these variables differed across the four life regions (i.e. Tropical, sub-Andean, Andean and Paramo). To do so, we used the functions 'aov' and 'TukeyHSD' from the stats R package (R 2023).

We performed logistic regression using the 'glm' function in R, specifying the statistical distribution family as "binomial." We then used the function 'pR2' from the pscl package (Jackman 2024) to compute McFadden's pseudo r-squared measure from the generalised linear models. This function involves a comparison of the log-likelihood for the fitted model against the log-likelihood of a null/restricted model with no predictors, normalised to run from zero to one as the fitted model provides a better fit to the data (providing a rough analogue to the computation of r-squared in a linear regression) (Jackman 2024).

3 Results

The sub-Andean region has the highest species richness (429 species), followed by the Andean (220 species) and then the Tropical region (190 species). The Paramo region has the lowest species richness from all regions (44 species). We found that 219 species were only present within one life region, and 243 and 18 species were found across two and three life regions, respectively (Supplementary Material Table S15.1). In terms of threat status within each region, the higher in elevation the region goes, the larger the percentage of threatened species which tend to be found. Species richness showed an increase and decrease relationship with elevation, with the Tropics and the Paramo (lowest and highest elevation sites, respectively) having lower species richness (Fig. 1a). The percentage of threatened species was 30%, 55%, 66% and 60% in the Tropics, sub-Andean, Andean and Paramo regions, respectively (Fig. 1b). As such, the Tropics is the only region with a significantly higher proportion of nonthreatened species ($X^2 = 24.9$, DF $= 1$, $p < 0.001$), whereas the

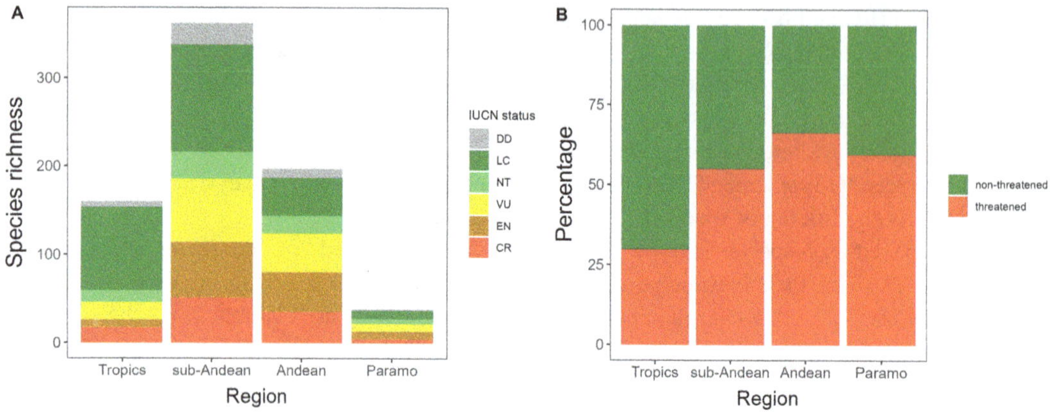

Fig. 1 Differences in species richness (a) and percentage of threatened and non-threatened species (**b**) across the four life regions. Colours in **a** refer to IUCN categories: DD data deficient, LC least concern, NT near threatened, VU vulnerable, EN endangered, CR critically endangered. Colours in b refer to threatened (red) and non threatened (green) species

Andean region, which is the region with the second-highest species richness, is the only one with a significantly higher proportion of threatened species compared to non-threatened ($X^2 = 20.5$, DF = 1, $p < 0.001$; Fig. 1b).

The species belonging to the Tropical region have a significantly larger range area compared to species from other regions ($F_{3,756} = 4.86$, $p < 0.001$; Fig. 2a), whereas species in the sub-Andean, Andean and Paramo regions have similar range sizes ($F_{3,756} = 4.86$, $p > 0.5$; Fig. 2a). Regarding human density, species belonging to the Paramo region experience similar levels of human density compared to species from both the sub-Andean and the Andean regions ($F_{3,756} = 11.86$, $p > 0.3$). Nevertheless, species from the Tropics, the sub-Andean region and the Andean region all experience a different degree of human density, with species in higher elevations experiencing higher human density than species from lower elevations ($F_{3,756} = 11.86$, $p < 0.01$; Fig. 2b).

Species inhabiting the Tropical region have a lower percentage of pasture area within their ranges compared to the rest of the regions ($F_{3,756} = 28.31$, $p < 0.001$), species from the sub-Andean region had a lower percentage of pasture area within their ranges compared to species from the Andean and the Paramo regions ($F_{3,756} = 28.31$, $p < 0.01$), while species from the Andean and Paramo regions have similar percentages ($F_{3,756} = 28.31$, $p = 0.95$; Fig. 2c). We found that all species have a similar percentage of cropland area within their ranges, irrespective of the region they inhabit ($F_{3,756} = 2.92$, $p > 0.1$; Fig. 2d).

The logistic regression analysis showed that, confounded with their threat category, the probability of threat is higher for species that occupy smaller geographical areas and lower for species with a large range size (z value = −9.2, DF = 714, $p < 0.001$, McFadden $R^2 = 0.23$) (Fig. 3a). Regarding human density, we found that while the probability of threat increased significantly with an increase of human density (z value = 2.68, DF = 714, $p < 0.01$), the overall model fit was very weak (McFadden $R^2 = 0.007$; Fig. 3b). Regarding the proportion of pasture area, while we found that the probability of threat increased significantly with a higher proportion of pasture area (z value = 3.13, DF = 714, $p < 0.001$), this model fit was also weak (McFadden $R^2 = 0.01$; Fig. 3c). Finally, we found that the threatened probability of species was similar irrespective of the proportion of cropland area within the range (z value = −0.5, DF = 714, $p = 0.61$; Fig. 3d).

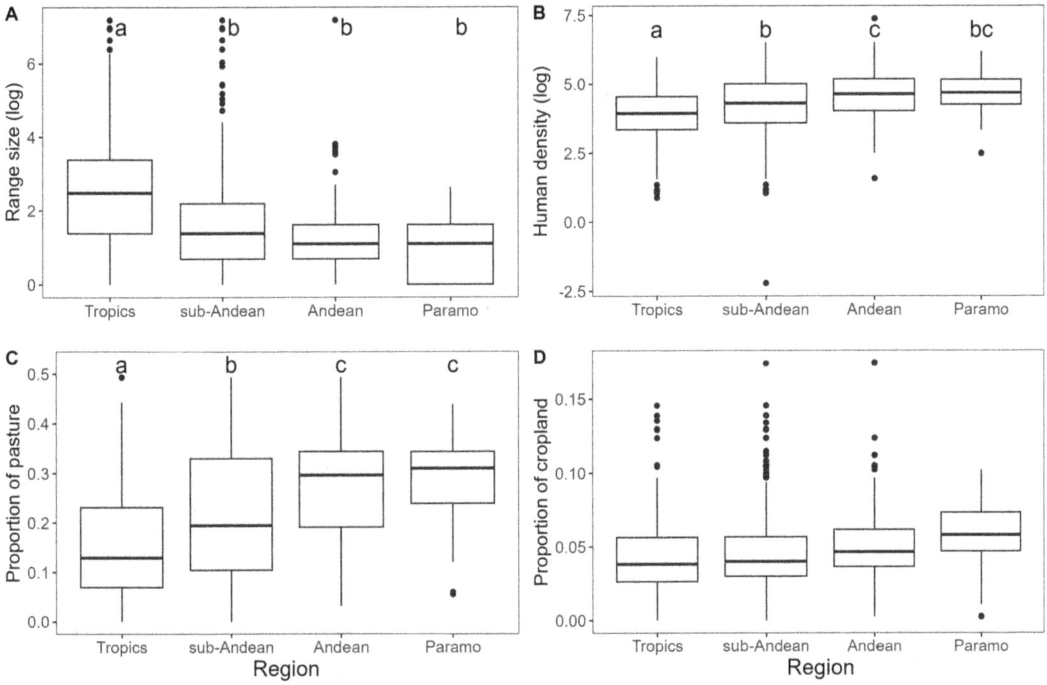

Fig. 2 Range size (**a**), human density (**b**), proportion of pasture area (**c**) and proportion of cropland area (**d**) across the Tropical, sub-Andean, Andean and Paramo regions. Different letters represent results deemed significant at $\alpha = 0.05$

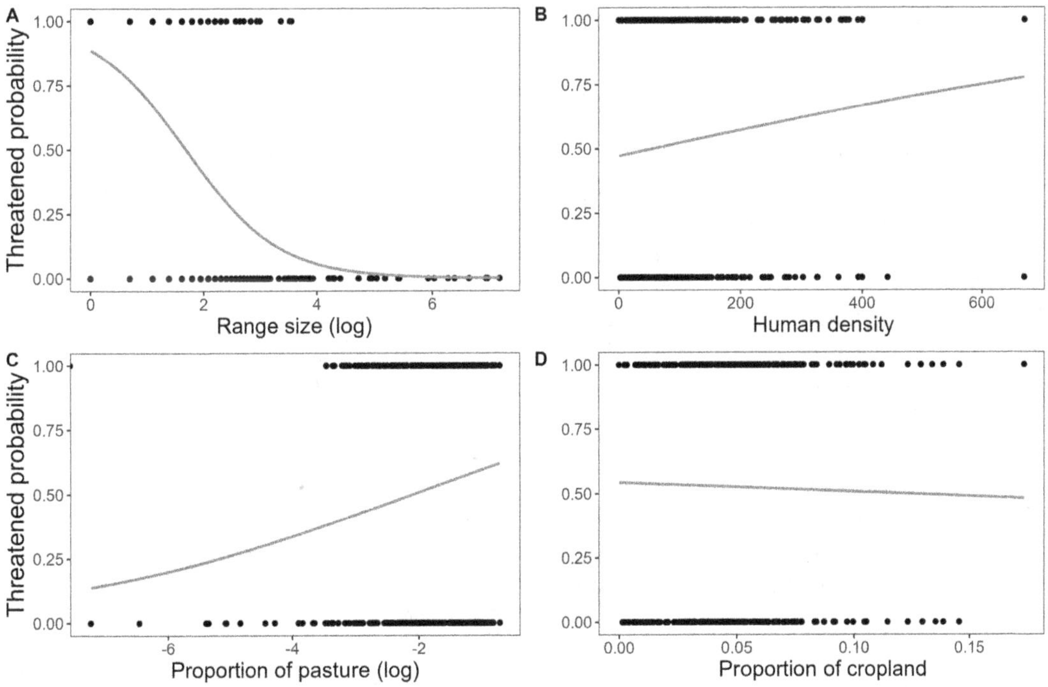

Fig. 3 Logistic regression of range size (**a**), human density (**b**), proportion of pasture area (**c**) and proportion of cropland area (**d**) across the probability of being threatened. Y-axis represents non-threatened and threatened species by 0 and 1, respectively

Table 1 Total beta diversity and contribution of turnover and nestedness (species richness) to total beta diversity and for each life region

	Total and percentage in parenthesis	Contribution per region (percentage)			
		Tropics	sub-Andean	Andean	Paramo
Beta diversity	0.43	0.26	0.21	0.23	0.28
Beta diversity from turnover/replacement	0.16 (0.38%)	0.49	0.09	0.39	0.03
Beta diversity from richness difference/ nestedness	0.26 (0.62%)	0.12	0.29	0.14	0.45

The overall beta diversity in our data is 0.43. Partitioning beta diversity components revealed that nestedness is the dominant component (>60% of beta diversity was due to nestedness). Yet the contribution of turnover to beta diversity was not negligible, contributing 38% to the total beta diversity. The contribution of each region to total beta diversity was on average 25% (range = 21–28%). When analysing the contribution of the regions to both components of beta diversity, we found a complex relationship, with all sites contributing significantly to either turnover or nestedness, but not both (Table 1).

4 Discussion

As expected, we found that species richness peaked at intermediate elevations (Hutter et al. 2013, 2017), with the sub-Andean region showing the highest and the Paramo region the lowest species richness. Although we found the Paramo region to show the lowest species richness (Hutter et al. 2017), recent evidence suggests the number of species that inhabit these regions may be much higher (up to 142 species) (Saboyá Acosta and Urbina-Cardona 2023). Most species have a broad altitudinal distribution (>500 m in altitude), with the majority of species being found across two or three life regions, and 219 found in only one life region. This is in contrast to that found by Bernal and Lynch (2008), where they showed that the majority of Andean Colombian species have narrow altitudinal ranges (<500 m in altitude), and only a few species have broad altitudinal distributions.

Our results show that the regions in the Colombian Andes with higher human density and agricultural intensification (sub-Andean, Andean and the Paramo regions) are those with a higher percentage of threatened species compared to non-threatened species. The Tropical region is the only one whose species have a significantly larger range area compared to the other regions, which may be related to the fact that the Tropics is also the only region with a higher percentage of non-threatened species compared to threatened species, the former of which tend to have larger range areas (IUCN 2023). Furthermore, there is a very strong tendency for the probability of threat to be greater if the range size of species is small. While human density and the proportion of pasture area within species' range areas showed that threatened probability increased with an increase of these variables, the models fit the data poorly, while the proportion of cropland area showed no association with threatened probability. These weak or null associations between human density, the proportion of pasture area and the proportion of cropland area with the threatened probability of a species was unexpected, since the most commonly documented threats to amphibians are types of habitat loss and degradation, with the top three being related to agriculture (77% of species impacted), while 40% of species are affected by infrastructure development (related to human density) (Luedtke et al. 2023). This weak relationship could be partially related to the shortfalls in taxonomy and spatial information in regions at higher elevation, such as the Paramo region (Saboyá Acosta and Urbina-Cardona 2023).

As expected, our results showed that amphibian species in the Colombian Andes showed a high beta diversity (De la Riva 2020). We found nestedness to be the dominant component of beta diversity for amphibians in the Colombian Andes. This means that most of the differences in communities are related to differences in richness. This could reflect large elevational ranges promoting speciation through habitat specialisation and isolation, which in turn increases endemism and consequently creates large differences in species richness (Kozak and Wiens 2007; Smith et al. 2007). Interestingly, nestedness being the dominant component of beta diversity is the opposite to that which was found for amphibians in a mountainous elevational gradient (Wang et al. 2022), in the Brazilian Atlantic forest (da Silva et al. 2014) and in a cloud forest and lowland rainforest in Peru (Villacampa et al. 2019). Nonetheless, the contribution of turnover to beta diversity was not irrelevant, contributing 38% to the total beta diversity. Furthermore, our altitudinal bands are very large, and if a more narrow altitudinal gradient is used, we could find further differences in beta diversity for amphibian communities in the Colombian Andes (Villacampa et al. 2019).

Our finding that each life region can have either high turnover and low nestedness values or the opposite is similar to (Baselga et al. 2012), who found that there is a global pattern for amphibians where regions show high turnover and low nestedness values, or low turnover and high nestedness values (Baselga et al. 2012). Our results regarding the contribution of each region to both components of beta diversity (turnover and richness) suggest that the contributions of the Tropics and the Andean regions to beta diversity are mostly due to turnover. This means that the species found in both regions are very distinct. In other words, there is a large replacement of some species by others, likely caused by environmental sorting or spatial and historical constraints (Qian et al. 2005). In contrast, the contributions of the sub-Andean and Paramo regions to beta diversity are mostly related to differences in species richness (nestedness), as a result of both these regions having the highest and lowest species richness across sites. These differences are due to gain (sub-Andean) and loss (Paramo) of species, and these values of nestedness indicate that some species are absent from some sites, yet we are at the moment unable to infer the underlying mechanism (i.e. extinction, differential dispersal capacity (Ulrich et al. 2009)).

Overall, the similar contribution to total beta diversity of each life region suggests that similar processes lead to high levels of differentiation in amphibians, while ecological and biogeographic factors influencing the different components of beta diversity vary across the life regions (McKnight et al. 2007). The complex patterns among components of amphibian beta diversity that we found could be the result of environment- and elevation-related biotic and abiotic factors making amphibian species respond differently to environmental and spatial changes due to differences in physiological and behavioural characteristics (da Silva et al. 2014)- e.g. temperature, precipitation (Wang et al. 2022). Indeed, mountain environmental conditions vary spatially, with communities therefore often exhibiting a high degree of change in community composition with elevation (Wang et al. 2022). Furthermore, to understand gradients of amphibian distribution, it is important to consider both environment and historical events (Buckley and Jetz 2007; Duarte et al. 2014). For example, while overall richness in the Colombian Andes is strongly determined by regional history, the patterns within the different life regions can be driven by water availability, temperature (Buckley and Jetz 2007) and species' ecophysiological constraints (González-del-Pliego et al. 2020, 2023). The Colombian Andes prove to be a very suitable study site as here is where we found the highest amphibian richness and proportion of endemic species across Neotropical cloud forests (Tobar-Suárez et al. 2022). Our study offers new insights into biodiversity patterns at regional levels and could inform conservation planning with regard to the optimal spatial arrangement of conservation areas.

Acknowledgements The authors thank A. Acosta-Galvis for the list of Andean amphibian species and D. Zugai for comments.

Competing Interests The authors have no conflicts of interest to declare that are relevant to the content of this chapter.

References

Acosta-Galvis AR (2023) Lista de los anfibios de Colombia: referencia en linea V.13.2023. Batrachia, Villa de Leyva, Boyacá

AmphibiaWeb (2024) AmphibiaWeb. University of California, Berkeley

Armenteras D, Gast F, Villareal H (2003) Andean forest fragmentation and the representativeness of protected natural areas in the eastern Andes, Colombia. Biol Conserv 113:245–256

Armesto LO, Señaris JC (2017) Anuros del norte de los andes: patrones de riqueza de especies y estado de conservación. Pap Avulsos Zool 57:491–526

Baselga A, Gómez-Rodríguez C, Lobo JM (2012) Historical legacies in world amphibian diversity revealed by the turnover and nestedness components of beta diversity. PLoS One 7:e32341

Basham EW, González del Pliego P, Acosta-Galvis AR, Woodcock P, Medina Uribe CA, Haugaasen T, Gilroy JJ, Edwards DP (2016) Quantifying carbon and amphibian co-benefits from secondary forest regeneration in the Tropical Andes. Anim Conserv 19:548–560

Bastazini VAG, Galiana N, Hillebrand H, Estiarte M, Ogaya R, Peñuelas J, Sommer U, Montoya JM (2021) The impact of climate warming on species diversity across scales: lessons from experimental meta-ecosystems. Glob Ecol Biogeogr 30:1545–1554

Bergamin RS, Bastazini VAG, Vélez-Martin E, Debastiani V, Zanini KJ, Loyola R, Müller SC (2017) Linking beta diversity patterns to protected areas: lessons from the Brazilian Atlantic Rainforest. Biodivers Conserv 26:1557–1568

Bernal MH, Lynch JD (2008) Review and analysis of altitudinal distribution of the Andean anurans in Colombia. Zootaxa 1826:1–25

Bernal MH, Paéz CA, Vejarano MA (2017) Composición y distribución de los anfibios de la cuenca del río Coello (Tolima), Colombia. Actual Biol 27:87–92

Buckley LB, Jetz W (2007) Environmental and historical constraints on global patterns of amphibian richness. Proc R Soc B Biol Sci 274:1167–1173

Chase JM, McGill BJ, McGlinn DJ, May F, Blowes SA, Xiao X, Knight TM, Purschke O, Gotelli NJ (2018) Embracing scale-dependence to achieve a deeper understanding of biodiversity and its change across communities. Ecol Lett 21:1737–1751

da Silva FR, Almeida-Neto M, Arena MVN (2014) Amphibian beta diversity in the Brazilian Atlantic forest: contrasting the roles of historical events and contemporary conditions at different spatial scales. PLoS One 9:e109642

De la Riva I (2020) Unexpected beta-diversity radiations in highland clades of Andean Terraranae frogs. In: Rull V, Carnaval AC (eds) Neotropical diversification: patterns and processes. Springer International Publishing, Cham, pp 741–764

Dray S, Bauman D, Blanchet G, Borcard D, Clappe S, Guenard G, Jombart T, Larocque G, Legendre P, Madi N, Wagner HH, Siberchicot A (2023) adespatial: multivariate multiscale spatial analysis. Version 0.3-23

Duarte LDS, Both C, Debastiani VJ, Carlucci MB, Gonçalves LO, Cappelatti L, Seger GDS, Bastazini VAG, Brum FT, Salengue EV, Bernardo-Silva JS (2014) Climate effects on amphibian distributions depend on phylogenetic resolution and the biogeographical history of taxa. Glob Ecol Biogeogr 23: 213–222

Etter A, Wyngaarden WV (2000) Patterns of landscape transformation in Colombia, with emphasis in the Andean region. AMBIO J Hum Environ 29:432–439

Etter A, McAlpine C, Wilson K, Phinn S, Possingham H (2006) Regional patterns of agricultural land use and deforestation in Colombia. Agric Ecosyst Environ 114: 369–386

Etter A, McAlpine C, Possingham H (2008) Historical patterns and drivers of landscape change in Colombia since 1500: a regionalized spatial approach. Ann Assoc Am Geogr 98:2–23

Fjeldså J, Álvarez MD, Juan Mario L, León B (2005) Illicit crops and armed conflict as constraints on biodiversity conservation in the Andes region. Ambio 34:205–211

Gilroy JJ, Prescott GW, Cardenas JS, Castañeda PGDP, Sánchez A, Rojas-Murcia LE, Medina Uribe CA, Haugaasen T, Edwards DP (2015) Minimizing the biodiversity impact of Neotropical oil palm development. Glob Chang Biol 21:1531–1540

Goldewijk KK, Beusen A, Doelman J, Stehfest E (2017) Anthropogenic land use estimates for the Holocene-HYDE 3.2. Earth Syst Sci Data 9:927–953

González-del-Pliego P, Freckleton RP, Edwards DP, Koo MS, Scheffers BR, Pyron RA, Jetz W (2019) Phylogenetic and trait-based prediction of extinction risk for data-deficient amphibians. Curr Biol 29:1557–1563. e1553

González-del-Pliego P, Scheffers BR, Freckleton RP, Basham EW, Araújo MB, Acosta-Galvis AR, Medina Uribe CA, Haugaasen T, Edwards DP (2020) Thermal tolerance and the importance of microhabitats for Andean frogs in the context of land use and climate change. J Anim Ecol 89:2451–2460

González-del-Pliego P, Freckleton RP, Scheffers BR, Basham EW, Acosta-Galvis AR, Medina Uribe CA, Haugaasen T, Edwards DP (2023) Phylogeny and morphology determine vulnerability to global warming in Pristimantis frogs. Land 12:130

Hekkala A-M, Roberge J-M (2018) The use of response measures in meta-analyses of land-use impacts on ecological communities: a review and the way forward. Biodivers Conserv 27:2989–3005

Hillebrand H, Blasius B, Borer ET, Chase JM, Downing JA, Eriksson BK, Filstrup CT, Harpole WS, Hodapp D, Larsen S, Lewandowska AM, Seabloom EW, Van de Waal DB, Ryabov AB (2018) Biodiversity change is uncoupled from species richness trends: consequences for conservation and monitoring. J Appl Ecol 55:169–184

Hutter CR, Guayasamin JM, Wiens JJ (2013) Explaining Andean megadiversity: the evolutionary and ecological causes of glassfrog elevational richness patterns. Ecol Lett 16:1135–1144

Hutter CR, Lambert SM, Wiens JJ (2017) Rapid diversification and time explain amphibian richness at different scales in the Tropical Andes, Earth's most biodiverse hotspot. Am Nat 190:828–843

IUCN (2023) The IUCN red list of threatened species. Version 2023-2. http://www.iucnredlist.org

Jackman S (2024) pscl: classes and methods for R developed in the political science computational laboratory. R package version 1.5.9. Sydney, Australia. https://github.com/atahk/pscl/

Kozak KH, Wiens JJ (2007) Climatic zonation drives latitudinal variation in speciation mechanisms. Proc Biol Sci 274:2995–3003

Lamoreux JF, Morrison JC, Ricketts TH, Olson DM, Dinerstein E, McKnight MW, Shugart HH (2006) Global tests of biodiversity concordance and the importance of endemism. Nature 440:212–214

Legendre P, Borcard D, Peres-Neto PR (2005) Analyzing beta diversity: partitioning the spatial variation of community composition data. Ecol Monogr 75:435–450

Luedtke JA, Chanson J, Neam K, Hobin L, Maciel AO, Catenazzi A, Borzée A, Hamidy A, Aowphol A, Jean A, Sosa-Bartuano Á, Fong GA, de Silva A, Fouquet A, Angulo A, Kidov AA, Muñoz Saravia A, Diesmos AC, Tominaga A, Shrestha B, Gratwicke B, Tjaturadi B, Martínez Rivera CC, Vásquez Almazán CR, Señaris C, Chandramouli SR, Strüssmann C, Cortez Fernández CF, Azat C, Hoskin CJ, Hilton-Taylor C, Whyte DL, Gower DJ, Olson DH, Cisneros-Heredia DF, Santana DJ, Nagombi E, Najafi-Majd E, Quah ESH, Bolaños F, Xie F, Brusquetti F, Álvarez FS, Andreone F, Glaw F, Castañeda FE, Kraus F, Parra-Olea G, Chaves G, Medina-Rangel GF, González-Durán G, Ortega-Andrade HM, Machado IF, Das I, Dias IR, Urbina-Cardona JN, Crnobrnja-Isailović J, Yang J-H, Jianping J, Wangyal JT, Rowley JJL, Measey J, Vasudevan K, Chan KO, Gururaja KV, Ovaska K, Warr LC, Canseco-Márquez L, Toledo LF, Díaz LM, Khan MMH, Meegaskumbura M, Acevedo ME, Napoli MF, Ponce MA, Vaira M, Lampo M, Yánez-Muñoz MH, Scherz MD, Rödel M-O, Matsui M, Fildor M, Kusrini MD, Ahmed MF, Rais M, Kouamé NGG, García N, Gonwouo NL, Burrowes PA, Imbun PY, Wagner P, Kok PJR, Joglar RL, Auguste RJ, Brandão RA, Ibáñez R, von May R, Hedges SB, Biju SD, Ganesh SR, Wren S, Das S, Flechas SV, Ashpole SL, Robleto-Hernández SJ, Loader SP, Incháustegui SJ, Garg S, Phimmachak S, Richards SJ, Slimani T, Osborne-Naikatini T, Abreu-Jardim TPF, Condez TH, De Carvalho TR, Cutajar TP, Pierson TW, Nguyen TQ, Kaya U, Yuan Z, Long B, Langhammer P, Stuart SN (2023) Ongoing declines for the world's amphibians in the face of emerging threats. Nature 622:308–314

McKnight MW, White PS, McDonald RI, Lamoreux JF, Sechrest W, Ridgely RS, Stuart SN (2007) Putting beta-diversity on the map: broad-scale congruence and coincidence in the extremes. PLoS Biol 5:e272

Moura MR, Ceron K, Guedes JJM, Chen-Zhao R, Sica YV, Hart J, Dorman W, Portmann JM, Gonzalez-del-Pliego P, Ranipeta A, Catenazzi A, Werneck FP, Toledo LF, Upham NS, Tonini JFR, Colston TJ, Guralnick R, Bowie RCK, Pyron RA, Jetz W (2024) A phylogeny-informed characterisation of global tetrapod traits addresses data gaps and biases. PLoS Biol 22:e3002658

Myers N, Mittermeier RA, Mittermeier CG, da Fonseca GAB, Kent J (2000) Biodiversity hotspots for conservation priorities. Nature 403:853–858

Orme CDL, Davies RG, Burgess M, Eigenbrod F, Pickup N, Olson VA, Webster AJ, Ding T-S, Rasmussen PC, Ridgely RS, Stattersfield AJ, Bennett PM, Blackburn TM, Gaston KJ, Owens IPF (2005) Global hotspots of species richness are not congruent with endemism or threat. Nature 436:1016–1019

Qian H, Ricklefs RE, White PS (2005) Beta diversity of angiosperms in temperate floras of eastern Asia and eastern North America. Ecol Lett 8:15–22

R (2023) R Core Team: a language and environment for statistical computing. R. F. f. S. Computing, Vienna. Version 4.2.3

Re: wild, Synchronicity Earth, IUCN SSC Amphibian Specialist Group (2023) State of the world's amphibians: the second global amphibian assessment. Re:wild, Austin

Roach NS, Urbina-Cardona N, Lacher TE (2020) Land cover drives amphibian diversity across steep elevational gradients in an isolated neotropical mountain range: implications for community conservation. Glob Ecol Conserv 22:e00968

Rodríguez N, Armenteras D, Rincón A, Morales M (2004) Forest biodiversity indicator in the Colombian Andes, San Diego

Saboyá Acosta LP, Urbina-Cardona JN (2023) Current state of knowledge of Páramo amphibians in Colombia: spatio temporal trends and information gaps to be strengthened for effective conservation. Trop Conserv Sci 16:19400829231169984

Sánchez-González LA, Navarro-Sigüenza AG (2009) History meets ecology: a geographical analysis of ecological restriction in the Neotropical humid montane forests avifaunas. Divers Distrib 15:1–11

Smith SA, De Oca ANM, Reeder TW, Wiens JJ (2007) A phylogenetic perspective on elevational species richness patterns in middle american treefrogs: why so few species in lowland tropical rainforests? Evolution 61: 1188–1207

Tobar-Suárez C, Urbina-Cardona N, Villalobos F, Pineda E (2022) Amphibian species richness and endemism in tropical montane cloud forests across the Neotropics. Biodivers Conserv 31:295–313

Ulrich W, Almeida-Neto M, Gotelli NJ (2009) A consumer's guide to nestedness analysis. Oikos 118: 3–17

Villacampa J, Whitworth A, Allen L, Malo JE (2019) Altitudinal differences in alpha, beta and functional diversity of an amphibian community in a biodiversity hotspot. Neotrop Biodivers 5:60–68

Wang X, Zhong M, Yang S, Jiang J, Hu J (2022) Multiple β-diversity patterns and the underlying mechanisms across amphibian communities along a subtropical elevational gradient. Divers Distrib 28:2489–2502

Experiences in Conservation Planning for Amphibians in the Andean Region

Ariadne Angulo, Claudio Azat, Diego F. Cisneros-Heredia, Margarita Lampo, Celsa Señaris, J. Nicolás Urbina-Cardona, and Marcos Vaira

1 Introduction

Conservation plans or strategies (CPs) are a key foundation for increasing the effectiveness of conservation efforts. They achieve this by defining goals, objectives, expected outcomes, identifying knowledge and conservation gaps, prioritizing actions, allocating resources, and fostering partnerships. In addition, CPs establish timelines and a monitoring framework to track implementation and measure impact (Wren et al. 2024). Conservation planning can take place at multiple geographical scales, ranging from global to regional, national and subnational levels (Wren et al. 2024). It is an inherently participatory process, requiring the engagement of relevant stakeholders to secure their buy-in and full commitment to the plan's implementation (Knight et al. 2006). Historically, amphibians have been overlooked in broader conservation planning efforts, including those focusing on protected areas (Nori et al. 2015). While early amphibian conservation action plans primarily targeted individual threatened species, more recent strategies tend to adopt a multispecies approach (Wren et al. 2024). This shift is crucial, as amphibians remain the most threatened vertebrate class, with nearly 41% of species threatened with extinction, according to the second Global Amphibian Assessment (GAA2) (Luedtke et al. 2023).

1.1 The Global Amphibian Conservation Action Plan: A Brief Overview

The global Amphibian Conservation Action Plan (ACAP), first published in 2007 (Gascon et al.

Supplementary Information: The online version contains supplementary material available at https://doi.org/10.1007/978-3-032-00074-3_16.

A. Angulo
IUCN SSC Amphibian Specialist Group, Toronto, ON, Canada

C. Azat
One Health Institute, Universidad Andrés Bello, Santiago, Chile

D. F. Cisneros-Heredia
Colegio de Ciencias Biológicas y Ambientales, Instituto de Biodiversidad Tropical IBIOTROP, Universidad San Francisco de Quito USFQ, Quito, Pichincha, Ecuador

M. Lampo
Fundación para el Desarrollo de las Ciencias Físicas, Matemáticas y Naturales (FUDECI), Caracas, Venezuela

C. Señaris
Estación Biológica de Doñana (EBD-CSIC), Seville, Spain

J. N. Urbina-Cardona
Pontificia Universidad Javeriana, Facultad de Estudios Ambientales y Rurales, Departamento de Ecología y Territorio. Carrera, Bogotá, Colombia

M. Vaira (✉)
IUCN SSC Amphibian Specialist Group, IUCN Species Survival Commission, Gland, Switzerland

Instituto de Ecorregiones Andinas (INECOA), CONICET – UNJu, San Salvador de Jujuy, Argentina
e-mail: marcos.vaira@conicet.gov.ar

J. N. Urbina-Cardona et al. (eds.), *Andean Herpetofauna*, Sustainable Development Goals Series,
https://doi.org/10.1007/978-3-032-00074-3_16

2007) and updated as a digital resource in 2015 (Wren et al. 2015) and in 2024 (Wren et al. 2024), provides a multipronged, unified global strategy for amphibian conservation. It was designed as a framework to develop unified strategies, to raise awareness of the amphibian plight, and to engage stakeholders from global to local levels.

As the most comprehensive version to date, the 2024 ACAP provides a synthesis of key developments in amphibian conservation, identifying knowledge gaps, challenges, needs, and opportunities for conservation action, particularly in the context of global environmental change. It was designed to guide conservation actions adopted by international and national conservation organizations, governments, institutions focused on biodiversity research and education, funding bodies, and other relevant stakeholders. This global action plan could provide a vehicle to integrate amphibian conservation into regional, national, or local strategies and to delineate conservation actions, research efforts, inform funding initiatives, policies, and increase public awareness. Implementing the ACAP is an ambitious task given the many challenges involved, but the effort is both timely and critically important to consolidate the work of organizations and governments in amphibian conservation.

1.2 Conservation Planning at Regional Levels

Given the limited resources available for amphibian conservation, multispecies approaches, particularly in biodiversity hotspots, will be increasingly necessary to bridge conservation gaps and more effectively tackle the global amphibian extinction crisis (see IUCN SSC Amphibian Specialist Group 2024). The Tropical Andes are recognized as the world's most species-rich biodiversity hotspot (Myers et al. 2000; Pennington et al. 2010), a designation that applies to amphibians in the region (Hutter et al. 2017, Tobar-Suárez et al. 2022). However, the Andes also harbor a high number of endemic and threatened amphibian species (Re:wild et al. 2023), many of which have highly restricted ranges and specific habitat needs, so many of them exist outside protected areas (Nori et al. 2015). One key outcome of the second Global Amphibian Assessment was the identification of Threatened Amphibian Landscapes (TAL), defined as areas with high concentration of threatened species facing multiple threats. Of the 50 TALs identified, 12 (24%) are located in Andean countries, with 8 (16%) specifically within the Andean mountain range (Re:wild et al. 2023).

In light of these findings, there is an urgent need for multispecies CPs for amphibians in Andean nations. Such an approach would be a more effective response to mitigate or reverse the escalating declines and extinctions facing these species. While several CPs have been developed for amphibians in Andean countries, there has been no targeted effort to assess the broader geographic and taxonomic scope of amphibian conservation planning at the regional level. Thus, we sought to evaluate multispecies amphibian CPs in the Andean region, focusing on their geographical scope, time frame, methodological approach, stakeholder involvement, funding, and postrelease follow-up actions. We complement our findings with personal experiences from CP initiatives and offer recommendations for future efforts, highlighting key knowledge areas to focus on, and suggesting the establishment of a feedback process aligned with the updated global ACAP.

2 Materials and Methods

Given that the dissemination and awareness of a conservation plan or strategy are fundamental for securing buy-in, adoption, and successful implementation, we focused the scope of the data-gathering component of this study on publicly available documentation related to amphibian CPs including goals, objectives, outcomes, knowledge gaps, priority actions, partnerships needed, and timelines. These documents ranged from workshop reports to fully published conservation plans and strategies. As such, we only included documents that were permanently available online. Although we are aware of draft planning documents developed over several years that have yet to be published, these were not included in the general analysis (but see below for further details).

To search for published documents, we used Google's search engine with various combinations of Spanish keywords related to amphibian conservation, action plans, strategies, amphibians, and the names of the seven Andean nations (Venezuela, Colombia, Ecuador, Peru, Bolivia, Chile, and Argentina). Our search covered international (two or more Andean countries) and national-level documents. For national-level documents, we also examined whether they were referenced in published official national biodiversity strategies and plans (NBSAPs). We only considered higher-level and multispecies CPs, that is, all plans and programs at the national or biome/ecosystem scales, excluding documents from departmental, state, municipal, or other political-administrative jurisdiction.

Regarding unpublished draft documents, we provide a separate discussion, as some of us were directly involved in their development. We discuss some of the challenges faced in developing CPs and offer recommendations for future amphibian conservation planning efforts.

Data on species richness and threat status are retrieved from The IUCN Red List of Threatened Species (IUCN 2024).

3 Results

We identified five published amphibian CPs, three with a national focus, one with a binational focus, and one with a multinational focus (see Fig. 1 for a sample of species covered in these documents). In addition, we found a nationally scoped book resulting from a project that was aimed at developing a national conservation plan for amphibians (Table 1).

Of the seven Andean countries, Venezuela, Bolivia, and Argentina have nationally scoped published amphibian CPs. In contrast, Colombia and Ecuador have draft planning documents that have been in development for several years, with plans for eventual publication. The Bolivian and Argentinean plans focus on threatened and native amphibians, respectively, while the Venezuelan CP covers all of the Amphibia class in the country. A book on the conservation of Chilean amphibians was produced as part of a project aimed at developing a national conservation plan, but despite a workshop to agree on conservation actions, the plan was never launched. To the best of our knowledge, Peru is the only Andean country without any form of nationally scoped amphibian CP.

There are two international CPs for the region, and both focus on specific genera. The Harlequin Toad Action Plan targets *Atelopus*, one of the most threatened genera globally. While the plan covers most Andean countries, it extends beyond the region to include all countries where the genus is found, encompassing 103 species (https://amphibiansoftheworld.amnh.org/Amphibia/Anura/Bufonidae/Atelopus). On the other hand, the Darwin Frogs Binational Conservation Strategy focuses on the genus *Rhinoderma*, represented by two species, and spans both Chile and Argentina.

The three nationally scoped published CPs are in Spanish. Both the Darwin Frogs and *Atelopus* CPs are also available in English, with the *Atelopus* CP additionally published in Portuguese.

The implementation time frames vary considerably among CPs. Of the five published CPs, one does not have an explicit time frame (Venezuela, but see under Sect. 4.1.5), two have a 5-year duration (Argentina and Bolivia), one has a 10-year duration (*Rhinoderma* strategy), and the *Atelopus* CP spans a 20-year duration. The Argentinean and Bolivian CPs have now lapsed; the *Rhinoderma* strategy is over halfway through, with a mid-term workshop held in January 2024, and the *Atelopus* plan is in the early phase of development. All of these documents are in their first version (i.e., no revised versions have been released, although an update of the Darwin's CP is being developed).

All CPs reviewed included workshops as a key methodological approach for stakeholder participation and engagement. However, the specific

Fig. 1 A sample of species covered in published Andean amphibian CPs. (**a**) *Gastrotheca chrysosticta.* (Photo credit: Mauricio Akmentins), (**b**) *Rhinella sclerocephala.* (Photo credit: Celsa Señaris), (**c**) *Agalychnis medinae.* (Photo credit: Celsa Señaris), (**d**) *Telmatobius culeus.* (Photo credit: Arturo Muñoz), (**e**) *Rhinoderma darwinii.* (Photo credit: Claudio Azat), and (**f**) *Atelopus cruciger.* (Photo credit: Jaime Culebras)

decision support framework employed (e.g., Conservation Planning Specialist Group systematic conservation planning, Open Standards for the Practice of Conservation) is not explicitly mentioned in the national documents, although it is noted in the international ones (both the Darwin's frog and *Atelopus* CPs followed the IUCN strategic conservation planning guidelines methodology). Open consultations were also used for the Venezuelan CP and the *Rhinoderma* strategy, with the latter including a symposium prior to the workshop.

Most CPs in Andean countries have been led by nongovernmental organizations (NGOs). Four of the five published CPs are led by members of the academic and NGO communities, and only one is led by government. The draft CP of Colombia is led by the national government and will be implemented by Colombian Association of Herpetology (ACh) and the IUCN SSC Amphibian Specialist Group—Colombia (ASG Colombia), while the draft CP of Ecuador was led by the national government and executed by NGOs and universities. Direct involvement of

Table 1 Published and draft higher-level CPs for Andean countries

	CPs					Other documents	Not yet published	
	Argentina	Bolivia	Chile and Argentina	Venezuela	Regional	Chile	Colombia	Ecuador
Action plan/ strategy name	Plan de Acción para la conservación de los Anfibios de la República Argentina (Vaira et al. 2018)	Plan de Acción para la conservación de los Anfibios Amenazados de Bolivia (Ministerio del Medio Ambiente y Agua 2012)	Estrategia Binacional de Conservación de las Ranitas de Darwin (UICN Grupo de Especialistas de Anfibios Chile 2018)	Anfibios de Venezuela: Estado del conocimiento y recomendaciones para su conservación (Molina et al. 2009)	Plan de Acción para la Conservación de la Rana Arlequín (*Atelopus*) (2021–2041) (Valencia and Fonte 2021)	Anfibios de Chile, un desafío para la conservación (Soto-Azat and Valenzuela-Sánchez 2012)	Programa Nacional para la Conservación de los Anfibios en Colombia. Plan de Acción 2025–2035 (Ministerio de Ambiente y Desarrollo Sostenible, Asociación Colombiana de Herpetologia, Wildlife Conservation Society, Grupo de Especialistas en Anfibios Comisión de Supervivencia de Especies—Unión Internacional para la Conservación de la Naturaleza. Bogotá, DC, Colombia 2025)	Plan de Acción para la conservación de Anfibios de Ecuador (Proyecto Conservación de la Biodiversidad de Anfibios Ecuatorianos y Uso Sostenible de sus Recursos Genéticos, Ministerio de Ambiente y Agua y Programa de las Naciones Unidas para el Desarrollo Ecuador. Quito, Ecuador)
Language(s)	Spanish	Spanish	Spanish and English	Spanish	Spanish, Portuguese and English	Spanish	Spanish	Spanish
URL	https://cuadernosdeherpetologia.com/index.php/CdH/issue/view/20/15	https://dgbap.mmaya.gob.bo/wp-content/uploads/2017/10/	https://www.estrategiarhinoderma.org/	https://www.iucn-amphibians.org/resources/publications/	https://www.atelopus.org/the-initiative	https://mma.gob.cl/wp-content/uploads/2018/	NA	NA

(continued)

Table 1 (continued)

	CPs					Other documents	Not yet published	
	Argentina	Bolivia	Chile and Argentina	Venezuela	Regional	Chile	Colombia	Ecuador
	https://www.amphibianark.org/wp-content/uploads/2021/03/Plan-de-Accion-Argentina-2018.pdf	PLAN-ACCION-ANFIBIOS-AMENAZADOS.pdf		action-plans/action-plans-by-regions/		08/Conservacion-de-Anfibios.pdf		
Geographic scope	National	National	Binational (Chile and Argentina)	National	International	National	National	National
Taxonomic scope	177 native species	54 threatened species	Genus *Rhinoderma* (2 species)	All amphibia	Genus *Atelopus*, 99 species	63 native species	All amphibia	All amphibia
Period covered	2018–2022	2012–2017	2018–2028	2009	2021–2041	2013	2025–2035	2020–2030
Plan duration (years)	5	5	10	NA	20	Not implemented	10	10
Plan versions	1	1	1 (second version is being prepared after mid-term workshop)	1	1	0	1	2 drafts
Methodology used	Workshops, remote follow-up	Workshop	Symposium, workshops, open consultation	Workshops, open consultation	Workshop	Workshop	Workshop	Workshop
Plan development leadership	Academia and NGO	National government	Academia and NGOs	Academia and NGOs	NGOs and academia	Academia and government	The national government convened and led a workshop in 2014, but a committee comprised of five representatives from academia and NGOs was tasked to write the National Program for the Conservation of Amphibians in	Led by the national government through a project (PARG) of the Ministry of Environment and Water and PNUD. An NGO executed the workshops' development and the plan's writing

							Colombia and its Action Plan	
Stakeholders involved in plan development	Representation ($N = 46$) from academia/scientific society (AHA); 35 workshop participants wrote the plan	Government and amphibian specialists	Representation from government, academia, and NGOs in the form of 300 attendees at symposium, 30 participants at workshop, and 300 attendees at the launch; amphibian specialists wrote the plan	Representation from two governmental agencies, five universities, and three NGOs; amphibian specialists wrote the plan	Representation ($N = 38$) from academia, NGOs, government, civil society, and donors covering 11 countries, seven of which are *Atelopus* range countries	Diverse stakeholders, including government, academia, NGOs, and private sector	Representation ($N = 32$) from NGOs, universities, museum, government agencies, National Police, INTERPOL, and the Regional Autonomous Corporations (CARs) at the workshop. The plan is authored by 21 herpetologists	Representation ($N = 32$) from NGOs (4), universities (13), zoos (1), Ministry of Environment and Water, including PARG (10), other governmental agencies (3), independent consultant (1)
Funding sources	None	NA	Academia, NGOs, and private (forestry) sector	NGOs, academia, and the private (mining) sector	NGOs, academia	Ministry of Environment of Chile	National government	Global Environmental Facility (GEF)
Follow-up	No formal follow-up	No evidence of follow-up	Academic paper published, mid-term workshop (January 2024) to assess progress	No formal follow-up	Regular newsletters to stakeholders; a second workshop was undertaken (November 2023) to assess progress (L. M. Fonte, personal communication)	No follow-up	A government-convened meeting with the main leads of the National Program and Action Plan (ASG, WCS and AcH) took place in February 2024 to define the implementing committee and the advisory group for specific topics/regions. In	No follow-up

(continued)

Table 1 (continued)

	CPs					Other documents	Not yet published	
	Argentina	Bolivia	Chile and Argentina	Venezuela	Regional	Chile	Colombia	Ecuador
							March 2024, ASG Colombia held a virtual symposium with 60 participants to discuss the context of the global ACAP and its relationship with the National Amphibian Conservation Program; preliminary working committees were established for the five regions of the country (Andes, Amazon, Caribbean, Chocó and Orinoco). In April 2024, AcH held a remote, nonconsultative meeting with its members to present the National Amphibian Conservation Program and that same day the adjusted Program	

							document was sent to the Ministry. In October 2024, within the framework of the COP16, the "Biodiversity Action Plan of Colombia to 2030" was published and within goal 3 is action #26: "Implementation of the National Strategy for the conservation of threatened amphibians" in order to advance towards Goal 4 of the Kunming-Montreal Global Biodiversity Framework	
Implementation mechanisms	None on record	Leads and collaborators were specifically identified for each action highlighted in the plan and a time frame allocated to each action; however, we have been unable to find any further details on	Creation of the Darwin's Frogs Alliance in 2018 and its ratification in 2024. This alliance is comprised of several entities of Chile, Argentina, the United States, the United Kingdom, and Germany. There are a total of 30 entities	None on record	Creation of the *Atelopus* Survival Alliance in 2019 and establishment of working groups with respective leads. This alliance is comprised of universities,	No implementation	Specific actions have been assigned to broad stakeholder categories (e.g., universities and NGOs); geographically scoped working committees have been set up to lead and monitor	No implementation

(continued)

Table 1 (continued)

	CPs					Other documents	Not yet published	
	Argentina	Bolivia	Chile and Argentina	Venezuela	Regional	Chile	Colombia	Ecuador
		plan implementation	spanning universities, NGOs, zoos, museums, private sector, and governmental entities of Chile and Argentina that support the strategy		NGOs, government agencies, civil society and donor organizations		plan implementation	
Notes	The implementation of the Plan did not generate alliances, commissions, networks or groups that were in charge of the development and fulfillment of the goals. The entire implementation was at the initiative of the groups of specialists who wrote the Plan and those who joined the proposed actions (e.g., drafting of the Manual of Techniques and Protocols for the Survey and Study of Amphibians in Argentina)	Proposed actions follow strategic lines delineated in the National Strategy of Threatened Vertebrates of Bolivia. A number of potential stakeholders are listed in the plan, but wording suggests that these stakeholders haven't been involved in the development of the plan				More than a plan, this is a book on the conservation of Chilean amphibians. It is the product of a project that intended to develop a national conservation plan for amphibians. A workshop was developed to agree on conservation actions, but the plan was never launched, and therefore not implemented		

NGO nongovernmental organization, *AHA* Asociación Herpetológica Argentina, *ASG* IUCN SSC Amphibian Specialist Group, *MADS* Ministerio de Ambiente y Desarrollo Sostenible, *WCS* Wildlife Conservation Society, *AcH* Asociación Colombiana de Herpetología

Table 2 Amphibian species richness and national amphibian CPs for each Andean country, and their relationship with National Biodiversity Strategies and Action Plans (NBSAPs)

Country	# of species[a]	# of threatened species	% of threatened species	National amphibian plan/ strategy?	Plan/ strategy's date	NBSAP periods	National plan referenced in NBSAP?
Venezuela	385	132	34.3	Yes	2009	2010–2020	No
Colombia	907	313	35	Draft	2025–2035	2016–2030	NA[b]
Ecuador	631	299	47.4	Draft	NA	2015–2030	NA[b]
Peru	647	152	23.5	No	NA	2014–2018; 2021	NA[b]
Bolivia	313	58	18.5	Yes	2012	2019–2030	No, but action planning efforts for *Telmatobius culeus* are mentioned
Chile	63	35	55.6	No	NA	2017–2030	No
Argentina	178	34	19.1	Yes	2018	2016–2020	NA, but a report on the species *Alsodes pehuenche* was included in the sixth national report to CBD COP

[a]Note that adding all species would not give an accurate total given that there are shared species among the different countries

[b]In these instances, amphibian action plans/strategies cannot be cited as they have yet to be developed or published, or were published after the respective NBSAP

national governments allows for greater institutional support in their adoption as public policy, but it also often results in significant delays in the publication process, without any guarantee of their implementation. Specifically, the Ministry of Environment included in the new "Colombian Biodiversity Action Plan 2030" (Ministerio de Ambiente y Desarrollo Sostenible 2024) action #26 "Implementation of the National Strategy for the conservation of threatened amphibians" to address the issue of threatened species mentioned in Goal 4 of the Kunming-Montreal Global Biodiversity Framework. While some CPs mention general sources of funding for events and document production, specific financial details are largely unavailable.

In terms of implementation monitoring, none of the national CPs had formal follow-up after their publication, with the exception of the two international and genus-focused CPs. As for implementation mechanisms, the two international CPs established alliances during their respective workshops, which now serve as umbrella and framework for the proposed actions.

Table 2 presents data on amphibian species richness, number and percentage of threatened species, the existence of national amphibian conservation planning document, and whether these plans are incorporated in national biodiversity plans or strategies. Collectively, the Andean countries are home to a large number of amphibian species, including many that are unique and threatened (percentages of threatened species per country range from 18.5% for Bolivia to 55.6% for Chile). However, despite both the high amphibian species richness and the significant proportion of threatened species, only three countries have publicly available CPs at the national level. In Colombia, the CP document has been officially disseminated by the Colombian Association of Herpetology (https://www.youtube.com/live/CLyI6mBtPl4?si=mEFmp2p1B-Al5R9V), and the Alexander von Humboldt Institute (https://www.youtube.com/

live/yjAHjbcIVzk?si=ytGPO-5MaBVC8Xdl&t=2030). Additionally, the new Colombian Biodiversity Action Plan 2030 (Ministerio de Ambiente y Desarrollo Sostenible et al. 2024) includes the "Implementation of the National Strategy for the conservation of threatened amphibians" in its action #27, but the document is not yet publicly available. Similarly, in Ecuador, the CP was publicly presented in August 2020 at an event organized by the Ministry of Environment and Water of Ecuador (https://www.undp.org/es/ecuador/noticias/ecuador-presenta-la-lista-roja-de-anfibios-y-su-plan-de-accion-para-la-conservacion), yet remains unpublished.

As parties to the Convention on Biological Diversity (CBD), all Andean countries are required to develop National Biodiversity Strategies and Action Plans or NBSAPs to guide the conservation and sustainable use of biodiversity (CBD 2024). While all countries in the region have a NBSAP, the integration of amphibian conservation plans (CPs) into these national strategies has been inconsistent. Of the three nationally scoped published class-wide amphibian CPs, those of Venezuela and Bolivia predate their respective NBSAP, whereas Argentina's CP was published within the time frame of its NBSAP, leaving no opportunity for its inclusion. Neither the Venezuelan nor the Argentinean CPs were referenced in their country's NBSAP, although the Argentinean Plan Nacional Extinción Cero, which was included in the sixth national report to CBD COP, referenced the species *Alsodes pehuenche*. In Bolivia, the NBSAP referenced conservation planning documents for species of Lake Titicaca, which included the Titicaca water frog (*Telmatobius culeus*), without delving into details of how this plan aligns with the broader NBSAP framework.

There are currently two nationally scoped class-wide draft CPs, one for Colombia and one for Ecuador. In the case of the Colombian document, which is foundational to the establishment of a national program for amphibian conservation, a nationally scoped workshop was convened by the Ministry of Environment and Sustainable Development (MADS) in July 2014, which was attended by 32 participants representing NGOs, academia, national and regional governments, and INTERPOL. In the context of this workshop, a committee consisting of five herpetologists representing stakeholders from academia and NGOs committed to writing the National Program for the Conservation of Amphibians in Colombia and its Action Plan. A first draft was shared with MADS in 2017, feedback and updates were incorporated, and the revised version was sent back to MADS in mid-2019. In spite of several prompts from participants, no further action was taken by MADS, until the Chair of the IUCN Species Survival Commission wrote a intervention letter to the Minister in 2023 to request an update on the program and plan. Since then, MADS has resumed work to publish and roll out the program and associated plan, convening key stakeholders to a meeting in early 2024 and making adjustments to the program. The program and plan were expected to be published in 2025. However, given that the new Biodiversity Action Plan of Colombia to 2030 (Ministerio de Ambiente y Desarrollo Sostenible 2024) has incorporated as a specific action of the National Goal #3 the "Implementation of the National Strategy for the conservation of threatened amphibians," it is expected that, by 2026, the country will have a published and adopted document to start its implementation via the IUCN SSC Amphibian Specialist Group—Colombia (ASG Colombia) and the Colombian Association of Herpetology. For the Ecuadorian CP, in 2019 and 2020, the Ministry of Environment and Water (MAAE) and its PARG project (Conservation of Ecuadorian Amphibians and Sustainable Use of their Genetic Resources) convened nationally scoped workshops, which were attended by 32 participants representing NGOs, academia, zoos, and MAAE. The draft action plan for amphibian conservation in Ecuador was written by consultants and NGO representatives (Centro Jambatu de Investigación y Conservación de Anfibios—Fundación Jambatu) and circulated by email for comments and endorsement (Coloma 2020). The PARG project organized a symposium on amphibian conservation in May–April 2019, where local, national, and international experiences were presented, including the Argentinean CP. The proceedings of this event mention the Ecuadorian CP as a guide for developing *ex situ* conservation actions (Coloma 2020). A local action plan (for conserving amphibians of the Abra de Zamora, a high Andean region in southern Ecuador) cites the Ecuadorian CP as its legal framework governing its scope of action (Ordóñez-Delgado et al. 2020). However, the Ecuadorian CP was never published. Despite several requests from participants and organizations over the years, further action has yet to be taken regarding the publication or dissemination of the Ecuadorian CP by the MAAE.

4 Discussion

Amphibian conservation planning in Andean countries has been relatively limited and largely driven by academic institutions and NGOs. To date, only Argentina, Bolivia and Venezuela have published nationally scoped CPs. Except for Bolivia (but see Sect. 4.1.5), these initiatives have been led by the academic and NGO communities. Even when the government serve as the public face of such efforts, as in Bolivia, experts from academia and NGOs often play a major role behind the scenes. Consequently, these sectors are likely to demonstrate a stronger long-term commitment to implementing conservation plans than government agencies.

There is a considerable lack of information regarding the implementation status of conservation plans developed by the Andean countries. Regular progress reviews, essential for evaluating the advances, identifying challenges and adjusting strategies as needed (see Andreone et al. 2012), are currently unavailable. To our knowledge, the only formal mid-term progress review is for the binational *Rhinoderma* strategy, which is currently underway. Additionally, a progress review of the *Atelopus* Action Plan was conducted during a workshop in November 2023, where thematic group leaders identified those actions that had already been implemented and identified priorities for the forthcoming years (L. F. M. Fonte, personal communication).

It would be highly valuable to conduct regular reviews as part of a CP implementation monitoring process. These evaluations would help to identify which actions are being effectively implemented and which are lagging, allowing for timely corrections (see, e.g., Andreone et al. 2012, who conducted a review of progress for "A Conservation Strategy for the Amphibians of Madagascar, ACSAM").

Conservation plans in Andean countries generally lack coordination with global initiatives. While the Argentinean and Venezuelan plans reference the Amphibian Conservation Action Plan (ACAP), they lack details on how their proposed actions align with the ACAP priorities. Strengthening the connection between local and global strategies can enhance the effectiveness of amphibian conservation efforts. The IUCN SSC Amphibian Specialist Group could play a key role in facilitating this alignment, demonstrated in the development of the Binational Conservation Strategy for Darwin's frogs, where the groups´ cochair and secretariat team provided key guidance.

With regards to official support, although the Chilean and Argentinean governments are a part of the binational Darwin's Frog Conservation Strategy, neither have officially endorsed it. However, there has been recent progress for this CP to be formally endorsed by the Chilean state through the tool of Recovery, Conservation and Management Plans (RECOGE) of the Ministry of Environment, which is foreseen to be adopted in the near future. Similarly, as *Rhinoderma darwinii* is only present in national parks in Argentina, the aim is for the plan to be recognized by the Administración de Parques Nacionales.

Although outside the scope of the current chapter, it is important to acknowledge the existence of various unofficial and localized amphibian initiatives. These efforts often focus on individual species, specific species assemblages or particular protected areas. However, they tend to be fragmented and may not always aligned with broader, more comprehensive plans. Like their national and international counterparts, these local efforts are often heavily reliant on the goodwill of NGOs and academic institutions.

4.1 Collective Experience

All chapter authors have been involved in amphibian conservation planning initiatives, collectively bringing a wealth of first-hand experience. Throughout our work, we have faced a number of challenges, many of which are common across the field. Wren et al. (2024) highlight several of these shared obstacles. In this chapter, we provide additional context and insights into the Andean regional experience. We also offer recommendations for addressing some of these challenges in the following sections.

4.1.1 Buy-In

Securing stakeholder buy-in is crucial for the successful implementation of a conservation plan and for achieving its conservation goals. Nearly all of the amphibian CPs examined in this study, and those we have been directly involved with, have aimed to engage relevant stakeholders. However, most of them have come from the academic and NGO communities, with some government participation, and limited involvement from the private sector. To gain broader support, it is essential to identify and cultivate amphibian conservation champions within government and other key stakeholders. By doing so, the groups they represent may become more engaged throughout the planning process and beyond, ensuring greater commitment to the long-term success of the initiative.

In addition, establishing coordination channels to integrate the goals outlined in CPs into the NBSAPs and promoting their adoption within strategic priorities of the respective national scientific councils can strengthen the involvement of governments, NGOs, and relevant stakeholders. Action plans can provide the framework for consolidating the efforts of governments and organizations to promote the inclusion of amphibian conservation issues into national biodiversity conservation strategies. However, the successful implementation of these action plans relies heavily on the level of coordination and commitment from a diverse range of stakeholders. Regional and national branches of the IUCN SSC Amphibian Specialist Group and national herpetological societies can play a pivotal role in connecting local conservation initiatives with academia, NGOs, and government institutions. For example, in Colombia, ASG Colombia brought together 89 herpetologists committed to implementing the national program for the conservation of amphibians, with 72% of them already engaged in conservation projects in the Andean region. Some of those projects were showcased during the first and second ASG Colombia symposiums held on March 1, 2024 and July 17 and 18, 2025, respectively. Currently, the Colombian Ministry of Environment has endorsed ASG Colombia and the Colombian Association of Herpetology as the primary implementing entities of the CP, with the support from governmental biodiversity research institutes (e.g., Alexander von Humboldt Institute, Sinchi Amazonian Institute for Scientific Research).

Although the Ecuadorian CP has not fully materialized, local action plans for amphibian conservation have been developed and implemented in southern Ecuador (Siavichay et al. 2016; Ordoñez-Delgado et al. 2020). These initiatives, led by local herpetologists and conservationists with support from NGOs, universities, and local governments, have often operated with more limited financial resources compared to the national CP. Despite these constraints, these local efforts have achieved great success, demonstrating the potential of localized, community-driven conservation actions. Many of the individuals involved in these local initiatives also contributed to the national CP, highlighting the importance of local expertise and collaboration in driving successful conservation outcomes.

An important challenge, particularly with regarding to government and private sector involvement, is that their representatives attending amphibian CP events often lack decision-making power. As a result, they are unable to commit to leading on specific actions, crucial for the implementation of these plans. An additional challenge with the private sector has been a limited willingness to engage. For example, for the Darwin's frog strategy, only a handful of invited forest companies responded. While it is very difficult for external parties to influence the selection of government and private sector representatives at CP events, it is critical that those who attend are either empowered to commit to specific actions and deliverables on behalf of their organization or possess enough influence within their hierarchy to secure commitments from decision-makers.

4.1.2 Publication

The publication of CPs is fundamental to making informed, streamlined, and cost-efficient decisions for the conservation of biodiversity; it further makes the information within available to the wider world, allowing to raise awareness, fundraise for the cause, and seek broader buy-in. Having a CP published in English in addition to Spanish may further help raise awareness for the CP's subject, as has been the case for the two internationally scoped CPs focusing on *Rhinoderma* and *Atelopus*.

Where the academic and NGO communities have ownership of the CP, funds permitting, it will often make it to the public sphere; however, where the process depends on the government, publication timelines will be unpredictable. While the majority of the documents that we are aware of have been published, two of the CPs that have been led by the government (i.e., Colombia and Ecuador) have had very long processes and have not yet been published. In the case of Colombia, publication of the respective document has been in the making for a full decade, whereas in Ecuador, the existence of a national action plan was announced in August 2020 (PNUD 2024), and the plan is still not available in the public domain. In the case of Chile, although the Ministry of Environment issued a call for a project seeking the development of a National Amphibian Conservation Plan in 2012, and the project was awarded and executed, inclusive of two workshops, the actual plan was never published, and the main result was the publication of the book "Anfibios de Chile, un desafío para la conservación" (Lobos et al. 2013). Extended delays in publication have an impact, as the information that was current at the time of the CP event may become outdated if too much time has passed between the event(s) and publication. It is therefore recommended that CP leads, including government, consider a schedule for publication to ensure that the document can be published in a timely manner.

4.1.3 Bad Practices

We have mentioned above how important publication is; however, unfortunately it is common for CPs to be published, only to end up as glossy documents on a shelf and not be adopted and implemented (Wren et al. 2024). Furthermore, CPs can be published and formally adopted by governments, but not necessarily implemented thereafter. For example, many of the plans and programs for the management of wild species of fauna and flora in Colombia are adopted by the Ministry of Environment but are not implemented (see related IUCN intervention letter in Supplementary Material S16.1); and since they do not prioritize goals or actions at national or regional levels, it is difficult to follow up and monitor specific actions. An issue specific to the Colombian Amphibian Conservation Program is that the government wants to promote commercial captive breeding and amphibian tourism despite poor practices by the only legal commercial captive breeding farm in Colombia (Castaño-Camacho and Zimmermann 2022), and the danger of increased illegal trafficking and the potential introduction of invasive species and amphibian diseases to sites with greater influx of tourism. A similar outcome could potentially occur with the Ecuadorian CP, where significant incentives are directed toward promoting commercial captive breeding. However, it is important to note that, following the legalization of the amphibian trade in the country, there has been an observed increase in illegal trafficking (Archives USFQ–TUERI Wildlife Hospital & Laboratory of Animal Health, unpublished data).

In the case of Venezuela, the plan was not officially adopted by the state (see Sect. 4.1.5

for details). In 2013, the government introduced the "Plan de la Patria," outlining the objective: "to cultivate a holistic conservation policy, promoting sustainable practices, safeguarding, and fostering scientific awareness of both biological diversity and the nation's water resources." However, this objective has become entangled in a broader political narrative emphasizing socialism as the sole antidote to capitalism. Consequently, political alignment has often dictated access to public funding, leading to exclusion of most of the scientific community and NGOs involved in this initiative.

In Bolivia, even though the national government convened experts and led the amphibian CP initiative, because the final document was not approved by official decree it was not formally adopted by the government itself (A. Muñoz, personal communication). Once the document was published, there was no further follow-up to ensure that the various actions highlighted within were implemented.

4.1.4 Linkages

Broadly speaking, there are internal and external linkages in relation to a CP: internal linkages are those that are needed to implement the various actions highlighted in the CP document, and which require coordination which would have ideally been discussed in the framework of the CP initiative; external linkages are those that emerge between the CP and other relevant initiatives that may not have been present or contemplated in the development of the CP. Identifying and establishing these external linkages would allow for better coordinated action where common issues and challenges are concerned, in this way expanding amphibian CPs to be included in other CP initiatives. This does, however, require additional human resources that may not be available beyond the remit of the CP.

For example, with over 20 national CPs targeting species at risk, Colombia exemplifies a committed approach to preserving biodiversity through detailed action plans that specify goals, objectives, and timelines, with plans facilitating coordinated and strategic conservation actions. However, it is unclear to what extent there may be cross-pollination between the different plans, something that is relevant considering that many of the threat drivers (e.g., pollution, deforestation, illegal trafficking, among others) are shared among biotic groups.

Where publication of amphibian CPs preceded NBSAPs, these have not been taken into consideration in the respective NBSAP. In the case of Argentina, although the national Amphibian Action Plan had not been published before the national 2016–2020 NBSAP, a previous initiative developed by a number of authors of the national plan (Categorization of the conservation status of amphibians in Argentina) is mentioned in the report (Secretaría de Gobierno de Ambiente y Desarrollo Sustentable de Presidencia de la Nación—República Argentina 2020).

As a community, we may need to rethink how we promote greater coordination for the implementation of CPs. In spite of our best efforts, we may still need to focus on developing close working relationships and coordination with government and NGOs, ensuring fluid communication, sharing of ideas, and joint fundraising for CP implementation. Potential stakeholders who can facilitate CP implementation might be jointly identified, and developing consensus-based prioritization criteria can aid in achieving progress and support. Some actions or goals of the CPs might be prioritized by their relevance and the urgency of implementation or to respond to emerging problems and new demands.

4.1.5 Human Resources and Changes in Leadership

Changes in leadership of a CP initiative, state agencies (national or local), or other key stakeholders can have a significant impact not only on the CP but also on what follows. When the idea for a CP emerged for the amphibians of Venezuela, it was a governmental entity that convened specialists for its development. At the time, there was a respected herpetologist working in the Ministry, and joining forces with the government, Conservation International Venezuela provided logistical and financial support for the development of workshops. When all events had been carried out and the final document was ready to be launched as an official national publication, the lead at the Ministry passed away and was replaced by another public servant who shunned alliances with North American institutions. This development, together with drastic and sudden changes in public policy, hampered the CP initiative, and in spite of multiple meetings seeking to lead it to fruition, the document was never adopted as an official government-endorsed publication and was thus published by the academic and NGO communities. Without government support, it was very difficult to move forward with implementation of the CP.

In Argentina, two executive government areas (environment and science) strongly linked to the implementation of the national strategy and programs for the conservation and sustainable use of biodiversity, went through drastic cycles of change in their structures in the last 10 years, and were reverted from the rank of National Ministries to Secretariats, thus reducing their proper functioning with the heads of office overseeing several tasks besides environment and science. These changes pose challenges to ongoing biodiversity and research programs.

In Ecuador, after the conclusion of the PARG project (Conservation of Ecuadorian Amphibians and Sustainable Use of their Genetic Resources), the parent initiative for the Ecuadorian CP's development, the government's political support to push for its publication and implementation has not been forthcoming.

In order to mitigate the possibility of similar future scenarios, it would be advisable to encourage the development of stable public policies, at least in the short- to mid-term periods. The establishment of regional or supranational strategies or alliances can strengthen the implementation of CP initiatives that are not being carried out individually at national or local levels. However, incorporating additional decision-making parties from different countries or governments can add complexity to the development of supranational CP initiatives.

4.1.6 Funding

A common thread that runs through amphibian CPs is that as hard as it is to obtain, funding may be available to undertake workshops and other steps necessary for the development of a CP, but obtaining funding to implement their actions is much more difficult and far more expensive, leaving stakeholders to their own devices for CP implementation. In our experience, unless there is a champion coordinating implementation and leading the monitoring process, it is very unlikely that a CP will be fully implemented. Thus, it is key that funding be contemplated for coordination and monitoring postproduction of the CP in funding proposals.

5 Conclusion

In the face of aggravated environmental issues and chronic lack of funding for biodiversity conservation, geographic and taxonomic higher-level amphibian CPs will be increasingly needed to better address and mitigate the amphibian crisis. The number of existing higher-level amphibian CPs is currently small, and the need for targeted goals and coordinated action is enormous. Although the number of these CPs is limited, great effort has gone into producing them and most have been led by the academic and NGO communities, raising the question: whose responsibility is it to plan for and implement conservation for amphibians in the Andean countries and elsewhere? We believe that this is a responsibility shared among all stakeholders, and governments in general need to take a much more proactive role to ensure that their amphibian biodiversity is conserved.

Acknowledgments We thank Arturo Muñoz and Luis Marin da Fonte for providing background information on the Bolivian CP initiative and the *Atelopus* CP initiative, respectively.

Competing Interests The authors have no conflicts of interest to declare that are relevant to the content of this chapter.

References

Andreone F, Carpenter AI, Copsey J et al (2012) Saving the diverse Malagasy amphibian fauna: where are we four years after implementation of the Sahonagasy Action Plan? Alytes 29(1–4):44–58

Castaño-Camacho A, Zimmermann ML (2022) ¿Están en peligro los tesoros de Colombia? Natural Press.

Available via https://www.naturalpress.ca/estan-en-peligro-los-tesoros-de-colombia/

Coloma A (2020) Plan de acción de conservación de anfibios del Ecuador. MAE, PNUD, GEF, PARG, Quito

Convention on Biological Diversity (CBD) (2024) What is an NBSAP? Available via https://www.cbd.int/nbsap/introduction.shtml. Accessed 3 July 2024

Gascon C, Collins JP, Moore RD, Church DR, McKay JE, Mendelson JRI (2007) Amphibian Conservation Action Plan: proceedings IUCN/SSC amphibian conservation summit (2005). IUCN/SSC Amphibian Specialist Group, Gland and Cambridge

Hutter CR, Lambert SM, Wiens JJ (2017) Rapid diversification and time explain amphibian richness at different scales in the Tropical Andes, Earth's most biodiverse hotspot. Am Nat 190(6):828–843. https://doi.org/10.1086/694319

IUCN (2024) The IUCN Red List of threatened species. Version 2023-1. Available via https://www.iucnredlist.org. Accessed 3 July 2024

IUCN SSC Amphibian Specialist Group (2024) Amphibian Conservation Action Plan (ACAP): a status review and roadmap for global amphibian conservation. In: Wren S, Borzée A, Marcec-Greaves R, Angulo A (eds) IUCN SSC Occasional Paper No 57. IUCN, Gland

Knight AT, Cowling RM, Campbell BM (2006) An operational model for implementing conservation action. Conserv Biol 20(2):408–419. https://doi.org/10.1111/j.1523-1739.2006.00305.x

Lobos G, Vidal M, Correa C, Labra A, Díaz-Páez H, Charrier A, Rabanal F, Díaz S, Tala C (2013) Anfibios de Chile, un desafío para la conservación. Ministerio del Medio Ambiente, Fundación Facultad de Ciencias Veterinarias y Pecuarias de la Universidad de Chile & Red Chilena de Herpetología, Santiago. https://mma.gob.cl/wp-content/uploads/2018/08/Conservacion-de-Anfibios.pdf

Luedtke JA, Chanson J, Neam K et al (2023) Ongoing declines for the world's amphibians in the face of emerging threats. Nature 622:308–314. https://doi.org/10.1038/s41586-023-06578-4

Ministerio de Ambiente y Desarrollo Sostenible (2024) Plan de Acción de Biodiversidad de Colombia al 2030. Ministerio de Ambiente y Desarrollo Sostenible, Bogotá. Available via https://www.cop16colombia.com/es/wp-content/uploads/2024/10/PLAN-DE-BIODIVERSIDAD-.pdf

Ministerio de Medio Ambiente y Agua (2012) Plan de Acción para la Conservación de los Anfibios Amenazados de Bolivia. PGD Impresiones, La Paz

Molina C, Señaris JC, Lampo M, Rial A (2009) Anfibios de Venezuela: Estado del conocimiento y recomendaciones para su conservación. Ediciones Grupo TEI. https://www.iucn-amphibians.org/resources/publications/action-plans/action-plans-by-regions/

Myers N, Mittermeier RA, Mittermeier CG, da Fonseca GAB, Kent J (2000) Biodiversity hotspots for conservation priorities. Nature 403:853–858. https://doi.org/10.1038/35002501

Nori J, Lemes P, Urbina-Cardona N, Baldo D, Lescano J, Loyola R (2015) Amphibian conservation, land-use changes and protected areas: a global overview. Biol Conserv 191:367–374. https://doi.org/10.1016/j.biocon.2015.07.028

Ordónez-Delgado L, Székely P, Székely D, Serrano F, Armijos-Ojeda D (2020) Plan de acción para la conservación de los anfibios del Abra de Zamora. Universidad Técnica Particular de Loja y Naturaleza y Cultura Internacional, Loja. https://doi.org/10.13140/RG.2.2.15950.08008

Pennington RT, Lavin M, Särkinen T, Lewis GP, Klitgaard BB, Hughes CE (2010) Contrasting plant diversification histories within the Andean biodiversity hotspot. PNAS 107(31):13783–13787. www.pnas.org/cgi/doi/10.1073/pnas.1001317107

PNUD (2024) Ecuador presenta la Lista Roja de Anfibios y su Plan de Acción para la Conservación. Available via https://www.undp.org/es/ecuador/noticias/ecuador-presenta-la-lista-roja-de-anfibios-y-su-plan-de-accion-para-la-conservacion. Accessed 3 July 2024

Re:wild, Synchronicity Earth, IUCN SSC Amphibian Specialist Group (2023) State of the world's amphibians: the second global amphibian assessment. Re:wild, Texas

Secretaría de Gobierno de Ambiente y Desarrollo Sustentable de Presidencia de la Nación – República Argentina (2020) Sexto Informe Nacional para la Conferencia de las Partes del Convenio Sobre La Diversidad Biológica (CBD). Available via https://www.argentina.gob.ar/interior/ambiente/biodiversidad/sexto-informe. Accessed 3 July 2024

Siavichay F, Maldonado G, Mejía Coronel D, Webster JF, Torres N, Costa K (2016) Plan para la conservación de los anfibios urbanos de Cuenca. Ilustre Municipalidad de Cuenca, BDE, CGA, Cuenca

Soto-Azat C, Valenzuela-Sánchez A (2012) Conservación de anfibios de Chile. Universidad Nacional Andrés Bello, Santiago, Chile. Available via https://mma.gob.cl/wp-content/uploads/2018/08/Conservacion-de-Anfibios.pdf. Accessed 3 July 2024

Tobar-Suárez C, Urbina-Cardona N, Villalobos F, Pineda E (2022) Amphibian species richness and endemism in tropical montane cloud forests across the Neotropics. Biodivers Conserv 31:295–313. https://doi.org/10.1007/s10531-021-02335-z

UICN Grupo de Especialistas de Anfibios Chile (2018) Estrategia Binacional de Conservación de las Ranitas de Darwin. Santiago, Chile. Available via https://www.estrategiarhinoderma.org/. Accessed 3 July 2024

Vaira M, Akmentins MS, Lavilla EO (eds) (2018) Plan de Acción para la Conservación de los Anfibios de la República Argentina. Cuad Herpetol 32(supl. 1): 5–56. https://doi.org/10.31017/CdH.2018.(2018-s1)

Valencia LM, Fonte LFM (2021) Plan de Acción para la Conservación de la Rana Arlequín (*Atelopus*)

(2021–2041). Iniciativa de Supervivencia *Atelopus*. Available via https://es.atelopus.org/the-initiative

Wren S, Baker A, Andreone F et al (2024) Chapter 9. Strategic planning: a basis for effective action. In: IUCN SSC Amphibian Specialist Group. Amphibian Conservation Action Plan (ACAP): a status review and roadmap for global amphibian conservation, pp 219–236

Wren S, Angulo A, Meredith H, Kielgast J, Dos Santos M, Bishop P (eds) (2015) Amphibian Conservation Action Plan. April 2015. IUCN SSC Amphibian Specialist Group. Available via https://www.iucn-amphibians.org/resources/acap/

Andean Snakes: Understanding Biodiversity Knowledge Shortfalls

Juan C. Diaz-Ricaurte, Juan D. Vásquez-Restrepo, Paola A. Carrasco, Filipe C. Serrano, Alejandro Navarro-Morales, Cristopher A. Antúnez-Fonseca, and Marcio Martins

1 Introduction

The Andes Mountain Range is the most extensive continental geological formation worldwide and the highest in America, extending along the western margin of South America, with a length of 8900 km and a maximum elevation of 6962 m (Graham 2009; Mendoza 2014). These geological features resulted from various events and formation processes throughout the mountain range between the Cretaceous and Cenozoic periods (120–40 Mya). This is why it is considered a relatively recent geological formation (Doan 2003; Cobbold et al. 2007; Graham 2009; Schellart 2017). A series of uplifts in the Andes has altered the landscape structure of South America, dividing it into three regions with distinct geological and climatic characteristics (Josse et al. 2009). In the southernmost part are the Southern Andes, considered the oldest, highest, and coldest areas (66 Mya), followed by the Central Andes with intermediate heights and little vegetation (50 Mya), and finally in the northernmost part, the Northern Andes (25 Mya), considered one of the most diverse and recent regions to form due to the branching of the mountain range in this tropical area (Van der Hammen 1974; Gregory-Wodzicki 2000; Lavenu 2006; Strecker et al. 2007; Graham 2009; Folguera et al. 2016; Restrepo et al. 2017; Heredia et al. 2018; Horton 2018).

All these characteristics reflect a wide diversity of factors such as elevation, climate, topography,

Supplementary Information: The online version contains supplementary material available at https://doi.org/10.1007/978-3-032-00074-3_17.

J. C. Diaz-Ricaurte (✉)
Departamento de Ecologia, Instituto de Biociências, Universidade de São Paulo, São Paulo, SP, Brazil

Grupo de investigación en Biodiversidad y Desarrollo Amazónico (BYDA), Semillero de Investigación en Ecofisiología y Biogeografía de Vertebrados, Centro de investigaciones Amazónicas Macagual – Cesar Augusto Estrada Gonzales, Universidad de la Amazonia, Florencia, Caquetá, Colombia
e-mail: juan.diaz@alumni.usp.br

J. D. Vásquez-Restrepo
Department of Ecology and Evolutionary Biology, Princeton University, Princeton, USA

P. A. Carrasco
Centro de Zoología Aplicada, Facultad de Ciencias Exactas, Físicas y Naturales, Universidad Nacional de Córdoba/Instituto de Diversidad y Ecología Animal (IDEA, CONICET-UNC), Córdoba, Argentina

F. C. Serrano · M. Martins
Departamento de Ecologia, Instituto de Biociências, Universidade de São Paulo, São Paulo, SP, Brazil

A. Navarro-Morales
Centro de Investigación de la Biodiversidad Andino Amazónica – INBIANAM, Universidad de la Amazonia, Florencia, Caquetá, Colombia

C. A. Antúnez-Fonseca
Ecology and Conservation, Universidade Federal de Mato Grosso do Sul, Campo Grande, Mato Grosso do Sul, Brazil

J. N. Urbina-Cardona et al. (eds.), *Andean Herpetofauna*, Sustainable Development Goals Series,
https://doi.org/10.1007/978-3-032-00074-3_17

and vegetation that, when combined, give rise to areas with high landscape heterogeneity and geographical isolation (Esquerré et al. 2018; Hazzi et al. 2018). These factors are considered the main processes of diversification in the Andes, contributing to it being known as a biodiversity hotspot (Myers et al. 2000; Rahbek et al. 2019). The Andes' recent geological formation, high diversity, and endemism, particularly in groups such as snakes, highlight its importance for biodiversity research (Duellman 1999; Graham 2009; Lagomarsino et al. 2016; Esquerré et al. 2018; Guedes et al. 2018; Rangel et al. 2018; Azevedo et al. 2020). The limited dispersion and ectothermic condition of snakes have made them especially susceptible to environmental pressures, leading to events of diversification, endemism, and adaptation in cis-Andean and trans-Andean areas (Graham 2009; Navarrete et al. 2009; Lynch 2012; Serrano et al. 2024).

While the Andes boast a rich diversity of snakes, significant knowledge gaps persist regarding these reptiles. Limited research scope and scale impede a comprehensive understanding of their distribution, ecology, and evolution (Rojas-Morales 2012, Angarita-Sierra and López-Hurtado 2020). These knowledge deficiencies, as defined by Hortal et al. (2015), encompass various biodiversity gaps, such as the Linnean, Wallacean, Prestonian, Darwinian, Raunkiaerian, Hutchinsonian, and Eltonian shortfalls.

The Linnean shortfall, highlighting the issue of undescribed or misidentified species (Hortal et al. 2015), is particularly evident in Andean snakes. Although 259 species are currently reported (see Sect. 2 for more details), this number likely underestimates true diversity due to taxonomic limitations (Pincheira-Donoso et al. 2013). For example, most species have been described using classical taxonomy, based mainly on external morphological characters (Padial et al. 2010; Noriega et al. 2015). Conversely, the application of molecular techniques, together with internal morphological analysis (e.g., CT-scan), has unveiled cryptic species and intricate phylogenetic relationships in many snake groups (Zaher et al. 2009; Ananjeva 2019; Angarita-Sierra et al. 2022; García-Cobos and Vásquez-Restrepo 2022).

The Wallacean shortfall highlights incomplete geographical distribution data across various scales (Hortal et al. 2015). This is particularly evident in the Andes, where knowledge gaps regarding the distribution of many species, especially rarer ones, are significant (Torres-Carvajal et al. 2019, 2020). This is largely due to the difficulty of accessing remote areas and the lack of systematic sampling (Maes et al. 2012; Villarreal-Rosas et al. 2022). Consequently, uncertainty in species distribution directly impacts conservation status assessments. Furthermore, the lack of distribution data leads to knowledge gaps regarding population states, a phenomenon known as the Prestonian shortfall (Hortal et al. 2015). Knowing the distribution of a species is fundamental to determine the size and stability of the population, essential information for the categorization of threat according to IUCN criteria. Unfortunately, about 30% of Andean snakes are in the Data Deficient (DD) category, reflecting the lack of information about their distribution and abundance (Maciel-Mata et al. 2015).

The Darwinian shortfall (Hortal et al. 2015) is notably apparent in Andean snake research (Azevedo et al. 2020). Despite increasing interest in the evolutionary and biogeographical aspects of these reptiles, the influence of the Andes' complex geological history, including orogeny and habitat fragmentation, remains poorly understood. Although studies have increased in recent decades, many remain localized and isolated (Albino 2011). This knowledge gap impedes the reconstruction of Andean snake biogeography, a crucial factor in understanding current distribution patterns (Wallacean shortfall). Furthermore, understanding their evolutionary history is essential for identifying conservation priorities and developing effective management strategies. Therefore, it is necessary to intensify efforts to understand the evolutionary processes that have shaped the diversity of snakes in the Andes.

Despite the critical importance of ecological aspects, substantial knowledge gaps persist regarding Andean snakes. This is particularly evident in the Raunkiaerian and Eltonian shortfalls, which highlight the limited understanding of these species' functional and ecological traits, their interactions within ecosystems, and their ecological roles (Hortal et al. 2015). For instance, the physiological adaptations of Andean snakes to the region's low temperatures, high rainfall, and climatic variability (Hutchinsonian shortfall) remain largely unknown. This information deficit impedes accurate vulnerability assessments to environmental changes and anthropogenic threats, hindering the establishment of effective conservation priorities (Benavides and Rocha 2012; Hurtado 2012; Martins 2024). It is crucial to conduct studies that address these gaps and provide detailed information on the physiological adaptations, ecological interactions, and functional roles of Andean snakes. For instance, studies on diet, thermoregulation, seasonal activity, and habitat use can provide valuable information to understand how these species interact with their environment and how they respond to environmental changes. Moreover, these studies can help identify the most vulnerable species and design more effective conservation strategies.

Based on the above, it is clear that there is a high level of ignorance regarding different aspects of this reptile group, considered one of the most threatened globally (Gibbons et al. 2000; Cox et al. 2022), despite being one of the most attractive groups for researchers. Lynch (2012) suggests that declines in South American snake diversity may be attributed to stigmatization-driven persecution, road mortality, illegal wildlife trade (Caceres-Martínez et al. 2017; Hierink et al. 2020), and anthropogenic habitat destruction, particularly deforestation due to agricultural, livestock, and urban expansion (Cordier et al. 2021). Additionally, emerging threats such as climate change, which can affect distribution patterns, activity, food availability, and disease proliferation, exacerbated by habitat loss, represent significant risks for snake populations (Reading et al. 2010; Lorch et al. 2016; Zipkin et al. 2020; Cox et al. 2022). For example, it is unknown how Andean snakes respond to extreme weather events (e.g., droughts or floods), or how they are affected by pollution from heavy metals and pesticides. Furthermore, conflicts between humans and snakes, driven by ignorance and fear, represent an additional threat to the conservation of these species.

This chapter aims to address taxonomic, phylogenetic, biogeographic, and conservation aspects of Andean snakes, thereby highlighting key knowledge gaps and proposing potential avenues for future research and conservation action. Furthermore, we discuss the relevance of integrating scientific and traditional local knowledge, along with promoting environmental education, for fostering peaceful human–snake coexistence.

2 Biogeography and Challenges Facing Andean Snakes: The Interplay of Wallacean and Linnean Shortfalls

Threats and their impacts are not randomly distributed around the world, but current threats, such as habitat loss due to deforestation, are concentrated in specific regions (Harfoot et al. 2021). Incorporating spatial knowledge of threats and species distribution is crucial for conservation efforts, reserve planning (Bini et al. 2006), and understanding macroecological patterns (Moura and Jetz 2021; Diniz-Filho et al. 2023). However, basic data on distribution remain scarce for most species. The Wallacean shortfall is especially evident in the Neotropical region (e.g., Nogueira et al. 2019; Lees et al. 2020; Serrano et al. 2023; Frateles et al. 2024), where there is also a rapid pace of conversion of natural habitats to agriculture and pastures (Harfoot et al. 2021). This has led to a sharp decrease in the abundance of vertebrates worldwide (Almond et al. 2022). However, on a continental scale, there are also biases regarding species distribution (Collen et al. 2008). This data scarcity is particularly pronounced in the Andes (Swenson et al. 2012), exacerbated by two key factors: high endemism, resulting in species with restricted ranges, and limited regional sampling. The latter is due to both the difficult terrain at high elevations

(Swenson et al. 2012), and insufficient investment in biodiversity science by neighboring countries (Himley 2009; Mathez-Stiefel et al. 2017; Clerici et al. 2019).

Snakes, with over 4100 currently recognized species (Uetz et al. 2025), represent a highly diverse component of terrestrial vertebrate assemblages. Although their abundance and diversity are greatest in tropical lowland regions globally (Roll et al. 2017), snakes have adapted to a broad spectrum of environments and ecosystems. However, challenges in determining the precise number of snake species within the South American Andean region arise from persistent information gaps, including those related to species distribution (Wallacean shortfall), phylogenetic placement (Darwinian shortfall), and taxonomic consensus (Linnean shortfall) (Hortal et al. 2015). Furthermore, the definition of an "Andean species" remains ambiguous. For instance, should the designation apply solely to species originating in the Andes, or also to those whose ranges encompass this region? If the latter, what proportion of a species' range must fall within the Andes for it to be considered Andean? Moreover, the delineation of the Andes itself, whether from the foothills or a specific elevational threshold, requires clarification.

While precise quantification is challenging, an approximation of snake species richness within the Andean region can be derived by overlaying species distribution ranges with the Andean boundary. For species distributions, we utilized polygons from the *Global Assessment of Reptile Distributions* v.1.7 [GARD] (http://www.gardinitiative.org), and for the Andean delineation, polygons from the Global Mountain Biodiversity Assessment [GMBA] (Snethlage et al. 2022). The GMBA's "mountain" zone delimitation employs an algorithm based on terrain roughness parameters, reducing subjective expert opinion (see Snethlage et al. (2022) for details). In the Andes, this encompasses elevations from sea level to approximately 6700 meters. Given that species distribution polygons are typically two-dimensional approximations, they may exhibit commission errors (area overestimation). To mitigate this, we applied a 10% tolerance threshold. Species with less than 10% of their total distribution area within the Andean boundary were considered marginal or nonrepresentative. This criterion not only compensates for polygon overestimation but also accounts for the species' evolutionary history within the Andes, recognizing the close link between evolutionary trajectory and distribution range.

Here, we map the richness of Andean snakes and assess potential temporal trends in the proportion of high-elevation species described. To generate a richness map, we used the distribution data of all species classified as "Andean" based on the above criteria, employing the GARD database (Roll et al. 2017) and the letsR package (Vilela and Villalobos 2015) in R 4.0.3 (R Core Team 2020). We then quantified the proportion of each species' distribution range within the Andean Mountain region. Richness was calculated using 1 degree grid cells (approximately 111 km × 111 km at the equator). Subsequently, we examined whether the proportion of a species' distribution range within the Andes increased with the year of taxonomic description, and whether more recently described species exhibited smaller distribution ranges. This analysis serves as a proxy for evaluating the interaction between Wallacean and Linnean shortfalls.

From this analysis, we derived an approximate count of 259 snake species with distribution entirely or partially restricted to the Andes (see Supplementary Material). This figure represents a conservative estimate; species may have been excluded due to selection criteria, data absence, or included due to taxonomic discrepancies (e.g., synonyms, subspecies). Of these, approximately 140 species exhibit over 80% of their distribution within the Andean region, suggesting they could be considered near-endemic, or at least highly restricted to the Andes, considering spatial delimitation uncertainties. These species are primarily concentrated in the tropical Andean zone, between 10° N and 20° S, following a richness gradient with a peak in the northern Andes of Colombia and Ecuador (Fig. 1a), exceeding 50 species per grid cell. This pattern is generally consistent across Colubridae/Dipsadidae, Elapidae, and Viperidae (Fig. 1b–e, g), with

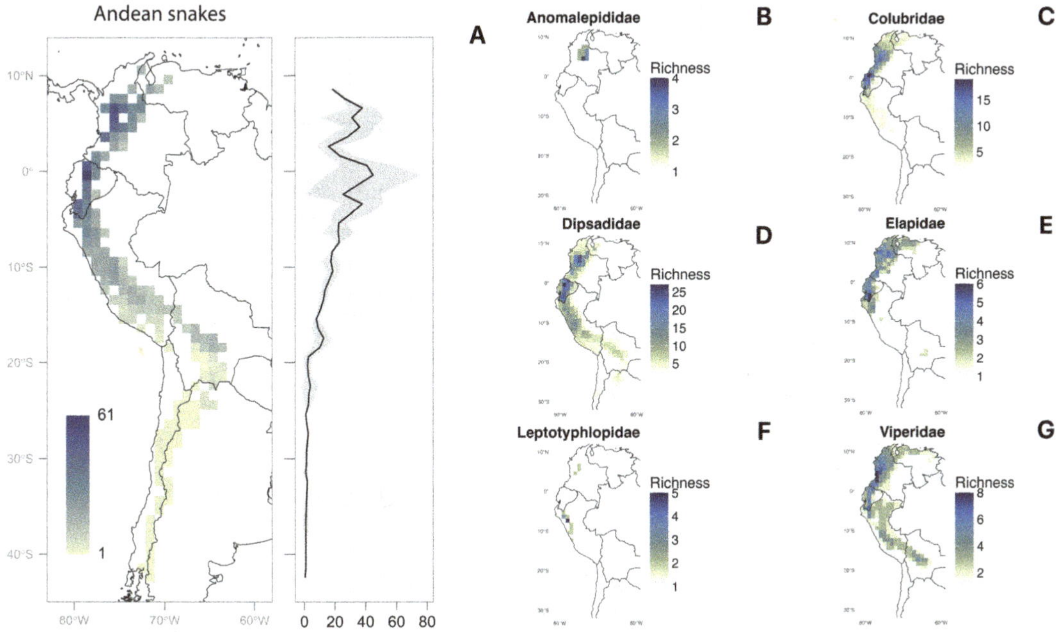

Fig. 1 Spatial distribution of snake species richness in the South American Andean region, delineated by major families: (**a**) all species, (**b**) Anomalepididae, (**c**) Colubridae, (**d**) Dipsadidae, (**e**) Elapidae, (**f**) Leptotyphlopidae, and (**g**) Viperidae

Dipsadidae extending further south into the Bolivian Andes (Fig. 1d). Leptotyphlopidae exhibits its highest richness in northern Peru (Fig. 1f).

The latitudinal diversity gradient, a widely documented phenomenon across major terrestrial vertebrate groups, including birds, amphibians, mammals, and reptiles (Hawkins et al. 2003; Buckley et al. 2010; Jenkins et al. 2013; Mittelbach et al. 2007; Roll et al. 2017; Saupe et al. 2019; Tobar-Suárez et al. 2022; Raz et al. 2024), remains a subject of ongoing inquiry. While the mechanisms underlying the heightened diversity of tropical regions are not fully elucidated, numerous eco-evolutionary hypotheses, encompassing evolutionary time, spatial heterogeneity, climatic stability, mutation rates, and competition, have been proposed (Pianka 1966; Fine 2015; Zhang et al. 2022). Concurrently, the Tropical Andes represent the most vulnerable region within the Andean mountain range, particularly in Ecuador and Colombia, and for groups such as reptiles (Bax and Francesconi 2019).

We also observed that, across snake families with at least five Andean species, the proportion of a species' distribution range within the Andes increased with the year of description (Fig. 2). This trend was particularly evident in Anomalepididae, while less pronounced in Dipsadidae, for which species continue to be described both within and outside the Andean range (Arteaga and Batista 2023). Furthermore, recently described species tended to exhibit smaller distribution areas, a pattern also documented in other reptiles (Guedes et al. 2024). This is often attributable to the lower accessibility and sampling intensity of high-elevation regions, which, combined with restricted distribution ranges, reduces the likelihood of species discovery and delays formal description (Guedes et al. 2024). Additionally, the increasing use of genetic data, supplementing traditional morphological analyses, contributes to a finer delimitation of biological entities, often

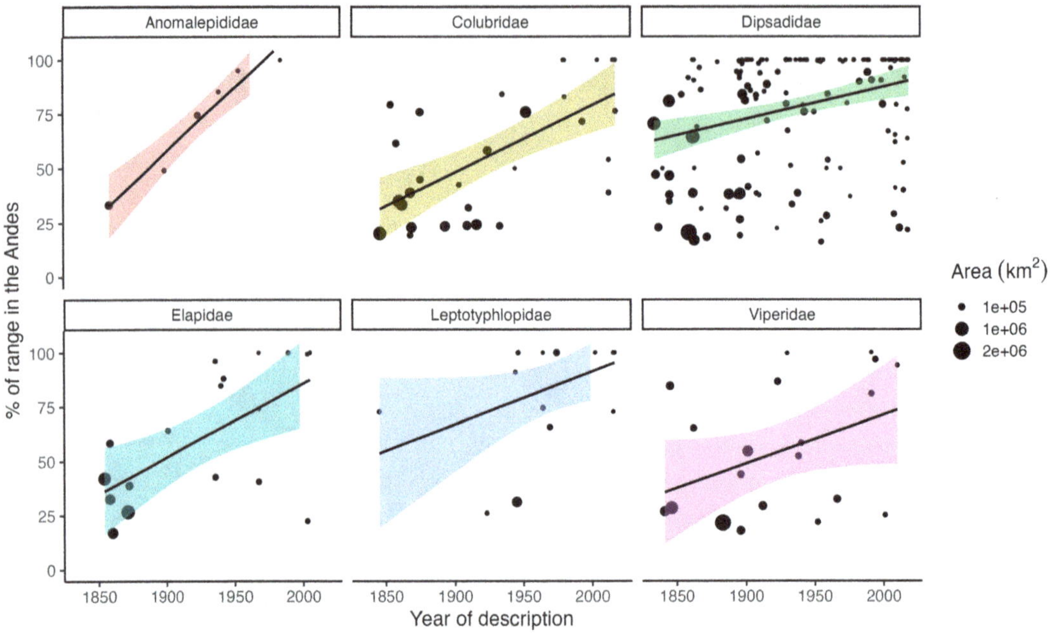

Fig. 2 Relationship between the year of description and the percentage of the distribution range within the Andean Mountain region (points), or families with at least five described species. Point size corresponds to distribution area (km^2)

resulting in smaller estimated species ranges (Agapow et al. 2004; Robuchon et al. 2019).

Overall, we observed that the spatial patterns of Andean snake species descriptions are skewed toward species with distributions not exclusively confined to the Andean mountain range. Andean endemic species, predominantly characterized by restricted distributions, have been primarily described in recent decades. While these are likely high-elevation species, data on minimum and maximum elevation remain limited. For example, in the recently published comprehensive reptile trait database by Oskyrko et al. (2024), 62.9% of the species considered herein (163 of 259) lacked elevation data, particularly within the Elapidae and Leptotyphlopidae families (21.1% and 17.6%, respectively). Furthermore, numerous species are known only from their type localities, notably within the species-rich genus *Atractus* (Roll et al. 2017, Serrano F.C., "unpublished data"). These findings underscore the Wallacean shortfall in Andean snakes, which may impede effective conservation strategies, particularly for high-elevation, range-restricted species potentially at increased risk from climate change (Young et al. 2011).

3 Linnean Shortfall: Taxonomic Aspects of Andean Snakes

The Linnaean system, the dominant classification scheme in biology for over 250 years, is hierarchical.[1] Whitin this framework, snakes are classified in the class Reptilia, order Squamata, suborder Serpentes, encompassing various families, genera, and species. Based on IUCN data, the Andes are inhabited by seven to eight families: Anomalepididae (2.3%), Boidae (0.4%), Colubridae/Dipsadidae (74.5%), Elapidae (7.3%), Leptotyphlopidae (6.6%), Tropidophiidae (0.8%), and Viperidae (8.1%).

[1] Within the zoological nomenclatural system, the endings of the names of any supergeneric category explicitly indicate the level to which that category belongs. For example, the ending -oidea indicates superfamilies (Colubroidea), -idae indicates families (Colubridae), -inae is used for subfamilies (Dipsadinae), and -ini is the ending used for tribes (Dipsadini).

These families comprise approximately 54 genera, 22 of which are monotypic, with five genera concentrating over 50% of the diversity: *Atractus* (ca. 71 spp.), *Dipsas* (ca. 22 spp.), *Micrurus* (ca. 19 spp.), *Bothrops* (ca. 12 spp.), and *Erythrolamprus* (ca. 11 spp.).

However, snake taxonomy often includes supergeneric groups beyond traditional taxonomic categories. These encompass subdivisions of Linnaean ranks and unranked clades. For instance, the suborder Serpentes is divided into the infraorders Scolecophidia and Alethinophidia. Scolecophidia, including Anomalepididae and Leptotyphlopidae, are characterized by simplified skulls, reduced eyes, and slender bodies, often referred to as "worm-shaped snakes" (from Greek *skṓlēkos* "worm" and *óphis* "snake"). Alethinophidia, meaning "true snakes" (from Greek *alēthinós* "true"), exhibit more kinetic skulls, specialized venom fangs in some cases, and enhanced vision. Andean Alethinophidia families include Boidae, Tropidophiidae, Colubridae/Dipsadidae, Elapidae, and Viperidae, with the first two forming Henophidia and the latter three forming Caenophidia ("advanced snakes"). Traditionally, snake taxonomy relied on external morphology, mainly meristic characters (countable features) such as scales and dentition (Peters and Orejas-Miranda 1970). Hemipenial morphology, though applicable only to males, has also been widely used in species delimitation (Dowling and Savage 1960; Zaher 1999). While some characters are useful for diagnosing higher taxa, species-level diagnoses require multiple characters due to limited morphospace and frequent overlap or convergence.

The taxonomic history of Andean snakes is extensive and complex, developed through contributions from numerous researchers over time, intertwined with the broader Neotropical herpetological research. Historically, this field has been dominated by foreign researchers, primarily men, with recent but still underrepresented contributions from women (Grosso et al. 2021; Rock et al. 2021). Thus, periods of substantial increase in Andean snake species descriptions occurred between 1860–1880, 1900–1920, 1940–1960, and 1990–2015 (Fig. 3).

Despite the extensive information available on snake natural history, biology, and evolutionary relationships, their high diversity makes large-scale studies difficult. This complexity leads to information gaps across groups and regions. For example, taxonomic and systematic studies in Colombian herpetology exhibit recurring gaps at medium-to-large scales (Vásquez-Restrepo 2021), often accompanied by research biases. Urbina-Cardona et al. (2023) found that Colombian herpetological literature prioritizes taxonomic listings, distribution expansions, and natural history over areas such as anatomy,

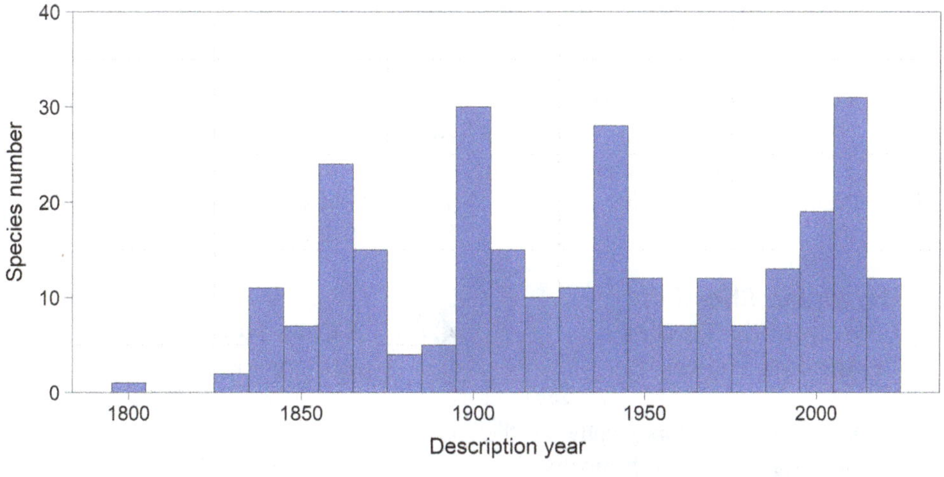

Fig. 3 Number of Andean snake species described between 1834 and 2015 (decadal intervals)

physiology, population ecology, and biogeography. In addition, the general state of knowledge about reptiles in the country reflects historical biases, favoring Testudines and Crocodylia over Squamata (Páez et al. 2006).

Shifts in systematics paradigms have redirected contemporary taxonomy from solely describing new species to revising and reorganizing existing ones (Padial and De la Riva 2020). Simultaneously, taxonomic outputs have transitioned from extensive monographs to scientific articles or focused sections within articles, particularly in phylogenetic systematics. This change has increased the volume of publications addressing snake taxonomy, yet simultaneously hindered their traceability, as taxonomic elements, often secondary to the main objective, may be inadequately indexed. For instance, a preliminary search using ("*Snake*" OR "*Serpentes*" OR "*Ophidi**") AND ("*Andes*" OR "*Andean*") AND ("*Taxonomy*" OR "*Systematics*") on platforms like Scopus (https://www.scopus.com) or Web of Science (https://www.webofscience.com) reveals fewer than 60 unrefined results. Nevertheless, resources such as The Reptile Database (http://www.reptile-database.org) have become indispensable for tracking taxonomic updates and associated literature.

However, the existence of databases that facilitate access to information does not make them the definitive authority on classification. Inevitably, authors differ in name usage and classification schemes. The dipsadines exemplify this, where, despite frequent monophyly recovery, some authors classify them as a family [= Dipsadidae] (Vidal et al. 2007, 2010; Zaher et al. 2009, 2019; Grazziotin et al. 2012), and others as a subfamily [= Dipsadinae] (Pyron et al. 2013; Zheng and Wiens 2016).

4 Darwinian Shortfall: Phylogenetic Studies in Andean Snakes

The complex orogeny and topography of the Andes have influenced the diversification process of various Neotropical organism groups, and snakes are no exception. Given the relatively low vagility of these animals, the geographic isolation seemingly facilitated by these orogenic processes has led to Andean snakes generally having relatively restricted distributions (Torres-Carvajal and Hinojosa 2020). Some of these Andean species belong to taxonomic groups with a wide distribution in South America for which it is known that the orogeny of the Andes had a significant influence on their diversification process, such as the case of the dipsadids of the Tachymenini tribe or vipers of the genus *Bothrops* (Hamdan et al. 2020; Trevine et al. 2022). Both to identify the distribution patterns of Andean snakes and to understand how the orogeny of the Cordillera has influenced their diversification, it is unavoidable to go through the path of phylogenetic analyses.

Phylogenetics is an essential tool of taxonomy for classifying organisms reflecting their evolutionary history. It is also a tool for the branch of biogeography that seeks to understand the historical processes that shaped the current distributions of species. In the case of Andean snakes, phylogenetic studies are not abundant, but a series of works carried out in the last decade have provided important information about their diversity and biogeographic history. In some cases, these phylogenies led to the identification of new species with Andean distribution, including representatives of Dipsadidae, like *Atractus* or *Synophis* (Pyron et al. 2015; Torres-Carvajal et al. 2015; Arteaga et al. 2017), and Viperidae, as is the case of *Bothrops monsignifer* (Timms et al. 2019). In others, phylogenies were used to estimate divergence times of lineages, demonstrating that the uplift of the Andes played an important role in the diversification of some groups, such as *Atractus*, *Leptophis*, Dipsadini, or Tropidophiidae (Torres-Carvajal and Terán 2021; Serrano et al. 2024; Zaher et al. 2024).

4.1 A Case Study: The *Bothrops-Bothrocophias* Complex

One of the groups of snakes perhaps with the greatest influence of the Andean orogeny in their diversification process is that of the vipers of the complex *Bothrops–Bothrocophias*, a group of 57 species (Uetz et al. 2025). These species occupy a wide spectrum of environments and display enormous morphological variability, which implies a complex systematics. The phylogenetic relationships within *Bothrops–Bothrocophias* were analyzed in several studies (e.g., Fenwick et al. 2009; Carrasco et al. 2012, 2023), and there is currently enough evidence that the species of the complex form a monophyletic group. There is also a consensus in recognizing species groups (Fig. 4a), clades that have been repeatedly recovered in different analyses. Most of these clades include one or more species with exclusively Andean distribution and/or species with some populations present in the Andes.

Bothrocophias is an essentially Andean genus whose species are present in montane forests on both sides of the Cordillera. The genus was proposed by Gutberlet Jr and Campbell (2001) and currently includes nine species, some of them previously assigned to *Bothrops* Wagler, 1824 (Uetz et al. 2025). In particular, the "*microphthalmus*" group presents distinctive morphological characteristics with respect to the species of *Bothrops*; however, the morphological variation of the group throughout its distribution is not well known. In a recent study, Angarita-Sierra et al. (2022) obtained a molecular-based phylogeny of the group that revealed the paraphyly of *Bothrocophias microphthalmus* with respect to *B. hyoprora*. Based on these results and a morphological review, the authors proposed recognizing two new species (*B. myrringae* and *B. tulitoi*) as the sister clade of *B. hyoprora*. Despite this taxonomic proposal, the paraphyly of *B. microphthalmus* remains. Angarita-Sierra et al. (2022) suggested that the populations of *B. microphthalmus* from Ecuador and Bolivia represent candidate species and indicated that the population from Peru (which includes the type locality) deserves the specific epithet *microphthalmus*. Another taxonomic evaluation required in *Bothrocophias* is that of the phylogenetic position of *B. lojanus*. This is an Andean species with a relatively restricted distribution, which inhabits dry montane forests in southern Ecuador and northern Peru (Carrasco et al. 2016).

The phylogenetic position of this species has been unstable and remains unclear. Historically the species was assigned to *Bothrops* until phylogenetic analyses based on molecular data of the species (e.g., Arteaga et al. 2016; Hamdan et al. 2020) recovered it as related to *Bothrocophias*, and currently, it is recognized as belonging to that genus (Uetz et al. 2025). However, when molecular data is analyzed in combination with morphological characters and/or different phylogenetic methodologies, the position of *B. lojanus* may become unstable (Fig. 4b). This indicates that the phylogenetic position of this Andean species is not sufficiently resolved; its generic assignment remains doubtful and the instability in its position can lead to the paraphyly of both *Bothrops* and *Bothrocophias* (Carrasco et al. 2023). This has important implications for the study of the evolutionary history of the group; resolving its position is crucial to understand how the divergence *Bothrops–Bothrocophias* occurred.

Regarding the phylogeny of *Bothrops*, among the most basal members are *B. pictus* and *B. ammodytoides* (Fig. 4a), two species with relatively wide distributions in Peru and Argentina, respectively. Both species are present in a wide altitudinal spectrum, from high regions in the Andes to regions near sea level (Campbell and Lamar 2004; Carrasco et al. 2010). A study by Fenker et al. (2014), who analyzed the phylogenetic diversity of the *Bothrops–Bothrocophias* complex, showed *B. ammodytoides* and *B. pictus* among the species with the greatest evolutionary distinctiveness. While the distributions of *B. pictus* and *B. ammodytoides* are separated by more than a thousand kilometers, and the distribution of one is trans- and the other cis-Andean, both share a similar dorsal coloration which is typical of desert pitvipers (Campbell and Lamar 2004). Phylogenetic analyses (e.g., Fenwick et al. 2009, Carrasco et al. 2012, 2023) indicate that this phenotypic similarity may be an inherited ancestral condition. In general, phylogenetic analyses recovered *B. pictus* as a taxon basal

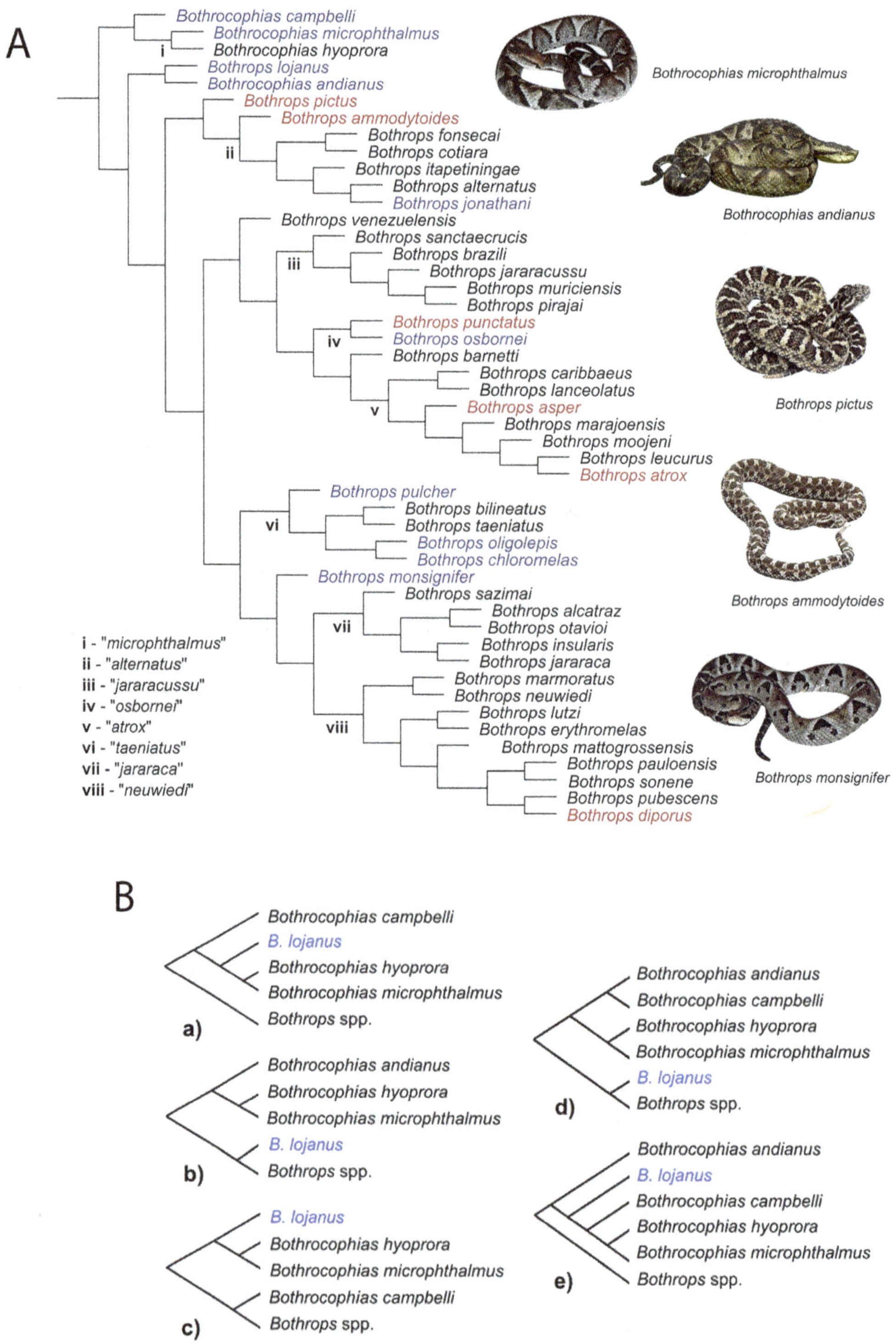

Fig. 4 (**a**) Phylogeny of the complex *Bothrops–Bothrocophias* (i–viii: names of the species groups). In blue: species with exclusively Andean distribution. In red: species that in their distribution include populations in the Andes. (Adapted from Carrasco et al. (2023)). (**b**) Alternative phylogenetic positions of *B. lojanus* obtained in different studies. (a) Arteaga et al. (2016), Alencar et al. (2016), Zaher et al. (2019); (b) Timms et al. (2019); (c) Figueroa et al. (2016); (d–e) Carrasco et al. (2023)

to the rest of the species of the genus; in some analyses that included morphological data, the species was recovered as closely related to the "*alternatus*" clade to which *B. ammodytoides* belongs. The "*alternatus*" clade includes another species with Andean distribution, *B. jonathani*, present on the western slopes of the Cordillera in southern Bolivia and northern Argentina (Carrasco et al. 2009). This is an enigmatic species that is not often found and of which there are very few specimens in scientific collections. Hamdan et al. (2020) used a calibrated molecular phylogeny to reconstruct the biogeographic history of *Bothrops* and inferred that the diversification of *B. pictus* and the "*alternatus*" group occurred during the Miocene and was related to the uplift of the Andes. It is likely that the ancestral population of *Bothrops* experienced an early vicariance event with the rise of the Andes, leading to an independent history of *B. pictus* west of the Cordillera and *B. ammodytoides* expanding southeastward to its present austral distribution (Carrasco et al. 2010). In view of these scenario, it is possible to affirm that the orogeny of the Andes played an important role in the early diversification of *Bothrops* during the Miocene (c. 11 mya).

Later orogenic events also seem to have influenced the diversification of other *Bothrops* groups, such as "*taeniatus*" and "*osbornei*" (Fig. 4a). According to the results of Hamdan et al. (2020), the rise of the Eastern Cordillera of the Colombian Andes would have played a role in the diversification of "*taeniatus*" during the Pliocene (c. 5.47 mya). This is a group of arboreal species that includes three Andean species present in humid montane forests in Colombia, Ecuador, Peru, and probably Bolivia (*B. chloromelas*, *B. oligolepis*, and *B. pulcher*). Similarly, the rise of the northern Andes during the Pliocene (c. 5.46 mya) would have influenced the diversification of "*osbornei*," a group that includes two semi-arboreal species that inhabit humid montane forests: *B. osbornei*, on the western slopes of the Andes in Ecuador, and *B. punctatus*, on the western slopes of Colombia and Ecuador reaching coastal areas and extending to Panama. Recent diversification events also involve *B. asper*, a species that belongs to the "*atrox*" group. This species has a wide distribution that extends from Mexico to northwestern Peru, and includes Andean lineages in Ecuador. Salazar-Valenzuela et al. (2019) studied the diversification of these lineages using genomic data, and their results suggest a colonization of Andean habitats facilitated by the presence of valleys. This is an example of how the topography of the Andes, once they reached their current physiognomy, promoted the diversification of snake fauna by generating environmental gradients and availability of new habitats.

The taxonomic status of near-Andean populations of some species of *Bothrops* with broad distributions requires an evaluation, and an example is *B. diporus*, a species with a predominantly Chacoan distribution. According to phylogeographic analyses (Machado et al. 2014; Carrasco et al. 2019), *B. diporus* is a paraphyletic species. Carrasco et al. (2019) hypothesized the existence of three lineages which they called *B. diporus*, *B.* aff. *pubescens*, and *B.* aff. *diporus,* which would require a taxonomic review; *B.* aff. *diporus* includes populations present at high elevations in the Precordillera of the Argentinian Andes.

5 Conservation of Andean Snakes

The Andes are a global biodiversity "hotspot" (Myers et al. 2000), yet they remain poorly protected (Bax and Francesconi 2019), with natural vegetation loss reaching 78% (Josse et al. 2011) and among the hotspots most vulnerable to deforestation-driven species loss (Brooks et al. 2002). Andean snakes, especially those at higher elevations, face extinction risk due to (i) habitat loss from deforestation and natural habitat transformation (Brooks et al. 2002); (ii) climate change that affects high-elevation species unable to track temperature shifts (Larsen et al. 2011);

and (iii) the prevalence of species with restricted ranges, increasing their vulnerability to stochastic events (Manes et al. 2021). Here, we reviewed population trends and threat status of Andean snakes from the IUCN Red List (Rodrigues et al. 2006) and related them to altitudinal distribution, assessing whether high-elevation species are disproportionately threatened. For this, we extracted minimum elevation data from ReptTraits (Oskyrko et al. 2024), a comprehensive global reptile trait database, to assess spatial conservation gaps. Because as we mentioned, 63% of Andean snake species in our dataset lacked information on minimum elevation data, we performed this analysis with the remaining 97 species.

In total, 37 species were unassessed by the IUCN, potentially due to the rapid recent description of Andean snakes, particularly within the genus *Atractus* (Arteaga et al. 2017; Melo-Sampaio et al. 2021). For the remaining 222 species, the majority of the evaluated Andean snake species exhibited unknown population trends, with only 23% showing a directional population trend (Table 1). This pattern was consistent across families, especially for Leptotyphlopidae and Tropidophiidae, where all species had unknown population trends. Although stable populations generally outnumbered declining populations within each family (Table 2), the high proportion of unknown trends may mask substantial declines in Andean snakes, as observed for most Neotropical vertebrates (Almond et al. 2022). In general, half of the species are classified as "Least Concern," with a decreasing number in progressively threatened categories. However, nearly a third of the species are "Data Deficient" (Table 3).

Most families have a majority of species classified as "Least Concern" category, while the proportions of nonthreatened (Least Concern and Near Threatened) and threatened (Vulnerable, Endangered, and Critically Endangered) species vary among families (Table 4). For example, Colubridae, Elapidae, and Viperidae have low percentages of "Data Deficient" species (6.5%, 15.8%, and 14.3%, respectively). This may stem from the higher research focus on the latter two families due to their medical significance, though this primarily relates to venom studies rather than comprehensive distributional assessments. Notably, Colubridae has a relatively high percentage of "Critically Endangered" species (12.9%).

Table 1 Absolute and relative number (% of total species) of species exhibiting each population trend

Population trend	Number of species	Proportion
Decreasing	20	9
Stable	31	14
Unknown	171	77

Table 2 Absolute and relative number (% of total of species) of species in each type of population trends among species

Family	Population trend	Number of species	Proportion
Anomalepididae	Stable	1	16.7
	Unknown	5	83.3
Boidae	Decreasing	1	100
Colubridae	Decreasing	3	11.1
	Stable	5	18.5
	Unknown	19	70.4
Dipsadidae	Decreasing	9	6.6
	Stable	18	13.2
	Unknown	109	80.1
Elapidae	Decreasing	4	21.1
	Stable	3	15.8
	Unknown	12	63.2
Leptotyphlopidae	Unknown	11	100
Tropidophiidae	Unknown	2	100
Viperidae	Decreasing	3	15
	Stable	4	20
	Unknown	13	65

Table 3 Absolute and relative number (% of total species) in each IUCN category

IUCN category	Number of species	Proportion
Data Deficient (DD)	66	29.7
Critically Endangered (CR)	8	3.6
Endangered (EN)	8	3.6
Vulnerable (VU)	16	7.2
Near Threatened (NT)	12	5.4
Least Concern (LC)	112	50.5

Table 4 Absolute and relative number (% of total species) of species in each IUCN category among families

Family	IUCN category	Number of species	Proportion
Anomalepididae	DD	5	83.3
Anomalepididae	LC	1	16.7
Boidae	LC	1	100
Colubridae	CR	4	12.9
Colubridae	DD	2	6.5
Colubridae	LC	21	67.7
Dipsadidae	CR	3	1.9
Dipsadidae	LC	48	29.8
Dipsadidae	EN	7	4.3
Dipsadidae	LC	60	37.3
Dipsadidae	NT	8	5
Dipsadidae	VU	10	6.2
Elapidae	CR	1	5.3
Elapidae	DD	3	15.8
Elapidae	EN	1	5.3
Elapidae	LC	11	57.9
Elapidae	NT	2	10.5
Elapidae	VU	1	5.3
Leptotyphlopidae	DD	5	29.4
Leptotyphlopidae	LC	4	23.5
Leptotyphlopidae	VU	2	11.8
Tropidophiidae	LC	1	50
Tropidophiidae	NT	1	50
Viperidae	DD	3	14.3
Viperidae	LC	13	61.9
Viperidae	NT	1	4.8
Viperidae	VU	3	14.3

However, some assessments may rely on outdated geographical occurrence data, potentially underestimating the true threat status of Andean snakes (Serrano et al. 2023). Therefore, species evaluations should consider not only distribution range size but also remaining suitable habitat and current threats, which are often overlooked or outdated (Meiri et al. 2023).

The altitudinal distribution of IUCN categories showed that nonthreatened species predominantly occur at lower elevations (Fig. 5), while threatened species are more common at higher elevations. Notably, "Critically Endangered" and "Data Deficient" species are primarily found above 1000 meters. This pattern may be attributed, in part, to the "sky island" effect of mountains, where species' distribution ranges tend to contract with increasing elevation.

Overall, our analysis reveals that the conservation status of many Andean snake species remains inadequately assessed and largely outdated, hindering effective conservation efforts. We confirm that population trend and threat status data for reptiles are often limited, particularly for recently described species (Meiri et al. 2023). Most families exhibit a predominance of species with unknown population trends and a substantial proportion classified as "Data Deficient." Furthermore, "Critically Endangered" and "Data Deficient" species are concentrated at higher elevations, further emphasizing their extinction vulnerability. While elevation information may

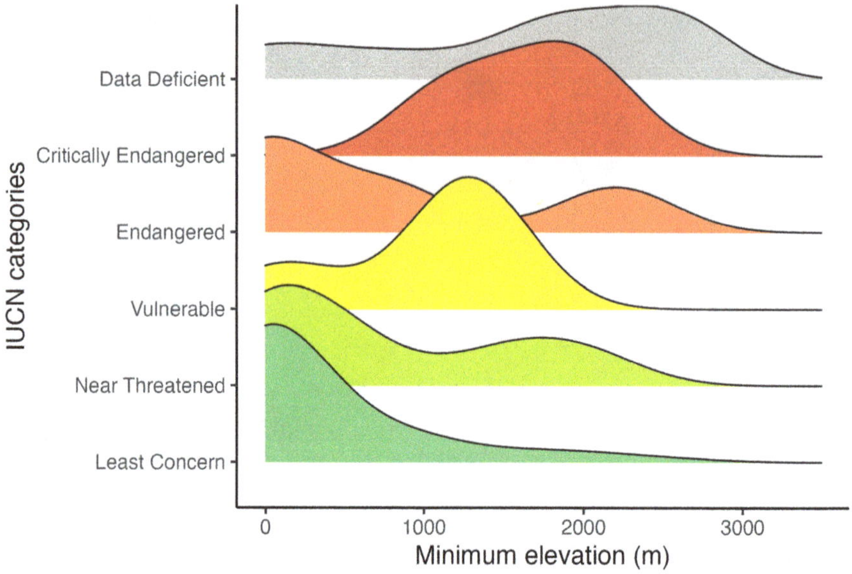

Fig. 5 Density plots of the minimum elevation for Andean snakes across the IUCN threat categories

be implicitly considered in "Critically Endangered" assessments, the vulnerability of "Data Deficient" species (nearly a third of evaluated species) is concerning, as this category does not confer legal protection (Trindade-Filho et al. 2012). Therefore, assessing the threat status of "Data Deficient" Andean snake species and determining population trends for the majority of species are imperative. This information is crucial for guiding spatially explicit conservation initiatives.

6 Conclusions, Limitations, and Future Perspectives

Comprehensive studies on the ecology, evolution, biology, and conservation of this vertebrate group are imperative, particularly in light of escalating threats, including climate change, emerging diseases, and their synergistic interactions with established pressures such as pollution, deforestation, and illegal trafficking (Davy et al. 2021; Allain et al. 2024). The results and approaches here, while concise, provide an overview of the current knowledge regarding Andean snakes and address some key biological gaps. The high proportion of recently described species underscores a deficit of detailed information across various aspects (as discussed in Sects. 1 and 2). We are, in essence, describing new species at a rate that outpaces our capacity to study them. A lack of integrative and interdisciplinary approaches represents a significant limitation in our scientific exploration of these issues.

6.1 Overcoming the Linnean and Wallacean Shortfalls: Toward More Accurate and Updated Classifications

As discussed, snake taxonomy has undergone significant transformations with advances in molecular systematics, moving from morphology-based classifications to those supported by evolutionary models and comprehensive evidence. However, taxonomic work remains influenced by inherent subjectivity, particularly in species delimitation. For Andean snakes, a substantial number of species lack genetic data (Fig. 6), presenting both challenges and opportunities for future research (e.g., addressing the Darwinian shortfall). Many of

Fig. 6 Phylogenetic relationships of Andean snake species according to Tonini et al. (2016). This tree includes 210 of the estimated 259 species. Note the prevalence of polytomies, indicating imputed species lacking genetic data

these species, absent from molecular phylogenies, are rare, known from few specimens, or have restricted distributions. This has contributed to taxonomic stability, albeit due to limited study. Advances in historical DNA extraction and sequencing from museum specimens offer a viable alternative in some cases (Hykin et al. 2015; Raxworthy and Smith 2021).

Reevaluating species limits is another pressing task in Andean snake taxonomy. A tendency to describe new species rather than to revise existing ones and synonymize where necessary is prevalent. This bias, as previously noted, is partly due to specimen availability and the shift from a biological to a phylogenetic species concept (Agapow et al. 2004; Isaac et al. 2004; Padial and de la Riva 2006; Robuchon et al. 2019). While estimates suggest numerous species remain undescribed (Wiens 2023), recognizing that new species descriptions are not always infallible is crucial. Diagnoses may reflect population variations, isolation by distance or sampling biases. Integrating taxonomy with phylogenetic systematics, supported by population genetics, is essential for resolving these inconsistencies.[2]

The description of cryptic species is a prominent trend in taxonomy. However, caution is warranted. While the unified species concept (de Queiroz 2005) suggests species need not be morphologically diagnosable, "cryptic diversity" should not be used as a simplistic justification for taxonomic revisions without robust morphological evidence (Korshunova et al. 2019). Cryptic diversity exists, but it is often defined within a narrow character space, neglecting other potentially informative traits.

Promoting collaborative networks for species taxonomy, especially for widely distributed groups, is vital for comprehensive studies. Geographical gaps (Wallacean shortfall) and taxonomic uncertainties (Linnean shortfall) persist, hindering accurate species identification. Regions with unknown species compositions often rely on historical classifications, even when they are known to be potentially inaccurate. Biological collections are invaluable sources of historical diversity information (e.g., Camacho-Rozo and Urbina-Cardona 2024). As noted by Vásquez-Restrepo (2021), a bias exists in amphibian and reptile systematics, where studies initially intended to be comprehensive often exclude samples from Colombia despite available material.

6.2 Darwinian Shortfall, Challenges, and Opportunities in the Genomic Era

The *Bothrops–Bothrocophias* example in Sect. 4.1 highlights the importance of phylogenetic and phylogeographic analyses for understanding Andean herpetofaunal diversity and evolutionary history. These studies face challenges related to genetic sample quantity and distribution, and the elusive nature of some species. Overcoming these obstacles will enable research to reveal poorly understood diversity in regions like inter-Andean valleys and the Eastern Precordillera. Integrating data from diverse sources, including morphology, is desirable. While morphological data collection is time-consuming, it can significantly contribute to phylogenetic inference and taxonomic conflict resolution.

6.3 Interdisciplinarity for Conservation

Future research should address morphological, behavioral, physiological, life history (Raunkiaerian shortfall), abundance (Prestonian shortfall), abiotic tolerance (Hutchinsonian shortfall), and biotic interaction (Eltonian shortfall) aspects, especially in the context of climate change. Interdisciplinary approaches are essential, fostering knowledge integration, and collaborative research (as mentioned in the Linnean and Darwinian shortfalls). This research can provide fundamental knowledge and guide biodiversity conservation strategies. Continued efforts to strengthen interdisciplinary science are crucial. Building solid foundations for integrating scientific, social, and political knowledge is increasingly necessary. Our results, approaches, and recommendations serve as a foundation for developing effective conservation strategies for Andean snakes in the face of global change.

[2] An example of this is *Atractus nicefori*, an endemic species from Colombia described by Amaral in 1930 and apparently known only from its type locality, in Jericó, Antioquia. The description of this species says that its dorsal color pattern is "lead gray"; however, its type material, of which there are photographs in the reptile collection of the Museum of Comparative Zoology at Harvard University, is far from this pattern. The actual color of this species is closer to reddish-brown, with darker transverse bands, which were not originally described. Thanks to this confusion, we have been looking for a species that does not exist as it was originally described for almost 100 years, which has also led to it being considered in a threat category due to its rarity. However, knowing its real lepidosis and coloration, a recent search in biological collections has revealed additional specimens from multiple locations (data not yet published). There is even a surprising similarity between this species and *Atractus manizalesensis*, of which it is not known if they could represent the same evolutionary entity.

Acknowledgments We thank the financial support provided by the São Paulo Research Foundation (FAPESP) under grants #2023/14087-2 and #2025/12960-6 (JCDR), and #2023/06999-1 (FCS). CAAF thank Coordenação de Aperfeiçoamento de Pessoal de Nível Superior—Brazil (CAPES)—Finance Code 001, and the Secretaría de Desarrollo Social (SEDESOL), Becas Solidarias—Honduras.

References

Agapow PM, Bininda-Emonds OR, Crandall KA, Gittleman JL, Mace GM, Marshall JC, Purvis A (2004) The impact of species concept on biodiversity studies. Q Rev Biol 79(2):161–179. https://doi.org/10.1086/383542

Albino AM (2011) Evolution of Squamata reptiles in Patagonia based on the fossil record. Biol J Linn Soc 103(2):441–457. https://doi.org/10.1111/j.1095-8312.2011.01691.x

Alencar LRV, Quental TB, Grazziotin FG, Alfaro ML, Martins M, Venzon M, Zaher H (2016) Diversification in vipers: phylogenetic relationships, time of divergence and shifts in speciation rates. Mol Biol Evol 105:50–62. https://doi.org/10.1016/j.ympev.2016.07.029

Allain SJ, Leech DI, Hopkins K, Seilern-Moy K, Rodriguez-Ramos Fernandez J, Griffiths RA, Lawson B (2024) Characterisation, prevalence and severity of skin lesions caused by ophidiomycosis in a population of wild snakes. Sci Rep 14(1):5162. https://doi.org/10.1038/s41598-024-55354-5

Almond REA, Grooten M, Juffe-Bignoli D, Petersen T (2022) Living planet report 2022 – building a nature positive society. WWF, Gland

Ananjeva NB (2019) Current state of the problems in the phylogeny of squamate reptiles (Squamata, Reptilia). Biol Bull Rev 9:119–128. https://doi.org/10.1134/S2079086419020026

Angarita-Sierra T, López-Hurtado CA (2020) Exploring the reproductive ecology of the tropical semifossorial snake Ninia atrata. Zool Res 41(2):157.https://cstr.cn/32248.14.zr.2020.015

Angarita-Sierra T, Cubides-Cubillos SD, Hurtado-Gómez JP (2022) Hidden in the highs: two new species of the enigmatic toadheaded pitvipers of the genus *Bothrocophias*. Vertebr Zool 72:971–996. https://doi.org/10.3897/vz.72.e87313

Arteaga A, Pyron RA, Peñafiel N, Romero-Barreto P, Culebras J, Bustamante L, Yánez-Muñoz MH, Guayasamin JM (2016) Comparative phylogeography reveals cryptic diversity and repeated patterns of cladogenesis for amphibians and reptiles in Northwestern Ecuador. PLoS One 11(4):1–40. https://doi.org/10.1371/journal.pone.0151746

Arteaga A, Mebert K, Valencia JH, Cisneros-Heredia DF, Peñafiel N, Reyes-Puig C, Vieira-Fernandes JL, Guayasamin JM (2017) Molecular phylogeny of *Atractus* (Serpentes, Dipsadidae), with emphasis on Ecuadorian species and the description of three new taxa. ZooKeys 661:91–123. https://doi.org/10.3897/zookeys.661.11224

Arteaga A, Batista A (2023) A consolidated phylogeny of snail-eating snakes (Serpentes, Dipsadini), with the description of five new species from Colombia, Ecuador, and Panama. ZooKeys 1143:1–49.https://doi.org/10.3897/zookeys.1143.93601

Azevedo JA, Guedes TB, Nogueira CDC, Passos P, Sawaya RJ, Prudente AL, Barbo FE, Strüssmann C, Franco FL, Arzamendia B, Giraydo AR, Argôlo AJS, Jansen M, Zaher H, Tonini JFR, Faurby S, Antonelli A (2020) Museums and cradles of diversity are geographically coincident for narrowly distributed Neotropical snakes. Ecography 43(2):328–339. https://doi.org/10.1111/ecog.04815

Bax V, Francesconi W (2019) Conservation gaps and priorities in the Tropical Andes biodiversity hotspot: implications for the expansion of protected areas. J Environ Manag 232:387–396. https://doi.org/10.1016/j.jenvman.2018.11.086

Benavides H, Rocha C (2012) Indicadores que manifiestan cambios en el sistema climático de Colombia (Años y décadas más calientes y las más y menos lluviosas). Instituto de Hidrología, Meteorología y Estudios Ambientales – IDEAM, Bogotá, DC. Available via DIALOG. http://www.ideam.gov.co/documents/21021/21138/Indicadores+de+-cambio+climatico+en+el+país.pdf/5ac540b8-e3f7-4076-91fe-d876f31101f9. Accessed 28 May 2024

Bini LM, Diniz-Filho JAF, Rangel TF, Bastos RP, Pinto MP (2006) Challenging Wallacean and Linnean shortfalls: knowledge gradients and conservation planning in a biodiversity hotspot. Divers Distrib 12(5):475–482. https://doi.org/10.1111/j.1366-9516.2006.00286.x

Brooks TM, Mittermeier RA, Mittermeier CG, Da Fonseca GA, Rylands AB, Konstant WR, Flick P, Pilgrim J, Oldfield S, Magin G, Hilton-Taylor C (2002) Habitat loss and extinction in the hotspots of biodiversity. Biol Conserv 16(4):909–923. https://doi.org/10.1046/j.1523-1739.2002.00530.x

Buckley LB, Davies TJ, Ackerly DD, Kraft NJB, Harrison SP, Anacker BL, Cornell HV, Damschen EL, Grytnes JA, Hawkins BA, McCain CM, Stephens PR, Wiens JJ (2010) Phylogeny, niche conservatism and the latitudinal diversity gradient in mammals. Proc Biol Sci 277(1691):2131–2138. https://doi.org/10.1098/rspb.2010.0179

Caceres-Martínez HC, Villamizar MP, Arias-Alzate FA (2017) Diagnóstico sobre el tráfico de fauna silvestre en el departamento de Norte de Santander, Colombia. Rev Biodivers Neotrop 7(3):189–199. https://doi.org/10.18636/bioneotropical.v7i3.652

Camacho-Rozo CP, Urbina-Cardona N (2024) Major knowledge shortfalls for Colombian Amazonian anurans: implications for conservation. Austral Ecol 49(8):e13564. https://doi.org/10.1111/aec.13564

Campbell JA, Lamar WW (2004) Venomous reptiles of Western Hemisphere. Cornell University Press, Ithaca

Carrasco PA, Harvey MB, Muñoz A (2009) The rare Andean pitviper *Rhinocerophis jonathani* (Serpentes: Viperidae: Crotalinae): Redescription with comments on its systematics and biogeography. Zootaxa 2283(1): 1–15. https://doi.org/10.11646/zootaxa.2283.1.1

Carrasco PA, Leynaud GC, Scrocchi GJ (2010) Redescription of the southernmost snake species *Bothrops ammodytoides* (Serpentes: Viperidae: Crotalinae). AMRE 31(3):323–338. https://doi.org/10.1163/156853810791769491

Carrasco PA, Mattoni CI, Leynaud GC, Scrocchi GJ (2012) Morphology, phylogeny and taxonomy of South American bothropoid pitvipers (Serpentes: Viperidae). Zool Scr 41(2):109–124. https://doi.org/10.1111/j.1463-6409.2011.00511.x

Carrasco PA, Venegas PJ, Valencia JH (2016) First confirmed records of the endangered Andean pitviper *Bothrops lojanus* Parker, 1930 (Viperidae: Crotalinae) from Peru. Herpetol Notes 9:297–301

Carrasco PA, Grazziotin FG, Farfán RSC, Koch C, Ochoa JA, Scrocchi GJ, Leynaud GC, Chaparro JC (2019) A new species of *Bothrops* (Serpentes: Viperidae: Crotalinae) from Pampas del Heath, southeastern Peru, with comments on the systematics of the *Bothrops neuwiedi* group. Zootaxa 4565(3):301–344. https://doi.org/10.11646/zootaxa.4565.3.1

Carrasco PA, Koch C, Grazziotin F, Venegas P, Chaparro JC, Scrocchi G, Salazar-Valenzuela D, Leynaud G, Mattoni C (2023) Total-evidence phylogeny and evolutionary morphology of New World pitvipers (Serpentes, Viperidae, Crotalinae). Cladistics 39(2): 71–100. https://doi.org/10.1111/cla.12522

Clerici N, Salazar C, Pardo-Díaz C, Jiggins CD, Richardson JE, Linares M (2019) Peace in Colombia is a critical moment for neotropical connectivity and conservation: save the northern Andes–Amazon biodiversity bridge. Conserv Lett 12(1):1–7. https://doi.org/10.1111/conl.12594

Cobbold PR, Rossello EA, Roperch P, Arriagada C, Gómez LA, Lima C (2007) Distribution, timing, and causes of Andean deformation across South America. Geol Soc Spec Publ 272(1):321–343. https://doi.org/10.1144/GSL.SP.2007.272.01.17

Collen B, Ram M, Zamin T, McRae L (2008) The tropical biodiversity data gap: addressing disparity in global monitoring. Trop Conserv Sci 1(2):75–88. https://doi.org/10.1177/194008290800100202

Cordier JM, Aguilar R, Lescano JN, Leynaud GC, Bonino A, Miloch D, Loyola R, Nori J (2021) A global assessment of amphibian and reptile responses to land-use changes. Biol Conserv 253:108863. https://doi.org/10.1016/j.biocon.2020.108863

Cox N, Young BE, Bowles P, Fernandez M, Marin J, Rapacciuolo G, Böhm M, Brooks TM, Hedges SB, Hilton-Taylor C, Hoffmann M, Jenkins RKB, Tognelli MF, Alexander GJ, Allison A, Ananjeva NB, Auliya M, Avila LJ, Chapple DG, Cisneros-Heredia DF, Cogger HG, Colli GR, Silva A, Eisemberg CC, Els J, Fong A, Grant TD, Hitchmough RA, Iskandar DT, Kidera N, Martins M, Meiri S, Mitchell NJ, Molur S, Nogueira CC, Ortiz JC, Penner J, Rhodin AGJ, Rivas GA, Rödel MO, Roll U, Sanders KL, Santos-Barrera G, Shea GM, Spawls S, Stuart BL, Tolley KA, Trape JF, Vidal MA, Wagner P, Wallace BP, Xie Y (2022) A global reptile assessment highlights shared conservation needs of tetrapods. Nature 605(7909):285–290. https://doi.org/10.1038/s41586-022-04664-7

Davy CM, Shirose L, Campbell D, Dillon R, McKenzie C, Nemeth N, Braithwaite T, Cai H, Degazio T, Dobbie T, Egan S, Fotherby H, Litzgus JD, Manorome P, Marks S, Paterson JE, Sigler L, Slavic D, Slavik E, Urquhart J, Jardine C (2021) Revisiting ophidiomycosis (snake fungal disease) after a decade of targeted research. Front Vet Sci 8:665805. https://doi.org/10.3389/fvets.2021.665805

De Queiroz K (2005) A unified concept of species and its consequences for the future of taxonomy. Proc Calif Acad 56(18):196–215

Diniz-Filho JAF, Jardim L, Guedes JJ, Meyer L, Stropp J, Frateles LEF, Pinto RB, Lohman LG, Tessarolo G, de Carvalho CJB, Ladle RJ, Hortal J (2023) Macroecological links between the Linnean, Wallacean, and Darwinian shortfalls. Front Biogeogr 15(2):1–11. https://doi.org/10.21425/F5FBG59566

Doan TM (2003) A south-to-north biogeographic hypothesis for Andean speciation: evidence from the lizard genus *Proctoporus* (Reptilia, Gymnophthalmidae). J Biogeogr 30(3):361–374. https://doi.org/10.1046/j.1365-2699.2003.00833.x

Dowling HG, Savage JM (1960) A guide to the snake hemipenis: a survey of basic structure and systematic characteristics. Zoologica (N Y) 45(2):17–28. https://doi.org/10.5962/p.203350

Duellman WE (1999) Patterns of distribution of amphibians: a global perspective. The Johns Hopkins University Press, Baltimore

Esquerré D, Brennan IG, Catullo RA, Torres-Pérez F, Keogh JS (2018) How mountains shape biodiversity: the role of the Andes in biogeography, diversification, and reproductive biology in South America's most species-rich lizard radiation (Squamata: Liolaemidae). Evolution 73(2):214–230. https://doi.org/10.1111/evo.13657

Fenker J, Tedeschi LG, Pyron RA, Nogueira CDC (2014) Phylogenetic diversity, habitat loss and conservation in South American pitvipers (Crotalinae: *Bothrops* and *Bothrocophias*). Divers Distrib 20(10):1108–1119. https://doi.org/10.1111/ddi.12217

Fenwick AM, Gutberlet RL Jr, Evans JA, Parkinson CL (2009) Morphological and molecular evidence for phylogeny and classification of South American pitvipers, genera *Bothrops*, *Bothriopsis*, and *Bothrocophias* (Serpentes: Viperidae). Zool J Linnean Soc 156(3):

617–640. https://doi.org/10.1111/j.1096-3642.2008.00495.x

Figueroa A, McKelvy AD, Grismer LL, Bell CD, Lailvaux SP (2016) A species-level phylogeny of extant snakes with description of a new colubrid subfamily and genus. PLoS One 11(9):1–31. https://doi.org/10.1371/journal.pone.0161070

Fine PV (2015) Ecological and evolutionary drivers of geographic variation in species diversity. Annu Rev Ecol Evol Syst 46(1):369–392. https://doi.org/10.1146/annurev-ecolsys-112414-054102

Folguera A, Naipauer M, Sagripanti L, Ghiglione MC, Orts DL, Giambiagi L (2016) An introduction to the Southern Andes (33–50°S): book structure. In: Folguera A, Naipauer M, Sagripanti L, Ghiglione MC, Orts DL, Giambiagi L (eds) Growth of the Southern Andes. Springer earth system sciences. Springer, Berlin, pp 1–7. https://doi.org/10.1007/978-3-319-23060-3_1

Frateles LEF, da Silva NJ Jr, Terribile LC, Diniz-Filho JAF (2024) Linnean shortfall and space-time patterns in species description of New World coralsnakes (Serpentes: Elapidae). Zool Scr 53(3):299–311. https://doi.org/10.1111/zsc.12644

García-Cobos D, Vásquez-Restrepo JD (2022) Notes on the distribution, morphological variation, and mechanoreceptors of *Tretanorhinus nigroluteus* (Serpentes: Dipsadidae) in Colombia. Biota Colomb 23(1):1–11. https://doi.org/10.21068/2539200X.943

Gibbons JW, Scott DE, Ryan TJ, Buhmann KA, Tuberville TD, Metts BS, Greene JL, Mills T, Leiden Y, Poppy S, Winne CT (2000) The Global Decline of Reptiles, Déjà Vu Amphibians: reptile species are declining on a global scale. Six significant threats to reptile populations are habitat loss and degradation, introduced invasive species, environmental pollution, disease, unsustainable use, and global climate change. BioScience 50(8):653–666. https://doi.org/10.1641/00063568(2000)050[0653:TGDORD]2.0.CO;2

Graham A (2009) The Andes: a geological overview from a biological perspective. Ann Missouri Bot Gard 96(3): 371–385. https://doi.org/10.3417/2007146

Grazziotin FG, Zaher H, Murphy RW, Scrocchi G, Benavides MAA, Zhang Y, Bonatto SL (2012) Molecular phylogeny of the New World Dipsadidae (Serpentes: Colubroidea): a reappraisal. Cladistics 28(5):437–459. https://doi.org/10.1111/j.1096-0031.2012.00393.x

Gregory-Wodzicki KM (2000) Uplift history of the Central and Northern Andes: a review. Geol Soc Am Bull 112(7):1091–1105. https://doi.org/10.1130/0016-7606(2000)112<1091:UHOTCA>2.0.CO;2

Grosso J, Fratani J, Fontanarrosa G, Pereyra M, Duport-Bru AS, Schneider RG, Casagranda MD, Ferraro DP, Vicente NS, Sálica MJ, Pereyra L, Medina RG, Bessa C, Semhan RV, Vera MC (2021) Male homophily in South American herpetology: one of the major processes underlying the gender gap in publications. AMRE 42(4):407–418. https://doi.org/10.1163/15685381-bja10063

Guedes TB, Sawaya RJ, Zizka A, Laffan S, Faurby S, Pyron RA, Bérnelis RS, Jansen M, Passos P, Prudente ALC, Cisneros-Heredia DF, Braz HB, Nogueira CC, Antonelli A (2018) Patterns, biases and prospects in the distribution and diversity of Neotropical snakes. Glob Ecol Biogeogr 27(1):14–21. https://doi.org/10.1111/geb.12679

Guedes JJ, Gomes de Lima HV, Mendonça LR, Chen-Zhao R, Diniz-Filho JAF, Moura MR (2024) Temporal trends in global reptile species descriptions over three decades. Syst Biodivers 22(1):2419832.https://doi.org/10.1080/14772000.2024.2419832

Gutberlet RL Jr, Campbell JA (2001) Generic recognition for a neglected lineage of South American pitvipers (Squamata: Viperidae: Crotalinae), with the description of a new species from the Colombian Chocó. Am Mus Novit 2001(3316):1–16. https://doi.org/10.1206/0003-0082(2001)316<0001:GRFANL>2.0.CO;2

Hamdan B, Guedes TB, Carrasco PA, Melville J (2020) A complex biogeographic history of diversification in Neotropical lancehead pitvipers (Serpentes, Viperidae). Zool Scr 49(2):145–158. https://doi.org/10.1111/zsc.12398

Harfoot MB, Johnston A, Balmford A, Burgess ND, Butchart SH, Dias MP, Hazin C, Hilton-Taylor C, Hoffmann M, Isaac NB, Iversen LL, Outhwaite CL, Visconti P, Geldmann J (2021) Using the IUCN Red List to map threats to terrestrial vertebrates at global scale. Nat Ecol Evol 5(11):1510–1519. https://doi.org/10.1038/s41559-021-01542-9

Hawkins BA, Porter EE, Diniz-Filho JAF (2003) Productivity and history as predictors of the latitudinal diversity gradient of terrestrial birds. Ecology 84(6): 1608–1623. https://doi.org/10.1890/0012-9658(2003)084[1608:PAHAPO]2.0.CO;2

Hazzi NA, Moreno JS, Ortiz-Movliav C, Palacio RD (2018) Biogeographic regions and events of isolation and diversification of the endemic biota of the tropical Andes. Proc Natl Acad Sci USA 115(31):7985–7990. https://doi.org/10.1073/pnas.1803908115

Heredia N, García-Sansegundo J, Gallastegui G, Farías P, Giacosa R, Hongn F, Tubía JM, Alonso JL, Busquets P, Charrier R, Clariana P, Colombo F, Cuesta A, Gallastegui J, Giambiagi L, González-Menéndez L, Limarino C, Martín-González F, Pedreira D, Quintana L, Rodríguez-Fernández LR, Rubio-Ordóñez A, Seggiaro RE, Serra-Varela S, Spalletti L, Cardó R, Ramos VA (2018) The pre-andean phases of construction of the southern Andes basement in neoproterozoic–paleozoic times. In: Folguera A, Contrera-Reyes E, Heredia N, Encinas A, Lannelli SF, Oliveros V, Dávila FM, Collo G, Giambiagi L, Maksymowicz A, Iglesia MP, Turienzo M, Naipauer M, Orts D, Litvak VD, Alvarez O, Arriagada C (eds) The evolution of the Chilean-Argentinean Andes. Springer International

Publishing, Switzerland, pp 111–131. https://doi.org/10.1007/978-3-319-67774-3_5

Hierink F, Bolon I, Durso AM, de Castaneda RR, Zambrana-Torrelio C, Eskew EA, Ray N (2020) Forty-four years of global trade in CITES-listed snakes: trends and implications for conservation and public health. Biol Conserv 248(2020):1–10. https://doi.org/10.1016/j.biocon.2020.108601

Himley M (2009) Nature conservation, rural livelihoods, and territorial control in Andean Ecuador. Geoforum 40(5):832–842. https://doi.org/10.1016/j.geoforum.2009.06.001

Hortal J, de Bello F, Diniz-Filho JAF, Lewinsohn TM, Lobo JM, Ladle RJ (2015) Seven shortfalls that beset large-scale knowledge of biodiversity. Annu Rev Ecol Evol Syst 46:523–549. https://doi.org/10.1146/annurev-ecolsys-112414-054400

Horton BK (2018) Sedimentary record of Andean mountain building. Earth Sci Rev 178:279–309. https://doi.org/10.1016/j.earscirev.2017.11.025

Hurtado G (2012) Análisis del comportamiento promedio y tendencias de largo plazo de la temperatura máxima media para las regiones hidroclimáticas de Colombia. Instituto de Hidrología, Meteorología y Estudios Ambientales – IDEAM, Bogotá, DC. Available via DIALOG. http://www.ideam.gov.co/documents/21021/21138/Analisis+del+Comportamiento+de+la+Temperatura+Maxima.pdf/2a2f247c-f457-45f3-ac9a-f2a481f8daa6. Accessed 20 May 2024

Hykin SM, Bi K, McGuire JA (2015) Fixing formalin: a method to recover genomic-scale DNA sequence data from formalin-fixed museum specimens using high-throughput sequencing. PLoS One 10(10):1–16. https://doi.org/10.1371/journal.pone.0141579

Isaac NJB, Mallet J, Mace GM (2004) Taxonomic inflation: its influence on macroecology and conservation. Trends Ecol Evol 19(9):464–469. https://doi.org/10.1016/j.tree.2004.06.004

Jenkins CN, Pimm SL, Joppa L (2013) Global patterns of terrestrial vertebrate diversity and conservation. Proc Natl Acad Sci USA 110(28):E2602–E2610. https://doi.org/10.1073/pnas.1302251110

Josse C, Cuesta F, Navarro G, Barrena V, Cabrera E, Chacón-Moreno E, Ferreira W, Peralvo M, Saito J, Tovar A (2009) Ecosistemas de los Andes del Norte y Centro. Bolivia, Colombia, Ecuador, Perú y Venezuela. Secretaría General de la Comunidad Andina, Lima

Josse C, Cuesta F, Navarro G, Barrena V, Becerra MT, Cabrera E, Chacón-Moreno E, Ferreira W, Peralvo M, Saito J, Tovar A, Naranjo LG (2011) Physical geography and ecosystems in the tropical Andes. In: Herzog SK, Martínez R, Jørgensen PM, Tiessen H (eds) Climate change and biodiversity in the tropical. InterAmerican Institute for Global Change Research (IAI) and Scientific Committee on Problems of the Environment (SCOPE), pp 152–169

Korshunova T, Picton B, Furfaro G, Mariottini P, Pontes M, Prkić J, Fletcher KC, Malmberg K, Lundin K, Martynov A (2019) Multilevel fine-scale diversity challenges the 'cryptic species' concept. Sci Rep 9(6732):1–23. https://doi.org/10.1038/s41598-019-42297-5

Lagomarsino LP, Condamine FL, Antonelli A, Mulch A, Davis CC (2016) The abiotic and biotic drivers of rapid diversification in Andean bellflowers (Campanulaceae). New Phytol 210(4):1430–1442. https://doi.org/10.1111/nph.13920

Larsen TH, Brehm G, Navarrete H, Franco P, Gomez H, Mena JL, Morales V, Argollo J, Blacutt L, Canhos V (2011) Range shifts and extinctions driven by climate change in the tropical Andes: synthesis and directions. In: Herzog SK, Martínez R, Jørgensen PM, Tiessen H (eds) Climate change and biodiversity in the tropical. InterAmerican Institute for Global Change Research (IAI) and Scientific Committee on Problems of the Environment (SCOPE), pp 47–67

Lavenu A (2006) Neotectónica de los Andes entre 1 N y 47 S (Ecuador, Bolivia y Chile): una revisión. Rev Asoc Geol Argent 61(4):504–524

Lees AC, Rosenberg KV, Ruiz-Gutierrez V, Marsden S, Schulenberg TS, Rodewald AD (2020) A roadmap to identifying and filling shortfalls in Neotropical ornithology. Auk 137(4):1–17. https://doi.org/10.1093/auk/ukaa048

Lorch JM, Knowles S, Lankton JS, Michell K, Edwards JL, Kapfer JM, Staffen RA, Wild ER, Schmidt KZ, Ballman AE, Blodgett D, Farrell TM, Glorioso BM, Last LA, Price SJ, Schuler KL, Smith CE, Wellehan JFX, Blehert DS (2016) Snake fungal disease: an emerging threat to wild snakes. Philos Trans R Soc Lond Ser B Biol Sci 371(1709):20150457. https://doi.org/10.1098/rstb.2015.0457

Lynch JD (2012) El contexto de las serpientes de Colombia con un análisis de las amenazas en contra de su conservación. Rev Acad Colomb Cienc Exact Fis Nat 36(140):435–449. https://doi.org/10.18257/raccefyn.36(140).2012.2495

Machado T, Silva VX, Silva MJJ (2014) Phylogenetic relationships within *Bothrops neuwiedi* group (Serpentes, Squamata): geographically highly-structured lineages, evidence of introgressive hybridization and Neogene/Quaternary diversification. Mol Phylogenet Evol 71:1–14. https://doi.org/10.1016/j.ympev.2013.10.003

Maciel-Mata CA, Manríquez-Morán N, Octavio-Aguilar P, Sánchez-Rojas G (2015) El área de distribución de las especies: revisión del concepto. Acta Univ 25(2):3–19. https://doi.org/10.15174/au.2015.690

Maes J, Egoh B, Willemen L, Liquete C, Vihervaara P, Schägner JP, Grizzetti B, Drakou EG, La Notte A, Zulian G, Bouraoui F, Paracchini ML, Braat L, Bidoglio G (2012) Mapping ecosystem services for policy support and decision making in the European Union. Ecosyst Serv 1(1):31–39. https://doi.org/10.1016/j.ecoser.2012.06.004

Manes S, Costello MJ, Beckett H, Debnath A, Devenish-Nelson E, Grey KA, Jenkins R, Ming T, Kiessling W,

Krause C, Maharaj SS, Midgley GF, Price J, Talukdar G, Vale MM (2021) Endemism increases species' climate change risk in areas of global biodiversity importance. Biol Conserv 257:1–22. https://doi.org/10.1016/j.biocon.2021.109070

Martins M (2024) Endangered reptiles. In: Scheiner SM (ed) Encyclopedia of biodiversity, vol 2, 3rd edn. Academic Press, New York, pp 280–297

Mathez-Stiefel SL, Peralvo M, Báez S, Rist S, Buytaert W, Cuesta F, Fadrique B, Feeley KJ, Groth AAP, Homejer J, Llambí LD, Locatelli B, López MF, Malizia A, Young KR (2017) Research priorities for the conservation and sustainable governance of Andean forest landscapes. Mt Res Dev 37(3): 323–339. https://doi.org/10.1659/MRD-JOURNAL-D-16-00093.1

Meiri S, Chapple DG, Tolley KA, Mitchell N, Laniado T, Cox N, Bowles P, Young BE, Caetano G, Geschke J, Böhm M, Roll U (2023) Done but not dusted: reflections on the first global reptile assessment and priorities for the second. Biol Conserv 278:109879. https://doi.org/10.1016/j.biocon.2022.109879

Melo-Sampaio PR, Passos P, Prudente AL, Venegas PJ, Torres-Carvajal O (2021) Systematic review of the polychromatic ground snakes *Atractus snethlageae* complex reveals four new species from threatened environments. J Zool Syst Evol 59(3):718–747. https://doi.org/10.1111/jzs.12453

Mendoza F (2014) Cordillera de Los Andes, una oportunidad para la integración y desarrollo de América del Sur. FAO, Santiago de Chile

Mittelbach GG, Schemske DW, Cornell HV, Allen AP, Brown JM, Bush MB, Harrison SP, Hulbert AH, Knowlton N, Lessios HA, McCaim CM, McCune AR, McDade LA, McPeek MA, Near TJ, Price TD, Ricklefs RE, Roy K, Sax DF, Schluter D, Sobel JM, Turelli M (2007) Evolution and the latitudinal diversity gradient: speciation, extinction and biogeography. Ecol Lett 10(4):315–331. https://doi.org/10.1111/j.1461-0248.2007.01020.x

Moura MR, Jetz W (2021) Shortfalls and opportunities in terrestrial vertebrate species discovery. Nat Ecol Evol 5(5):631–639. https://doi.org/10.1038/s41559-021-01411-5

Myers N, Mittermeier RA, Mittermeier CG, Da Fonseca GA, Kent J (2000) Biodiversity hotspots for conservation priorities. Nature 403(6772):853–858. https://doi.org/10.1038/35002501

Navarrete LF, López-Johnston JC, Balza LM, Blanco A (2009) Guía de las serpientes de Venezuela. Biología, Venenos, Conservación y Checklist. Gráficas ACEA, Caracas

Nogueira CC, Argolo AJ, Arzamendia V, Azevedo JA, Barbo FE, Bernils RS, Bolochio BE, Borges-Martins-M, Brasil-Godinho M, Braz H, Buononato MA, Cisneros-Heredia DF, Colli GR, Costa HC, Franco FL, Giraudo A, Gonzales RC, Guedes T, Hoogmoed MS, Marques OAV, Montingelli GG, Passos P, Prudente ALC, Rivas GA, Sanchez PM, Serrano FC, Silva NJ, Strüssmann C, Vieira-Alencar JPS, Zaher H, Sawaya RJ, Martins M (2019) Atlas of Brazilian snakes: verified point-locality maps to mitigate the Wallacean shortfall in a megadiverse snake fauna. South Am J Herpetol 14(sp1):1–274. https://doi.org/10.2994/SAJH-D-19-00120.1

Noriega JA, Santos AM, Aranda SC, Calatayud J, De Castro I, Espinoza VR, Hórreo JL, Medina NG, Peláez ML, Hortal J (2015) ¿Cuál es el alcance de la crisis de la Taxonomía? Conflictos, retos y estrategias para la construcción de una Taxonomía renovada. Revista IDE@ – SEA 9:1–16

Oskyrko O, Mi C, Meiri S, Du W (2024) ReptTraits: a comprehensive dataset of ecological traits in reptiles. Sci Data 11(1):1–6. https://doi.org/10.1038/s41597-024-03079-5

Padial JM, de la Riva I (2006) Taxonomic inflation and the stability of species lists: the perils of Ostrich's behavior. Syst Biol 55(5):859–867. https://doi.org/10.1080/10635150600081588

Padial JM, Miralles A, De la Riva I, Vences M (2010) The integrative future of taxonomy. Front Zool 7(16):1–14. https://doi.org/10.1186/1742-9994-7-16

Padial JM, de la Riva I (2020) A paradigm shift in our view of species drives current trends in biological classification. Biol Rev 96(2):731–751. https://doi.org/10.1111/brv.12676

Páez VP, Arredondo JC, López C, Martínez LM, Molina C, Restrepo A (2006) Reptiles de Colombia, diversidad y estado de conocimiento. In: Chaves E, Santamaria M (ed) Instituto de investigaciones Biológicas Alexander von Humboldt. Bogotá, D. C., pp 118–131

Pianka ER (1966) Latitudinal gradients in species diversity: a review of concepts. Am Nat 100(910):33–46. https://doi.org/10.1086/282398

Peters JA, Orejas-Miranda B (1970) Catalogue of neotropical squamata. Part I. Snakes. Bull US Natl Mus 297:1–347

Pincheira-Donoso D, Bauer AM, Meiri S, Uetz P (2013) Global taxonomic diversity of living reptiles. PLoS One 8(3):e59741. https://doi.org/10.1371/journal.pone.0059741

Pyron RA, Burbrink FT, Wiens JJ (2013) A phylogeny and revised classification of Squamata, including 4161 species of lizards and snakes. BMC Evol Biol 13(1): 1–54. https://doi.org/10.1186/1471-2148-13-93

Pyron RA, Guayasamin JM, Peñafiel N, Bustamante L, Arteaga A (2015) Systematics of *Nothopsini* (Serpentes, Dipsadidae), with a new species of *Synophis* from the Pacific Andean slopes of southwestern Ecuador. ZooKeys 541:109–147. https://doi.org/10.3897/zookeys.541.6058

R Core Team (2020) R: a language and environment for statistical computing. R Foundation for Statistical Computing. Available from: http://www.Rproject.org/ Accessed 20 Apr 2024

Rahbek C, Borregaard MK, Colwell RK, Dalsgaard B, Holt BG, Morueta-Holme N, Nogues-Bravo D,

Whittaker R, Feldså J (2019) Humboldt's enigma: ¿What causes global patterns of mountain biodiversity? Science 365(6458):1108–1113. https://doi.org/10.1126/science.aax0149

Rangel TF, Edwards NR, Holden PB, Diniz-Filho JAF, Gosling WD, Coelho MTP, Cassemiro FAS, Colwell RK (2018) Modeling the ecology and evolution of biodiversity: biogeographical cradles, museums, and graves. Science 361(6399):1–13. https://doi.org/10.1126/science.aar5452

Raxworthy CJ, Smith BT (2021) Mining museums for historical DNA: advances and challenges in museomics. Trends Ecol Evol 36(11):1049–1060. https://doi.org/10.1016/j.tree.2021.07.009

Raz T, Allison A, Avila LJ, Bauer AM, Böhm M, Caetano GDO, Colli G, Doan TM, Doughty P, Grismer L, Itescu Y, Kraus F, Martins M, Morando M, Murali G, Nagy ZT, Nogueira CC, Novosolov M, Oliver PM, Passos P, Pincheira-Donoso D, Sindaco R, Slavenko A, Torres-Carvajal O, Uetz P, Wagner P, Zimin A, Roll U, Meiri S (2024) Diversity gradients of terrestrial vertebrates–substantial variations about a common theme. J Zool 322(2):126–140. https://doi.org/10.1111/jzo.13130

Reading CJ, Luiselli LM, Akani GC, Bonnet X, Amori G, Ballouard JM, Filippi E, Naulleau G, Pearson D, Rugiero L (2010) Are snake populations in widespread decline? Biol Lett 6(6):777–780. https://doi.org/10.1098/rsbl.2010.0373

Restrepo A, Molina-Zuluaga C, Hurtado JP, Marín CM, Daza JM (2017) Amphibians and reptiles from two localities in the northern Andes of Colombia. Check List 13(4):203–237. https://doi.org/10.15560/13.4.203

Robuchon M, Faith DP, Julliard R, Leroy B, Pellens R, Robert A, Thévenin C, Véron S, Pavoine S (2019) Species splitting increases estimates of evolutionary history at risk. Biol Conserv 235:27–35. https://doi.org/10.1016/j.biocon.2019.03.041

Rock KN, Barnes IN, Deyski MS, Glynn KA, Milstead BN, Rottenborn ME, Andre NS, Dekhtyar A, Dekhtyar O, Taylor EN (2021) Quantifying the gender gap in authorship in herpetology. Herpetologica 77(1):1–13. https://doi.org/10.1655/0018-0831-77.1.1

Rodrigues AS, Pilgrim JD, Lamoreux JF, Hoffmann M, Brooks TM (2006) The value of the IUCN Red List for conservation. Trends Ecol Evol 21(2):71–76. https://doi.org/10.1016/j.tree.2005.10.010

Rojas-Morales JA (2012) Snakes of an urban-rural landscape in the central Andes of Colombia: species composition, distribution, and natural history. Phyllomedusa 11(2):135–154. https://doi.org/10.11606/issn.2316-9079.v11i2p135-154

Roll U, Feldman A, Novosolov M, Allison A, Bauer AM, Bernard R, Böhm M, Castro-Herrera F, Chirio L, Collen B, Colli GR, Dabool L, Das I, Doan TM, Grismer LL, Hoogmoed M, Itescu Y, Kraus F, LeBreton M, Lewin A, Martins M, Maza E, Meirte D, Nagy ZT, Nogueira CC, Pauwels OSG, Pincheira-Donoso D, Powney GD, Sindaco R, Tallowin OJS, Torres-Carvajal O, Trape JF, Vidan E, Uetz P, Wagner P, Wang Y, Orme CDL, Greyner R, Meiri S (2017) The global distribution of tetrapods reveals a need for targeted reptile conservation. Nat Ecol Evol 1(11):1677–1682. https://doi.org/10.1038/s41559-017-0332-2

Salazar-Valenzuela D, Kuch U, Torres-Carvajal O, Valencia JH, Gibbs HL (2019) Divergence of tropical pitvipers promoted by independent colonization events of dry montane Andean habitats. J Biogeogr 46(8):1826–1840. https://doi.org/10.1111/jbi.13661

Saupe EE, Myers CE, Townsend A, Soberón J, Singarayer J, Valdes P, Qiao H (2019) Spatio-temporal climate change contributes to latitudinal diversity gradients. Nat Ecol Evol 3(10):1419–1429. https://doi.org/10.1038/s41559-019-0962-7

Schellart WP (2017) Andean mountain building and magmatic arc migration driven by subduction-induced whole mantle flow. Nat Commun 8(1):1–13. https://doi.org/10.1038/s41467-017-01847-z

Serrano FC, dos Santos Vieira-Alencar JP, Díaz-Ricaurte JC, Valdujo PH, Martins M, de Campos Nogueira C (2023) The Wallacean Shortfall and the role of historical distribution records in the conservation assessment of an elusive Neotropical snake in a threatened landscape. J Nat Conserv 72:126350. https://doi.org/10.1016/j.jnc.2023.126350

Serrano FC, Pontes-Nogueira M, Sawaya RJ, Alencar LRV, Nogueira CC, Grazziotin FG (2024) There and back again: when and how the world's richest snake family (Dipsadidae) dispersed and speciated across the Neotropical region. J Biogeogr 51(5):878–893. https://doi.org/10.1111/jbi.14790

Snethlage MA, Geschke J, Ranipeta A, Jetz W, Yoccoz NG, Körner C, Spehn EM, Fischer M, Urbach D (2022) A hierarchical inventory of the world's mountains for global comparative mountain science. Sci Data 9(1):149. https://doi.org/10.1038/s41597-022-01256-y

Strecker MR, Alonso RN, Bookhagen B, Carrapa B, Hilley GE, Sobel ER, Trauth MH (2007) Tectonics and climate of the southern central Andes. Annu Rev Earth Planet Sci 35(1):747–787. https://doi.org/10.1146/annurev.earth.35.031306.140158

Swenson JJ, Young BE, Beck S, Comer P, Córdova JH, Dyson J, Embert D, Encarnación F, Ferreira W, Franke I, Grossman D, Hernandez P, Herzog SK, Josse C, Navarro G, Pacheco V, Stein BA, Timaná M, Tovar A, Tovar C, Vargas J, Zambrana-Torrelio CM (2012) Plant and animal endemism in the eastern Andean slope: challenges to conservation. BMC Ecol 12(1):1–19. https://doi.org/10.1186/1472-6785-12-1

Timms J, Chaparro JC, Venegas PJ, Salazar DV, Scrocchi G, Cuevas J, Leynaud GC, Carrasco PA (2019) A new species of pitviper of the genus *Bothrops* (Serpentes: Viperidae: Crotalinae) from the Central Andes of South America. Zootaxa 4656(1):99–120. https://doi.org/10.11646/zootaxa.4656.1.4

Tobar-Suárez C, Urbina-Cardona N, Villalobos F, Pineda E (2022) Amphibian species richness and endemism in tropical montane cloud forests across the Neotropics. Int J Biodivers Conserv 31(1):295–313. https://doi.org/10.1007/s10531-021-02335-z

Tonini JFR, Beard KH, Ferreira RB, Jetz W, Pyron RA (2016) Fully-sampled phylogenies of squamates reveal evolutionary patterns in threat status. Biol Conserv 204:23–31. https://doi.org/10.1016/j.biocon.2016.03.039

Torres-Carvajal O, Hinojosa KC (2020) Hidden diversity in two widespread snake species (Serpentes: Xenodontini: *Erythrolamprus*) from South America. Mol Phylogenet Evol 146:1–8. https://doi.org/10.1016/j.ympev.2020.106772

Torres-Carvajal O, Venegas PJ, Nunes PMS (2020) Description and phylogeny of a new species of Andean lizard (Gymnophthalmidae: Cercosaurinae) from the Huancabamba depression. South Am J Herpetol 18(1):13–23.https://doi.org/10.2994/SAJH-D-18-00069.1

Torres-Carvajal O, Terán C (2021) Molecular phylogeny of Neotropical Parrot Snakes (Serpentes: Colubrinae: *Leptophis*) supports underestimated species richness. Mol Phylogenet Evol 164:1–7. https://doi.org/10.1016/j.ympev.2021.107267

Torres-Carvajal O, Echevarría LY, Venegas PJ, Chávez G, Camper JD (2015) Description and phylogeny of three new species of *Synophis* (Colubridae, Dipsadinae) from the tropical Andes in Ecuador and Peru. ZooKeys 546:153–179. https://doi.org/10.3897/zookeys.546.6533

Torres-Carvajal O, Echevarría LY, Lobos SE, Venegas PJ, Kok PJ (2019) Phylogeny, diversity and biogeography of Neotropical sipo snakes (Serpentes: Colubrinae: *Chironius*). Mol Phylogenet Evol 130:315–329. https://doi.org/10.1016/j.ympev.2018.10.022

Trevine VC, Grazziotin FG, Giraudo A, Sallaberry-Pincheira N, Vianna JA, Zaher H (2022) The systematics of *Tachymenini* (Serpentes, Dipsadidae): an updated classification based on molecular and morphological evidence. Zool Scr 51(6):643–663. https://doi.org/10.1111/zsc.12565

Trindade-Filho J, de Carvalho RA, Brito D, Loyola RD (2012) How does the inclusion of Data Deficient species change conservation priorities for amphibians in the Atlantic Forest? Biodivers Conserv 21:2709–2718. https://doi.org/10.1007/s10531-012-0326-y

Uetz P, Freed P, Aguilar R, Reyes F, Kudera J, Hošek J (2025) The reptile database. Available from: http://www.reptile-database.org. Accessed 24 Jan 2025

Urbina-Cardona N, Saboyá LP, Camacho-Rozo CP, Acosta AR, Arenas-Rodríguez A, Albarracín-Caro JF, Moreno-Cabal M, Novoa-Salamanca NM, Camacho-Duran MJ, Echeverry G, Hernández-Gallego MJ, Pirateque L, Aldana V, Echeverry D, Zabala-Forero FA (2023) Producción científica sobre la herpetología en Colombia: perspectivas desde los temas de investigación hacia la conservación biológica. Caldasia 45(1):4–20. https://doi.org/10.15446/caldasia.v45n1.97216

Van der Hammen T (1974) The Pleistocene changes of vegetation and climate in tropical South America. J Biogeogr 1(1):3–26. https://doi.org/10.2307/3038066

Vásquez-Restrepo JD (2021) Inclusion is not representativeness: the context of Colombian samples in the taxonomic and systematic mid-large herpetological literature. Rev Latinoam Herpetol 4(2):236–247. https://doi.org/10.22201/fc.25942158e.2021.02.281

Vidal N, Delmas A, David P, Cruaud C, Couloux A, Hedges SB (2007) The phylogeny and classification of caenophidian snakes inferred from seven nuclear. C R Biol 330(2):182–187. https://doi.org/10.1016/j.crvi.2006.10.001

Vidal N, Dewynter M, Gower DJ (2010) Dissecting the major American snake radiation: a molecular phylogeny of the Dipsadidae Bonaparte (Serpentes, Caenophidia). C R Biol 333(1):48–55. https://doi.org/10.1016/j.crvi.2009.11.003

Vilela B, Villalobos F (2015) letsR: a new R package for data handling and analysis in macroecology. Methods Ecol Evol 6(10):1229–1234. https://doi.org/10.1111/2041-210X.12401

Villarreal-Rosas J, Wells JA, Sonter LJ, Possingham HP, Rhodes JR (2022) The impacts of land use change on flood protection services among multiple beneficiaries. Sci Total Environ 806:150577. https://doi.org/10.1016/j.scitotenv.2021.150577

Wiens JJ (2023) How many species are there on Earth? Progress and problems. PLoS Biol 21(11):1–4. https://doi.org/10.1371/journal.pbio.3002388

Young B, Young KR, Josse C (2011) Vulnerability of tropical Andean ecosystems to climate change. In: Herzog SK, Martínez R, Jørgensen PM, Tiessen H (eds) Climate change and biodiversity in the tropical. InterAmerican Institute for Global Change Research (IAI) and Scientific Committee on Problems of the Environment (SCOPE), pp 170–181

Zaher H (1999) Hemipenial morphology of the South American xenodontine snakes: with a proposal for a monophyletic Xenodontinae and a reappraisal of colubroid hemipenes. Bull Am Mus Nat Hist 240:1–168. http://hdl.handle.net/2246/1646

Zaher H, Grazziotin FG, Cadle JE, Murphy RW, De Moura-Leite JC, Bonatto SL (2009) Molecular phylogeny of advanced snakes (Serpentes, Caenophidia) with an emphasis on South American Xenodontines: a revised classification and descriptions of new taxa. Pap Avulsos Zool 49(11):115–153. https://doi.org/10.1590/S0031-10492009001100001

Zaher H, Murphy RW, Arredondo JC, Graboski R, Machado-Filho PR, Mahlow K, Montingelli GG, Bottallo-Quadros A, Orlov NL, Wilkinson M, Zhang Y, Grazziotin FG (2019) Large-scale molecular phylogeny, morphology, divergence-time estimation, and the fossil record of advanced caenophidian snakes (Squamata: Serpentes). PLoS One 14(5):1–82. https://doi.org/10.1371/journal.pone.0217959

Zaher H, Trusz C, Koch C, Entiauspe-Neto OM, Battilana J, Grazziotin FG (2024) Molecular phylogeny and biogeography of the dwarf boas of the family Tropidophiidae (Serpentes: Alethinophidia). Syst Biodivers 22(1):2319289. https://doi.org/10.1080/14772000.2024.2319289

Zhang Y, She J, Long X, Zhang M (2022) Spatio-temporal evolution and driving factors of eco-environmental quality based on RSEI in Chang-Zhu-Tan metropolitan circle, central China. Ecol Indic 144:109436. https://doi.org/10.1016/j.ecolind.2022.109436

Zheng Y, Wiens JJ (2016) Combining phylogenomic and supermatrix approaches, and a time-calibrated phylogeny for squamate reptiles (lizards and snakes) based on 52 genes and 4162 species. Mol Phylogenet Evol 94: 537–547. https://doi.org/10.1016/j.ympev.2015.10.009

Zipkin EF, Di Renzo GV, Ray JM, Rossman S, Lips KR (2020) Tropical snake diversity collapses after widespread amphibian loss. Science 367(6479):814–816. https://doi.org/10.1126/science.aay5733

Herpetofauna of the Andes: A Critical Synthesis

Carlos A. Navas and J. Nicolás Urbina-Cardona

1 Introduction

This book focuses on the herpetofauna of the Andean cordillera, a mountain range extending over 8900 km and traversing seven countries, creating latitudinal and elevational gradients that, added to geological and microclimatic factors, create highly diverse ecological settings and bioclimatic clines that promote biodiversity. As demonstrated throughout the four parts of this volume, the complex climatic heterogeneity of the Andes has driven the diversification of a remarkably rich herpetofauna, including frogs, lizards, snakes, and the less studied salamanders and, particularly, caecilians, turtles, and crocodiles Despite their lower diversity, Caecilians and lungless salamanders hold evolutionary significance, particularly in the northern Andes, where they represent independent transitions to fossoriality and amphibian adaptation to montane environments. With contributions from researchers across national and disciplinary boundaries, the book presents a comprehensive overview of Andean herpetology, portraying the Andes as a dynamic engine of speciation, turnover, and endemism. This synthesis draws on those contributions to articulate overarching themes, identify critical knowledge gaps, and reflect on unresolved tensions that remain in the field.

2 Andean Herpetology in Four Parts

This book opens with a prologue (Chapter "Prologue") by the esteemed Andean herpetologist Nora Ibargüengoytía. Chapter "The Andes: The Physical Setting" addresses the geographic and physical foundations of the Andes, with particular emphasis on tectonic uplift, climatic zonation, and environmental heterogeneity. These abiotic factors shape distinct microhabitats—from cloud forests to puna and steppe—and influence species richness and trait diversity across the herpetofauna.

Part I, *Diversity and Distribution Patterns* (Chapters "Beyond Species Richness: A Behavioural Perspective of the Anuran Diversity in the Andes of Colombia", "Functional Diversity of Amphibians and Reptiles in Tropical Dry Forest Remnants and Anthropogenic Cover Types in the Lower Bogotá River Sub-Basin", "Biogeographic Patterns of Richness and Endemism in Liolaemidae: Identifying Hotspots for the Most Diverse Andean Reptile Radiation", and "The Climatic Niche Evolution of South American

C. A. Navas (✉)
Department of Physiology, Biosciences Institute, University of São Paulo, São Paulo, SP, Brazil
e-mail: navas@usp.br

J. N. Urbina-Cardona
Pontificia Universidad Javeriana, Facultad de Estudios Ambientales y Rurales, Departamento de Ecología y Territorio. Carrera, Bogotá, Colombia

J. N. Urbina-Cardona et al. (eds.), *Andean Herpetofauna*, Sustainable Development Goals Series,
https://doi.org/10.1007/978-3-032-00074-3_18

Squamates (Reptilia: Squamata)". Chapter "Beyond Species Richness: A Behavioural Perspective of the Anuran Diversity in the Andes of Colombia" explores the behavioral diversity of Colombian Andean anurans, highlighting variation in communication, reproduction, and parental care, especially in underdocumented direct-developing species. Chapter "Functional Diversity of Amphibians and Reptiles in Tropical Dry Forest Remnants and Anthropogenic Cover Types in the Lower Bogotá River Sub-Basin (Department of Cundinamarca, Colombia)" focuses on the functional diversity of amphibians and reptiles in dry forest remnants, demonstrating how land cover transformation filters trait combinations and disproportionately affects specialized taxa. Chapter "Biogeographic Patterns of Richness and Endemism in Liolaemidae: Identifying Hotspots for the Most Diverse Andean Reptile Radiation" maps richness and endemism in Liolaemidae lizards, identifying biogeographic hotspots and linking them to Andean uplift. Chapter "The Climatic Niche Evolution of South American Squamates (Reptilia: Squamata)" assesses climatic niche evolution in squamates, showing repeated transitions between lowland and highland zones and asymmetries in thermal tolerance. Together, these chapters illustrate how geography, dispersal barriers, ecological traits, and climate history shape the distribution and diversification of Andean herpetofauna.

Part II, *Evolutionary Histories and Life Strategies* (Chapters "The Shared Evolutionary History of South America's Mountain Biotas: Genetic Evidence from Amphibians and Reptiles", "Anurans of the Patagonian Andes: A Little-Known Variety of Reproductive and Developmental Strategies", "Unveiling Patagonian Lizards: A Three-Decade Journey Through Ecology, Evolution, and Adaptation in a Changing World", and "Drivers of Physiological Responses and Range Distribution in the Context of Rapaport's Rule: The Case of East Andean Liolaemidae Lizard Species"), explores the mechanisms of adaptation that have shaped reproductive and developmental strategies in Andean species. Chapter "The Shared Evolutionary History of South America's Mountain Biotas: Genetic Evidence from Amphibians and Reptiles" examines how viviparity and extended larval development are evolutionary responses to the thermal and hydric constraints of high elevations, with notable variations across clades and altitudinal bands. Chapter "Anurans of the Patagonian Andes: A Little-Known Variety of Reproductive and Developmental Strategies" discusses reproductive strategies and the role of temperature in shaping developmental and thermoregulatory adaptations in Patagonian anurans. Chapter "Unveiling Patagonian Lizards: A Three-Decade Journey Through Ecology, Evolution, and Adaptation in a Changing World" examines the reproductive physiology and thermal adaptations of Patagonian lizards, focusing on viviparity and evolutionary plasticity in response to cold, high-altitude environments, with an emphasis on thermal tolerance and metabolic flexibility. Finally, Chapter "Drivers of Physiological Responses and Range Distribution in the Context of Rapaport's Rule: The Case of East Andean Liolaemidae Lizard Species" investigates the physiological drivers of range distribution and thermal tolerance in Liolaemidae lizards, with a focus on how metabolic rate, diet, and thermal adaptations interact with environmental variability to influence species' geographic range and life-history strategies. Together, these chapters provide a comprehensive view of how evolutionary histories, life-history strategies, and physiological adaptations are intrinsically linked to the ecological challenges of the Andes, including thermal tolerance, reproductive timing, and the role of environmental variability in shaping species' life cycles.

Part III, *Ecological and Physiological Adaptations* (Chapters "Physiological Ecology of Andean Amphibians Associated with Freezing Temperatures", "Thermal Biology of Squamate Reptiles in the High Mountains of the Northern Andes", and "Thermal Adaptations in Tropical Andean Amphibians: Exploring Spatial Variability and Global Change Impacts"), deepens our understanding of how environmental gradients intersect with metabolism, performance, and thermal ecology. The chapters discuss fine-scale behavioral regulation, freeze tolerance, and plasticity in thermal limits among amphibians and squamates. Chapter "Physiological Ecology of Andean Amphibians Associated with Freezing Temperatures" examines cold tolerance mechanisms, including freeze resistance in tropical highland frogs. Chapter "Thermal Biology of Squamate Reptiles in the High Mountains of the Northern Andes" presents comparative data on thermal biology in Andean squamates. Chapter "Thermal Adaptations in Tropical Andean Amphibians: Exploring Spatial Variability and Global Change Impacts", using primarily larval data, explores anuran thermal tolerance across elevations. Temperature emerges as a pervasive driver of behavioral, physiological, and morphological evolution, with remarkable cases of adaptation across taxa.

Part IV, *Conservation Challenges and Emerging Threats* (Chapters "Emerging Infectious Diseases of Amphibians in the Andean Region", "Diversity and Extinction Risk of Colombian Andean Amphibians Across Life Regions", "Experiences in Conservation Planning for Amphibians in the Andean Region", "Andean Snakes: Understanding Biodiversity Knowledge Shortfalls", and "Herpetofauna of the Andes: A Critical Synthesis"), offers a multifaceted perspective on threats to Andean herpetofauna. Chapter "Emerging Infectious Diseases of Amphibians in the Andean Region" synthesizes knowledge on infectious diseases, focusing on well-studied pathogens while noting uneven sampling and diagnostic coverage and underrepresented diseases. Chapter "Diversity and Extinction Risk of Colombian Andean Amphibians Across Life Regions" applies spatial modeling to assess extinction risk in Colombian amphibians, identifying vulnerable life zones. Chapter "Experiences in Conservation Planning for Amphibians in the Andean Region" presents regional experiences in amphibian conservation planning—from methodological advances to institutional tensions. It underscores the importance of amphibian-specific data and transboundary cooperation. Collectively, these chapters call for coordinated efforts that integrate ecology, disease, biogeography, and governance. Chapter "Andean Snakes: Understanding Biodiversity Knowledge Shortfalls" highlights profound taxonomic and data shortfalls for Andean snakes, calling for basic descriptive and ecological work. These chapters offer an overview on conservation issues, from basic science to active engagement.

While collectively rich in insight, the chapters also reveal persistent thematic and geographic gaps. Several Andean regions remain understudied in this book, notably the central Andes of Peru and Bolivia, the Mérida Cordillera in Venezuela, the Chilean sector of Andean Patagonia, as well as dry inter-Andean valleys and urban or peri-urban landscapes. These areas, despite their ecological, evolutionary, and species conservation significance, are largely absent from the book's core analyses. Taxonomically, caecilians and salamanders, and most snakes, continue to receive limited attention (with notable exceptions in chapter "Andean Snakes: Understanding Biodiversity Knowledge Shortfalls"). Beyond taxonomy, empirical data on amphibian tadpoles are entirely lacking, and demographic or population structure studies remain scarce for the vast majority of species. This paucity of population-level information is particularly concerning, as such data are foundational for assessing extinction risk and implementing effective, field-based conservation actions.

Table 1 provides a comparative overview of the systematic groups covered across the chapters, emphasizing the lineages most frequently studied in the volume. The data reveal that 53% of the chapters focus on the order Anura, followed by 46.6% addressing Sauria. Snakes are discussed in 26% of the chapters, while 20% mention salamanders, though without dedicated research on this order. Notably, only one-chapter reports on turtles and crocodiles, and no work specifically addresses Gymnophiona (caecilians).

Table 1 Systematic groups addressed in the chapters of this volume and their key lineages

	Class: Amphibia			Class: Reptilia				
Chapter/Taxonomic (Sub)Order	Anura	Caudata	Gymnophiona	Squamata: Sauria	Squamata: Serpentes	Crododylia	Testudines	Herpetofauna (Super)Families
Chapter "The Andes: The Physical Setting", Benavides and Arenas-Bautista								
Chapter "Beyond Species Richness: A Behavioural Perspective of the Anuran Diversity in the Andes of Colombia", Vargas-Salinas et al.	✓							Dendrobatoidea, Hylidae, Centrolenidae
Chapter "Functional Diversity of Amphibians and Reptiles in Tropical Dry Forest Remnants and Anthropogenic Cover Types in the Lower Bogotá River Sub-Basin (Department of Cundinamarca, Colombia)", Veloza and Urbina-Cardona	✓			✓	✓	✓	✓	Bufonidae, Dendrobatidae, Hylidae, Leptodactylidae, Microhylidae, Alligatoridae, Crocodylidae, Podocnemididae, Boidae, Colubridae, Viperidae, Corytophanidae, Dactyloidae, Dipsadidae, Gekkonidae, Gekkonidae, Iguanidae, Phyllodactylidae, Scindidae, Sphaerodactylidae, Teiidae
Chapter "Biogeographic Patterns of Richness and Endemism in Liolaemidae: Identifying Hotspots for the Most Diverse Andean Reptile Radiation", Esquerré and Skeels				✓				Liolaemidae
Chapter "The Climatic Niche Evolution of South American Squamates (Reptilia: Squamata)", Velasco et al.				✓	✓			Amphisbaenidae, Aniliidae, Anomalepididae, Anguidae, Boidae, Colubridae, Dipsadidae, Elapidae, Gekkonidae, Gymnophthalmidae, Hoplocercidae, Leiosauridae, Leptotyphlopidae, Liolaemidae, Phyllodactylidae, Dactyloidae, Scincidae, Sphaerodactylidae, Teiidae, Tropidophiidae, Tropiduridae, Typhlopidae, Viperidae
Chapter "The Shared Evolutionary History of South America's Mountain Biotas: Genetic Evidence from Amphibians and Reptiles", Prates et al.	✓			✓				

Chapter "Anurans of the Patagonian Andes: A Little-Known Variety of Reproductive and Developmental Strategies", Ubeda et al.	✓							Calyptocephalellidae, Rhinodermatidae, Batrachylidae, Alsodidae, Leptodactylidae, Bufonidae
Chapter "Unveiling Patagonian Lizards: A Three-Decade Journey Through Ecology, Evolution, and Adaptation in a Changing World", Piantoni et al.				✓				Liolaemidae, Gekkonidae,
Chapter "Drivers of Physiological Responses and Range Distribution in the Context of Rapaport's Rule: The Case of East Andean Liolaemidae Lizard Species", Cruz et al.				✓				Liolaemidae
Chapter "Physiological Ecology of Andean Amphibians Associated with Freezing Temperatures", Carvajalino-Fernández and Navas	□✓							Bufonidae, Strabomantidae, Hylidae
Chapter "Thermal Biology of Squamate Reptiles in the High Mountains of the Northern Andes", Calderón-Espinosa et al.				□✓	✓			Agamidae, Anguidae, Colubridae, Dactyloidae, Lacertidae, Liolaemidae, Phrynosomatidae, Teiidae, Tropiduridae, Xantusiidae
Chapter "Thermal Adaptations in Tropical Andean Amphibians: Exploring Spatial Variability and Global Change Impacts", Pintanel et al.	✓	✓						Hemiphractidae, Hylidae, Strabomantidae
Chapter "Emerging Infectious Diseases of Amphibians in the Andean Region", Flechas et al.	✓							Bufonidae, Calyptocephalellidae, Craugastoridae, Hylidae, Leptodactylidae, Pipidae, Telmatobiidae
Chapter "Diversity and Extinction Risk of Colombian Andean Amphibians Across Life Regions", González-del-Pliego and Bastazini	✓	✓						
Chapter "Experiences in Conservation Planning for Amphibians in the Andean Region", Angulo et al.	✓	✓						Atelopus, Gastrotheca, Rhinella
Chapter "Andean Snakes: Understanding Biodiversity Knowledge Shortfalls", Díaz-Ricaurte et al.					✓			Anomalepididae, Boidae, Colubridae, Elapidae, Leptotyphlopidae, Tropidophiidae, Viperidae

In total, the chapters mention 47 families, with several families appearing in multiple chapters. Specifically, Hylidae, Liolaemidae, Bufonidae, Colubridae, Gekkonidae, and Tropiduridae are mentioned across four to five chapters. Table 2 provides a comparative summary of the methodological approaches employed throughout the chapters, highlighting the diversity and imbalances in research strategies. The data reveal that 68.5% of the chapters included Field Surveys, 81.2% employed Statistical Modeling, 43% utilized Experimental Work, and 43% incorporated GIS/Remote Sensing. Notably, only 31% of the chapters involved Molecular Techniques or addressed Systematics/ Phylogenetics, while Policy/Stakeholder Analysis was covered in just one chapter. Together, these tables offer a concise synthesis of the book's taxonomic and methodological landscape, while also highlighting clear opportunities for future research.

3 Emerging Research Gaps and Thematic Imbalances

As editors, we recognize knowledge gaps in Andean herpetology from two main vantage points. The first comes from the chapters themselves: throughout the volume, authors explicitly identify research shortfalls that constrain both understanding and conservation of Andean amphibians and reptiles. Taxonomic and distributional gaps are particularly acute for reptiles—especially snakes (Chapter "Andean Snakes: Understanding Biodiversity Knowledge Shortfalls")—and for amphibians in dry forests (Chapter "Functional Diversity of Amphibians and Reptiles in Tropical Dry Forest Remnants and Anthropogenic Cover Types in the Lower Bogotá River Sub-Basin (Department of Cundinamarca, Colombia)") and high-elevation environments (Chapters "Anurans of the Patagonian Andes: A Little-Known Variety of Reproductive and Developmental Strategies", "Physiological Ecology of Andean Amphibians Associated with Freezing Temperatures", and "Thermal Biology of Squamate Reptiles in the High Mountains of the Northern Andes"), and across entire regions such as the central Andes of Peru and Bolivia, the Mérida Cordillera in Venezuela, the Chilean portion of Andean Patagonia, and the dry inter-Andean valleys. Urban and peri-urban settings also remain largely overlooked, despite their growing ecological and conservation relevance. Many species still lack phylogenetic resolution, reliable occurrence records, and standardized trait data (Chapters "Biogeographic Patterns of Richness and Endemism in Liolaemidae: Identifying Hotspots for the Most Diverse Andean Reptile Radiation", "The Shared Evolutionary History of South America's Mountain Biotas: Genetic Evidence from Amphibians and Reptiles", and "Andean Snakes: Understanding Biodiversity Knowledge Shortfalls"), impeding both basic ecological analysis and conservation planning. Life history and functional and physiological traits—such as body size, diet, reproductive mode, and thermal tolerance—are unevenly documented, which limits our capacity to assess ecological vulnerability across gradients (Chapters "Functional Diversity of Amphibians and Reptiles in Tropical Dry Forest Remnants and Anthropogenic Cover Types in the Lower Bogotá River Sub-Basin (Department of Cundinamarca, Colombia)", "Drivers of Physiological Responses and Range Distribution in the Context of Rapaport's Rule: The Case of East Andean Liolaemidae Lizard Species", and "Thermal Adaptations in Tropical Andean Amphibians: Exploring Spatial Variability and Global Change Impacts"). Reproductive and behavioral ecology remain underexplored in direct-developing species and in larval stages, particularly in Patagonian and tropical montane systems (Chapters "Beyond Species Richness: A Behavioural Perspective of the Anuran Diversity in the Andes of Colombia" and "Anurans of the Patagonian Andes: A Little-Known Variety of Reproductive and Developmental Strategies"). In the realm of thermal ecology, several chapters point to a lack of field-based temperature data (Chapter "Thermal Biology of Squamate Reptiles in the High Mountains of the Northern Andes") and to poor integration between organismal physiology and

Table 2 Methodological approaches applied in the chapters of this volume

Chapter	Field surveys	Experimental work	GIS/ remote sensing	Statistical modeling	Molecular techniques	Systematics/ phylogenetics	Policy/ stakeholder analysis	Archival/ literature review
Chapter "The Andes: The Physical Setting", Benavides and Arenas-Bautista								✓
Chapter "Beyond Species Richness: A Behavioural Perspective of the Anuran Diversity in the Andes of Colombia", Vargas-Salinas et al.	✓	✓		✓				✓
Chapter "Functional Diversity of Amphibians and Reptiles in Tropical Dry Forest Remnants and Anthropogenic Cover Types in the Lower Bogotá River Sub-Basin (Department of Cundinamarca, Colombia)", Veloza and Urbina-Cardona	✓		✓	✓				✓
Chapter "Biogeographic Patterns of Richness and Endemism in Liolaemidae: Identifying Hotspots for the Most Diverse Andean Reptile Radiation", Esquerré and Skeels			✓	✓	✓	✓		✓
Chapter "The Climatic Niche Evolution of South American Squamates (Reptilia: Squamata)", Velasco et al.			✓	✓	✓	✓		✓
Chapter "The Shared Evolutionary History of South America's Mountain Biotas: Genetic Evidence from Amphibians and Reptiles", Prates et al.			✓	✓	✓	✓		✓
Chapter "Anurans of the Patagonian Andes: A Little-Known Variety of Reproductive and Developmental Strategies", Ubeda et al.	✓	✓						✓
Chapter "Unveiling Patagonian Lizards: A Three-Decade Journey Through Ecology, Evolution, and Adaptation in a Changing World", Piantoni et al.	✓	✓		✓				✓
Chapter "Drivers of Physiological Responses and Range Distribution in the Context of Rapaport's Rule: The Case of East Andean Liolaemidae Lizard Species", Cruz et al.	✓	✓	✓	✓		✓		✓
Chapter "Physiological Ecology of Andean Amphibians Associated with Freezing Temperatures", Carvajalino-Fernández and Navas	✓	✓		✓				✓
Chapter "Thermal Biology of Squamate Reptiles in the High Mountains of the Northern Andes", Calderón-Espinosa et al.	✓	✓		✓				✓
	✓	✓		✓				✓

(continued)

Table 2 (continued)

Chapter	Field surveys	Experimental work	GIS/ remote sensing	Statistical modeling	Molecular techniques	Systematics/ phylogenetics	Policy/ stakeholder analysis	Archival/ literature review
Chapter "Thermal Adaptations in Tropical Andean Amphibians: Exploring Spatial Variability and Global Change Impacts", Pintanel et al.								
Chapter "Emerging Infectious Diseases of Amphibians in the Andean Region", Flechas et al.	✓			✓	✓			✓
Chapter "Diversity and Extinction Risk of Colombian Andean Amphibians Across Life Regions", González-del-Pliego and Bastazini	✓		✓	✓				✓
Chapter "Experiences in Conservation Planning for Amphibians in the Andean Region", Angulo et al.							✓	✓
Chapter "Andean Snakes: Understanding Biodiversity Knowledge Shortfalls", Díaz-Ricaurte et al.	✓		✓	✓	✓	✓		✓

environmental variables (Chapters "Physiological Ecology of Andean Amphibians Associated with Freezing Temperatures" and "Thermal Adaptations in Tropical Andean Amphibians: Exploring Spatial Variability and Global Change Impacts"). Disease ecology in the Andes has largely focused on chytridiomycosis, while research on other pathogens, including bacterial infections, helminths, protozoa, and viruses, remains sparse; furthermore, diagnostic efforts are unevenly distributed across the region, with significant geographical gaps in the study of these diverse pathogens (Chapter "Emerging Infectious Diseases of Amphibians in the Andean Region"). On the conservation front, amphibian-specific data are still poorly integrated into spatial planning tools (Chapters "Diversity and Extinction Risk of Colombian Andean Amphibians Across Life Regions" and "Experiences in Conservation Planning for Amphibians in the Andean Region"), governance structures are often fragmented (Chapter "Experiences in Conservation Planning for Amphibians in the Andean Region"), and there is a marked absence of frameworks that combine ecological, climatic, and land-use information in a meaningful way (Chapters "Diversity and Extinction Risk of Colombian Andean Amphibians Across Life Regions" and "Andean Snakes: Understanding Biodiversity Knowledge Shortfalls"). Taxonomic bias also emerges: caecilians, turtles, crocodyles, aquatic snakes, and microendemic lizards are rarely mentioned or entirely absent, despite likely facing severe threats undergoing environmental change (Table 1). And little is known about most fossorial forms. In particular, caecilians and lungless salamanders are entirely absent from empirical studies, despite their occurrence in cloud forests and paramo ecosystems; fossorial snakes, although present in several Andean countries, are unrepresented across ecological, physiological, and conservation chapters.

A second layer of gaps arises from our editorial experience. When this book was still in its early stages, the editorial team envisioned an "ideal" set of chapters to represent the diversity and complexity of Andean herpetology. However, it quickly became clear that such comprehensive coverage would not be achievable. This was not due to a lack of interest or effort, but rather to structural asymmetries in Andean herpetological research—including the uneven distribution of research groups, funding disparities, and limited opportunities for training and institutional support. As a result, several important topics remain unaddressed, despite their scientific relevance and initial consideration for inclusion. Critical gaps in Andean herpetology remain largely unexplored, despite their central importance to life-history evolution, extinction risk, and conservation policy. These include questions about the functional diversity of páramo amphibians and how it would respond to species extinctions, as well as the phylogeographic history and conservation status of highly diverse genera like *Pristimantis*. The ecological relationships between Andean herpetofauna and other mountain systems are also insufficiently studied, along with the temporal changes in species threat categories and shifting landscape composition and configurations. Furthermore, important conservation questions remain unaddressed, such as the effectiveness of protected area networks in representing amphibian and reptile diversity, and analyses of regional threat heterogeneity. The ecophysiological tolerances of Andean tadpoles, particularly in response to climatic and hydrological changes, are also notably absent in literature. These omissions point to broader knowledge gaps that hinder our understanding of regional herpetological dynamics and conservation outcomes.

Finally, some gaps became apparent only after completing the book—not because we failed to include them, but because we did not fully anticipate their absence. One example is the functional role of amphibians and reptiles in Andean ecosystems. Other cross-cutting themes such as phenology, species interactions (e.g., competition, mutualism), and the role of herpetofauna in trophic networks and ecosystem processes remain underdeveloped, limiting our capacity to fully assess ecological functioning and resilience.

While diversity and distribution are central themes of the book, little is said about the trophic positions, energy flow, or nutrient cycling roles of herpetofauna. We also came to realize how little is known about urban and peri-urban dynamics. Urban expansion is a dominant force in many Andean regions, yet we found no research assessing its impacts on herpetofauna (but see chapter "Functional Diversity of Amphibians and Reptiles in Tropical Dry Forest Remnants and Anthropogenic Cover Types in the Lower Bogotá River Sub-Basin (Department of Cundinamarca, Colombia)"), despite known effects from urban heat islands, pollution, and habitat fragmentation elsewhere, and initiatives converging urban and biodiversity studies. Likewise, ethnoherpetology—the cultural knowledge, symbolic roles, and resource uses of herpetofauna among Indigenous, local, and campesino communities—remains absent, despite its potential value for biocultural conservation.

Lastly, this volume does not address biological invasions, overharvesting, or the illegal wildlife trade, despite the widespread presence of feral predators, invasive species, and introduced commercial species such as trout in the Andes. The illegal wildlife trade remains a significant, yet underexplored, threat to Andean amphibians and reptiles, with species being extracted for the international pet trade, medicinal use, and cultural rituals. Driven by high global demand, this trade has contributed to population declines, but its ecological impacts are poorly understood, and regulations are weakly enforced. Invasive species, such as the American bullfrog and introduced trout, disrupt native ecosystems by altering habitats, outcompeting native species, and spreading diseases like chytridiomycosis. The ecological consequences of these species are substantial, requiring more research on their management and eradication. Overharvesting, both legal and illegal, particularly for ornamental and medicinal purposes, has further endangered several Andean species, and while the full extent of this threat remains unclear, stricter regulations and enforcement are essential to safeguard these species. Pollution, particularly water contamination and habitat degradation, continues to exacerbate the threats from diseases and invasions, further jeopardizing amphibian and reptile populations. Despite growing awareness of the risks posed by invasive species and disease, the impacts of pollution—especially water contamination and agrochemical runoff—remain understudied. Addressing pollution through improved waste management, water quality monitoring, and agrochemical regulation is vital for preserving critical habitats.

Taken together, these layered absences are not oversights in editorial judgment—they reflect the current contours and limitations of Andean herpetological research. By acknowledging these silences, this chapter complements the strengths of the volume and helps set an agenda for the next generation of studies: one that is more integrative, more inclusive of different knowledge systems, and more attuned to the full ecological and cultural complexity of the Andes.

4 Unresolved Tensions

We use the term *unresolved tensions* to refer to subtle asymmetries in how Andean herpetology is researched, interpreted, and applied—areas where conceptual framing, methodological emphasis, or thematic integration remain unsettled and admit different views. These tensions are not contradictions, but rather indicators of structural imbalances in research effort and scope.

One such tension emerges from a bibliometric pattern that became evident as we prepared this volume. In January 2025, a Google Scholar search for scientific articles published since 2000 and combining the terms "Herpetology" and "Andes" (or "Andean") in the title yields a surprisingly small number of results. However, when searches are performed using systematic terms—such as "frogs," "lizards," or "snakes"—the output is much larger: over 150 articles on frogs, more than 100 on lizards, nearly 40 on snakes, fewer than 10 on salamanders, and only one on Gymnophiona. This lack of attention is compounded by the limited integration of approaches that bridge taxonomy with physiology, biogeography, and conservation—a gap especially evident in taxa such as snakes, where evolutionary, ecological, and thermal data remain disjointed. This discrepancy suggests that much of the existing research is narrowly taxonomic in scope and does not engage with broader, integrative questions in herpetology as a discipline. The Andes appear to be a region where herpetological research is abundant in parts but fragmented in whole.

To assess whether this pattern is typical of mountain systems or specific to the Andes, we conducted a comparison with another major range: the Rocky Mountains. A similar search logic, now replacing "Andes" with "Rocky Mountains," and not restricted to title, revealed more than twice as many articles addressing general herpetological themes in the latter. Andean hits do duplicate Himalaya and Alps, yet broad herpetological research seems scarce for the exceptional biodiversity and ecological complexity that the Andean cordillera represents. Whether this reflects historical research patterns, institutional asymmetries, or reveals conceptual framing remains an open question. A related imbalance is also visible in conservation literature. Approximately 5000 articles link herpetology, Andes, and conservation. Yet, when refined to "conservation planning," the number drops to around 1000. Fewer than 800 articles reference land use, and only about 350 addresses both land use and climate change. This suggests a disproportionate focus on climate change, which may be driven more by global funding trends and discursive prominence than by the actual balance of threats in the region. Land-use change—through deforestation, agriculture, mining, and urban expansion—is a well-documented driver of biodiversity loss in the Andes, yet it is underrepresented in the literature.

Further asymmetries emerge when examining how climate change is approached in herpetological research. Among approximately 10,000 climatological scientific papers with the explicit intention to assess climate change, fewer than 20 focus on the Andes, and even fewer address their tropical region. This scarcity raises concerns about the empirical foundations of climate-related claims in studies on Andean herpetofauna. While many biological assessments reference global projections—such as average planetary temperature—there is little clarity on how these global trends are actually expressed across the elevational gradients and bioclimatic mosaics of the Andes. For instance, climatic anomalies in the Northern Andes remain virtually uncited in herpetological literature—signaling a persistent disconnect between climatology and herpetological biodiversity in the region. While generalization to Earth's averages may offer contextual framing but likely obscures fine-scale physiological responses and local adaptations of montane ectotherms, which are influenced by altitudinally constrained microclimates and topography-induced thermal mosaics. Importantly, this eventual lack of more precise reference does not result by omission, but because such data does not exist or is not yet publicly available. This tension critically limits our ability to assess how Andean herpetofauna are actually experiencing and responding to climate change, particularly at individual and populational levels of organization.

Finally, a separation of climate and land-use change including urbanization as fully independent threats obscures their interactive effects. Deforestation, agricultural expansion, city growth, and infrastructure development reshape habitats in ways that amplify climate pressures—altering thermal regimes, moisture availability, community structure, and species interactions. Yet, integrative studies that explore these synergies remain scarce for Andean herpetofauna. Addressing them will require a shift from sectoral approaches to holistic frameworks that bridge climate science, landscape ecology, trait-based ecology, physiology, and conservation governance. This volume, summed to these final reflections, underscores that strengthening Andean herpetological research is not just a matter of generating more data. It is about developing conceptual and methodological frameworks that capture the full complexity of environmental change in the region. To the many outstanding local initiatives, many of which appear in this

Book, we must add broader scientific enterprises, necessarily international and transdisciplinary.

5 Integrative Approaches and Future Directions

This volume makes clear that Andean herpetology has growth over decades to conform many well-established and already traditional areas of research. Yet, perhaps a pivotal moment has been reached. While the region harbors some of the world's richest amphibian and reptile assemblages, the science that supports their understanding and conservation remains fragmented—systematically, thematically, methodologically, nationally, and institutionally. The field must support and value the many disciplinary initiatives that produce science of international quality, while promoting and embracing more integrative approaches that bridge gaps across ecological scales, biological disciplines, and geopolitical boundaries. Therefore, several priorities emerge. First, there is an urgent need to develop regionally grounded climate research that is open to science and can inform about context- and species-specific vulnerability assessments. Strengthening partnerships between herpetologists and climatologists, and investing in fine-scale thermal data and monitoring infrastructure, seems a consequential next step. Beyond thermal ecology, other key physiological dimensions—such as locomotor performance and water balance—remain largely unexplored, despite their importance for understanding species' survival, dispersal potential, and resilience under changing environmental conditions.

Second, research as a process must enhance cross-scalar frameworks that link organismal physiology, trait evolution, and habitat change. This is happening already, and some chapters in this book provide such evidence. However, an extremely ambitious, yet realistic, goal would be integrating field-based thermal exposure, hydric constraints, and ecological interactions with larger-scale models of species distribution and community turnover from a macroecological perspective. Also, the field would benefit from comparative studies across taxa and ecosystems, for example, those considering amphibians and reptiles together along shared environmental gradients. Such studies are useful to reveal convergent adaptations, but also divergences in adjustment type, in turn reflecting complementary and distinct vulnerabilities and therefore, conservation needs.

Despite significant advances in ecological and physiological understanding, several critical aspects of Andean herpetofauna remain underexplored, including ecological interactions, phenological patterns, and altitudinal ecological connectivity in fragmented landscapes. Key research gaps include the lack of empirical studies on the eco-physiology, thermal ecology, locomotor performance, and hydric physiology of caecilians and salamanders, as well as the absence of demographic analyses and population structure studies for both amphibians and reptiles. Additionally, there is a notable lack of empirical data on Andean tadpoles, despite their ecological significance and vulnerability to climatic and hydrological changes. The ecological impacts of invasive species, such as trout and bullfrogs, and the integration of eco-physiological and acoustic traits into functional diversity frameworks are also critical gaps. These omissions hinder our ability to predict species responses to climate change and landscape transformation. To address these challenges, future research must embrace integrative eco-evolutionary frameworks that link local microadaptations with broader biogeographical patterns and the functional roles of species in ecosystems. Bridging these gaps will be essential for effective conservation planning, ensuring the persistence of Andean herpetofauna in a rapidly changing world. This will require targeted investment, interdisciplinary collaboration, and more effective articulation between ecological research and policy development, especially in countries with limited institutional capacity.

Finally, we believe that greater attention must be directed toward the human dimensions of Andean herpetology. As mentioned earlier, urban studies and ethnoherpetology offer crucial perspectives, to which we now add the value of herpetology in education. Looking ahead, the future of Andean herpetology could be positively and exponentially transformed if researchers invest in durable and equitable networks of transnational collaboration. Bridging regional data asymmetries—particularly by strengthening research networks in underrepresented regions such as the central Andes—will be essential to redress current biases in geographic and taxonomic coverage. Such initiatives include strengthening transdisciplinarity, fostering long-term partnerships and academic exchanges, implementing capacity-building programs that enhance institutional and human resource development, expanding open-access data platforms, and creating mechanisms to reduce disparities in training and funding opportunities across Andean countries.

Conservation outcomes will depend not only on scientific insight but also on anchoring these efforts in institutional continuity. Addressing the outstanding gaps identified throughout this volume will demand not only stronger science, but also sustained collaboration, and equitable knowledge exchange. This implies confronting systemic challenges—including underfunding, institutional fragility, gender and regional inequities, academic isolation, and the growing disconnection between science and local realities. Building a healthier and more inclusive herpetological community will depend on the establishment of a diverse Andean network of researchers committed to data sharing for meta-analyses, collaborative projects across countries, student and faculty exchange programs, and the cocreation of national conservation strategies where they are still lacking. This network must also push governments to adopt and adequately fund amphibian and reptile conservation plans, while amplifying the voices of researchers from historically marginalized institutions and communities. By aligning research agendas with these principles, Andean herpetology can evolve from a collection of extremely valuable studies that take place in the Andes into a fully articulated research field. If this transformation occurs, science may become a more effective force in understanding and conserving one of the planet's most complex mountain systems. We hope this volume contributes to building that future.

GPSR Compliance
The European Union's (EU) General Product Safety Regulation (GPSR) is a set of rules that requires consumer products to be safe and our obligations to ensure this.

If you have any concerns about our products, you can contact us on

ProductSafety@springernature.com

In case Publisher is established outside the EU, the EU authorized representative is:

Springer Nature Customer Service Center GmbH
Europaplatz 3
69115 Heidelberg, Germany

www.ingramcontent.com/pod-product-compliance
Lightning Source LLC
Chambersburg PA
CBHW080636160426
43267CB00005B/60
* 9 7 8 3 0 3 2 0 0 0 7 3 6 *